INDUCTION MACHINES
Their Behavior and Uses

PHILIP L. ALGER
Consulting Professor of Electrical Engineering
Rensselaer Polytechnic Institute
Troy, New York

SECOND EDITION
Completely Revised and Updated

NEW INTRODUCTION
by M. Harry Hesse

Gordon and Breach Publishers
Australia • Austria • Belgium • China • France • Germany • India • Japan • Luxembourg • Malaysia •
Netherlands • Russia • Singapore • Switzerland • Thailand • United Kingdom • United States

Copyright © 1965, 1970 by Gordon and Breach Science Publishers, Inc. Second Edition with additions published by OPA (Overseas Publishers Association) Amsterdam B.V., under license by Gordon and Breach Science Publishers SA.

First Edition published under the title *The Nature of Induction Machines*.

Second Edition completely revised and updated.

All rights reserved.

First published 1965
Third printing with additions 1995

No part of this book may be reproduced or utilized in any form or by any means, electronic or mechanical, including photocopying and recording, or by any information storage or retrieval system, without permission in writing from the publisher. Printed in the United States of America.

Gordon and Breach Science Publishers SA
Postfach
4004 Basel
Switzerland

Library of Congress Catalog Card No. 64-18799

To
HOWARD MAXWELL
Manager of Engineering and Manufacturing
Induction Motor Department
General Electric Company
1918–1944

Mr. Maxwell's quizzical reply, "Have you tried it?" when told of any new theory, and the great freedom to experiment that he gave his young engineers, made working for him a wonderfully rewarding experience.

TO TRUTH SEEKERS

High heart and flashing eye,
And all the world beneath the sky
      To conquer with the rush of youth.
For man in glorious mold is made—
Undaunted courage, unafraid
      To seek, and then proclaim the truth.

And if one says that blue's the sky,
Another grey—do not deny
      The other man a vision sees.
Perhaps you are the color blind;
Perhaps in your own humble mind
      Green is the sky and blue the trees.

But if to your ideal you hold,
And that ideal is high and bold,
      What matter if externals pass?
The tides know perpetual change,
Flowers go, and seasons range
      Like shadows on the summer grass.

O, give to us serene, the art
Of keeping in a quiet heart
      Desire insistent for the truth
      —High heart and flashing eye,
And all the world beneath the sky
      To conquer with the rush of youth.

                                              C.J.A.

PREFACE

The Sons of Martha

   It is their care in all the ages to take the buffet and cushion the shock.
   It is their care that the gear engages; it is their care that the switches lock.
   It is their care that the wheels run truly; it is their care to embark and
      entrain,
   Tally, transport, and deliver duly the Sons of Mary by land and main.

RUDYARD KIPLING

The purpose of this book is to give its readers an understanding and a visual perception of, and some familiarity with, the behavior and uses of induction machines.

Like most things familiar and utterly reliable, the induction motor has long been taken for granted by industry, and by the young engineer with his eye on the future. Yet these three hundred million electric motors that operate our homes and farms, and give the power of ten horses to every American factory worker, deserve our thoughtful attention. They, and the ten million more such aids to "the sons of Martha" that we employ each year, present a world of challenging problems to the student and the professor, as well as to their designers and their users. And, too, as the most important of all types of electric machinery, dollar-wise, their study is richly rewarding to all who aspire to be power engineers.

After giving a brief statement of basic principles, a clear perception of each phenomenon is provided, by means of equivalent circuits, energy-flow, harmonic analysis, and rotating field concepts. Equations are derived for numerical calculation of each element of performance in its simplest form. The secondary aspects are considered separately, enabling the final result to be obtained as the sum of a converging infinite series, rather than by a single complex formula. Holding the several elements of each problem in separate view by such means enables those skilled in the art to make rapid and accurate approximations, and gives them the ability to choose wisely between alternatives.

The generalized machine concepts of Gabriel Kron are used to develop the theory of various special and unsymmetrical connections of induction machines, and to show the relations between induction and other types of rotating machines. Thus, the book is intended to provide a foundation for the understanding of all members of the rotating electric machine family.

The first edition of this book, published in 1951 by John Wiley & Sons, Inc., and copyrighted by the General Electric Company, dealt with polyphase machines only. New material has been added dealing with single-phase machines, reactors and Saturistors, and methods of speed and torque control. Especial attention is given to the newly-developed series type of hysteresis motor, that employs hard magnetic materials such as Alnico in the flux leakage paths of the secondary winding. This "Saturistor" principle is suitable for motors of industrial sizes, in marked contrast to the long-familiar shunt type of hysteresis motor, which employs hard magnetic materials in the magnetizing flux path of the secondary winding, and is useful in fractional-horsepower sizes only. Much new information has been added also, to bring the other chapters up to date.

The author is deeply indebted to the early teaching of Comfort A. Adams and of Waldo V. Lyon, whose papers written many years ago are still classics in the field of electric machinery. Also, he owes a great deal to the support and friendly counsel of many former associates in the General Electric Company, especially Howard Maxwell, H. R. West, Marvin L. Schmidt, W. M. Schweder, G. Angst, and the late Sven R. Bergman and Charles J. Koch. The original genius and unfailing enthusiasm of Gabriel Kron have provided great aid and inspiration, and the many constructive suggestions of Professor L. V. Bewley have been most helpful.

The author is greatly indebted to the General Electric Company for their generous permission to use the copyrighted material in the Wiley edition.

Finally, the author wishes to express his deep appreciation of the invaluable assistance provided throughout by Miss Ruth Shaver.

PHILIP L. ALGER

Schenectady, New York

PREFACE TO SECOND EDITION

In this second edition I have brought up to date the tables of motor frame sizes, torques and temperature rises provided in American Standards, and added recent data on the performance and possible applications of the Saturistor principle. Also, a large number of references to recent publications have been added. Most of the equations and formulas have been changed to use newtons instead of pounds for force and centimeters instead of inches for lengths, to accord with now accepted practice.

I wish again to express my appreciation of the help provide by Miss Ruth Shaver in editing.

PHILIP L. ALGER

Schenectady, New York
April, 1969

INTRODUCTION

From time to time there are cogent reminders that we are but dwarfs standing on the shoulders of giants. Philip Langdon Alger was one of those giants. In 1951 Dr. Alger published *The Nature of Polyphase Induction Machines*, John Wiley & Sons, and in 1965 *The Nature of Induction Machines*, Gordon and Breach Publishers. In 1970 he prepared the completely revised and updated second edition of the 1965 book. These books have become the classic standard reference works on induction machines the world over and have earned him the name "Mr. Induction Motor."

Twenty-five years have elapsed since the second edition of the book was published, and the publishers felt that the time had come for a revised third edition of his book. For several reasons I felt that the main body of the text should not be changed. The publishers then suggested a new printing with the addition of an introduction justifying this view of a classic work.

Philip Langdon Alger was born on January 27, 1894, the son of a Navy captain and professor of mathematics at St. John's College (now the United States Naval Academy) in Annapolis, Maryland, at a time when Charles Proteus Steinmetz, Michael Pupin, and others engaged in extensive discussions before the A.I.E.E. on the merits of the induction motor that Nikola Tesla had given to the world only a few years earlier.

In 1912 Alger received a B.S. degree in mathematics from St. John's College. Three years later he received a B.S.E.E. degree from the Massachusetts Institute of Technology and undertook a research assistantship to study dynamo electric machinery. In 1920 he received an M.S.E.E. degree from Union College, Schenectady, New York.

During World War I Alger became a lieutenant in the Ordinance Department of the United States Army, where he distinguished himself by becoming interested in the cause of erratic trajectory shells from a 6-inch Navy gun being tested under his direction. His studies led to successful improvements that were published in the 1919 volume of the *Coast Artillery Journal*. After the war he served as captain and then major in the Ordinance Department Reserve Corps from 1920 to 1943.

Alger's career in the General Electric Company began in 1919 when he started work on the design of induction motors. He made an intensive study of the various components of reactance in induction motors, the results of which were fundamental to the company's contributions to the development of the induction motor drives for battleships and aircraft carriers of the United States

Navy, as well as to the improved design of synchronous generators.

In 1923, as a result of problems with shaft currents, particularly in the Alexanderson high-frequency alternators, he determined the origin of and means for eliminating such shaft currents. In 1924 he took on a new assignment in General Electric's A-C Engineering Department to work on synchronous motors. Based upon his earlier experience, he developed a new amortisseur winding for synchronous motors that permitted them to be started directly across the line without starting compensators. He published the A.I.E.E. paper "The Calculation of Armature Reactance of Synchronous Machines" in 1928, which remains a classic in the annals of rotating electric machinery.

Alger was appointed to the staff of the vice-president of engineering in 1929 to sponsor and coordinate developments in electric apparatus throughout General Electric. His chief interest still lay in motors. Attention now turned to more effective use of materials in the design of motors, to make motors physically smaller and lighter in weight for a given output. This required the critical examination of many traditions in design engineering, application engineering, and among the motor users. He recognized that men are creatures of habit; and he realized that one must have the wisdom to recognize what is sound and what is simply tradition, and then have the courage to break with tradition when necessary. He developed a philosophy of "rules are made to be broken— in a nice way." He exercised strong leadership in committees and working groups of the A.I.E.E. and A.S.A. to bring about new N.E.M.A. standards for the motors in the 1940s. The new motors weighed less than one-third of their predecessors of the late 1920s, were quieter, and performed as well or better and at less cost.

In the 1940s many new synthetic insulation materials began to appear that could be effectively used in electric machinery, but their application was hampered by existing standards. As chairman of an A.I.E.E. working group, he inspired the work that resulted in the preparation of the first test procedure for the evaluation of insulation systems for motors, which, in turn, required a complete review of A.I.E.E. Standard No. 1, "General Principles upon which Temperature Limits Are Based in the Rating of Electric Equipment."

Alger returned to the induction motor department as consulting engineer in 1950. He turned his attention to the causes of noise in induction motors. By inspiring others, and by his own contributions, the entire picture of noise relationships in electric motors was unfolded. His 1954 A.I.E.E. paper "The Magnetic Noise of Polyphase Induction Motors," is one of the most significant of the many contributions to this subject.

It was during this period that Alger drew upon his vast experience and published his first book on polyphase induction motors. His subsequent two books contain improvements that he found desirable. These books therefore reflect Philip Langdon Alger and the methods that he considered best in the design of induction motors. This is perhaps my strongest argument against any alterations of this book.

It must be recognized that during the early years of his career, design calculations meant *hand* calculations. A great deal of effort went into mathematical formulations that were a compromise between accuracy and expediency. Today, with high-speed digital computers, design calculations are considerably more detailed and accurate, and, more importantly, are changing continually. Furthermore, in a free-enterprise system, practical design computer programs are proprietary information, and it is doubtful that any two are alike. To introduce such new concepts into his book is, therefore, neither practical nor desirable. Such information is best found in the contemporary technical literature.

I repeat that this book represents the methods that Alger deemed best for designing induction motors. This was an evolutionary process spanning perhaps four decades, and presumably forms the backbone of General Electric's design philosophy. Other manufacturers, through years of tradition and accumulation of patents, will have a somewhat different design philosophy. One single book could not do justice to all.

Perhaps the single addition that could be made to Alger's book, which would not tamper with the main text but rather enhance the appreciation of his work and thoughts, is a list of his publications (p. 515). It will be quickly observed that many of his papers are shared with co-authors, which is a tribute to his generosity and his inspirational leadership as an educator and developer of men. The Discussions (p. 519) not only reflect his continual alertness to what was happening around him, but will enable the reader to gain some insight into Philip Alger, the man.

There is an aspect of Philip Alger, the man, which is not nearly as well known as his professional career. Having known him for many years, I offer some recollections that may be of interest to the reader. In 1959 he retired from General Electric and accepted an appointment as adjunct professor of electrical engineering at Rensselaer Polytechnic Institute. The previous decade had produced profound changes in the philosophy of engineering education. In electrical engineering, the traditional "rotating machinery" subjects had offered the most stubborn resistance to change. Although now generally referred to as "energy conversion," no well-defined center of gravity for the course content had been established. Alger embarked on a new course and developed lecture notes entitled "Dynamic Circuit Theory and Rotating Electric Machines," which he subsequently modified. Being an avid experimentalist himself, he also developed an innovative electric machinery projects laboratory. Not only did he inspire students to take an interest in electrical machines, but with his dedication there resulted a number of senior and master's theses, several of which were published as A.I.E.E. papers.

INTRODUCTION

When I joined Rensselaer's Center for Electric Power Engineering in 1970, I shared a room with Philip Alger, which was much more of a workplace than an office. It was filled with all sorts of machine parts, stator and rotor laminations, and so on. The laboratory contained a variety of motors of different sizes and configurations. Perhaps the machine that impressed me most was an induction motor with inverted construction, having the squirrel cage on the stator. This was a machine that he designed and had constructed at his own expense. With more than a decade of successful "hands-on" laboratory experience, he was a most valuable resource in the design of the Faraday's Law Machinery Laboratory concept, which was implemented at Rensselaer in 1971.

By 1972 Alger was seventy-eight years old and he retired as adjunct professor. However, for several years thereafter, until his death in 1979, he continued to maintain an active interest in what we were doing, not only in the machines laboratory but also in the professional societies, especially in the area of electric power engineering education. The opportunity to visit with him at his home in Schenectady, New York, was always a pleasure.

M. Harry Hesse
Professor of Electric Power Engineering
Rensselaer Polytechnic Institute

CONTENTS

CHAPTER	PAGE
1. BASIC PRINCIPLES—REACTORS—SATURISTORS	1

1. The Nature of Rotating Electric Machines *(3)* 2. The Simple Series Circuit *(5)* 3. Resistance Calculations *(10)* 4. Magnetic Calculations *(11)* 5. Induced Voltage Calculations *(14)* 6. Magnetic and Electrostatic Energy Storage *(17)* 7. Magnetic Forces *(19)* 8. Reactance Calculations *(23)* 9. Reactors *(26)* 10. Saturistors *(28)*

2. TRANSFORMERS—CIRCUIT ANALYSIS 35

1. Transformers *(37)* 2. The Equivalent Circuit *(41)* 3. The Importance of the Equivalent Circuit *(42)* 4. Per Unit Notation *(43)* 5. Steady-State Circuit Analysis *(43)* 6. Circle Diagram *(44)* 7. Direct-Current Transient Circuit Analysis *(46)* 8. Constant Linkage Theorem *(49)* 9. Alternating-Current Transient Circuit Analysis *(49)* 10. The Superposition Theorem *(52)* 11. General Solution of Transient Equations, Employing the Null-Unit Function *(54)* 12. Transient Performance *(57)* 13. Dimensional Analysis *(60)* 14. Energy Flow *(64)*

3. THE ROTATING MAGNETIC FIELD 69

1. Idealized Magnetic Structure *(71)* 2. Current and Magnetomotive Force Waves *(72)* 3. Torque *(73)* 4. Power Output *(75)* 5. Armature Windings *(76)* 6. Generated Voltages *(80)* 7. Production of the Rotating MMF *(82)* 8. The MMF Produced by a Single Coil Carrying Direct Current *(83)* 9. Change from Direct to Alternating Current *(86)* 10. Resolution into Rotating MMFs *(86)* 11. Addition of MMFs due to Coils in the Same Phase Winding *(87)* 12. Addition of MMFs due to Different Phases *(89)* 13. Magnetizing Current *(93)* 14. Winding Analysis *(96)* 15. Symmetrical Components *(100)*

4. THE POLYPHASE INDUCTION MOTOR 105

1. Principle of Operation and Utility *(107)* 2. Electromagnetic Structure *(108)* 3. Torque and Slip *(109)* 4. Phasor Diagram *(112)* 5. Circle Diagram *(114)* 6. Starting Performance *(116)* 7. No-Load Tests *(117)* 8. Load Tests *(122)* 9. Stray-Load Loss Measurement *(123)* Appendix—Behrend and Tesla *(126)*

5. POLYPHASE INDUCTION MOTOR PERFORMANCE CALCULATIONS 129

xviii   CONTENTS

CHAPTER   PAGE

    1. The Equivalent Circuit *(131)*   2. Circuit Calculations *(133)*   3. Generalized Circuit Calculations *(140)*   4. Calculations in the Region of Standstill *(142)*   5. Calculations in the Region of Full Load *(144)*   6. Operation as a Generator *(148)*   7. Power-Factor Determination *(148)*   8. Calculations in the Region of Maximum Torque *(153)*   9. Dynamic Braking *(156)*   10. Summary of Formulas *(159)*

6. DESIGN OF INDUCTION MACHINES   163

    1. The Designing Process *(165)*   2. The Frame Structure *(166)*   3. Standard Frame Dimensions *(168)*   4. Ventilation and Temperature Rise *(171)*   5. Shafts and Bearings *(172)*   6. Insulation *(174)*   7. The Magnetic Core *(177)*   8. The Windings *(179)*   9. Magnetizing Current Calculation *(181)*   10. Loss Calculations *(188)*

7. REACTANCE CALCULATIONS   197

    1. Elements of Reactance *(199)*   2. The Magnetizing Reactance *(201)*   3. The Primary Slot Reactance *(201)*   4. The Zero-Phase Sequence Reactance *(206)*   5. The Secondary Slot Reactance *(208)*   6. The Differential Reactance *(209)*   7. The Zigzag Reactance *(216)*   8. Overlap Method of Calculating Zigzag Reactance *(222)*   9. Effect of Skew on Reactance *(227)*   10. The Belt-Leakage Reactance *(228)*   11. Peripheral Air-Gap Leakage *(232)*   12. Coil End Leakage of Armature Windings *(233)*   Appendix A—Sine-Wave Linkages of an Open-Slot Stator *(244)*   Appendix B—The Flux Distribution and the Associated Energy Flow in an Annular Core *(246)*

8. SPEED-TORQUE-CURRENT RELATIONS   259

    1. Methods of Speed and Torque Control *(261)*   2. Energy Losses and Heating during Acceleration *(262)*   3. High-Impedance Rotor *(263)*   4. Deep-Bar Rotor *(265)*   5. Idle-Bar Rotors *(272)*   6. Double Squirrel-Cage Rotor *(277)*   7. The Series Hysteresis Motor (Idle Alnico Bars) *(283)*   8. Reduced Voltage Starting *(286)*   9. Part-Winding Starting *(288)*   10. Split-Winding Starting *(296)*   11. Pole Changing and Concatenation *(302)*   12. Wound Rotor with Rheostatic Control *(306)*   13. Wound Rotor with Saturistor *(309)*   14. Speed Control with Auxiliary Machines or Frequency Converters *(318)*

9. CRAWLING, LOCKING, AND STRAY LOSSES   323

    1. Effects Produced by Harmonic Fields *(325)*   2. Permeance Waves *(328)*   3. Asynchronous Crawling *(331)*   4. Stray Losses *(339)*   5. Standstill Locking *(346)*   6. Synchronous Crawling *(355)*   7. Unbalanced Magnetic Pull *(356)*

CONTENTS

Chapter	Page
10. Magnetic Noise and Voltage Ripples	363

1. The Nature of Magnetic Noise *(365)*   2. Stator Frame Vibration *(367)*   3. Sound Radiation *(370)*   4. Noise Calculations *(372)*   5. Resonant Frequencies *(376)*   6. Slot Frequency Force Waves *(377)*   7. Effects of Dissymetry *(383)*   8. Induced Voltage Ripples *(383)*   9. Effects of Harmonics in the Impressed Voltage *(388)*

11. Kron's Generalized Machine   391

1. Vistas *(393)*   2. The Generalized Machine *(394)*   3. Direct-and Quadrature-Axis Circuits *(395)*   4. Effect of Rotation *(400)*   5. Torque Equations *(401)*   6. Limitations of the General Equivalent Circuit *(402)*   7. Transformations of Reference Axes (403)   8. Symmetrical Components *(404)*   9. Three-Phase to Two-Phase Transformations *(405)*   10. Matrix Transformations *(407)*   11. Impedance Matrix of the Generalized Machine *(411)*   12. Equivalent Circuits *(416)*   Appendix—Gabriel Kron *(421)*

12. Single-Phase Induction Motors   427

1. Principal Features *(429)*   2. Stator Windings *(430)*   3. The Capacitor Motor *(431)*   4. Speed Control of Capacitor Motors *(440)*   5. The Pure Single-Phase Motor *(445)*   6. The Dilemma of Single-Phase Motor Theory *(447)*   7. Resistance-Split Phase Motors *(457)*   8. Shaded-Pole Motors *(458)*   9. Repulsion-Induction Motors *(460)*

13. Special Types and Connections of Induction Machines   465

1. Unbalanced Two-Phase Motors *(467)*   2. Unbalanced Three-Phase Motors *(469)*   3. Dual-Winding Motors *(476)*   4. Transient Currents *(480)*   5. Reluctance Motors *(486)*   6. The Permasyn Motor *(488)*   7. Hysteresis Motors *(489)*   8. Synchronous Induction Motors *(490)*   9. A-C Excited Induction Motors *(490)*   10. Linear Induction Motors *(491)*

14. Rating and Application of Induction Motors   497

1. The Purpose of a Rating System *(499)*   2. General-Purpose Motors *(500)*   3. Starting Current Limitations *(503)*   4. Temperature Limitations *(503)*   5. Special-Purpose Motors *(508)*

Papers by Philip Langdon Alger and with Co-Authors   515

Discussions by Philip Langdon Alger   519

Index   525

CONTENTS—CHAPTER 1

Basic Principles—Reactors—Saturistors

		PAGE
1	The Nature of Rotating Electric Machines	3
2	The Simple Series Circuit	5
3	Resistance Calculations	10
4	Magnetic Calculations	11
5	Induced Voltage Calculations	14
6	Magnetic and Electrostatic Energy Storage	17
7	Magnetic Forces	19
8	Reactance Calculations	23
9	Reactors	26
10	Saturistors	28

1

BASIC PRINCIPLES—REACTORS—SATURISTORS

1 THE NATURE OF ROTATING ELECTRIC MACHINES

Electric motors and generators are machines for converting electric into mechanical power, or vice versa, making use of five natural laws:

1. A voltage e acting in a closed conducting circuit produces a current i proportional to the net circuit voltage and inversely proportional to the circuit impedance (Sect. 1.2). The power consumed in the circuit is equal to i^2R, where R is the effective circuit resistance (Sect. 1.3).

2. An electric current i flowing in a circuit of inductance L produces a magnetic flux linking the circuit proportional to Li, and a stored magnetic energy equal to $Li^2/2$ (Sect. 1.4).

3. A varying magnetic flux ϕ linking a conducting circuit produces a voltage e in the circuit proportional to $d\phi/dt$ (Sect. 1.5).

4. A voltage e impressed across a dielectric with a capacitance C creates a dielectric flux proportional to Ce, and a stored energy equal to $Ce^2/2$ (Sect. 1.6).

5. When the stored magnetic (or dielectric) energy of a circuit is changed as the result of the motion of a current-carrying conductor (or a charge carrying plate), a force is produced on the conductor (or the plate) proportional to the rate of change of the stored energy with the distance moved (Sect. 1.7).

A typical electric motor of the polyphase squirrel-cage induction type is illustrated in Fig. 1.1. The stator consists of a laminated steel core carrying insulated windings in slots, and supported in a frame so designed as to provide protection, adequate ventilation, and good appearance. In this type of motor, the rotor carries a short-circuited winding in nearly closed slots, separated from the stator by a small radial air gap. The rotor is supported by bearings in end shields, or brackets, which complete the enclosure and assure accurate positioning of the rotor in the stator bore.

The art of designing these rotating machines consists in (1) laying out a field structure and magnetic circuit best adapted to carry the necessary number of webers of magnetic flux with low magnetomotive force consumption and minimum hysteresis and eddy-current losses; (2) providing primary and secondary windings so disposed as to give the necessary current-carrying capacity and insulation strength, with a minimum of resistance loss, leakage inductance, and space occupied; (3) providing a mechanical structure adapted to transmit the torques produced to the load, and to operate quietly at the desired speed with optimum margins of safety and a minimum of windage losses; and (4) providing adequate cooling

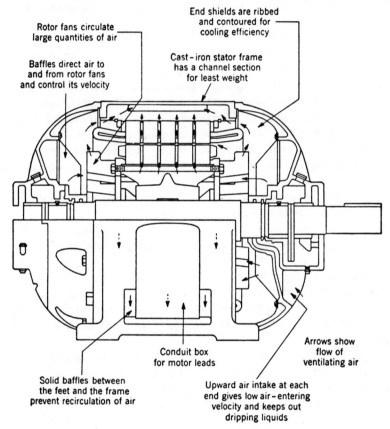

FIG. 1.1. Typical Polyphase Induction Motor with Squirrel-Cage Rotor and Sleeve Bearings.

means to remove the heat produced by the power losses without exceeding a safe temperature rise. To do all these things well, and at the same time to secure minimum costs, speedy production through properly standardized parts, good appearance, and close matching of the machine to the application, require engineering skills and ingenuity of a high order.

The designer must be able to calculate the performance of a given machine, once the dimensions are fixed. His real task, however, is to choose the materials and dimensions to give the desired performance.

There is only one answer to the first problem, but there are infinitely many answers to the second problem. Truly, the designer is an artist, who embodies his ideas in copper, steel, and insulation. He has the artistic opportunities of the sculptor and a wide scope for the creation of dynamic forms of beauty as well as usefulness.

With the continuing development of new and better materials, the unceasing demand of industry for new and better ways of driving generators and of applying motors, and the intense pressure of competition for lower costs, great advances are still ahead in the designing art. Table 1.1 illustrates this continuing trend of progress in design, showing the steady increase in the horsepower rating at 1800 rpm obtained from a given size of motor frame (NEMA frame 404) over the past seventy years.

Table 1.1

Years	Normal Horsepower Rating
1898–1903	7.5
1903–1905	10
1905–1914	15
1914–1924	20
1924–1929	25
1929–1940	30
1940–1956	40
1956–1961	50
1961–1966	60
1966–	100

2 The Simple Series Circuit

If an electric circuit consists of a resistance R ohms, an inductance L henries, and a capacitance C farads in series, the current of i amperes that flows in response to an impressed instantaneous voltage of e volts is determined at each instant of time, t, by the equation:

$$(1.1) \qquad e = iR + L\frac{di}{dt} + \frac{q}{C},$$

where q is the instantaneous charge of the capacitance, equal to:

$$(1.2) \qquad q = \int_0^t i\, dt.$$

The equations are most easily solved if the way that e varies with time is expressed by representing e as a Fourier's series of sinusoidal terms of different frequencies, $0, f_1, f_2, f_3, \ldots$:

$$e = E_0 + E_1 \sin 2\pi f_1 t + E_2 \sin 2\pi f_2 t + E_3 \sin 2\pi f_3 t + \ldots,$$

or:

$$(1.3) \quad e = E_0 + E_1 \sin(\omega_1 t - \theta_1) + E_2 \sin(\omega_2 t - \theta_2)$$
$$+ E_3 \sin(\omega_3 t - \theta_3) + \ldots.$$

In most practical cases, f_2, f_3, etc. are multiples of the fundamental frequency, f_1, as they all originate in the speed of rotation of the generator that is the source of voltage. The separate terms are then called the "harmonics" of the voltage wave.

Whatever the actual variation of e with time, it is always possible to find values of E_0, E_1, E_2, f_1, f_2, etc., to satisfy Eq. 1.3. Then, each of the terms of Eq. 1.3 can be substituted separately in Eq. 1.1 to find the corresponding current harmonics, and the actual current will be equal to their sum, provided that R, L, and C are independent of the current and frequency.

Considering only a single harmonic of Eq. 1.3, $E \sin \omega t$, and substituting this in Eq. 1.1, we have, under steady-state conditions:

$$(1.4) \qquad E \sin 2\pi f t = iR + L\frac{di}{dt} + \frac{1}{C}\int_0^t i\, dt.$$

For Eq. 1.4 to be true, if R, L, and C are constants, the steady-state value of i must also vary sinusoidally at the frequency f:

$$(1.5) \qquad i = I \sin(2\pi f t - \theta).$$

Substituting this in Eq. 1.4, and taking the initial value of q to be zero, the steady-state equation becomes:

$$(1.6) \qquad E \sin 2\pi f t = IR \sin(2\pi f t - \theta) + I\left(2\pi f L - \frac{1}{2\pi f C}\right)\cos(2\pi f t - \theta).$$

As both voltage and current vary sinusoidally, they can be represented by phasors† displaced by a fixed phase angle, θ, and revolving uniformly at the rate of f revolutions per second. Their instantaneous values are the projections of the phasors on the vertical (j) axis, at the time t, assuming E to be horizontal when $t = 0$, Fig. 1.2. The $\sin(2\pi f t - \theta)$ term of Eq. 1.6 is

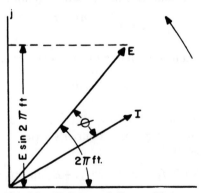

FIG. 1.2. Voltage and Current Phasors.

† A phasor is a quantity whose value is expressed by a complex number, $x + jy$. See *A.S.A.* Z10.8, paragraph 9.

θ in phase *behind* the $\sin 2\pi ft$ term, since it is a maximum $\theta/2\pi$ revolutions *after* $\sin 2\pi ft$ is a maximum. Likewise, the $\cos(2\pi ft - \theta)$ term is 90° *ahead* of $\sin(2\pi ft - \theta)$, since it is a maximum one-quarter revolution *before* $\sin(2\pi ft - \theta)$ is a maximum.

It is customary to represent a 90° ahead rotation in phase by the symbol $j = \sqrt{-1}$, since two such rotations correspond to a multiplication by $j^2 = -1$, which is a rotation through 180°. A forward rotation of θ is represented alternatively by the symbol $\underline{/\theta}$, or by the multiplier, $\cos\theta + j\sin\theta$. Counterclockwise motion is taken as positive.

Therefore, we can write:

$$\sin(2\pi ft - \theta) = (\cos\theta - j\sin\theta)\sin 2\pi ft,$$

and $\quad \cos(2\pi ft - \theta) = j\sin(2\pi ft - \theta) = (\sin\theta + j\cos\theta)\sin 2\pi ft.$

Substituting these in Eq. 1.6, and omitting the common factor, $\sin 2\pi ft$:

(1.7) $$E = I\left[R + j\left(2\pi fL - \frac{1}{2\pi fC}\right)\right](\cos\theta - j\sin\theta),$$

whence $\quad I = \dfrac{E}{|Z|}, \quad \cos\theta = \dfrac{R}{\sqrt{(R^2 + X^2)}} = \dfrac{R}{|Z|} = $ power factor,

and $\quad \sin\theta = \dfrac{X}{\sqrt{(R^2 + X^2)}},$

where $\quad X = \left(2\pi fL - \dfrac{1}{2\pi fC}\right)$

$\quad\quad\quad$ = the net circuit reactance at the frequency f,

and $\quad Z = R + jX = $ the corresponding circuit impedance.

The current in a simple (static) circuit is equal to the sum of the separate harmonics of the impressed voltage, each divided by the corresponding impedance. Each current harmonic lags behind the corresponding harmonic of the impressed voltage by its own "power-factor angle", $\theta_n = \cos^{-1}(R_n/Z_n)$. The effective, or rms, value of any single harmonic component is its peak value divided by $\sqrt{2}$, since the average value of $(I\sin\theta)^2$ over a cycle is $I^2/2$. The total rms current in the circuit is equal to the square root of the sum of the squares of the rms values of the separate current harmonics.

The power, or energy flow, in the circuit has two components. These are the transmitted watts, which form the real power, or active energy flow, on the one hand; and the wattless, or reactive, power, which merely oscillates to and fro, on the other hand.

The power flowing in the circuit at any instant is the product of voltage times current:

(1.8) $$W = (E \sin 2\pi ft)[I \sin(2\pi ft - \theta)]$$
$$= (EI/2)\cos\theta - (EI/2)\cos(4\pi ft - \theta)$$

(Active power)　　　　(Reactive power)
$$= (EI/2)\cos\theta(1 - \cos 4\pi ft) - (EI/2)\sin\theta \sin 4\pi ft.$$

This consists of two parts: $(EI/2)\cos\theta(1 - \cos 4\pi ft)$ is a pulsating "active" power flowing always in one direction, and having an average value equal to rms volts × rms amperes × power factor; $(EI/2) \sin \theta \sin 4\pi ft$

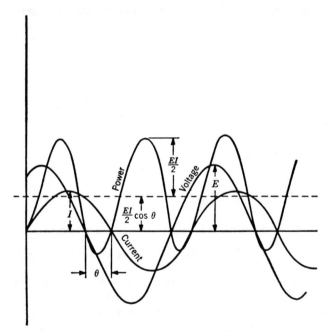

FIG. 1.3. Current, Voltage, and Power in a Single-Phase Circuit.

is an alternating "reactive" power which flows to and fro at double frequency, $2f$, and whose maximum amplitude is equal to rms volts × rms amperes × $\sqrt{(1-pf^2)}$. The sum of the two gives a total instantaneous power which pulsates at double frequency from a maximum value in one direction equal to $(1+\cos\theta) \times$ rms volts × rms amperes to a maximum value in the other direction equal to $(1-\cos\theta) \times$ rms volts × rms amperes, as indicated in Fig. 1.3.

In a balanced polyphase circuit, each separate phase carries both active and alternating power as above described, but at each instant the alternating powers of the n phases are displaced $360/n°$ in time phase, giving a zero sum, so that the only *net* power flowing in the combined circuits is the active power, distributed equally among the phases. There is no *net* alternating power flowing in the lines at any instant.

Some background of physical conceptions such as these is required if operations involving power determination are to be correctly carried through with the use of complex algebra. For, if a current is represented as $a+jb$, and a voltage as $c+jd$, referred to a common reference axis, their product $(ac-bd)+j(bc+ad)$ is equal to the single-phase *double-frequency* alternating power and not to the (active power)$+j$(reactive power) as might be supposed. The product has a phase angle equal to the sum of the phase angles of the current and voltage, and, therefore, represents a phasor rotating at twice the speed, or frequency, of its two factors. Being of double frequency, the phase angle of this product with reference to the current or voltage has no useful meaning. To obtain the true, or active, power by multiplication, it is necessary first to reverse the sign of the j term of either voltage or current, so that the product becomes $(ac+bd)\pm j(bc-ad)$. This has a phase angle equal to the difference of the phase angles of current and voltage, and, therefore, has zero speed of rotation, or is unidirectional. The real term now represents the active power $(EI/2)\cos\theta$, and the j term the reactive power $(EI/2)\sin\theta$.

The usual convention is to consider the magnetizing volt-amperes of an inductive load as positive reactive power. Thus, a synchronous condenser, or a capacitor, which draws a leading current from the line, is considered a generator of reactive power; and an induction motor, or a reactor, which draws its (lagging) magnetizing current from the power source, is a consumer of reactive power. This requires that the sign of the j term of the *current* be reversed when current and voltage are multiplied together, to obtain the correct sign for the reactive power. Thus, if the rms voltage impressed on an inductive load is E, the lagging current of I rms amperes flowing *into* the load is $I(\cos\theta-j\sin\theta)$, the active power delivered to the load is $EI\cos\theta$ watts, and the reactive power delivered to the load is *plus* $EI\sin\theta$ watts.

This process of deriving the total power from the product of I and E, expressed as phasors, is represented symbolically by the equation:

$$W = EI^* = \text{(active power)} + j\text{(reactive power)}$$

where I^* is the conjugate of the phasor I. (If $I = a-jb$, $I^* = a+jb$.)

3 Resistance Calculations

The resistivity of pure copper at 75 C† is:

(1.9) $\quad\quad\quad\quad \rho_{75} = 2.10 \times 10^{-6}$ ohms per cm cube

$\quad\quad\quad\quad\quad\quad = 0.827 \times 10^{-6}$ ohms per in. cube.

Since the temperature coefficient of resistance of copper at 20 C is 0.00393 per degree C, the resistivity at another temperature, $T°C$, can be found by the equation:

(1.10) $\quad\quad\quad\quad \rho_T = \rho_{75}\left(\dfrac{235+T}{310}\right).$

A typical transformer winding consists of a large number of coils of copper wire, interleaved with turn and section insulation to withstand transient voltage stresses. The resistance of the winding is equal to:

(1.11) $\quad R = \dfrac{\text{resistivity} \times \text{mean length of turn} \times \text{no. of turns in series}}{\text{area of one conductor}}$

$\quad\quad = \dfrac{\text{resistivity} \times \text{mean length of turn} \times (\text{turns in series})^2}{\text{conductor area} \times \text{turns in series}}$

$\quad\quad = \dfrac{2.10 N^2 L_t}{10^6 C}$ ohms at 75 C,

where $N =$ number of turns in series,

$\quad\quad L_t =$ mean length of turn in cm, and

$\quad\quad C =$ total copper cross-section of winding in square cm. (one coil side only).

The coefficient 2.10 becomes 0.827 if dimensions are in inches.

This formula makes it easy to compare widely different designs, with minimum chance of numerical error. If another metal than copper is used, the equivalent copper area should be used in Eq. 1.11 instead of the actual area. For aluminum $C = 0.61$ times the aluminum area, since the conductivity of hard aluminum is 0.61 that of copper. (This ratio varies from 0.62 for pure metal down to between 0.45 and 0.55 for typical casting alloys employed in squirrel-cage rotors.)

At high frequencies, stray leakage fluxes passing through the copper generate local voltages and eddy currents, which distort the current distribution and increase the effective resistance. These effects can be

† The temperature assumed in efficiency calculations according to American Standards.

calculated by well-known laws (Sect. 8.4), but they are small in well-designed machines under operating conditions.

As Eq. 1.11 indicates, with a given copper section, the winding resistance varies in proportion to the square of the number of turns. For a given core size, or magnetic flux capacity, the number of turns is proportional to the impressed voltage. For a given power, or kva rating, the current and, therefore, the needed conductor section are inversely proportional to the voltage.

Hence, the resistance drop at full-load current, expressed as a fraction of rated voltage, is a constant, substantially independent of the voltage, for a given core size, copper space, and kva rating:

$$(1.12) \quad \frac{IR}{E} = \frac{\text{kva} \times R}{E^2} \alpha \left[\frac{\text{kva} \times (\text{turns})^2 \times \text{turn length}}{(\text{turns})^2 \times \text{copper section}} = \frac{\text{kva} \times \text{turn length}}{\text{copper section}} \right].$$

Equation 1.12 shows that the I^2R losses of a machine are dependent on the kva rating and the physical dimensions, but are independent of the voltage rating, except in so far as the space required for high-voltage insulation requires a longer turn length or reduced copper section.

4   Magnetic Calculations

A current of i amperes flowing in N turns of a conductor creates a magnetomotive force:

$$(1.13) \quad M = Ni \text{ ampere turns}$$
$$= 0.4\pi \, Ni \text{ gilberts}.$$

Mmf intensity is measured usually in terms of ampere turns per cm., or in oersteds, where:

(1.14)   1 oersted = 1 gilbert per cm. = 0.787 ampere turns per cm.

(2.02 ampere turns per inch)

An mmf of Ni ampere turns linking a uniform magnetic path of area A square meters and length G meters creates a magnetic flux $\phi = BA$ webers, where:

$$(1.15) \quad B = \mu\mu_0 Ni/G = 4\pi\mu Ni(10^{-7})/G \text{ webers per square meter.}$$

μ is the relative permeability of the flux path material, and μ_0 is the permeability of free space, $4\pi(10^{-7})$ henries per meter.

For air and other non-magnetic materials, μ is one. For steel, μ varies in the manner shown by Fig. 1.4. If G is measured in cm.:

$$(1.16) \quad B = 1.258\mu Ni(10^{-8})/G \text{ webers per sq cm.}$$

If the flux path is not uniform, it should be divided into sections that are substantially uniform, and the mmf required to produce the same flux in each should be calculated. The sum of these separate mmfs will be the total mmf required for the complete magnetic circuit.

Magnetic flux densities of the order of 1.4 to 1.7 webers per square meter normally are employed in rotating electric machines. Even at these densities, the mmfs required for the magnetic paths in iron normally are

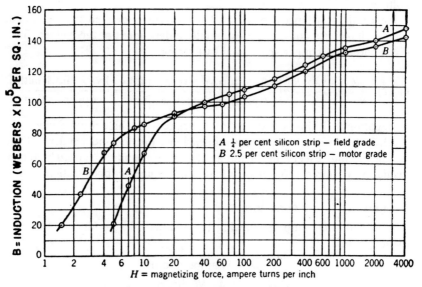

FIG. 1.4. Magnetization Curves of Silicon Steel.

less than about 30 per cent of that required for the air gap between stator and rotor.

It is customary to represent a direction of current, flux, or voltage away from the observer (into the plane of the paper) by a cross, and a similar flow toward the observer by a dot.

The positive direction of emf and of current flow in d-c circuits is taken to be that away from a positive source, as from the cathode to the anode of a battery. In a zinc–acid–copper cell, the copper electrode is positive, and the zinc is negative, so that the positive direction of current flow in the external circuit is from the copper to the zinc; or, in a lead–sulphuric acid storage battery, it is from the lead peroxide plate to the sponge lead plate. This convention was adopted many years ago before the electron theory was developed. It is now considered that a current consists of a flow of electrons, or units of negative charge. The electrons, being negative, move in the direction opposite to that taken as the positive direction of current.

From the viewpoint of electronic theory, therefore, an opposite sign convention would be required to correspond with physical reality.

The positive direction of magnetic flux in the external magnetic circuit is taken to be that away from the north pole of a magnet. The voltages, and corresponding currents, induced by any motion are always in a direction to oppose the motion, in accordance with Lenz's law and the conservation of energy.

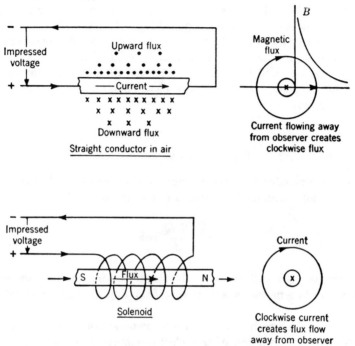

FIG. 1.5. Magnetic Flux and Current Relations.

These relations of flux and current are indicated in Fig. 1.5, where a current from left to right is seen to produce an upward flux above the conductor (dots), and a downward flux below it (crosses). Or a current flowing away from the observer creates a flux encircling it in a clockwise direction. And a clockwise flow of current in a coil creates a magnetic flux through the coil, away from the observer.

If a straight conductor in air, with the return path far away, carries a current i amperes, a circular magnetic field will be created around the conductor, Fig. 1.5. At any radius R (outside the conductor), the flux density will be, by Eq. 1.15:

(1.17) $\quad B = 4\pi i (10^{-7})/2\pi R = 2i(10^{-7})/R$ webers per square meter.

if R is in meters; or $0.508i(10^{-8})/R$ webers per sq in., if R is in inches.

For an actual case, the resultant flux density at any point is found by summing the contributions made by all the current-carrying conductors present.

Equation 1.17 is useful for estimating the inductance of free conductors and coil ends. It should be particularly noted that the flux density is a maximum at the surface of the conductor, and decreases inversely as the radius, as indicated at the upper right of Fig. 1.5. Hence, if a closed single wire loop is formed in air, the flux through the loop will be concentrated in a narrow strip along the periphery, and the flux density in the center of the loop will be small by comparison. If the student forms the habit of visualizing magnetic fields in this way, he will find designing easier and will avoid mistakes.

5 Induced Voltage Calculations

When a time-varying magnetic flux of ϕ webers links N turns of an electric circuit, a voltage e is induced in the circuit:

$$(1.18) \qquad e = -\frac{N\,d\phi}{dt} \text{ volts.}$$

The negative sign indicates that a current flowing in response to the induced voltage produces a magnetic field opposing the original field (Lenz's law).

In terms of electronic theory, every metallic conductor contains an "electron gas", composed of free electrons moving about in a random manner with velocities dependent on the temperature. The potential barrier due to atomic forces existing at the surface of the metal prevents the electrons from escaping from the conductor, unless their escape is facilitated by high temperature, electron bombardment, or other effects. A magnetic field cutting the conductor creates a force on each electron, causing a drift of the entire body of electrons along the conductor. The resulting current is measured by the net number of electrons passing a given point per second. A current of one ampere (one coulomb per second) represents a flow of 3×10^9 electrostatic units of charge (statcoulombs) or 6.25×10^{18} electrons per second, since the charge of one electron is 4.8×10^{-10} electrostatic units.

Equation 1.18 may be expressed alternatively in terms of the rate of flux cutting when a conductor of length l meters moves at a velocity of V meters per second transversely across a fixed magnetic field of B webers per square meter:

(1.19) $$E = BlV \text{ volts.}\dagger$$

The product BlV must be integrated around the entire circuit in which the induced voltage E is to be measured, so as to include all the changes in flux linkages that are occurring.

Whether the induced voltage is calculated by Eq. 1.18, as a "transformer voltage", or by Eq. 1.19, as a "speed voltage", is immaterial, since the two expressions are equivalent ways of calculating the rate of change of flux linkages of the circuit. Which expression is used in any given case is largely a matter of convenience, but it is important that all the changes in linkages be taken into account. To emphasize this, it is well to state the law of induction more fully, as given by L. V. Bewley:

> The electromotive force induced in a closed circuit C is equal to (the time rate of change in magnitude of that flux in space which is linked at the given instant with the circuit C) + (the sum for all the elements around the circuit of the triple product *length of the moving element × component of velocity perpendicular to the length of the element × component of flux density normal to the plane of motion of the element*).

If the flux, ϕ, linking an N-turn coil varies sinusoidally in time, at a frequency f cycles per second:

(1.20) $$\phi = \phi_M \sin 2\pi f t \text{ webers,}$$

then $$E = -\frac{N\,d\phi}{dt} = -2\pi f N \phi_M \cos 2\pi f t \text{ volts;}$$

so that the root-mean-square value of E is:

(1.21) $$E_{rms} = \frac{2\pi f N \phi_M}{\sqrt{2}}$$
$$= 4.443 f N \phi_M \text{ volts,}$$

and its time phase lags 90° behind that of the inducing flux.

This is the basic equation for induced voltage used in designing all types of a-c machines. At 60 cycles, a maximum flux of 3.75×10^{-3} webers is required to induce one rms volt per turn.

In a given closed circuit, the flux can not increase indefinitely, because of magnetic saturation and the practical limits of available mmf. Therefore, the time average voltage given by Eq. 1.18 over a considerable period must approach zero as a limit. The only way that a continuous unidirectional

---

† In older texts, Eqs. 1.18 and 1.19 had a factor of 10^{-8} on the right-hand side, because the maxwell was then taken as the unit of flux. The modern use of the weber = 10^8 maxwells as the flux unit eliminates this 10^{-8} factor.

voltage can be produced electromagnetically is by means of a switching device, which continually substitutes a new circuit element, or, in effect, changes the value of N in Eq. 1.18. A commutator, for example, is a device for short-circuiting each coil of an armature winding in sequence, as soon as the coil has turned to a position of minimum induced voltage (or maximum flux), and then reconnecting the coil into a parallel circuit in the reversed direction. Thus, all coils with decreasing flux are constantly being connected in parallel reversed with all the coils having increasing flux, and the voltage of the combined circuit is sustained in one direction.

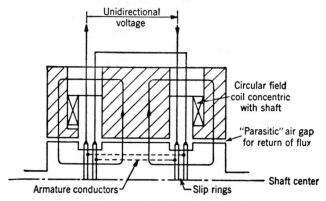

FIG. 1.6. Homopolar Generator.

Many inventors have tried to circumvent the need for a commutator by devising generators in which the magnetic flux cuts the armature conductors in only one direction. All these machines can be reduced to some form of homopolar, or acyclic generator, as in Fig. 1.6.

In a homopolar generator, all the magnetic flux flows across the air gap in one direction from one section of the core length, and, after flowing along the shaft, returns across the air gap to another section of the core. The rotor conductors (in the center section of the core) are continuously cut by flux in one direction only. By bringing them out to slip rings at both ends of the rotor, a continuous voltage is obtained corresponding to one turn linking the entire flux. To produce a voltage corresponding to more than one turn, separate series-connected slip rings must be provided for each conductor.

This machine forms the limiting case of a one-segment commutator. Homopolar machines are limited to low voltages, and require high speeds if much power is to be obtained. B. G. Lamme's classic article describing the development of a 2,000-kw, 1,200-rpm generator of this type, and the attendant difficulties with high-speed current collection using brushes running on metal slip rings, is very well worth reading. Recently, however, the development of liquid metal current collection, employing sodium–potassium mixtures in a nitrogen atmosphere, has permitted the collection of currents up to the order of a million amperes at speeds above 3,000 rpm, so that direct-current power outputs of many thousands of kilowatts now can be obtained from acyclic generators.

6 Magnetic and Electrostatic Energy Storage

It is important to visualize the way that energy is stored in a magnetic field. If i amperes flowing through N turns link a magnetic flux path of permeability μ, area A square meters, and length G meters, the flux produced, by Eq. 1.15 is:

$$(1.22) \qquad \phi = BA = \frac{4\pi\mu A N i \, 10^{-7}}{G} \text{ webers.}$$

If, initially, the current is zero, and a voltage is suddenly applied, the current and flux will build up over a period of time. The voltage induced in the circuit, opposing the impressed voltage, from Eq. 1.18, will be:

$$(1.23) \qquad e = -N \, d\phi/dt = \frac{-4\pi\mu N^2 A \, 10^{-7}}{G} \, di/dt \text{ volts.}$$

Thus, the amount of energy delivered to the magnetic field during the time t, while the current is building up to i amperes, will be:

$$(1.24) \qquad W_{mag} = -\int_0^{t_1} e i \, dt = \frac{4\pi\mu N^2 A \, 10^{-7}}{G} \int_0^i i \, di = \frac{2\pi\mu N^2 A \, 10^{-7} i^2}{G}$$

$$= \frac{B^2 A G \, 10^7}{8\pi\mu} \text{ wsec (joules)}$$

so that the energy per cubic meter is $B^2 10^7 / 8\pi\mu$ wsec.

For example, if the flux density is 1 weber per square meter (10,000 maxwells per sq cm), in an air path, the stored magnetic energy is:

$$(1.25) \qquad (10^7/8\pi)(10^{-6}) = 0.398 \text{ wsec* per cu cm}$$

$$= 6.53 \text{ wsec per cu in.}$$

$$= 4.81 \text{ ft lb per cu in.}$$

We do not know that the energy is actually located in the volume of the field, but this assumption gives the right answer, and it is very helpful in visualizing the effects of any changes in the field. If the field is not uniform, the total energy is found by integrating $10^7 B^2 / 8\pi\mu$ times each element of

---

*1 watt-second = 1 newton-meter = 0.737 ft lb.

volume over the entire space. Since μ for steel is of the order of 100 or more, the energy stored in the iron portions of a magnetic path is very small. Therefore, in a rotating electric machine, the air spaces between stator and rotor, in the slot openings, and near the end conductors are the seat of nearly all the magnetic energy. Since the machine is used to convert and transmit energy, the air spaces may be considered analogous to the steam passages in a turbine. Just as the turbine blades and diaphragms are shaped to enable the steam flow to transform its thermal energy into mechanical energy with the least loss from eddies or surface friction, so the teeth and air-gap dimensions of an electric machine are chosen to enable the magnetic energy whirling through the air spaces to be converted into mechanical energy with the least loss from stray fields or wasted mmf.

In visualizing the production of a magnetic field by a current, it must be remembered that the field does not appear instantaneously, but is propagated through free space at the velocity of light. Thus, there is a time lag between the appearance of a current and that of its self-inductive flux in air, just as there is a lag in the building up of the current in a circuit of high inductance. In each case, the lag is due to the necessity of storing energy in the magnetic field, this energy being equal to the product of the current by the in-phase component of induced voltage, integrated over the period of field build-up.

Energy is also stored in the electric field produced by a voltage acting across a gap. If the voltage is E, the gap length is d cm, its relative dielectric constant is K, and the cross-sectional area of the gap is S sq cm, the electric stress is:

$$E/d \text{ volts per cm.}$$

The electrostatic capacity of the gap is:

(1.26) $$C = SK/4\pi d \text{ statfarads}$$

$$= \frac{SK}{36\pi d \times 10^{11}} \text{ farads;}$$

and the corresponding charge on each surface is:

(1.27) $$Q = CE = \frac{SKE}{36\pi d \times 10^{11}} \text{ coulombs.}$$

The stored energy is the product of the charge by the average voltage during build-up:

(1.28) $$W = \tfrac{1}{2}QE = \tfrac{1}{2}CE^2 = \frac{SKE^2}{72\pi d\, 10^{11}} \text{ wsec.}$$

The energy per unit volume is:

(1.29) $$\frac{W}{Sd} = \frac{K}{72\pi \times 10^{11}}\left(\frac{E}{d}\right)^2 \text{ wsec per cu cm}$$

$$= 0.442K \times 10^{-13}\left(\frac{E}{d}\right)^2 \text{ wsec per cu cm.}$$

For air, the relative dielectric constant is 1, and the limiting dielectric stress before breakdown occurs is of the order of 30,000 volts per cm at atmospheric pressure. Therefore, the maximum electrostatic energy that it is feasible to store in room air is $0.442 \times 10^{-13}(30,000)^2 = 0.398 \times 10^{-4}$ wsec per cu cm.

Comparing this with Eq. 1.25, it is seen that the energy that it is feasible to store in an air gap with a magnetic field is some 10^4 times greater than can be stored with an electrostatic field. This explains why electromagnetic rather than electrostatic machines are universally used for power purposes.

7 Magnetic Forces

The fact that energy is stored in a magnetic (or electrostatic) field implies that a mechanical force is exerted by that field.

Suppose that a magnetic field with flux density B is created in a gap of length G and area A, and then that mechanical force is applied to increase G to $G+g$, while at the same time the mmf applied is varied so as to maintain a constant flux. Then, since ϕ does not change, there is no voltage induced in the circuit, and no power is supplied or removed electrically. However, the total magnetic energy is increased, from Eq. 1.24, by an amount:

(1.30) $$\Delta W = Ag(B^2 10^7/8\pi\mu) \text{ wsec, or newton-meters.}$$

This extra energy must have been supplied by the work done in increasing G by the amount g, against the resistance of the magnetic force, F. Hence:

(1.31) $$Fg = \Delta W, \quad \text{or} \quad F = AB^2 10^7/8\pi\mu \text{ newtons,}$$

where A is in square meters and B in webers per square meter.

Thus, a magnetic field of density B webers per square meter, acting across an air gap, exerts an attractive force equal to:

(1.32) $$F = B^2\, 10^7/8\pi \text{ newtons per square meter}$$

if B is in webers per square inch,

$$F = 1.387 B^2\, 10^8 \text{ pounds per square inch.}$$

A typical value of B in the air gap of an electric motor or generator is 0.8 weber per square meter, which corresponds to a force of $80/\pi = 25.4$ newtons per square cm, or 36.8 lb per sq in. The feasible flux density in the air gap is limited by the saturation density of steel to less than 2 webers per square meter, giving not over 230 lb per sq in., which places an upper limit on the forces that can be produced in practical machines.

It should be remembered that no magnetic force is exerted on a wooden pole, for example. Therefore, the force on a steel, or magnetic, pole is only due to the flux that exists over and above that which would exist if the pole were replaced by air. The actual force on the magnetic pole, or keeper, due to the flux leaving it, is proportional to the difference between the absolute F values in the air and in the iron, or:

$$(1.33) \qquad F_{\text{effective}} = \frac{B^2 \, 10^7}{8\pi} \left(1 - \frac{1}{\mu}\right) \text{ newtons per square meter.}$$

As a check on the analysis, we may consider the case of a magnetic field produced by a constant current, I, while the length of flux path is reduced by mechanical means from a very large value to G. The power supplied by the electric circuit will be:

$$(1.34) \qquad W_{\text{elect}} = -\int_0^t eI \, dt = -4\pi\mu N^2 A I^2 \, 10^{-7} \int_\infty^G d(1/g)$$
$$= 4\pi\mu N^2 A I^2 \, 10^{-7}/G \text{ wsec.}$$

This is twice the total energy stored in the magnetic field, by Eq. 1.24. However, the work done in decreasing the gap to G is:

$$(1.35) \qquad W_{\text{mech}} = -\int_\infty^G F \, dg = -\frac{AB^2 \, 10^7}{8\pi\mu} \, dg$$
$$= -2\pi\mu N^2 A I^2 \, 10^{-7} \int_\infty^G dg/g^2 = 2\pi\mu N^2 A I^2 \, 10^{-7}/G \text{ wsec.}$$

This is the same as the magnetic energy, Eq. 1.24. Hence, the power supplied electrically goes half into the magnetic field, and half into mechanical work. This is in accord with the basic law, due to Lord Kelvin, that:

> When, in a singly excited magnetic circuit, without saturation, a deformation takes place, at a constant current, the energy supplied from the line is divided into two equal parts, one half increasing the stored energy of the magnetic circuit, the other half being converted into mechanical work.

The principle of virtual displacement used in finding Eq. 1.31 can be used to find the magnetic force in any other situation. Suppose a conductor of length l meters, carrying a current of I amperes, is located in a cross magnetic field of density B webers per square meter, as shown in Fig. 1.7.

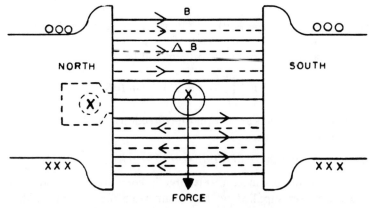

FIG. 1.7. Current in Magnetic Field.

Assuming that the lines of magnetic force go straight across the two air gaps, above and below the conductor, the flux density produced by the current above the conductor, from Eq. 1.15 will be:

(1.36) $\qquad \Delta B = 2\pi I\, 10^{-7}/G$ webers per square meter.

Below the conductor, the flux density will be the same, except that it is negative. Therefore, the net density above will be $B_0 + \Delta B$, and that below will be $B_0 - \Delta B$. If the conductor moves down a distance x, the total magnetic energy will be increased by an amount:

1.37) $\qquad W = lGx(10^7/8\pi)\left[(B_0+\Delta B)^2 - (B_0-\Delta B)^2\right]$

$\qquad\qquad = lGx(10^7/8\pi)(4B_0\,\Delta B) = B_0\, lIx$ newton meters.

Hence, the force acting on the conductor is $F = W/x$, or:

(1.38) $\qquad\qquad F = B_0\, lI$ newtons.

The force is the same whether the conductor is in the air gap, as shown in Fig. 1.7, or in a slot at one side, as shown in dotted lines. The force acts on the conductor itself when the conductor is in the air-gap space, but it acts chiefly on the iron of the pole face and slot opening if the conductor is slot-embedded. This simple equation, (1.38), is the usual basis for calculating the torques developed by all types of rotating machines.

To find the force between two parallel conductors, we shall assume them to carry equal and opposite currents, I, and that their length, l, is so great that end effects can be neglected. If r_0 is the radius of each conductor, and R their distance apart in meters, the (external) inductance of each, including flux out to a distance D meters, by Eq. 1.17, will be:

$$(1.39) \quad L = \int_{r_0}^{D} lB\,dr/I = 2l\,10^{-7} \int_{r_0}^{D} dr/r = 2l\,10^{-7} \ln D/r_0 \text{ henries.}$$

The mutual inductance between them will be:

$$(1.40) \quad M = 2l\,10^{-7} \int_{R}^{D} dr/r = 2l\,10^{-7} \ln D/R \text{ henries.}$$

Since the two currents are in opposite directions, the flux-induced voltage around the loop will be:

$$(1.41) \quad e = 2(L-M)\,di/dt = 4l\,10^{-7} \ln R/r_0 \text{ volts,}$$

since the $\ln D$ terms cancel if D is large in comparison with R.

And the power supplied to the circuit while the current builds up to I amperes will be:

$$(1.42) \quad W = \int e i\,dt = 2(L-M) \int_0^I i\,di = (L-M)I^2$$
$$= 2I^2 l\,10^{-7} \ln R/r_0 \text{ newton meters.}$$

If R is increased to $R+\Delta R$ by mechanical means, the increase in magnetic energy will be:

$$(1.43) \quad \Delta W = 2I^2 l\,\Delta R\,10^{-7}/R \text{ newton meters.}$$

And, the force aiding the change is this divided by ΔR:

$$(1.44) \quad F = 2I^2 l\,10^{-7}/R \text{ newtons.}$$

This is the same force derivable from Eqs. 1.17 and 1.31 by substituting for B_0 in Eq. 1.31 the flux density produced at the center of one conductor by the current in the other, as given by Eq. 1.17. We may generalize and say that the force between two long parallel conductors of length l, carrying currents I_1 and I_2, and spaced R meters apart, is:

$$(1.45) \quad F = 2I_1 I_2\,10^{-7} l/R \text{ newtons}$$
$$= 4.50 I_1 I_2\,10^{-8} l/R \text{ lb.}$$

In general, magnetic forces on current-carrying conductors act in a direction to increase the inductance of the circuit. Thus, a loop carrying a high current tends to become a circle. For this reason, heavy cables lying on a floor, and the end windings of large machines, should be firmly tied down, to avoid possible injury from their motion under short-circuit current forces.

8 Reactance Calculations

The resistance component, R, of the impedance of any electromagnetic structure for power purposes must be small, to hold down the I^2R losses, and so obtain good efficiency. The capacitance component, $1/2\pi fC$, is small at power frequencies (200 cycles or less), because the electrostatic energy storage is very small, as shown in Sect. 1.6. Hence, for most rotating machines and transformers, the reactance, $2\pi fL$, is by far the largest component of the impedance.

The inductance, L, is a measure of the amount of magnetic energy that can be stored, and of the force that can be produced by a given current. And, it is an inverse measure of the amount of current that a given applied voltage will produce in the machine winding. Hence, designers and users of electromagnetic devices need to be thoroughly familiar with inductances, and how to calculate them.

The inductance, L, is defined as the ratio of the volts induced by a changing flux to the rate of change of the current producing the flux:

(1.46) $\quad L = e/di/dt = N\,d\phi/dt/di/dt = N\,d\phi/di$ henries

$\qquad\quad = N\phi/i$, if there is no saturation.

Fig. 1.8 shows a C-core reactor with two windings whose structure is typical of many electromagnetic devices. Consider first the upper winding

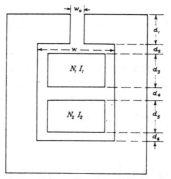

FIG. 1.8. C-Core Reactor with Air Gap.

with N_1 turns and carrying a current I_1 amperes. From Eq. 1.15, the flux it produces across the d_1 and d_2 sections of the slot opening is:

(1.47) $\quad \phi_1 = 4\pi N_1 I_1\,10^{-7}l[(d_1/w_0)+(d_2/w)]$ webers

if all dimensions are in meters.

If the ampere turns required for the iron portions of the magnetic circuit are negligible, the corresponding value of inductance for winding 1 is, from Eq. 1.46:

(1.48) $\quad L_1 = N_1\phi_1/I_1 = 4\pi N_1^2\,10^{-7}l[(d_1/w_0)+(d_2/w)]$ henries.

I_1 does not produce any flux across the d_4, d_5, or d_6 sections of the core window, or slot, because these are magnetically short-circuited by the iron core, with zero reluctance. However, there is some flux across d_3, which adds to the inductance. The flux density at the top of d_3 is the same as it is across d_2, and it is zero at the bottom of d_3, varying linearly between. The flux at the top links all the N_1 turns, that at the bottom does not link any of the turns, with the number linked varying linearly between. Since the inductance is proportional to flux times turns linked, the total inductance will be proportional to the integral:

$$(1.49) \qquad \int_0^1 (N_1 x)(d_3 x)\, dx = N_1 d_3/3$$

where x is the ratio of the distance up from the bottom of d_3 to its top. Hence, the total reactance of the slot-embedded part of winding 1 is:

$$(1.50) \qquad L_1 = 4\pi N_1^2 10^{-7} l\,[(d_1/w_0) + (d_2/w) + (d_3/3w)] \text{ henries.}$$

Of course, there are additional terms for the flux linking the return conductors of the winding, outside the slot, but these will not be considered here.

In the same way the inductance of the slot-embedded portion of winding 2 is:

$$(1.51) \qquad L_2 = 4\pi N_2^2 10^{-7} l\,\{(d_1/w_0) + [(d_2 + d_3 + d_4)/w] + (d_5/3w)\} \text{ henries.}$$

There is no flux across d_6. Evidently the d_1 and d_2 terms are common to both windings, or are mutual. The flux across d_3 due to winding 1 links all the turns of winding 2. However, the flux density here varies linearly from full value at the top of d_3 to zero at the bottom. The average density is, therefore, half the maximum, and the corresponding term in the mutual inductance is $d_3/2w$. Hence, the total mutual inductance between windings 1 and 2 is:

$$(1.52) \qquad L_m = 4\pi N_1 N_2 10^{-7} l\,[(d_1/w_0) + (d_2/w) + (d_3/2w)] \text{ henries.}$$

The flux density across d_3 due to winding 2 is uniform over the whole height, but the flux at the top links all the turns of winding 1, while that at the bottom does not link any. Therefore, the average linkages are proportional to $d_3/2w$, as before, and the mutual inductance is found to be the same from whichever winding it is viewed, as expected.

In most machines, the d_1/w_0 term is higher than the others, so that the leakage inductances, $L_1 - L_m$ and $L_2 - L_m$ are much smaller than the mutual inductance, when referred to the same number of turns in each winding.

REACTANCE CALCULATIONS

If the air gap, w_0, is made very small or zero, the ampere turns consumed in the iron portions of the magnetic circuit become large with respect to the gap ampere turns, and the value of the inductance will vary with the current, depending on the flux density in the iron, as indicated by Fig. 1.4. We must then decide on how to define the inductance of a saturated magnetic circuit. The answer depends on the purpose in view.

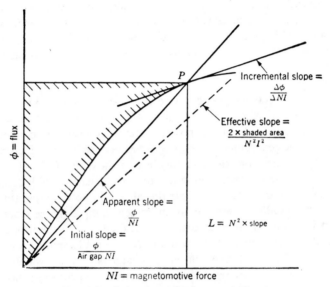

FIG. 1.9. Inductance of a Saturated Circuit.

Fig. 1.9 indicates four alternative definitions. lowest value is the incremental inductance, defi.. magnetization curve at the final flux value, or:

(1.53) $$L = \frac{N \Delta\phi}{\Delta I}.$$

The next lowest is the effective inductance, which defines the stored energy in the magnetic field:

$$L = -\frac{2}{I^2} \int ei\,dt,$$

or, from Eq. 1.18:

(1.54) $$L = \frac{2}{I^2} \int_0^\phi Ni\,d\phi = \frac{2 \times \text{shaded area in Fig. 1.9}}{I^2}.$$

The apparent inductance, defined by the final values of flux and current, ϕ/NI, and the initial inductance, defined by the air-gap line of the magnetization curve, are successively higher. Still other values may be derived to give the correct time of current build-up or decay between any two limiting values.

To avoid such ambiguities as these, the reactance calculations throughout this book are derived as $2\pi f N\phi/I$ values, on the apparent inductance basis. In rotating machines, nearly all the flux paths have series air gaps, so that saturation effects can usually be allowed for by simple corrections to the apparent inductance values. In the design of transformers, contactors, and other apparatus with closed magnetic circuits without air gaps, however, these saturation effects must be given more careful consideration.

9 Reactors

The simplest electromagnetic device is a reactor, consisting of a single coil linking a magnetic circuit, such as Fig. 1.8 (with its two windings connected in series aiding). In power systems, reactors are connected in series with large generators to limit short-circuit currents, in which case the air gap may be very large, or there may be no iron at all. The coil may then be cast in concrete to resist high-current forces. Reactors with completely closed iron cores are used to regulate the current in power apparatus, or to maintain the stability of arcs, as in fluorescent lamps. These reactors are provided with well-laminated cores, so that the losses in the iron are very small, and the resistance of the winding is also made small, to minimize power losses. The field coils and electromagnets that are widely used in direct-current circuits are reactors also, usually having air gaps and relative motion.

If there is no air gap in the core, the magnetic flux builds up to nearly its saturation value, even with small currents. Thus, when a sinusoidal voltage wave is applied, the flux also builds up sinusoidally, but the current wave is peaked, as in Fig. 1.10 (a). If the current waveform is maintained sinusoidal, the flux wave will be flat-topped, and the voltage will be peaked, as in Fig. 1.10 (b).

In most power circuits, the current harmonics are suppressed, as by using a three-phase Y-connection, or because the reactor forms only a small portion of the total impedance. On this basis, the closed core reactor voltage is nearly equal to that due to a sine wave flux with a peak value equal to the saturation limit for $\sqrt{2}$ times the rms current. And, the effective reactance varies nearly in proportion to the -0.75 power of the current. This type of reactor is widely used in the secondary circuits of wound rotor induction motors.

A saturable reactor is a closed core reactor with a d-c control winding. By varying the control current, as in response to an error signal, the core can be saturated to varying degrees, thereby varying the effective reactance of the a-c winding, so obtaining precise circuit regulation. The control winding has many turns, so that it can be excited with a small current and convenient low voltage (equal to the *IR* drop). To prevent high voltages

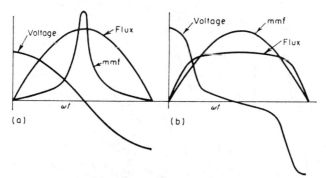

Fig. 1.10. Wave Forms for Saturated Magnetic Circuits.

from being induced in the control winding, it is essential that the flux in it should not vary during the a-c cycle. For this reason, saturable reactors are provided with three legs, as in Fig. 1.11. The center leg that carries the control winding has twice the cross-sections of the outer legs, which carry duplicate a-c coils. Normally, the a-c coils are connected in parallel aiding,

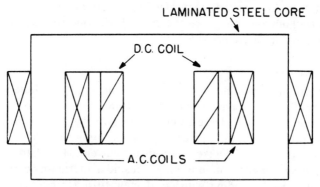

Fig. 1.11. Saturable Reactor.

so that they carry equal fluxes, and there is no a-c flux in the center leg. The d-c control flux divides equally between the two outer legs, reducing their flux capacity in proportion to the control current.

Fig. 1.12 shows curves of a-c voltage versus current for different values of control current, for a typical saturable reactor. With the parallel connection of the a-c coils, any change in the control current induces

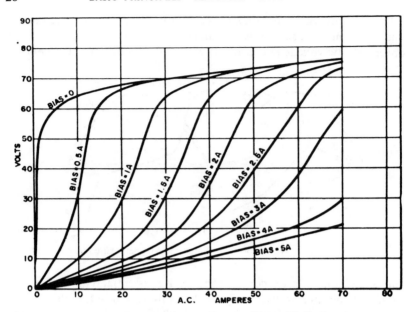

Fig. 1.12. Voltage-Current Curves of Saturable Reactor.

opposing currents in the two a-c coils, making the response slow. To avoid this, the two a-c coils may be connected in series, but then any dissymmetry may induce high a-c voltages in the control winding, so that fewer turns and more insulation are needed.

10 SATURISTORS

If a hard magnetic material, such as Alnico, is substituted for the laminations in the core of a reactor, and an alternating voltage is impressed, there will be a large hysteresis loss in the Alnico, so that the reactor impedance will have a large resistive component. Fig. 1.13 shows B-H curves for Alnico 5 taken with direct current. Fig. 1.14 shows the corresponding flux-time curve when a sinusoidal alternating current flows in the "Saturistor"† winding, with a peak magnetizing force of 800 oersteds. The flux wave zero point is delayed until the current has reached 75 per cent of its peak value, corresponding to a time lag of 48 electrical degrees, and giving an apparent power factor of sin 48° or 0.75.

Fig. 1.15 shows actual R, X, and Z curves for a Saturistor with Alnico 5 in the core legs, taken at 60 cycles. These curves show that a Saturistor has three interesting and useful properties:

† Trademark of the General Electric Company.

SATURISTORS

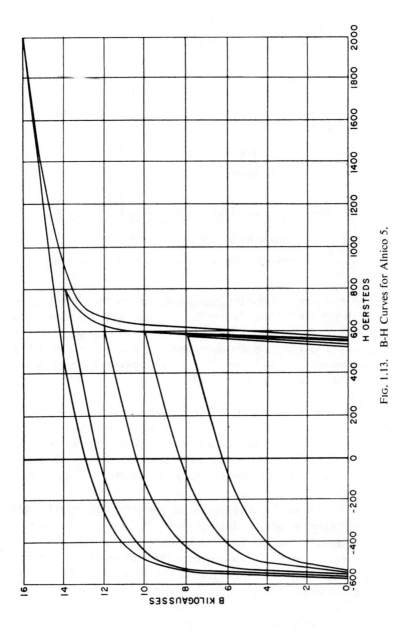

Fig. 1.13. B-H Curves for Alnico 5.

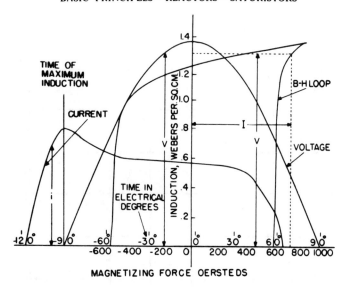

FIG. 1.14. Voltage-Current Wave Forms for Alnico 5 Saturistor.

1. The resistance, and also the impedance, are very low when the current is small. Below the "pick-up" current, corresponding to the coercive force of the Alnico, the impedance is the same as that of the winding with a magnetic core having a permeability equal to about 3.4 times that of air, and no hysteresis loss. At higher currents, the impedance rises rapidly to a maximum at about twice the "pick-up" current value, and at still higher currents it declines gradually.

Thus, the Saturistor impedance increases when the current increases—a unique property that is the inverse of the magnetic saturation effect.

2. The power factor of the Saturistor is 0.7 to 0.8 over a considerable range of current. And, the power factor is substantially independent of frequency, because the hysteresis loss, like the reactance, is proportional to frequency.

3. If a control winding is provided on the core, varying the control current will vary the effective impedance of the Saturistor, just as in the case of a saturable reactor. Figure 1.16 shows a three-legged Saturistor made of thick laminations (to increase the eddy-current loss without reducing the flux capacity at 60 cycles), with Alnico 5 blocks in the two outer legs enclosed by duplicate a-c windings, and a laminated center leg with a d-c winding. Figures 1.17 and 1.18 show the variations of effective 60-cycle resistance and reactance per phase with different control currents, when three of these units were connected in Y.

Therefore, the Saturistor is actually a saturable resistor, whence the

SATURISTORS

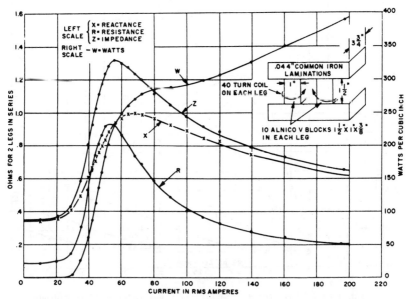

FIG. 1.15. 60-Cycle Impedance-Current Curves of Alnico 5 Saturistor.

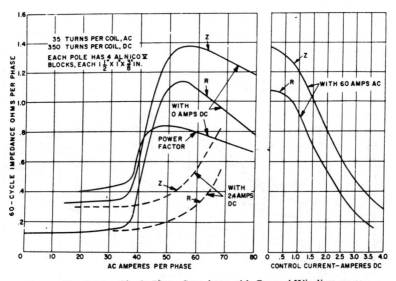

FIG. 1.16. Single-Phase Saturistor with Control Winding.

32   BASIC PRINCIPLES—REACTORS—SATURISTORS

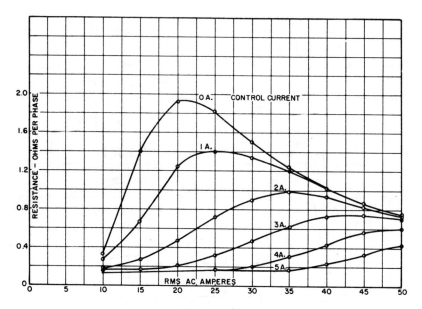

Fig. 1.17. 60-Hz Resistance-Current Curves for Three-Unit Saturistor, with Various Control Currents.

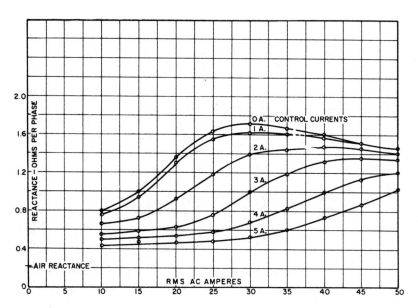

Fig. 1.18. 60-Hz Reactance-Current Curves for Three-Unit Saturistor, with Various Control Currents.

name—Satur(able Res)istor. It is a resistance that can be varied in response to a signal, without the use of any contactors, brushes, sliding contacts, or diodes. At current values greater than about 20 amperes, the Alnico is saturated, so that the flux and the hysteresis loss remain nearly constant. Therefore, at high currents the effective reactance varies nearly inversely as the first power of the current, and the resistance inversely as the square of the current. Thus, the power factor of a Saturistor falls rapidly when the current increases above the saturation value.

As will be shown later (Sect. 8.13), the Saturistor is uniquely useful in the secondary circuits of induction motors. It has marked limitations, due to the limited amount of hysteresis loss in available materials, the need for dissipating the high losses without excessive temperature rise, the brittle character of the hard magnetic materials, and their relatively high cost. Nevertheless, the Saturistor principle is expected to be widely useful.

BIBLIOGRAPHY

Reference to Section

1. *The General Theory of Electrical Machines*, B. Adkins, John Wiley, 1957.   1
2. "Saturistors and Low Starting Current Induction Motors," P. L. Alger, G. Angst, and W. M. Schweder, *I.E.E.E. Trans.*, Power Apparatus and Systems, Vol. 82, June 1964, pp. 291–98.   10
3. "Speed Control of Wound Rotor Motors with SCRs and Saturistors," Philip L. Alger, William A. Coelho, and Mukund R. Patel, *I.E.E.E. Trans. on Industry and General Applications*, Vol. IGA-4, No. 5, September/October 1968, pp. 477-485.
4. "Energy Conversion in Magnetic Devices," Philip L. Alger and Edward Erdelyi, *Electro-Technology*, June 1962.   6
5. "The Development of Electrical Machinery in the United States," P. L. Alger and F. D. Newbury, *Proc. Int. Elec. Congress*, Paris, 1932, pp. 485-532.   1
6. "Flux Linkages and Electromagnetic Induction," L. V. Bewley, *A.I.E.E. Trans.*, Vol. 48, 1929, pp. 327-37.   5
7. *Alternating Current Machinery*, L. V. Bewley, The Macmillan Co., 1949.   1
8. *Electrical Engineering Science*, P. R. Clement and W. C. Johnson, McGraw-Hill, 1960.   1
9. *Electromagnetism and Relativity*, E. G. Cullwick, Longmans Green and Co., London, 1957.   6
10. *Electric Machinery*, A. E. Fitzgerald and Charles Kingsley, Jr., Second Edition, McGraw-Hill, 1961.   1
11. "Improved Starting Performance of Wound Rotor Motors Using Saturistors," C. E. Gunn, *I.E.E.E. Trans.*, Power Apparatus and Systems, Vol. 82, June 1964, pp. 298-302.
12. *Scientific Basis of Electrical Engineering*, J. M. Ham and G. R. Slemon, John Wiley, 1961.   2
13. "A New Look at Acyclic Generation," L. Marshall Harvey and J. Reid Burnett, *General Electric Report G.E.A.-7265*.
14. *Electric Motors, Their Theory and Construction*, Third Edition, Vol. II, "Polyphase Current," H. M. Hobart, Issac Pitman & Sons, Ltd., London, 1923.   1

	Reference to Section

15. "Design Aspects of a Homopolar Inductor Alternator," G. C. Jain, *I.E.E.E. Trans.*, Vol. 83, No. 10, October 1964, pp. 1009-1015.
16. *The Magnetic Circuit*, V. Karapetoff, McGraw-Hill, 1911. — 4
17. "Development of a Successful Direct-Current, 2000 kw. Unipolar Generator," B. G. Lamme, *A.I.E.E. Trans.*, Vol. 31, Part II, 1912, pp. 1811-35. — 5
18. "The Story of the Induction Motor," B. G. Lamme, *A.I.E.E. Jour.*, Vol. 40, 1921, pp. 203-23. — 1
19. *Theory of Alternating Current Machinery*, Second Edition, A. S. Langsdorf, McGraw-Hill, 1955. — 7
20. "The Development of Torque in Slotted Armatures," A. S. Langsdorf, *I.E.E.E. Trans.*, Power Apparatus and Systems, Vol. 82, April 1963, pp. 82-8. — 7
21. "Some Aspects of Inductance When Iron Is Present," L. T. Rader and E. C. Litscher, *A.I.E.E. Trans.*, Vol. 63, 1944, pp. 133-39. — 8
22. *The Performance and Design of Alternating Current Machines*, Third Edition, M. G. Say, Isaac Pitman & Sons, Ltd., 1958. — 1
23. *The Physics of Electricity and Magnetism*, W. T. Scott, John Wiley, 1959. — 4
24. *Magnetoelectric Devices, Transducers, Transformers, and Machines*, Gordon R. Slemon, John Wiley, 1966.
25. *Fractional Horsepower Electric Motors*, C. G. Veinott, Second Edition, McGraw-Hill, 1948. — 1

CONTENTS—CHAPTER 2

Transformers—Circuit Analysis

		PAGE
1	Transformers	37
2	The Equivalent Circuit	41
3	The Importance of the Equivalent Circuit	42
4	Per Unit Notation	43
5	Steady-State Circuit Analysis	43
6	Circle Diagram	44
7	Direct-Current Transient Circuit Analysis	46
8	Constant Linkage Theorem	49
9	Alternating-Current Transient Circuit Analysis	49
10	The Superposition Theorem	52
11	General Solution of Transient Equations, Employing the Null-Unit Function	54
12	Transient Performance	57
13	Dimensional Analysis	60
14	Energy Flow	64

2
TRANSFORMERS—CIRCUIT ANALYSIS

1 Transformers

A transformer is an electromagnetic structure that enables alternating current power to be transformed with high efficiency from one voltage level to another at the same frequency. It is a multiple winding closed core reactor, consisting of a laminated magnetic circuit linked by two or more coils, Fig. 2.1. A current, I_1, flowing in the "primary" winding of N_1 turns produces $N_1 I_1$ ampere turns, which create a magnetic flux, ϕ_1, in the core

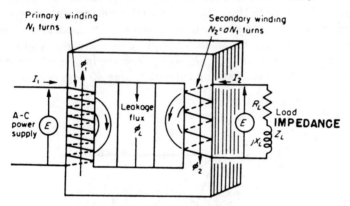

Fig. 2.1. Transformer.

As I_1 varies, ϕ_1 varies in proportion, the two always remaining in time phase at no load, except for the slight time lag of the flux due to hysteresis and eddy current losses in the iron. If E_1 is impressed on the primary, enough current will flow to produce the flux needed to create a counter voltage equal and opposite to E_1, except for the small voltage required to overcome the impedance drop of the magnetizing current. If a load impedance, Z_L, is connected across the secondary terminals, a current, I_2, will flow in the N_2 turns of the secondary windings, producing $N_2 I_2$ opposing ampere turns on the magnetic circuit, and tending to reduce the flux. This will reduce the counter voltage, allowing a greater primary current to flow. When steady state is reached, the primary current, less the magnetizing current, will be equal to $I_1 = N_2 I_2/N_1 = aI_2$; exactly balancing the ampere turns of I_2, and the flux will be restored to its original value, less the amount of leakage flux passing between the primary and secondary coils, Fig. 2.1.

The transformer performance can be visualized with the aid of the phasor diagram of Fig. 2.2. All the phasors in the figure, assumed to be

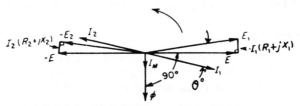

FIG. 2.2. Phasor Diagram of Transformer.

rotating at synchronous speed in a counterclockwise direction, have lengths proportional to their rms values, and angular positions representing their relative displacements in time phase. The induced voltage, E, common to both windings (on an equal turn basis) is found by subtracting the primary impedance drop, $I_1(R_1+jX_1)$ from the impressed voltage, E_1. The flux, ϕ, and the magnetizing current, I_M, lag 90 deg. behind E. The

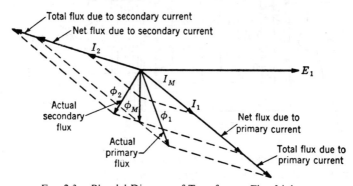

FIG. 2.3. Blondel Diagram of Transformer Flux Linkages.

secondary terminal voltage, E_2, is found by subtracting the secondary impedance drop, $I_2(R_2+jX_2)$ from $-E$. E_2 is entirely consumed by the load impedance drop, $I_2(R_L+jX_L)$. At the instant when the primary current, I_1 in Fig. 2.2, reaches its crest value, the line voltage, E_1, is $\theta°$ past its crest, and has an instantaneous value $\cos \theta$ times its maximum, while the magnetic flux is a little more than $90°-\theta$ from its crest, and is rising.

As an aid in visualizing the actual flux linkages in the transformer, the Blondel diagram of Fig. 2.3 is helpful. This is similar to Fig. 2.2 except that each current phasor is prolonged to represent the flux linkages it would produce if acting alone. Two phasors for each current are shown: one for the total, and the other for the net flux produced, the difference between the two representing the amount of leakage flux which does not link the other winding.

By adding vectorially the total primary flux to the net secondary flux, the actual flux linking the primary winding, ϕ_1, is found. Likewise, by combining the net primary flux with the total secondary flux, the actual secondary flux, ϕ_2, is obtained. And the combination of the two values of net flux gives the net flux that is mutual to the two windings, ϕ_M.

Although for convenience, we speak of the main flux as ϕ_M, and the differences between ϕ_1, ϕ_2, and ϕ_M as being superposed leakage fluxes, the fluxes physically existing in the two windings are actually ϕ_1 and ϕ_2.

This exact physical picture is an important aid to understanding, but it will be recognized that some of the leakage flux links only a part of either winding. ϕ_1 and ϕ_2, therefore, represent total *linkages*, rather than the fluxes at any particular place in the core.

For calculation purposes, it is convenient to use algebraic expressions rather than diagrams. Dealing usually with a definite frequency, it is convenient also to express the inductive properties of windings in terms of their mutual (or magnetizing) and their leakage reactances. Let:

$a = N_2/N_1 =$ ratio of secondary to primary winding turns

$R_1, R_2 =$ resistances of primary and secondary windings (ohms), referred to the primary number of turns (Sect. 1.3),

$M =$ mutual inductance of primary and secondary windings (henries),

$L_1, L_2 =$ total self-inductances of the primary and secondary windings (henries), respectively, referred to the primary number of turns (Sect. 1.8),

$f =$ frequency of sinusoidal applied voltages in cycles per second,

$F =$ ratio of actual frequency, f, to normal frequency, f_0, for which reactances are calculated,

$I_1, I_2 =$ primary and secondary currents in rms amperes, both referred to the primary number of turns,

$E_1, E_2 =$ primary and secondary voltages, both referred to the primary number of turns,

$Z_L = R_L + jX_L =$ impedance of secondary load circuit referred to primary turns.

Then:

$X_1 = 2\pi f_0 (L_1 - M) =$ leakage reactance of primary winding (ohms),
$X_2 = 2\pi f_0 (L_2 - M) =$ leakage reactance of secondary winding (ohms),
$X_M = 2\pi f_0 M =$ magnetizing reactance (ohms).

Neglecting core loss:

(2.1) $$E_1 = I_1(R_1+jX_1)+j(I_1-I_2)X_M,$$

(2.2) $$E_2 = j(I_1-I_2)X_M - I_2(R_2+jX_2).$$

With the secondary winding open, $I_2 = 0$. If the resistance is neglected:

(2.3) $$I_1 = I_M = \frac{E_1}{j(X_1+X_M)} = \frac{-jE_1}{X_M+X_1}, \quad \text{and} \quad E_2 = \frac{X_M E_1}{X_M+X_1};$$

or, in terms of total inductance coefficients at no load:

(2.4) $$I_1 = I_M = \frac{-jE_1}{2\pi f_0 L_1}, \quad \text{and} \quad E_2 = \frac{ME_1}{L_1}.$$

With the secondary winding short-circuited, $E_2 = 0$, and, neglecting resistance:

(2.5) $$I_2 = \frac{X_M I_1}{X_M+X_2}, \quad I_1 = \frac{-j(X_M+X_2)E_1}{(X_1+X_2)X_M+X_1 X_2};$$

or, substituting $2\pi f_0 L_1$ for $X_1 + X_M$, etc., and dropping the phasor notation of $-j$:

(2.6) $$I_2 = \frac{MI_1}{L_2}, \quad I_1 = \frac{L_2 E_1}{2\pi f_0(L_1 L_2 - M^2)}.$$

The ratio

(2.7) $$K = \frac{M}{\sqrt{(L_1 L_2)}} < 1$$

is often called the "coupling coefficient" of the transformer, in which case $K\sqrt{(L_1 L_2)}$ may be substituted for M in Eq. 2.6, giving on short circuit:

(2.8) $$I_1 = \frac{E_1}{2\pi f L_1(1-K^2)},$$

and

(2.9) $$I_2 = KI_1 \sqrt{\left(\frac{L_1}{L_2}\right)} = \frac{K^2 E_1}{2\pi f_0 M(1-K^2)}.$$

In practice, transformers are designed to minimize the leakage reactance, to obtain good performance and low cost. For this reason, the primary and secondary windings usually are made concentric, with half of each winding on each leg, instead of placing them on separate legs as shown in Fig. 2.1. Since K is nearly 1, engineers use X_1 and X_2 in their calculations rather than L_1 or L_2, and for many purposes X_M is taken to be ∞.

2 The Equivalent Circuit

The performance of a transformer is simply and accurately represented by an equivalent electric circuit, Fig. 2.4. This circuit has the great virtue that it is adapted to either numerical calculations or measurement on an electrical network. Also, it can be combined with the circuits representing generators, transmission lines, etc., making possible complete system analyses. It is very important, therefore, that the student obtain a clear perception of the physical meaning and properties of this circuit.

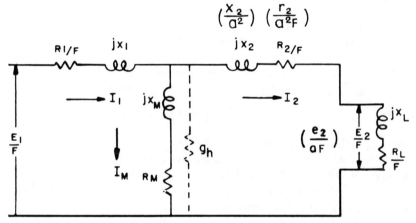

Fig. 2.4. Equivalent Circuit of Transformer.

It is derived from the actual circuit shown in Fig. 2.1 by converting all secondary values to primary terms, and recognizing that the magnetizing current and induced voltage are then common to both windings.

In the circuit, R_1 and R_2 represent the resistances, and X_1 and X_2 the leakage reactances of the primary and secondary windings, both referred to primary terms. That is, if r_2 is the measured ohmic resistance of the secondary winding, and $a = N_2/N_1$ is the effective ratio of secondary to primary turns, the ohmic value of R_2 to use in the circuit is $R_2 = r_2/a^2$. This relation is valid because (1) the product $N_1 I_1$ must equal $N_2 i_2 = a N_1 i_2 = N_1 I_2$, when the mmfs are equal and opposite, giving $I_2 = a i_2$ (i_2 being the actual secondary current); and (2) the energy consumed in resistance losses (or stored in the leakage magnetic field) must be the same for the same mmf and the same coil dimensions independent of a, that is:

(2.10) $$i_2^2 r_2 = \left(\frac{I_2}{a}\right)^2 (a^2 R_2) = I_2^2 R_2.$$

For given coil dimensions (and copper space factor), the resistance varies as the square of the number of series turns in the coil. If there are n parallel circuits, each with N series turns, each circuit has N times the length of one turn, and a section $1/Nn$th of the entire coil, so that the combined resistance of the n paths in parallel between terminals varies as $N(Nn)(1/n) = N^2$.

X_M is the magnetizing reactance, or ratio of the primary induced voltage to the magnetizing current, and R_M is the resistance in series with X_M representing the hysteresis and eddy-current losses in the magnetic circuit.

In so far as these core losses are due to eddy currents, induced in the laminations, core bolts, and other structural parts, it is more correct to represent them by a separate branch of the circuit, in parallel with the magnetizing current branch, as shown dotted in Fig. 2.4. For the simple analysis outlined here, however, it is convenient to lump the core loss with the magnetizing branch. The value of R_M is accordingly:

$$(2.11) \qquad R_M = g_h X_M^2, \text{ approximately,}$$

where g_h is the conductance of the core-loss branch.

It is desirable to retain the normal frequency values of the reactances in calculations, whatever the actual frequency. For this reason, all the actual f frequency values of voltage and of resistance in the circuit shown are divided by F, making the circuit valid for any frequency.

Ordinarily, $F = 1$, as $f = f_0$. If d-c is applied to the circuit, for example, $F = 0$, making the current simply $(E/F)/(R_1/F) = E/R_1$, as expected.

3 The Importance of the Equivalent Circuit

The equivalent circuit is the means of expressing in an agreed technical language both the requirements imposed by an electrical system on a given machine, and the performance that the machine is predicted to give. The circuit, therefore, is the meeting point of the application and the design engineers, serving to interpret the ideas of one to the other.

The designing engineer deals with the relations between the available materials, the dimensions chosen, and the resulting machine performance. He expresses his conclusions in the values of the equivalent circuit "constants". The application engineer deals with the load capacity, voltage regulation, short-circuit currents, speed of response, and other aspects of system behavior. He expresses the desired machine performance by assigning values to the equivalent circuit constants.

By calculations similar to those outlined in Sects. 1.3 and 1.8, all the constants of the transformer equivalent circuit of Fig. 2.4 can be found for

THE IMPORTANCE OF THE EQUIVALENT CIRCUIT

a given design; or, conversely, the needed dimensions can be determined to give desired values of the circuit constants.

4 Per Unit Notation

It is convenient and customary to express the values of R_1, X_1, etc., in the equivalent circuit in per unit terms. That is, instead of saying a transformer winding has a resistance of R ohms, we say it has a "per unit resistance" equal to IR/E, where I is the rated current and E the rated voltage of the winding. Unit resistance is simply the ratio of full-load voltage to full-load current, E/I, or the resistance that represents full-load output. A per unit resistance value of 0.01 means that there is 1 per cent voltage drop in the winding resistance under rated load conditions, or the resistance loss is 1 per cent of the rated volt-amperes. A chief advantage of the per unit notation is that the numerical values of the circuit constants are of the same order of magnitude for windings of all voltage and current ratings, whereas the ohmic values vary widely. For usual power transformers, the total resistance is of the order of 1 per cent or less, the reactance from 7 to 15 per cent, and the magnetizing current 5 per cent, more or less. Typical values may, therefore, be:

$$R_1 = R_2 = 0.005;\ X_1 = X_2 = 0.04;\ \text{and}\ X_M = 20.$$

The core loss is usually less than 1 per cent, so that, with $X_M = 20$, R_M might be 2.

5 Steady-state Circuit Analysis

In Fig. 2.4, if the secondary winding is open, $X_L = \infty$, and $I_2 = 0$, so that the open-circuit impedance (at no load) is:

$$(2.12) \qquad Z_{oc} = (R_1 + R_M) + j(X_1 + X_M).$$

If the secondary is short-circuited, $Z_L = 0$, and the circuit impedance is:

$$(2.13) \qquad Z_{sc} = (R_1 + jX_1) + \frac{(R_2 + jX_2)(R_M + jX_M)}{(R_M + R_2) + j(X_M + X_2)}.$$

Since X_M is normally very large compared with R_2, X_2, and R_M, this may be expanded in powers of Z/X_M, giving:

$$(2.14) \qquad Z_{sc} = R_1 + jX_1 + (R_2 + jX_2)\left(1 - \frac{R_2 + jX_2}{R_M + jX_M} + \ldots\right),$$

or, since, $R_2 \ll X_2$ and R_M is very small compared with X_M:

$$(2.15) \qquad Z_{sc} = \left[R_1 + R_2\left(1 - \frac{2X_2}{X_M}\right)\right] + j\left[X_1 + X_2\left(1 - \frac{X_2}{X_M}\right)\right].$$

approximately.

The per unit short-circuit current (steady state) is, therefore, nearly equal to the reciprocal of the total per unit impedance, or, practically, of the reactance. At the first instant after the line switch is closed, however, much higher currents may flow momentarily, depending on the initial phase angle of the voltage and the amount of remanent magnetic flux in the core, as discussed in Sect. 2.9.

To visualize the performance of the transformer under normal loads, it is helpful to consider the relation of secondary voltage and output to the load impedance, neglecting the magnetizing current. Let $Z = R + jX$ represent the total transformer impedance (sum of primary and secondary values), and $Z_L = R_L + jX_L$ equal the load impedance. The per unit voltage at the secondary terminals will then be:

$$(2.16) \quad \frac{E_2}{E_1} = \frac{Z_L}{Z + Z_L}$$

$$= \frac{R_L(R + R_L) + X_L(X + X_L) + j(RX_L - XR_L)}{(R + R_L)^2 + (X + X_L)^2}.$$

If the load is non-inductive, $X_L = 0$, and Eq. 2.16 reduces to:

$$(2.17) \quad \frac{E_2}{E_1} = \frac{R_L[(R + R_L) - jX]}{(R + R_L)^2 + X^2},$$

whose numerical magnitude, since R and X are small in comparison to R_L, is approximately:

$$(2.18) \quad \frac{E_2}{E_1} = 1 - \frac{R}{R_L} + \frac{2R^2 - X^2}{2R_L^2} + \ldots.$$

Or, if the load is purely reactive, $R_L = 0$, and, by symmetry:

$$(2.19) \quad \frac{E_2}{E_1} = 1 - \frac{X}{X_L} + \frac{2X^2 - R^2}{2X_L^2} + \ldots, \text{ approximately.}$$

That is, for resistance loads, the voltage regulation, or per unit drop in voltage from no load to full load, is nearly equal to the total per unit resistance; and for inductive loads it is nearly equal to the per unit reactance of the transformer.

6 Circle Diagram

If, with constant primary voltage, the load resistance, R_L, is decreased indefinitely, the current will increase, the secondary voltage will fall, and the power supplied to the load will reach a maximum and then decrease toward zero.

CIRCLE DIAGRAM

The ratio of the secondary current to primary voltage is:

(2.20) $I_2/E_1 = 1/(R+jX+R_L+jX_L) = \dfrac{(R+R_L)-j(X+X_L)}{(R+R_L)^2+(X+X_L)^2}.$

If, with R_L varying, we plot the component of current in phase with E_1 vertically, and the $-j$ component horizontally to the right, Eq. 2.20 forms a semicircle, as shown in Fig. 2.5. If R and R_L are both zero, the current is $-j/(X+X_L)$, which fixes the diameter of the circle. With X and X_L

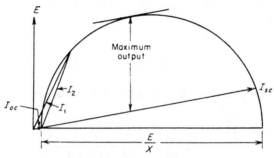

FIG. 2.5. Circle Diagram.

fixed, as R_L varies, the current phasor traverses the perimeter of the half circle, becoming equal to the magnetizing current when $R_L = \infty$. When $R_L = 0$, the current reaches its short circuit value, as given by Eq. 2.15.

The value of load resistance for maximum power may be found by differentiating the expression for the per unit power output with respect to R_L, and equating it to zero.

The per unit voltage across the load resistance, by analogy with Eq. 2.17, is:

(2.21) $\dfrac{E_{R_L}}{E_1} = \dfrac{R_L[(R+R_L)-j(X+X_L)]}{(R+R_L)^2+(X+X_L)^2},$

and the per unit power output is:

(2.22) $\dfrac{E_{R_L}^2}{R_L E_1^2} = \dfrac{R_L}{(R+R_L)^2+(X+X_L)^2}.$

The derivative of Eq. 2.22 with respect to R_L is:

$$\dfrac{(R+R_L)^2+(X+X_L)^2-2R_L(R_L+R)}{[(R+R_L)^2+(X+X_L)^2]^2}.$$

Equating this to zero, we find:

(2.23) $R_L = \sqrt{[R^2+(X+X_L)^2]}$

as the condition for maximum power output, and from Eq. 2.22:

$$\text{(2.24)} \quad \text{Per unit maximum output} = \frac{1}{\sqrt{[2\{R+\sqrt{[R^2+(X+X_L)^2]}\}]}}.$$

If a transformer has a total per unit resistance, R, of 0.02, and a reactance, X, of 0.0775, therefore, the maximum power it can deliver to a non-inductive load is five times its full-load rating, and the corresponding per unit load resistance is 0.10.

The magnitude of the voltage across the load resistance at maximum output is, from Eq. 2.21:

$$\text{(2.25)} \quad \frac{E_{R_L}}{E_1} = \frac{R_L}{\sqrt{[(R+R_L)^2+(X+X_L)^2]}}.$$

Substituting Eq. 2.23 in this, at maximum output:

$$\text{(2.26)} \quad \frac{E_{R_L}}{E_1} = \frac{1}{\sqrt{2}\sqrt{\left\{1+\frac{R}{\sqrt{[R^2+(X+X_L)^2]}}\right\}}}$$

$$= 0.707\left[1-\frac{R}{2(X+X_L)}\right], \text{ approximately,}$$

since R is small compared with $(X+X_L)$. That is, at maximum output, the voltage across the load resistance will be a little less than 70 per cent of the no-load voltage (with constant voltage on the primary).

Only a few per cent drop in voltage can be tolerated in normal service. Voltage regulation requirements, therefore, limit the permissible value of a power transformer's output to a load that is only a small fraction of the (maximum) output possible.

As Eqs. 2.19, 2.24, and 2.26 indicate, the performance limits of a transformer, and indeed of all types of a-c machines, are set by the leakage reactance value. The reactance is, in fact, an inverse measure of the (short-time) useful output obtainable from the machine.

The transformer circle diagram is representative of similar diagrams that apply to transmission lines, induction and synchronous machines, and other forms of apparatus. It is a useful aid in visualizing the performance derived from the equivalent circuit.

7 Direct-Current Transient Circuit Analysis

Consider a simple circuit with resistance R and inductance L in series. If a steady voltage, E, is suddenly applied, at time $t=0$, the initial current being zero, the circuit voltage relations are given by the equation:

$$E = Ri + \frac{d(Li)}{dt}, \qquad (2.27)$$

which, if L is constant, may be solved by substituting:

$$i = C_1 + C_2 e^{-at}. \qquad (2.28)$$

From energy considerations (Sect. 2.8), the current remains zero momentarily after the time $t = 0$, so that $C_2 = -C_1$. Substituting this in Eq. 2.27, we have:

$$E = C_1(R - Re^{-at} + aLe^{-at}).$$

When $t = \infty$, the exponential terms drop out, giving $C_1 = E/R$. Also, when $t = 0$, the sum of the exponential terms must be zero, so that $a = R/L$. Therefore:

$$i = \frac{E}{R}(1 - e^{-Rt/L}) = EI(t), \qquad (2.29)$$

where $I(t)$ represents the "response" or "indicial admittance" of the circuit to a suddenly applied unit voltage (Sect. 2.10).

The current consists of a steady-state component, $I = E/R$, and a transient component, which is initially $-E/R$, but dies away to zero along

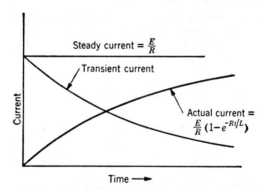

FIG. 2.6. Direct-Current Rise in an Inductive Circuit.

a logarithmic decrement curve (Fig. 2.6). The time constant, T_0, of the circuit, defined as the time in which the transient falls to $1/e$, or 0.368 times its initial value, is:

$$T_0 = \frac{L}{R} \text{ sec}. \qquad (2.30)$$

The energy stored in the magnetic field, while the current builds up from zero to its steady value, is:

(2.31) $$W_L = \int_0^\infty i\left(L\frac{di}{dt}\right)dt = \int_0^I Li\,di = \tfrac{1}{2}LI^2.$$

The energy stored in the magnetic field in watt-seconds is, therefore, equal to one-half the product of the inductance in henries by the square of the current in amperes (Eq. 1.24).

The energy input to the circuit resistance is, from Eqs. 2.27 and 2.29:

(2.32) $$W_R = \int_0^T Ri^2\,dt = RI^2\int_0^T (1 - 2e^{-Rt/L} + e^{-2Rt/L})\,dt$$

$$= RI^2T - 2LI^2(1 - e^{-RT/L}) + \frac{LI^2}{2}(1 - e^{-2RT/L}).$$

When t is large, this becomes:

(2.33) $$W_R = RI^2T - \frac{3LI^2}{2} \text{ wsec,}$$

so that the inductive delay in establishing the steady-state current causes a decrease in the total I^2R losses by an amount three times the energy stored in the magnetic field. The net energy drawn from the power supply is equal to the steady-state I^2R loss, starting from the instant the switch is closed, less twice the energy of the magnetic field.

If the winding is suddenly short-circuited, after the steady-state current I has been established, the subsequent value of the current is found by substituting $E = 0$ in Eq. 2.27, and making $i = I$ when $t = 0$, whence:

(2.34) $$i = Ie^{-RT/L}.$$

The total energy dissipated in the resistance by this current is:

(2.35) $$W_R = \int_0^\infty Ri^2\,dt = RI^2\int_0^\infty e^{-2Rt/L}\,dt,$$

$$= \tfrac{1}{2}LI^2.$$

That is, the current is maintained at a gradually decreasing value by the energy stored in the magnetic field until that energy is exhausted. If the circuit is opened before the flux has died away, the voltage across the terminals will rise to a high value, maintaining the current across the opening until the arc energy and resistance losses have dissipated the magnetic energy. For this reason, it is dangerous to interrupt the current of any inductive winding, without safety provisions such as discharge resistors or remote operation.

8 Constant Linkage Theorem

These relations illustrate the fact that energy is neither created nor destroyed, which gives rise to the constant linkage theorem.

Referring to Eq. 2.27, E and R are equal to zero for a closed circuit of zero resistance, so that:

$$\frac{d}{dt}(Li) = 0. \tag{2.36}$$

That is:

If the resistance of a closed circuit is zero, the algebraic sum of the magnetic linkages of the circuit remains constant.

Although, in practice, the circuit resistance is not zero, the linkages remain the same at the first instant after any sudden change in the applied voltage, since time is required to dissipate energy in resistance losses.

In any transformer, or rotating machine, therefore, when a magnetic flux links a winding, and that winding is suddenly short-circuited, the flux linkages of the winding will remain constant, except for the subsequent change due to the resistance decrement, no matter what happens. If the magnetic circuit is pulled apart, creating an air gap, more current will flow in the winding, supplying the necessary magnetomotive force to maintain the flux. If another mmf is impressed on the magnetic circuit, more current will flow in the winding to balance it. Remembering this principle will be helpful in understanding transient phenomena in all types of electric machines.

9 Alternating-Current Transient Circuit Analysis

If the voltage suddenly applied to the circuit of Sect. 2.7 is alternating, and the switch is closed at the instant when the voltage is $\theta°$ before its zero value, the current-voltage relations are given by:

$$E \sin(2\pi ft - \theta) = Ri + L\frac{di}{dt}. \tag{2.37}$$

This may be solved by substituting:

$$i = C \sin(2\pi ft - \alpha) + C_1 e^{-at}. \tag{2.38}$$

Before closing the switch, and momentarily afterward, i is zero. Putting $t = 0$ and $i = 0$ in this expression for i gives:

$$C_1 = C \sin \alpha,$$

and

$$i = C[\sin(2\pi ft - \alpha) + e^{-at} \sin \alpha]. \tag{2.39}$$

Substituting in Eq. 2.37:

(2.40) $\quad E\sin(2\pi ft - \theta) = RC\sin(2\pi ft - \alpha) + RCe^{-a}\sin\alpha$
$$+ 2\pi fLC\cos(2\pi ft - \alpha) - aLCe^{-at}\sin\alpha.$$

When $t = \infty$, the exponential terms vanish, whence:

(2.41) $\quad E\sin(2\pi ft - \theta) = C[R\sin(2\pi ft - \alpha) + 2\pi fL\cos(2\pi ft - \alpha)],$

giving (from Eq. 1.7):

$$C = \frac{E}{Z}, \quad Z = \sqrt{[R^2 + (2\pi fL)^2]}, \quad \alpha = \theta + \sin^{-1}\frac{2\pi fL}{Z},$$

and, when $t = 0$:

$$-E\sin\theta = C(2\pi fL\cos\alpha - aL\sin\alpha),$$

or: $\quad Z\sin\theta = \dfrac{aL}{Z}(R\sin\theta + 2\pi fL\cos\theta) - \dfrac{2\pi fL}{Z}(R\cos\theta - 2\pi fL\sin\theta),$

whence: $\quad a = R/L.$

Thus, finally:

(2.42) $\quad i = \dfrac{E}{Z}[\sin(2\pi ft - \alpha) + e^{-Rt/L}\sin\alpha].$

This "conventional" method of solving the differential Eq. 2.37 requires that the four constants, C, C_1, a, and α, be determined separately for each type of applied voltage-time curve. Also, some prescience was required to write down in advance the form of the solution.

More general methods of solving such equations are given in Sects. 2.10 and 2.11.

As in the case of Eq. 2.29, the current consists of two components, a steady-state alternating current equal to the applied voltage divided by the circuit impedance, and a transient direct current initially equal and opposite to the initial value of the steady-state current, but dying away exponentially to zero. If the switch is closed at the instant when the alternating current ought to be at its negative crest value

$$\left(\theta = 90° - \sin^{-1}\frac{2\pi fL}{Z}\right),$$

the direct component will have its maximum possible value, equal and opposite to the crest of the alternating current. In this case, $\alpha = 90°$, and from Eq. 2.42:

(2.43) $\quad i = \dfrac{E}{Z}[\sin(2\pi ft - 90°) + e^{-Rt/L}].$

If, further, the rate of decrement is slow (R/L small), the transient component will still be large when the steady-state current reaches its maximum in the reversed direction, making the peak current nearly twice the crest of the steady-state value (Fig. 2.7).

The presence of magnetic saturation reduces the value of L at high flux densities, and may greatly increase the peak current values under transient conditions, even though the secondary circuit is open.

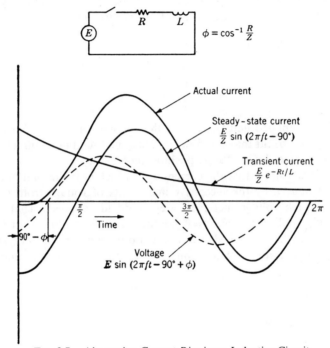

FIG. 2.7. Alternating-Current Rise in an Inductive Circuit.

If there is no residual flux, and the switch is closed when the voltage is zero ($\theta = 0$), the magnetic flux should be at its crest value to accord with steady-state conditions. The flux will then start from zero and increase continuously throughout the first half cycle, reaching a peak value nearly double normal flux density, if the decrement rate is small. If there is a residual flux, left the last time the circuit was opened, and this is in the same direction as the new flux produced when the voltage is reapplied, the flux may reach a peak value even more than twice normal, after a half cycle.

At such densities, the magnetic saturation is extremely high, and the magnetizing current required may reach thirty or more times full-load current. The copper loss during the half cycles when the flux density is above normal is much greater than that during the alternative half cycles when the flux density is below normal, creating an unsymmetrical voltage drop that restores magnetic symmetry after a few cycles. The residual flux will be lowered if a capacitance is permanently connected across the transformer windings, of the right value to cause several current reversals during its decrement period after the line switch is opened.

These peak currents due to magnetic offset are of interest because of their possible effects on relaying and on coil distortion, but their very short duration makes their heating effects of no importance.

10 The Superposition Theorem

Equation 2.29 divided by E, gives the "response" of the circuit,

$$I(t) = \frac{1}{R}(1 - e^{-Rt/L}),$$

to a suddenly applied unit voltage. If the applied voltage is a function of time, $e(t)$, the current that will flow, $i(t)$, may be found more directly than in the last section, by the "superposition theorem".

This theorem states that the current at any instant is the sum of all the currents produced by the successive increments of voltage from $t = 0$ to $t = t$, as if each voltage increment were suddenly applied. That is, the voltage $e(0)$ applied at time $t = 0$ produces a current $e(0)I(t)$ at time t. Also, a voltage $\Delta e(x)$ applied at a time x produces a current $\Delta e(x)I(t-x)$ at time t. At the limit, the additional voltage that is applied at time x is

$$\frac{de(x)}{dx} dx = e'(x) dx.$$

Hence, the total current at time t is:

(2.44) $$i(t) = e(0)I(t) + \int_0^t e'(x)I(t-x)\,dx.$$

This equation can be rearranged in many different forms. For our purposes, the most convenient form is that obtained by integrating the last term of Eq. 2.44 by parts, giving:

(2.45) $$i(t) = e(0)I(t) + e(x)I(t-x)\Big]_0^t - \int_0^t e(x)I'(t-x)\,dx,$$

$$= e(t)I(0) - \int_0^t e(x)I'(t-x)\,dx.$$

This form of the equation can be visualized by considering that the total current is the sum of the currents formed by suddenly applying successive spikes of voltage $e(x)$, each for a time dx, over the time period from $t = 0$ to $t = t$. If the voltage $e(x)$ is applied at the time x, and then a reversed voltage $-e(x)$ is applied at a time $x + dx$, the result will be to produce a net current, whose value will be:

(2.46) $$e(x)[I(t-x) - I(t-x-dx)] = -e(x)I'(t-x)\,dx.$$

The sign of the term is negative, because the differential of $(t-x)$ is $-dx$.

Considering the same circuit as in Sect. 2.7, for example, and taking the suddenly applied voltage to be sinusoidal, as in Eq. 2.37:

(2.47) $$e(t) = E\sin(2\pi ft - \theta),$$

Eq. 2.45 gives:

(2.48) $$i(t) = \frac{Ee^{-Rt/L}}{L}\int_0^t e^{Rx/L}\sin(2\pi fx - \theta)\,dx,$$

since $I(0) = 0$, and $I'(t-x) = -\frac{1}{L}e^{-R(t-x)/L}$ from Eq. 2.29.

The solution of Eq. 2.48, obtained from a table of integrals, or by integrating by parts, is:

(2.49) $$i(t) = Ee^{-Rt/L}e^{Rx/L}\left(\frac{R\sin(2\pi fx - \theta) - 2\pi fL\cos(2\pi fx - \theta)}{R^2 + (2\pi fL)^2}\right)\Bigg]_{x=0}^{x=t}$$

$$= \frac{E}{Z}e^{-Rt/L}e^{Rx/L}\sin(2\pi fx - \alpha)\Bigg]_{x=0}^{x=t}$$

$$= \frac{E}{Z}[\sin(2\pi ft - \alpha) + e^{-Rt/L}\sin\alpha],$$

where $\quad \alpha = \theta + \sin^{-1}\dfrac{2\pi fL}{Z}$

which is the same as Eq. 2.42.

Thus, by separating the solution for the transient current into two steps, first finding the circuit response to unit voltage, $I(t)$, and then employing the superposition theorem to find the response to $e(t)$, the process is simplified and adapted to solving more difficult problems.

However, in deriving Eqs. 2.44 and 2.45, we carried the integration only from $t = 0$ to $t = t$, assuming that nothing that transpired between $t = -\infty$ and $t = 0$ need be considered. We did include a constant term $e(0)I(t)$, or $I(0)e(t)$ outside the integral, recognizing that the integration does not include the current created by the initial value of the voltage. In practice, especially when multiple integrations are required, or when discontinuities occur at other times than $t = 0$, this procedure may lead to difficulties. A way of overcoming these difficulties is presented in the next section.

11 GENERAL SOLUTION OF TRANSIENT EQUATIONS, EMPLOYING THE NULL-UNIT FUNCTION

It will be noted that Eqs. 2.27 and 2.37 express the conditions in a circuit *after the switch has been closed*, and that a different equation, obtained by making $E = 0$, applies *before the switch is closed*. Conventional methods of solving transient problems, which follow this practice of using different equations before and after a circuit change occurs, have the disadvantage that constants of integration, such as C_1 and C_2 in Eq. 2.28, must be found. The process of finding these involves not only the initial conditions, but also a knowledge of physical laws. For example, in solving Eqs. 2.27 and 2.37, we brought in the fact that the current is the same immediately after the switch is closed as it was before, as required by the principle of the conservation of energy.

There is a more direct method of finding the current due to a unit applied voltage, which gives the constants of integration directly. The procedure that will be given here has been put forward by J. J. Smith, following Heaviside's ideas, and is validated by the work of the French mathematician, Laurent Schwartz.

The essential feature of the method is the use of the Heaviside null-unit function, $H(t, t_0)$, as a multiplier to represent the discontinuities introduced by the sudden closing of a switch or other abrupt circuit change. $H(t, t_0)$ is defined by the equation (Fig. 2.8 (*a*)):

(2.50)
$$H(t, t_0) = 0 \text{ if } t < t_0$$
$$= 1 \text{ if } t > t_0.$$

Figure 2.8 (*b*) and (*c*) show the two independent equations usually employed to describe the relations:

(2.51)
$$y = 0 \text{ when } t < t_0$$
$$= t \text{ when } t > t_0,$$

and Fig. 2.8 (*d*) shows the single equation valid over the complete time range that results when $H(t, t_0)$ is employed as a multiplier.

Evidently the derivative of H obeys the following laws:

(2.52)
$$H'(t, t_0) = \frac{dH(t, t_0)}{dt} = 0 \text{ if } t \neq t_0,$$

and
$$\int_{t_0 - \varepsilon}^{t_0 + \varepsilon} H'(t, t_0) \, dt = 1,$$

where ε is an indefinitely small interval of time.

GENERAL SOLUTION OF TRANSIENT EQUATIONS 55

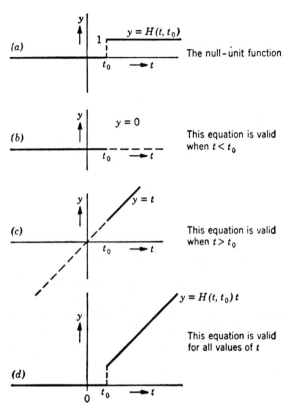

Fig. 2.8. The Use of the Null-Unit Function to Form a Single Equation Valid before and after an Abrupt Circuit Change.

In the strict mathematical sense, H' does not exist at $t = t_0$; nevertheless, as Schwartz has shown, it can be usefully employed in the manner following:

For the circuit of Sect. 2.7, for example, taking the time of closing the switch as $t_0 = 0$, and writing simply H for $H(t, 0)$, Eq. 2.27 may be rewritten:

$$(2.53) \qquad EH = RIH + L\frac{d(IH)}{dt}.$$

This equation is valid before as well as after the voltage is applied, since $H = 0$ when $t < 0$. Thus, the solution of this equation will give an expression for the current that is valid before as well as after the switch is closed, and so will embody the integration constants directly.

The solution for this will be of the form, similar to Eq. 2.28:

(2.54) $$IH = (C_1 + C_2 e^{-at})H.$$

Substituting this in Eq. 2.53, we have:

(2.55) $$EH = RH(C_1 + C_2 e^{-at}) - aC_2 LH e^{-at} + LH'(C_1 + C_2 e^{-at}),$$

where $H' = dH/dt$.

When $t \neq 0$, $H' = 0$, and, therefore, Eq. 2.55 must remain true after the H' terms are removed. Hence, also, the terms in H' must separately be equal to zero, at the instant when $t = 0$. Therefore:

$$[C_1 + C_2 e^{-at}]_{t=0} = 0,$$

or

(2.56) $$C_2 = -C_1.$$

Substituting this in the H terms of Eq. 2.55:

(2.57) $$EH = RHC_1(1 - e^{-at}) + aC_1 LHe^{-at}.$$

These equations are the same as those found in Sect. 2.7 by the conventional method of solving Eq. 2.27, and, therefore, the same final answer, Eq. 2.29 is obtained, except that H remains present as a multiplier:

(2.58) $$IH = \frac{EH}{R}(1 - e^{-Rt/L}).$$

What we have accomplished by introducing the null-unit function, H, is first, to make the equations valid over the entire range of time; second, to make differentiation and integration fully reversible; and, third, to eliminate any need for bringing in extraneous physical principles in solving the equations. This holds true for fractional-order as well as whole-order differentiation and integration. By using $H(t, t_0)$, $H(t, t_1)$, etc., corresponding to independent operations at different times, t_0, t_1, etc., the method can be extended to solve problems that involve a number of independent discontinuities.

From the superposition theorem, Eq. 2.44, if the voltage $e(x)$ is suddenly applied at the time $x = t_0$, the increment of voltage applied at the time x is:

(2.59) $$\frac{d[H(x, t_0)e(x)]}{dx} dx = e(x)H'(x, t_0) dx + H(x, t_0)e'(x) dx.$$

The current produced at time t by a unit voltage applied at the time x is:

$$H(t, x)I(t - x).$$

Hence, Eq. 2.44 for the total current at time t should be written:

$$(2.60) \quad H(t,t_0)i(t) = \int_{-\infty}^{t} H(t,x)I(t-x)\left[H'(x,t_0)e(\ ) + H(x,t_0)e'(x)\right]dx.$$

But $H'(x,t_0)$ is zero at all times except $x = t_0$. Therefore, all the multipliers of H' may be replaced by their single values at $t = t_0$, and, from Eq. 2.52, Eq. 2.60 becomes:

$$(2.61) \quad H(t,t_0)i(t) = H(t,t_0)I(t-t_0)e(t_0) + H(t,t_0)\int_{t_0}^{t} I(t-x)e'(x)\,dx,$$

which is the same as Eq. 2.44, multiplied by $H(t,t_0)$. The product $H(t,x)H(x,t_0)$ is equal to unity at each value of x throughout the integration from $t = t_0$ to $t = t$, and is equal to 0 when $t < t_0$. Hence, it is replaced by the factor $H(t,t_0)$ outside the integral, and by changing the lower limit of integration from $-\infty$ to t_0.

It will be observed that the initial term, outside the integral, now appears as a result of the discontinuity at $t = t_0$, and that it has come out as an automatic result of the integration procedure.

There is no space here to go into the many interesting aspects of this subject. The chief point of mathematical interest is the use of the derivative H' (often called the Dirac function, δ), in spite of the fact that it does not exist at $t = 0$ in a strict mathematical sense. The validity of this procedure has been proved by Laurent Schwartz. What he does is, whenever an infinite expression appears, to replace it by a "functional representation", which is finite; then perform the desired operations; and finally transform the resulting new functional representation back into the function it represents. Mathematicians have developed other ways of solving such problems, by use of the Laplace transformation, but, to the author at least, these methods appear less direct and less clearly visualizable.

It is as if engineers were canoeists going down a river, who, coming to some rapids, were accustomed to portage around by a "conventional" route. Heaviside, relying on intuition, simply "shot the rapids". The mathematicians then said that this is too uncertain a method, and pointed out instead a route via Laplace. Schwartz has now shown that Heaviside's method of going straight through is safe after all.

12 Transient Performance

Under switching surge conditions, the rate of change of current is so high that the magnetic flux in a transformer core hardly changes (because of opposing eddy circuits in the core laminations), and only the flux in air need be considered. The capacitance elements of the circuit, as indicated in Fig. 2.9, are then all-important. However, the transients that occur in normal operation can be calculated from the equivalent circuit of Fig. 2.4, without considering the capacitance, using the methods followed in deriving Eqs. 2.29 and 2.40.

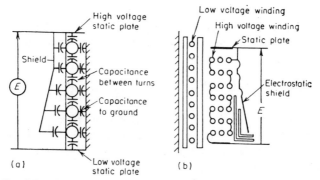

FIG. 2.9. Capacitance and Shielding of High-Voltage Transformer.

Under transient conditions, the wave forms are no longer sinusoidal, so that we must use $L\,di/dt$ for the voltage across an inductance, instead of $j\omega Li$. The change is accomplished by substituting $-jp$ for F, where $p = d/(\omega\,dt)$. When the circuit values are multiplied through by F, the voltage across a reactance becomes

(2.62) $$jFXi = j\omega FLi = \omega pLi = L\,di/dt.$$

The ω in the denominator of p is introduced in order to retain the normal reactance values in the circuit, instead of using inductances. The same result would be obtained if we measured t in radians of angular motion at synchronous speed, instead of in seconds, and omitted the ω.

From inspection of the circuit, Fig. 2.4, the steady-state current is given by

(2.63) $$\frac{E}{F} = I_1\left[\frac{R_1}{F} + jX_1 + \frac{1}{\dfrac{1}{jX_M} + \dfrac{1}{(R_2+R_L)/F + j(X_2+X_L)}}\right].$$

The same equation will apply to transient conditions, if we put $-jp$ for F and multiply through by F

(2.64) $$e(t) = i(t)\left[R_1 + pX_1 + \frac{pX_M[R_2+R_L+p(X_2+X_L)]}{R_2+R_L+p(X_2+X_L+X_M)}\right].$$

The expressions $e(t)$ and $i(t)$ are used instead of E and I to indicate that the voltage and current are now functions of time. For simplicity, let

$$R_3 = R_2 + R_L, \qquad X_3 = X_2 + X_L, \qquad X = X_1 + X_3$$

and

$$m = R_1(X_M + X_3) + (X_M + X_1)R_3.$$

Then Eq. 2.64 becomes

(2.65) $$i(t) = \frac{[R_3 + p(X_3 + X_M)]e(t)}{p^2(XX_M + X_1 X_3) + mp + R_1 R_3}.$$

This can be separated into two components by partial fractions, after finding the two roots p_1 and p_2 of the algebraic equation

(2.66) $$p^2(XX_M + X_1 X_3) + mp + R_1 R_3 = 0.$$

Since m is large compared with the other terms, the two roots are

(2.67) $$p_1 = \frac{-R_1 R_3}{m} \text{ approx.}$$

and

(2.68) $$p_2 = \frac{-m}{(XX_M + X_1 X_3)} \text{ approx.}$$

Equation 2.65 can be solved by resolving it into two partial fractions:

(2.69) $$i(t) = i_1 + i_2$$

where, replacing p by $d/(\omega\, dt)$,†

(2.70) $$di_1/dt - \omega p_1 i_1 = \frac{\omega[R_3 + p_1(X_M + X_3)]e(t)}{m + 2p_1(XX_M + X_1 X_3)}$$

and

(2.71) $$di_2/dt - \omega p_2 i_2 = \frac{\omega[R_3 + p_2(X_M + X_3)]e(t)}{m + 2p_2(XX_M + X_1 X_3)}.$$

The solutions of these equations are given by Eq. 2.29, 2.34 or 2.40, depending on the form of the applied voltage, $e(t)$. Under usual load conditions, R_L is so large that the transient currents are small, and they decay rapidly. The two cases of interest, therefore, are short circuit, when $R_L = X_L = 0$, and open circuit, when $R_L = \infty$. In the first case, $R_L = X_L = 0$, so that, from Eq. 2.67:

(2.72) $$p_1 = \frac{-R_1 R_2}{R_1(X_M + X_2) + R_2(X_M + X_1)}$$

$$= \frac{-R_1 R_2}{X_M(R_1 + R_2)} \text{ approx.}$$

(2.73) $$p_2 = \frac{-[R_1(X_M + X_2) + R_2(X_M + X_1)]}{XX_M + X_1 X_2}$$

$$= \frac{-(R_1 + R_2)}{X_1 + X_2} \text{ approx.}$$

† If $p_1, p_2, \ldots, p_n$ are the roots of the equation $F(p) = 0$, then
$$\frac{f(p)}{F(p)} = \frac{f(p_1)}{(p - p_1)F'(p_1)} + \frac{f(p_2)}{(p - p_2)F'(p_2)} + \cdots + \frac{f(p_n)}{(p - p_n)F'(p_n)}.$$

Substituting these values in Eqs. 2.70 and 2.71,

$$di_1/dt + \frac{\omega R_1 R_2 i_1}{X_M(R_1+R_2)} = \frac{\omega R_2^2 e(t)}{X_M(R_1+R_2)^2} \text{ approx.} \tag{2.74}$$

and

$$di_2/dt + \frac{\omega(R_1+R_2)i_2}{X_1+X_2} = \frac{\omega e(t)}{X_1+X_2} \text{ approx.} \tag{2.75}$$

The solutions for i_1 and i_2 can now be written by inspection, substituting the coefficients of i and e in each case for the R/L and $1/L$ coefficients in Eq. 2.37.

Similarly, in the case of $R_L = \infty$, we find

$$p_1 = -R_1/(X_M+X_1) \tag{2.76}$$

and $p_2 = -\infty$, so that $i_2 = 0$ and

$$di_1/dt + R_1 i_1/(X_M+X_1) = e(t)/(X_M+X_1). \tag{2.77}$$

In practice, the value of X_M during these transients may be only a small fraction of its normal value, because the offset current, Eq. 2.42, makes the magnetic flux in the core rise above normal, creating extreme saturation. When the residual core flux left when the circuit was last opened is in the same direction as the new flux, the saturation is even more extreme. Thus, the initial current when the transformer is put on the line with the secondary open can reach 30 or more times normal, depending on the instant at which the switch is closed.

13 Dimensional Analysis

It is desirable to visualize the changes in the transformer equivalent circuit constants produced by a change in dimensions, as this provides a perspective that is very helpful in design.

Assume that all the dimensions of a given transformer are multiplied by k; the number of winding turns and the flux and current densities being held constant. The following conditions will then hold:

1. The core area, the flux capacity, and, therefore, the voltage will increase by a factor k^2.

2. The current capacity, assuming adequate ventilation, will increase in proportion to the area of the conductor, or by a factor k^2.

3. The weight of the transformer will increase by a factor k^3.

Therefore:

4. The copper cross-section will increase by a factor k^2, but the conductor length will increase by k, so the resistance will decrease by a factor k.

5. The copper loss will increase by the factor k^3, since the current squared goes up k^4 times, while the resistance goes down k times.

6. The volume of core and, therefore, the iron losses will increase by a factor k^3.

7. The surface area for cooling will increase by k^2, but the total losses increase by k^3, so the heat to be dissipated per unit of surface goes up k times.

8. The areas of leakage flux paths will increase as k^2, but their lengths will increase as k, so the leakage reactance will increase by the factor k.

9. The area of the magnetic circuit will increase by the factor k^2, and its length by the factor k, so the magnetizing reactance will increase by the factor k.

10. The apparent kva rating will increase by the product of the voltage and current factors, or by k^4.

11. The kva per pound, or specific rating of the transformer, will apparently increase by a factor k.

12. The per unit resistance loss will decrease by a factor k.

13. The per unit iron losses will decrease by a factor k.

14. The per unit magnetizing current will decrease by the factor k.

15. The per unit leakage reactance will increase by the factor k.

16. The time constant of current decrement in transients (ratio of inductance to resistance) will increase by the factor k^2.

17. The surface temperature will increase by the factor k.

18. The frequency of mechanical resonance will decrease by the factor k.

These results are indicated in Fig. 2.10, which shows the actual circuit constants for a transformer k times larger than for Fig. 2.4, with $F = 1$. Fig. 2.11 shows the corresponding circuit in per unit terms.

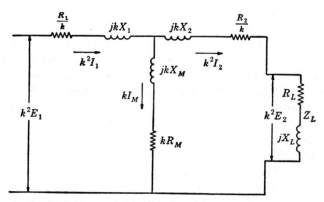

FIG. 2.10. Ohmic Equivalent Circuit of Transformer with all Dimensions k Times Larger than in Fig. 1.4.

This dimensional analysis shows why it is that transformer efficiencies and weight ratios improve as the size and rating increase. In practice, the use of higher voltages, requiring more insulation, and the need for relatively more ventilating space to hold down the temperatures, partially offset these gains. The larger transformer must be reproportioned to offset the increase in per unit leakage reactance inherent with increased size. Thus, a large transformer will necessarily have different characteristics from those of a small-scale model.

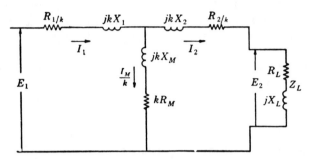

FIG. 2.11. Per Unit Equivalent Circuit of Transformer with all Dimensions k Times Larger than in Fig. 1.4.

Clearly, good performance requires that all the proportions of a transformer be varied as the physical size increases. Just how to do this to the best advantage depends on the relative importance of iron and copper losses, reactance, and temperature rise for the particular application.

Although this analysis has dealt with transformers only, the conclusions apply to all types of electric machines. In every case, the equivalent circuit representing the machine includes resistance and reactance elements whose relative values change with the dimensions in the same way as those of the transformer.

As a further preparation for the study of equivalent circuits representing machine performance, it is desirable to review the relations between the various systems of units and their fundamental "dimensions", which are used in electrical, magnetic, mechanical, and other calculations. The equations governing any of these systems may be represented by an equivalent electric circuit, so that it is helpful to be able to translate quickly the results of circuit calculations into any desired system of units.

In a mechanical system, all the physical quantities can be expressed as combinations of three basic units, such as mass, length, and time. To include electromagnetic phenomena, a fourth unit must be added, usually taken as electric charge. However, any four independent units may be chosen, each selection being convenient for a particular type of problem.

DIMENSIONAL ANALYSIS

Table 2.1 lists the various quantities usually encountered in the analysis of electric machines, and their dimensional values in four different systems of units.

TABLE 2.1

ELECTROMAGNETIC-MECHANICAL SYSTEM QUANTITIES

I	II	III	IV	V	VI
		Mechanical Systems		Electrical System	Magnetic System
Quantity	Symbol	Dimensions			
Units		L, M, Q, T	V, F, Q, T	I, E, L, T	ϕ, I, L, T
Length	L	L	VT	L	L
Mass	M	M a	$FV^{-1}T$ a	$EIL^{-2}T^3$	$\phi IL^{-2}T^2$
Time	T	T	T	T	T
Velocity	V	LT^{-1} x	V x	LT^{-1}	LT^{-1}
Force	F	LMT^{-2} e	F e	$EIL^{-1}T$	ϕIL^{-1}
Acceleration	A	LT^{-2}	VT^{-1}	LT^{-2}	LT^{-2}
Momentum	H	LMT^{-1}	FT	$EIL^{-1}T^2$	$\phi IL^{-1}T$
Damping constant	D	MT^{-1} b	FV^{-1} b	$EIL^{-2}T^2$	$\phi IL^{-2}T$
Spring constant	K	MT^{-2} c	$FV^{-1}T^{-1}$ c	$EIL^{-2}T$	ϕIL^{-2}
Moment of inertia	I	L^2M	FVT^3	EIT^3	ϕIT^2
Power	P	L^2MT^{-3} ex	FV ex	EI ex	ϕIT^{-1}
Energy	W	L^2MT^{-2}	FVT	EIT	ϕI ex
Charge	Q	Q	Q	IT	IT
Inductance	L	L^2MQ^{-2}	$FVQ^{-2}T^3$	$EI^{-1}T$ a	ϕI^{-1}
Capacity	C	$L^{-2}M^{-1}Q^2T^2$	$F^{-1}V^{-1}Q^2T^{-1}$	$E^{-1}IT$ 1/c	$\phi^{-1}IT^2$
Current, or mmf	I	QT^{-1}	QT^{-1}	I x	I e
Voltage	E	$L^2MQ^{-1}T^{-2}$	$FVQ^{-1}T$	E e	ϕT^{-1}
Impedance	$R, X,$ or Z	$L^2MQ^{-2}T^{-1}$	$FVQ^{-2}T$	EI^{-1} b	$\phi I^{-1}T^{-1}$	1/a
Resistivity	ρ	$L^3MQ^{-2}T^{-1}$	$FV^2Q^{-2}T^3$	LEI^{-1}	$L\phi I^{-1}T^{-1}$
Permeability	μ	LMQ^{-2}	$FQ^{-2}T^2$	$L^{-1}EI^{-1}T$	$L^{-1}\phi I^{-1}$
Permittivity	K	$L^{-3}M^{-1}Q^2T^2$	$F^{-1}V^{-2}Q^2T^{-2}$	$L^{-1}E^{-1}IT$	$L^{-1}\phi^{-1}IT^2$
Flux	ϕ	$L^2MQ^{-1}T^{-1}$	$FVQ^{-1}T^2$	ET	ϕ x
Reluctance	$\mathcal{R}$	$L^{-2}M^{-1}Q^2$	$F^{-1}V^{-1}Q^2T^{-3}$	$E^{-1}IT^{-1}$	$\phi^{-1}I$ b

a, b, c, e, x are corresponding parameters in the analogous equivalent circuits of each system, as governed by the differential equation:

$$e = a\frac{dx}{dt} + bx + c\int x\,dt.$$

The typical equation of motion of a mechanical system is:

$$(2.78) \qquad F = \frac{M\,d^2L}{dt^2} + D\frac{dL}{dt} + KL,$$

or, more generally:

$$(2.79) \qquad e = a\frac{dx}{dt} + bx + c\int x\,dt.$$

The correspondence of M to a, D to b, etc., is indicated in column III of the table, in which all quantities are expressed in terms of the familiar units of mass, length, time, and charge. However, since x corresponds to

dL/dt, or V, calculations with this equation and its equivalent circuit are more conveniently made if V, F, Q and T are taken as the basic units, as in column IV.

For an electrical system, Eq. 2.79 becomes (Eq. 1.1):

$$(2.80) \qquad e = L\frac{di}{dt} + Ri + \frac{1}{C}\int i\,dt.$$

The choice of current, voltage, length, and time as the basic units gives relations most convenient for use with this equation, as indicated in column V.

For an electromagnetic system, Eq. 2.79 becomes:

$$(2.81) \qquad i = \frac{1}{R}\frac{d\phi}{dt} + \mathcal{R}\phi.$$

This process of electric circuit representation and analysis is equally applicable to a very wide range of mechanical, hydraulic, and field phenomena, so long as currents are used to represent velocity or flow; voltages to represent forces or pressures; inductance to represent inertia; resistance to represent friction; and capacitance to represent elasticity. The electric circuit, therefore, provides a basic technical "language" that is common to all sorts of engineering problems.

A principal use for Table 2.1 is in checking the validity of equations that include several terms expressed in different parameters. Every term in such an equation, when reduced to its simplest equivalents, derived from the table, must contain identically the same powers of the basic parameters. For example, an energy equation may be written:

$$W = m_1 FL + m_2 I^2 RT + m_3 CE^2 + m_4 \frac{\phi^2}{\mu L} + m_5 MV^2,$$

where m_1, m_2, etc., are numerical coefficients.

This equation is valid, because each of the terms is reducible, by means of the equivalents in Table 2.1, to an expression of the form:

$$mL^2MT^{-2}.$$

14 Energy Flow

At this point, it is well to "take stock", and try to visualize the energy flow through a circuit.

To speak in homely analogies, it is convenient to think of the flow of electric current in a wire as similar to that of water in a pipe. When the switch is closed, applying voltage to a conductor, free electrons rush along the wire, building up a pressure of electric potential against the insulating material along the way, and a magnetic field around the conductor, until they come to the end. If the end is open, the electrons meet an elastic wall of insulation, or potential barrier, which stops them, and throws them back against the oncoming electrons, producing a wave of reflected current and voltage, which returns to the source. In this way, after an infinite series of reflections, the system settles down to a steady state, with the conductor

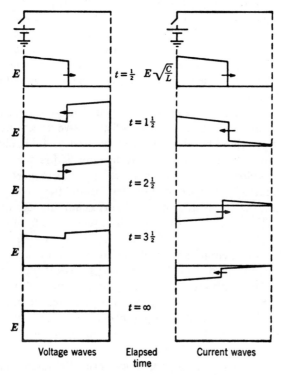

FIG. 2.12. Transient Waves when a D-C Voltage is Suddenly Impressed on an Open Line.

full of electrons pressing against the enclosing wall of insulation, ready to break through at any point of weakness.

The wave (of electron pressure, not electron motion) is transmitted along the conductor at almost the speed of light, 186,000 miles per second. The actual velocity is $1/\sqrt{(LC)}$, where L is the inductance in henries, and C the capacitance in farads, per unit length of line. The presence of iron or of insulation lowers the velocity, since the product LC is increased by a permeability or a dielectric constant greater than one.

Figure 2.12 shows the process of building up voltage on an open line, when a d-c voltage is suddenly impressed. The current wave is reflected at an open end with a change of sign, and at the closed end without a change of sign, the opposite being true for the voltage. If there were no resistance or dielectric losses, the waves would continue to oscillate back and forth indefinitely, but in practice the losses damp them out after only a few reflections.

If the conductor is closed at the far end, a similar process of successive wave reflections occurs, until a steady state is built up, with the voltage pressure used up along the line in overcoming the resistance and reactance drop. In the same way, when water is let into a pipe, the steady state is reached after a series of oscillations, or "water-hammer", and the applied pressure is balanced by the friction and inertia forces required by the flow that is established.

The concept of the building up of the currents and voltage distribution in a transmission line or winding, as a result of successive reflections (Fig. 2.12), has been well presented by Woodruff and by Bewley. It should always be remembered that the electrons initially flowing across the path formed when a switch is closed cannot tell what they are going to meet as they move along the conductor. The current at the first instant, therefore, is identically the same, and is measured by the surge admittance of the line, $[I = E \sqrt{(C/L)}]$, irrespective of whether the circuit at its far end is open or short-circuited. Thus, during the first few instants, the current and voltage relations in a circuit are widely different from those that exist after steady state is established. The theory of these transient effects is important in connection with lightning and switching surges, and the various impulse test methods that are used to verify machine performance under transient conditions. The whole building-up process is normally over in a few microseconds, however, so that it may be considered entirely apart from the normal operation of machines.

The oncoming wave of electrons, which forms the initial current in the conductor, is impeded by collisions with atoms, or friction, as measured by the conductor resistance, just as water is slowed down by friction in a pipe. Also, if there is any leakage or capacitance current through the insulation, the front of the current wave is frittered away, in much the same way that a wave of water running over dry sand is diminished by seepage into the porous surface.

Thus, the number of electrons passing over the conductor in a given time under a given voltage pressure is reduced by high resistance and by leakage, the corresponding energy being consumed in heating of the conductor, external losses, or dielectric energy.

Similarly, the electrons are delayed by the energy they must supply to create the magnetic field, as they move along the conductor. The energy thus stored in the magnetic field is exactly similar to the kinetic energy of a moving water column. With direct current, the inductance, or inertia, delays the building up of the current, but does not limit its final value. With alternating current (the back emf due to the inductance being proportional to the rate of rise of the current), the rise of current is caused to lag behind the rise of voltage, just as the velocity of a water column lags behind the applied pressure that accelerates it.

The energy that is stored in the inductive field may be thought of as kinetic energy, which creates an emf or back voltage, opposed to any change in the current. Thus, all the magnetic forces produced by currents, and all the electric voltages produced by changes in current, have their source in the energy stored in the magnetic field. There are, of course similar effects due to the energy stored in the electrostatic field, dependent on the voltage. At usual machine voltages, however, the energy, $(1/2)CV^2$, stored in the dielectric field, is insignificant by comparison with the magnetic energy in the air-gap field (Sect. 1.6). In theory, an electrostatic motor could be built, with the torque produced by the rate of change of the energy stored in the dielectric field. But the dielectric stresses, and the resulting insulation difficulties, required to obtain a useful power output are so great that the electrostatic machine has remained of scientific rather than practical interest.

For these reasons, it is fair to say that the real key to understanding an electric machine is to visualize clearly and calculate accurately the distribution of magnetic flux through the machine. The flux linkages measure the reactances of the windings. The rate of change of the stored magnetic energy as the rotor moves measures the forces and torques produced (Sect. 1.7 and Appendix B, Chapter 7).

BIBLIOGRAPHY

		Reference to Section
1.	*Mathematics for Science and Engineering*, P. L. Alger, McGraw-Hill, Second Edition, 1969	9
2.	"Traveling Waves Initiated by Switching," L. V. Bewley, *A.I.E.E. Trans.*, Vol. 58, 1939, pp. 18–26.	12
3.	*Transformer Engineering*, L. F. Blume, A. Boyajian, *et al.*, Second Edition, John Wiley, 1951.	1
4.	"Transformer Magnetizing Inrush Currents and Influence on System Operation," L. F. Blume, G. Camilli, *et al.*, *A.I.E.E. Trans.*, Vol. 63, 1944, pp. 366–75.	12
5.	"A Simplified Method of Analyzing Short-Circuit Problems," R. E. Doherty, *A.I.E.E. Trans.*, Vol. 42, 1923, pp. 841–8.	8
6.	*Mathematics of Modern Engineering*, Vol. 1, R. E. Doherty and E. G. Keller, Vol. 2 by E. G. Keller, John Wiley, 1936 and 1942.	9

TRANSFORMERS—CIRCUIT ANALYSIS

	Reference to Section
7. "Short-Circuit Current of Induction Motors and Generators," R. E. Doherty and E. T. Williamson, *A.I.E.E. Trans.*, Vol. 40, 1921, pp. 509–39.	12
8. *Transients in Linear Systems*, M. F. Gardner and J. L. Barnes, John Wiley, 1942.	10
9. *Mathematical and Physical Principles of Engineering Analysis*, W. C. Johnson, McGraw-Hill, 1944.	13
10. "Equivalent Circuits to Represent the Electromagnetic Field Equations," G. Kron, *The Physical Rev.*, Vol. 64, Aug. 1 and 15, 1943, pp. 126–8.	3
11. *Transient Circuit Analysis*, Y. H. Ku, Van Nostrand, 1961.	9
12. *Complex Variables and the Laplace Transform for Engineers*, W. R. LePage, McGraw-Hill, 1961.	9
13. "A Derivation of Heaviside's Operational Calculus Based on the Generalized Functions of Schwartz," J. J. Smith and P. L. Alger, *A.I.E.E. Trans.*, Vol. 68, Part II, 1949, pp. 939–44.	11
14. "Magnetizing-Inrush Phenomena in Transformer Banks," W. K. Sonnemann, C. L. Wagner, and G. D. Rockefeller, *A.I.E.E. Trans.*, Power Appar. & Systems, Vol. 77, Part III, Oct. 1958, pp. 884–92.	12
15. *Linear Transient Analysis*, E. Weber, John Wiley, 1954.	9

CONTENTS—CHAPTER 3

The Rotating Magnetic Field

		PAGE
1	Idealized Magnetic Structure	71
2	Current and Magnetomotive Force Waves	72
3	Torque	73
4	Power Output	75
5	Armature Windings	76
6	Generated Voltages	80
7	Production of the Rotating MMF	82
8	The MMF Produced by a Single Coil Carrying Direct Current	83
9	Change from Direct to Alternating Current	86
10	Resolution into Rotating MMFs	86
11	Addition of MMFs due to Coils in the Same Phase Winding	87
12	Addition of MMFs due to Different Phases	89
13	Magnetizing Current	93
14	Winding Analysis	96
15	Symmetrical Components	100

3

THE ROTATING MAGNETIC FIELD

1 IDEALIZED MAGNETIC STRUCTURE

A rotating electric machine consists in essence of two concentric cylinders, each mounting a set of magnetic poles, of alternate north and south polarity, as in Fig. 3.1. The poles may be permanent magnets, or direct-current-excited electromagnets. Or they may be the resultants of electric currents flowing in coils distributed around the periphery. In any

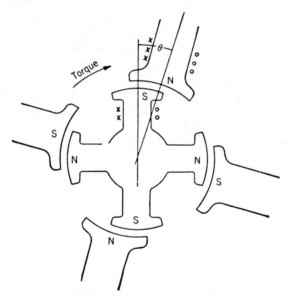

Fig. 3.1. Polar Structure of Rotary Electric Machine.

case, the structure of Fig. 3.1 provides a convenient way of visualizing the forces that are produced.

If there are P pairs of equally spaced poles on each cylinder, and an inner south pole is displaced $P\theta$ electrical degrees from the nearest outer north pole, the interpolar magnetic forces will produce a torque in a direction tending to increase the total flux (Sect. 1.7). The torque, therefore, will tend to bring the opposite poles in line, so that their mmfs will add; that is, θ will tend to become zero. If both cylinders are free to turn, and one is driven from an external power source, the other cylinder will follow its rotation at the same speed, with an angle of lag between poles just sufficient to develop the needed turning effort. In this case, the machine will be a synchronous clutch. If the magnetic torque is insufficient, or the second cylinder is held stationary, the poles will slip by each other, and the torque will be alternately positive and negative, going through P cycles for each revolution of relative motion, or slip. The net torque will be zero, except for power losses. The clutch is then said to be out of step.

If the outer magnets are removed and replaced by a stationary cylinder of laminated steel, carrying a slot-embedded winding, the rotation of the inner magnets will produce alternating voltages in the winding. When the winding is closed, alternating currents will be induced in it, which will oppose the changes in flux linkages imposed by the moving magnets. These induced currents will themselves create an mmf wave, opposing the impressed mmf of the revolving poles, and turning synchronously with the rotor. And the magnetic poles that would be produced by the induced currents acting alone will exert magnetic forces on the rotor poles, just as if they were magnets themselves.

If the poles revolve at a speed of N revolutions per minute, each rotor pole will pass a stationary point $N/60$ times a second. Two consecutive poles of opposite polarity will produce a complete cycle of voltage alternation in a stationary coil, so that the relation between frequency and speed is:

$$(3.1) \qquad f = \frac{PN}{60}, \quad \text{or} \quad N = \frac{60f}{P},$$

where f = frequency in cycles per second, P = number of pairs of poles, and N = revolutions per minute.

2 Current and Magnetomotive Force Waves

To calculate the forces produced by these $2P$ pole revolving fields, it is convenient to replace all the magnets by sinusoidally distributed waves of mmf. The resultant flux density in the air gap will then correspond to the resultant mmf acting at each point, found by summation of the rotor and stator mmfs.

A convenient starting point in the calculations is the rms value of the sinusoidally distributed ampere conductors per meter of periphery, usually called the "current loading", and designated by Δ (Eq. 3.36). The effective ampere conductors at any point on the periphery are then given by the expression:

$$(3.2) \qquad \text{Ampere conductors} = \sqrt{2}\,\Delta \cos P\theta \text{ amperes per meter.}$$

TORQUE

The ampere turns acting across the air gap at any point are equal to the total number of ampere conductors between that point and the nearest peak of the current density wave, or:

$$(3.3) \quad \text{Air-gap mmf} = \sqrt{2}\,\Delta \int_0^\theta \cos P\theta \, d\left(\frac{D\theta}{2}\right)$$

$$= \frac{\sqrt{2}\,D\Delta}{2P} \sin P\theta \text{ ampere turns,}$$

where D is the air-gap diameter (stator bore) in meters.

As Fig. 3.2 indicates, the mmf wave is 90° ahead of the current density wave in space phase.

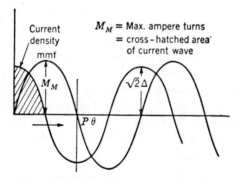

FIG. 3.2. Magnetomotive Force and Current Waves.

The flux density is proportional to the mmf at each point. By Eqs. 1.15 and 3.3:

$$(3.4) \quad B = \frac{4\pi(\text{mmf})}{G10^7} = \frac{8.886 D\Delta}{PG10^7} \sin P\theta = B_M \sin P\theta \text{ webers per sq m,}$$

if G is the effective air-gap length in meters.

3 TORQUE

The torque produced by interaction of the rotor and stator fields may be calculated either by Eq. 1.38, giving the torque as the product of field strength of the rotor by current loading of the stator; or by the principle of virtual displacement, giving the torque as the rate of change of stored magnetic energy with angular displacement.

By the first method, the effective stator current density is, by Eq. 3.2:

$$\sqrt{2}\,\Delta_1 \cos P\theta \, .$$

and, assuming a displacement of α mechanical degrees between stator and rotor fields, the air-gap flux density is, by Eq. 3.4:

$$B = B_1 \sin P\theta - B_2 \sin P(\theta + \alpha) \text{ webers per sq m.}$$

Only the second term of this equation produces any net torque with Δ_1. Therefore, by Eq. 1.38, the torque is:

$$\text{Torque} = \text{radius} \times \text{force}$$

$$= \frac{Dl}{2} \times \text{current} \times \text{flux density},$$

integrated around the periphery,

(3.5) $$\text{Torque} = \frac{\sqrt{2}\, DlB_2 \Delta_1}{2} \int_0^{2\pi} \cos P\theta \sin P(\theta + \alpha) \, d\left(\frac{D\theta}{2}\right)$$

$$= \frac{\pi \sqrt{2}\, D^2 l B_2 \Delta_1 \sin P\alpha}{4} \text{ newton meters.}$$

Thus, the torque is proportional to the square of the rotor diameter, the rotor core length, the air-gap flux density, the current loading, and the sine of the angular displacement between rotor and stator poles.

By the second method, the resultant air-gap flux destiny, due to the sum of rotor and stator mmfs, is:

(3.6) $$B = B_1 \sin P\theta - B_2 \sin P(\theta + \alpha).$$

The total stored energy in the air gap is, therefore, by Eq. 1.24:

(3.7) $$W = \frac{\text{volume of air gap} \times \text{average value of } B^2 \times 10^7}{8\pi} \text{ wsec}$$

$$= \frac{DlG10^7}{8}\left(\frac{P}{\pi}\right) \int_0^{\pi/P} B^2 \, d\theta.$$

Substituting Eq. 3.6 in Eq. 3.7:

$$W = \frac{PDlG10^7}{8\pi} \int_0^{\pi/P} [B_1^2 \sin^2 P\theta - 2B_1 B_2 \sin P\theta \sin P(\theta + \alpha)$$

$$+ B_2^2 \sin^2 P(\theta + \alpha)]$$

$$= \frac{PDlG10^7}{8\pi} \left[\frac{\pi}{2P}(B_1^2 - 2B_1 B_2 \cos P\alpha + B_2^2)\right] \text{ wsec}$$

$$= \frac{DlG10^7}{16} \text{ (maximum resultant air-gap flux density)}^2 \text{ wsec.}$$

POWER OUTPUT

The derivative of this with respect to α gives the torque:

$$\text{Torque} = \frac{dW}{d\alpha} = \frac{PDlG10^7 B_1 B_2 \sin P\alpha}{8} \text{ newton meters.}$$

Substituting the maximum value of B_1, in terms of Δ_1, from Eq. 3.4:

(3.5) $$\text{Torque} = \pi \sqrt{2} \frac{D^2 l B_2 \Delta_1 \sin P\alpha}{4} \text{ newton meters,}$$

where D and l are in meters, B_2 in webers per square meter, and Δ_1 in rms amperes per meter of periphery; which is the same as Eq. 3.5.

The torque in newton meters must be multiplied by 0.7373 to obtain the torque in pound-feet, or

(3.8) $$\text{Torque} = 0.819 D^2 l B_2 \Delta_1 \sin P\alpha \text{ lb-ft,}$$

if D and l are in inches, B_2 in webers per square inch, and Δ_1 in rms amperes per inch of periphery.

4 Power Output

The power output in watts is equal to the torque in newton meters times the speed of rotation in radians per second, or:

(3.9) $$\text{Power output} = \frac{\pi^2 \sqrt{2} D^2 l N B_2 \Delta_1 \sin P\alpha}{120} \text{ watts,}$$

where N = revolutions per minute. The coefficient is $\pi^2 \sqrt{2}/120 = 0.116$, and is the same whether the dimensions are in meters or in inches.

For example, if a maximum air-gap flux density of 5×10^{-4} webers per sq in. and an rms current loading of 2,000 amperes per in. at the maximum torque point are employed, the expected output at the angle of maximum torque ($P\alpha = 90°$) is:

$$0.116 D^2 l N (5 \times 10^{-4})(2,000) = 0.116 D^2 l N \text{ watts}$$

with D and l in inches.

Allowing a 3 to 1 ratio of maximum to rated output, this gives an expected motor rating in horsepower of:

(3.10) $$\text{Horsepower rating} = 3.2 D^2 l N,$$

if D and l are in meters, or:

$$5.2 D^2 l N 10^{-5},$$

if D and l are in inches.

An 1,800-rpm motor with 8-in. air-gap diameter and 5-in. core length could on this basis be rated about 30 hp. In practice, the decreasing per unit resistance and increasing per unit reactance with increase in dimensions, as outlined in Sect. 2.13, enable somewhat larger values of B and Δ to be used, as D is increased. Therefore, the actual output obtainable from a consistent line of small motors varies nearly as $(D^2 l)^{3/2}$, instead of simply as $D^2 l$. For the same ratio of l to D, the output varies as $D^{4.5}$, approximately, in motor sizes below about 50 hp.

It is convenient to remember the formula for power output in English units:

$$(3.11) \quad \text{Power output}^* = \frac{2\pi NT}{33{,}000} = \frac{NT}{5{,}252} \text{ hp} = \frac{NT}{7{,}040} \text{ kw,}$$

where N = revolutions per minute and T = torque in foot-pounds. This is derived by remembering that 1 hp (746 watts) by definition is equal to 33,000 ft-lb per min; and that the work done in one 360° revolution of a shaft against a torque of T lb-ft is $2\pi T$ ft-lb. (In Europe the Cheval-vapeur is taken to be 736 watts.)

5 Armature Windings

To understand rotating electric machines, a thorough comprehension is required of the way in which alternating currents flowing in a distributed coil winding produce a rotating magnetic field.

Figure 3.3 shows a cross section of a typical small induction motor, with a 4-pole, 3-phase, 36-slot stator winding. As previously explained, the winding is embedded in slots, in order to permit use of a very small air gap between rotor and stator, and to reduce the forces acting on the conductors. To be effective in producing the four-pole flux, each coil is given a span, or pitch, approximately equal to one-quarter of the periphery. The pitch is preferably a little less than a pole span, because this shortens the coil end length with only a small reduction in the useful flux. For economy, all coils are made identical. Thus, 36 coils, each enclosing 8 teeth, and so having 88.9 per cent pitch, are used in this case.

With 36 coils, and 36 slots in all, a two-layer winding results, having 2 coil sides in each slot. In manufacture, the top sides of the first 8 coils, forming the "jump" of the winding, are held up until the bottom sides of the last 8 coils are placed in the slots below them, as indicated in Fig. 3.4. To avoid this somewhat awkward process, the first coils are sometimes inserted with both sides in slot bottoms, and the last coils then have both sides in the tops of slots. This creates a slight dissymmetry, however, and is possible only when the coils consist of many turns of small wire, and are, therefore, very flexible.

Using only 18 coils, with 1 coil side per slot, appears at first sight to be simpler, especially as it eliminates the between-coil slot insulation. Such

*If T is in newton-meters, the factor 7040 becomes 9550.

single-layer windings were formerly used quite generally, and are still employed in small single-phase motors because of their convenience in coil assembly. In a polyphase winding, however, the coil ends must cross over each other, as indicated in Fig. 3.5. Since each coil side fills the full depth of the slot, the ends must be reshaped in flattened form, or must be bent

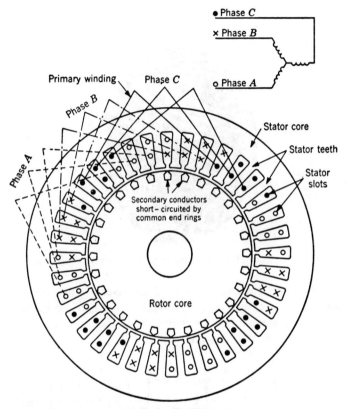

FIG. 3.3. Cross Section of Typical Three-Phase Induction Motor.

radially away from the air gap, to give room for the cross-overs. On this account, the coils in a single-layer polyphase winding must be of several different shapes, a factor which more than offsets their fewer number, so far as cost is concerned.

The double layer has the further advantages over the single-layer winding of lower leakage reactance, better waveform, and better heat dissipation, for the single-layer winding has half as many coil end bundles, each having twice as many conductors, as the double-layer winding. Since the reactance of a concentrated coil varies as the square of the turns, to a first approximation, the double-layer winding might be expected to have only half as much end turn reactance $[(N)^2 + (N)^2 = \frac{1}{2}(2N)^2]$. This is partially offset by the shorter length of the inside turns of the single-layer winding. The two-layer coils can have fractional pitch, reducing their end lengths and giving a

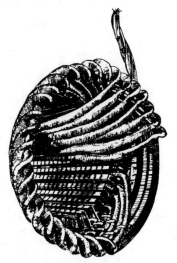

FIG. 3.4. Coils of Three-Phase Winding Partially Assembled.

more favorable mmf pattern, while the single-layer winding has the same mmf pattern as a full-pitch two-layer winding. The exposed surface of the coil ends will be greater for the double-layer than for the single-layer winding, giving better heat dissipation.

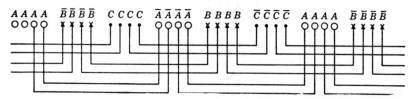

FIG. 3.5. Single-Layer, Two-Pole, Three-Phase Winding.

As pointed out in Sect. 2.13, the reactance and ventilation are less important, the smaller the size of machine. For small single-phase machines, the advantage in ease of winding the single-layer concentric winding more than offsets the disadvantages cited. In this case, the inner coils of each pole are wound with fewer turns, giving the effects of fractional pitch.

ARMATURE WINDINGS

If the four-pole magnetic field is truly sinusoidal in waveform, it will generate sinusoidal voltage waves in each coil as it rotates, the time phase of each coil voltage being displaced from the voltage of the preceding coil. In Fig. 3.3, as there are 9 slots per pole, the electrical angle between slots is 20°, and this will be the phase displacement of successive coil voltages. With uniform speed of field rotation, therefore, the space-phase angles on

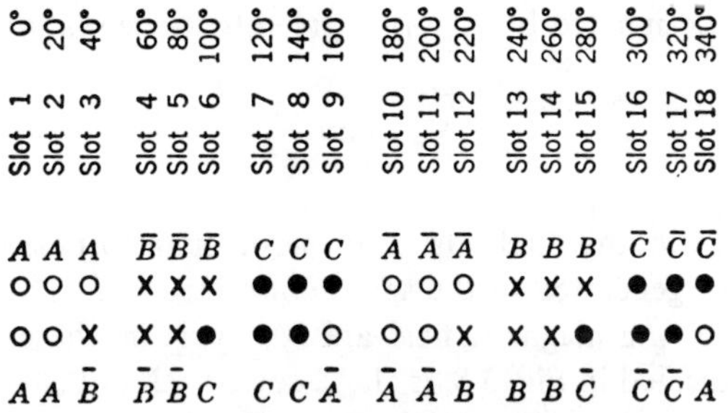

Slot 1 0°	Slot 2 20°	Slot 3 40°	Slot 4 60°	Slot 5 80°	Slot 6 100°	Slot 7 120°	Slot 8 140°	Slot 9 160°	Slot 10 180°	Slot 11 200°	Slot 12 220°	Slot 13 240°	Slot 14 260°	Slot 15 280°	Slot 16 300°	Slot 17 320°	Slot 18 340°
A	A	A	$\bar{B}$	$\bar{B}$	$\bar{B}$	C	C	C	$\bar{A}$	$\bar{A}$	$\bar{A}$	B	B	B	$\bar{C}$	$\bar{C}$	$\bar{C}$
○	○	○	×	×	×	●	●	●	○	○	○	×	×	×	●	●	●
○	○	×	×	×	●	●	●	○	○	○	×	×	×	●	●	●	○
A	A	$\bar{B}$	$\bar{B}$	$\bar{B}$	C	C	C	$\bar{A}$	$\bar{A}$	$\bar{A}$	B	B	B	$\bar{C}$	$\bar{C}$	$\bar{C}$	A

FIG. 3.6. Double-Layer, Two-Pole, Three-Phase Winding; 60° Phase Belts—8/9 Pitch.

the periphery are interchangeable with time-phase angles in the electric circuits, and the same phasor diagram may represent either space or time relations.

The arrangement in phases of the winding of Fig. 3.3 is indicated in Fig. 3.6, which shows one half the winding, or 2 poles and 18 slots. Considering the top coil sides only, the first 3 coils are connected in series to form a "phase belt" of phase A. The next 3 are connected in series in phase B reversed, and the third similarly are in phase C. The next 9 coils

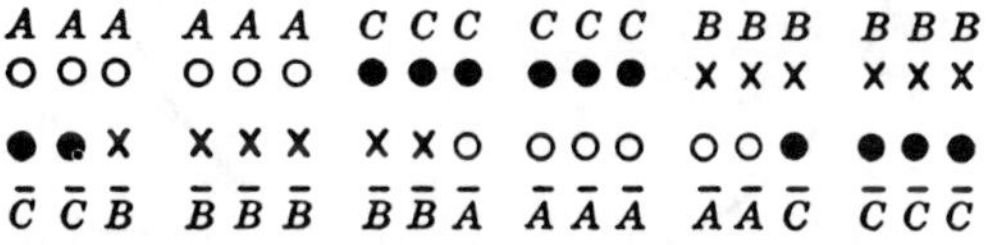

A A A	A A A	C C C	C C C	B B B	B B B
○ ○ ○	○ ○ ○	● ● ●	● ● ●	× × ×	× × ×
● ● ×	× × ×	× × ○	○ ○ ○	○ ○ ●	● ● ●
$\bar{C}$ $\bar{C}$ $\bar{B}$	$\bar{B}$ $\bar{B}$ $\bar{B}$	$\bar{B}$ $\bar{B}$ $\bar{A}$	$\bar{A}$ $\bar{A}$ $\bar{A}$	$\bar{A}$ $\bar{A}$ $\bar{C}$	$\bar{C}$ $\bar{C}$ $\bar{C}$

FIG. 3.7. Double-Layer, Two-Pole, Three-Phase Winding; 120° Phase Belts—8/9 Pitch.

repeat this sequence, with the directions of current flow reversed. There are in all 6 phase belts in 2 pole pitches; each 3 slots, or 60° wide. The winding is, therefore, called a 60° phase-belt winding. The bottom coil sides are exactly similar, but are displaced 8 slots, or 160° in phase, from the top coil sides, and are reversed in direction of current flow, giving a net phase difference of 20°. If the winding had 100 per cent pitch, the top and bottom coil sides in each slot would carry identical in-phase currents. As it is, 2 of the 3 slots in each phase belt carry coil sides in the same phase winding, and 1 carries coil sides of different phases.

Fig. 3.7 shows a three-phase, 120° phase-belt winding.

6 GENERATED VOLTAGES

Since all the bottom coil side voltages generated by the sinusoidal rotating magnetic field are displaced 20° from the top coil side voltages, the resultant coil voltage will be cos 10°, or 0.985, times the arithmetic sum of the two coil side voltages taken separately. This factor is called the pitch factor, and is designated by K_p. It is given by the equation:

$$K_p = \sin\frac{p\pi}{2}, \tag{3.12}$$

where p is the winding pitch, or ratio of coil span to pole pitch.

The voltages generated in the series-connected coils 1, 2, and 3 by the sinusoidal revolving magnetic field are 20° apart in phase. The resultant voltage, as indicated in Fig. 3.8, is $(1+2\cos 20°)/3$, or 0.960, times as great

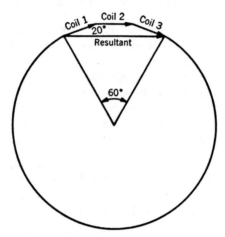

FIG. 3.8. Summation of Fundamental Voltages in 60° Phase Belt.

as it would be if all three were in time phase. This ratio is called the winding distribution factor, designated by K_d. If there were only one coil in the phase belt, K_d would be unity. If the number of slots per phase belt is large, the value of K_d approaches as a limit the ratio of the chord to the arc of the phase-belt angle. For a 60° phase belt, this is:

$$K_d = \frac{\sin 30°}{\pi/6} = \frac{3}{\pi} = 0.9549. \tag{3.13}$$

GENERATED VOLTAGES

Values of K_d for various slot numbers and phase-belt widths are given in Table 3.1, in accordance with the equation:

$$(3.14) \qquad K_d = \frac{\sin\dfrac{\pi}{2q}}{n \sin\dfrac{\pi}{2nq}},$$

where q is the number of phase belts per pole, and n the number of slots per phase belt.

The resultant voltage generated by the winding is equal to $K_p K_d$ times the arithmetic sum of all the series-connected conductor voltages.

TABLE 3.1

WINDING DISTRIBUTION FACTOR K_d (EQ. 3.14)

Slots per Phase Belt	60°	90°	120°	180°
1	1.000	1.000	1.000	1.000
2	0.966	0.924	0.866	0.707
3	0.960	0.911	0.844	0.667
4	0.958	0.906	0.837	0.654
5	0.957	0.904	0.833	0.648
∞	0.955	0.900	0.827	0.636

In designing an armature winding, the desirable number of turns, N, is found from Eq. 1.21, when the impressed voltage, frequency, and maximum flux are known. By varying the winding pitch, or changing a three-phase winding from Y to Δ connection, or connecting the phase belts under successive poles in parallel instead of series, it is possible to adjust the winding to make N a convenient whole number. Equation 1.21 may then be rewritten:

$$(3.15) \qquad E = 4.44 f N K_p K_d \phi_M \text{ volts},$$

where N is the number of turns in series, and E is the rms voltage across a single phase of the winding. The voltage across the terminals of a three-phase winding is equal to E, if it is Δ connected, or to $\sqrt{3}\,E$, if it is Y connected.

If the air-gap flux is not perfectly sinusoidal, and contains harmonic components with more than the fundamental number of poles, the voltage induced by any particular harmonic can be found in the same way as for the fundamental voltage, using the appropriate values of pitch and distribution factors.

For example, if a fifth (space) harmonic is present in the air-gap flux, this having five times as many poles as the fundamental, the angle of phase difference between the voltages induced in successive coils will be five times

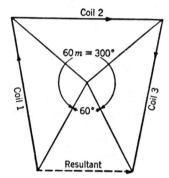

FIG. 3.9. Phase-Belt Voltage Due to Fifth Harmonic; 60° Phase Belts—Three Slots per Phase Belt.

as great as for the fundamental. The phasor diagram of voltages induced in a single phase belt will, therefore, be as shown in Fig. 3.9, and the expression for the mth harmonic distribution factor will be, from Eq. 3.14:

$$(3.16) \qquad K_{d_m} = \frac{\sin\dfrac{m\pi}{2q}}{n \sin\dfrac{m\pi}{2nq}}.$$

Similarly, the harmonic pitch factor will be:

$$(3.17) \qquad K_{p_m} = \sin\frac{pm\pi}{2}.$$

Equation 3.15 will then give the fifth harmonic voltage, if Eqs. 3.16 and 3.17 are used instead of Eqs. 3.14 and 3.12. It should be noted that, for the frequency of the voltage induced by the fifth harmonic to be the same as for the fundamental, the harmonic field must rotate at one-fifth the speed of the fundamental flux wave. For this reason, any space harmonics in the flux distribution produced by a field winding produce no useful voltages in the stator, and vice versa.

7 Production of the Rotating MMF

The calculation of the effective ampere turns of an armature winding, and thence the magnitude of the air-gap flux it produces, is the reverse of the procedure for calculating the induced voltage. It is important to visualize the way in which the resultant mmf is produced by the phasor addition of the mmfs made by the individual coils. A convenient procedure for doing this is to:

(a) Calculate the field produced by a single coil carrying a direct current, and resolve it into a series of space harmonic components.

(b) Replace the direct current by an alternating one, causing all the harmonic components of the field to pulsate sinusoidally at the line frequency, while remaining stationary in space.

(c) Resolve each pulsating harmonic component into two equal, constant, oppositely revolving fields.

(d) Add vectorially the similar harmonic fields produced by the adjacent series-connected coils in the same phase.

(e) Add vectorially the separate harmonic fields produced by the different phases.

8 The MMF Produced by a Single Coil Carrying Direct Current

Fig. 3.10 shows the developed air-gap surface of two pole pitches, or 360 electrical degrees, and a single armature coil with a pitch of $180p$ degrees, or p times full pitch. A direct current flowing in this coil will produce a

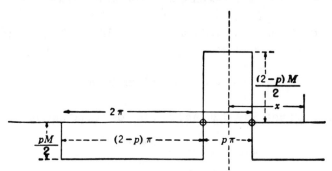

Fig. 3.10. Air-Gap Field of a Single Fractional-Pitch Coil.

uniform magnetic flux across the air gap over the coil width of $p\pi$, which returns over the remaining space of $(2-p)\pi$. Neglecting reluctance of the iron on both sides of the gap, and considering the coil mmf to be concentrated at slot openings of negligible width, the flux wave will consist of two dissimilar rectangles of equal area, as shown in Fig. 3.10. Since the same total flux crosses the gap in each direction, the flux density inside the coil is $(2-p)B$, and that outside the coil is pB, where B is the flux density that would be produced if the coil were full pitch ($p = 1$). The total flux is, therefore, $p(2-p)$ times the flux that would be produced by the full-pitch coil.

THE ROTATING MAGNETIC FIELD

By the well-known Fourier method of analysis, the air-gap mmf distribution of Fig. 3.10 may be represented as the sum of a series of sine waves.

Thus, the mmf, m, at any distance x from the center of the coil is:

(3.18) $\quad m = a\cos x + b\cos 2x + c\cos 3x + \ldots + n\cos kx + \ldots$

But:

(3.19) $\quad m = -\dfrac{pM}{2} \quad \text{from } x = -\pi \text{ to } x = -\dfrac{p\pi}{2},$

(3.20) $\quad = \dfrac{(2-p)M}{2} \quad \text{from } x = -\dfrac{p\pi}{2} \text{ to } x = \dfrac{p\pi}{2},$

(3.21) $\quad = -\dfrac{pM}{2} \quad \text{from } x = \dfrac{p\pi}{2} \text{ to } x = \pi,$

where M = ampere turns of the single coil. If $p = 1$, $M/2$ ampere turns act across the gap in each direction.

Multiplying Eq. 3.18 by $\cos kx$ and integrating from $x = -\pi$ to $x = \pi$,

(3.22) $\quad \displaystyle\int_{-\pi}^{\pi} m\cos kx\, dx = n\int_{-\pi}^{\pi}\cos^2 kx\, dx = n\pi,$

since all the terms on the right side of Eq. 3.18 vanish, except the term containing $\cos kx$, because:

$$\int_{-\pi}^{\pi} \cos nx \cos kx\, dx = 0 \text{ unless } n = k.$$

Multiplying Eqs. 3.19, 3.20, and 3.21 by $\cos kx$ and integrating:

(3.23) $\quad \displaystyle\int_{-\pi}^{\pi} m\cos kx\, dx = -\dfrac{pM\sin kx}{2k}\bigg]_{-\pi}^{-p\pi/2}$

$$+ \dfrac{(2-p)M\sin kx}{2k}\bigg]_{-p\pi/2}^{p\pi/2}$$

$$- \dfrac{pM\sin kx}{2k}\bigg]_{p\pi/2}^{\pi}$$

$$= \dfrac{2M}{k}\sin\dfrac{kp\pi}{2}.$$

Equating Eq. 3.23 to 3.22:

(3.24) $\quad n = \dfrac{2M}{\pi k}\sin\dfrac{kp\pi}{2}.$

Substituting Eq. 3.24 in 3.18:

(3.25) $$m = \frac{2M}{\pi}\left(\sin\frac{p\pi}{2}\cos x + \frac{1}{2}\sin\frac{2p\pi}{2}\cos 2x \right.$$
$$\left. + \frac{1}{3}\sin\frac{3p\pi}{2}\cos 3x + \ldots + \frac{1}{k}\sin\frac{kp\pi}{2}\cos kx + \ldots\right).$$

The fundamental sine wave of mmf, therefore, has a peak value equal to $(2M/\pi)\sin(p\pi/2)$. The average of the fundamental mmf over the pole pitch is $2/\pi$ times this, or is $(4M/\pi^2)\sin(p\pi/2)$. The average of the rectangular wave of total mmf, from Fig. 3.10, is $p(2-p)M/2$. Therefore, the ratio of the fundamental, or useful, flux to the total flux produced by the single coil is:

(3.26) $$\frac{\phi_1}{\phi_{\text{total}}} = \frac{\frac{8M}{\pi^2}\sin\frac{p\pi}{2}}{p(2-p)M} = \frac{8\sin\frac{p\pi}{2}}{\pi^2 p(2-p)}.$$

For $p = 1$, this is 0.811.

Figure 3.11 shows the principal harmonics obtained in this way for a full-pitch winding ($p = 1$).

It will be noted that the pitch factor, $\sin(kp\pi/2)$, given by Eq. 3.17 for calculating the induced voltage is also the pitch factor given by Eq. 3.25

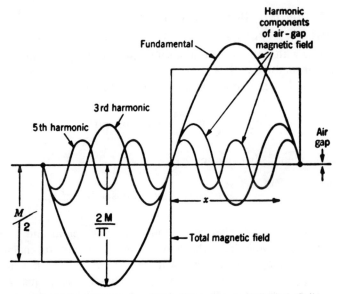

FIG. 3.11. Harmonics of Flux Wave Due to Full-Pitch Coil.

for the calculation of the harmonic components of flux produced by a coil.

There are both even and odd harmonics in the series of Eq. 3.25. It will be found later that the even harmonics represent the dissymmetry of the field, and that they cancel out when there is a whole number of symmetrical phase belts per pole (q is an integer), as in Fig. 3.3. They are of some importance, however, in fractional slot and 120° phase-belt windings.

9 Change from Direct to Alternating Current

If the current in the coil is alternating instead of direct, the mmf will pulsate sinusoidally in time, and all the harmonic components of the field will pulsate in synchronism. The effect of this is to multiply all the terms of Eq. 3.25 by $\cos \omega t$, where $\omega = 2\pi f$:

$$(3.27) \quad m = \frac{2M \cos \omega t}{\pi} \left(\sin\frac{p\pi}{2} \cos x + \frac{1}{2} \sin\frac{2p\pi}{2} \cos 2x + \ldots \right.$$

$$\left. + \frac{1}{k} \sin\frac{kp\pi}{2} \cos kx + \ldots \right).$$

Here, M is the peak ampere turns of the single coil, or $\sqrt{2}\, I$ times turns per coil, if I is the rms alternating current.

10 Resolution into Rotating MMFs

Considering the kth harmonic of Eq. 3.27 by itself, Fig. 3.11, we have an mmf wave with k times the fundamental number of poles, sinusoidally distributed around the periphery, and alternating sinusoidally in time at a frequency f cycles per second. At any point on the periphery, the mmf is:

$$(3.28a) \quad m_k = \frac{2M}{\pi k} \sin\frac{kp\pi}{2} \cos kx \cos 2\pi ft.$$

But this is identically equal to:

$$(3.28b) \quad m_k = \frac{M}{\pi k} \sin\frac{kp\pi}{2} [\cos(kx - 2\pi ft) + \cos(kx + 2\pi ft)].$$

That is, the stationary alternating mmf is at every point and at every instant indistinguishable from the sum of two equal constant-magnitude mmfs, revolving in opposite directions at the same speed, of $2f$ pole pitches per second (Fig. 3.12). Since the kth harmonic has k times as many poles as the fundamental, and as all mmfs move $2f$ pole pitches per second, the kth harmonic will evidently move at one kth the speed of the fundamental, or N/k revolutions per minute, if N is the synchronous speed of the fundamental.

From Eq. 3.28, at the point $x = 0$ (and also at $x = 2\pi/k, 4\pi/k, 6\pi/k, \ldots$), representing the center of a coil, at the time $t = 0$ (and also at $t = 1/f, 2/f, 3/f, \ldots$), the forward- and backward-revolving mmfs are coincident, giving their resultant its crest value of $(2M/\pi k) \sin (pk n/2)$. At the same time, $t = 0$, but at the point $x = \pi/2k$ (and also at $x = 3\pi/2k, 5\pi/2k, 7\pi/2k, \ldots$), the two fields are opposite, making their resultant equal to zero.

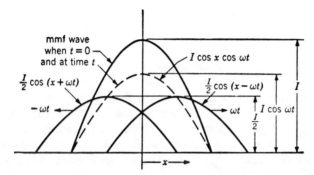

Fig. 3.12. Resolution of Alternating Wave into Two Constant-Magnitude Waves Revolving in Opposite Directions.

Starting at time 0, when the two kth harmonic mmfs are coincident, they will be 180 electrical degrees apart at the end of a quarter cycle, $t = 1/4f$, and will be 360° apart, or will again coincide, after the lapse of a half cycle. At every instant, the forward and backward mmfs will be equally distant on opposite sides of the coil center, so their sum will be stationary and alternating. The two forms of Eq. 3.28 are truly interchangeable expressions, but the great advantages of the second form in visualizing the summation of the mmfs due to the other coils make it far more useful.

11 Addition of MMFs Due to Coils in the Same Phase Winding

If every coil in the winding is identical, each one will produce a similar pair of oppositely revolving mmf waves for each harmonic. Each phase belt, composed of n consecutive coils carrying the same current, produces a kth harmonic mmf equal to the phasor sum of the n identical fields spaced consecutively one slot pitch apart. This resultant is found by phasor addition of the chords of a circular arc, as in Fig. 3.8 or Fig. 3.9. The corresponding distribution factor, or ratio of the resultant to the numerical sum of the mmfs, is given by Eq. 3.16. The mmf of a single phase belt of n coils is found by combining Eqs. 3.27, 3.28, and 3.16:

$$3.29) \quad m = \frac{Mn}{\pi}\left[K_p K_d \cos(x-\omega t) + \frac{1}{2}K_{p2}K_{d2}\cos(2x-\omega t) + \ldots \right.$$
$$\left. + \frac{1}{k}K_{pk}K_{dk}\cos(kx-\omega t) + \ldots \right]$$
$$+ \frac{Mn}{\pi}\left[K_p K_d \cos(x+\omega t) + \frac{1}{2}K_{p2}K_{d2}\cos(2x+\omega t) + \ldots \right.$$
$$\left. + \frac{1}{k}K_{pk}K_{dk}\cos(kx+\omega t) + \ldots \right].$$

The first bracketed term in Eq. 3.29 represents all the forward-revolving mmfs and the second one the backward-revolving mmfs. The values of K_{dk} and K_{pk} are given by Eqs. 3.16 and 3.17.

A symmetrical polyphase winding has q phase belts per pole, all physically identical, but spaced $360/2q$ electrical degrees apart, and carrying currents spaced $360/2q$ degrees apart in time phase. In the normal case, such as Fig. 3.3, when q is an integer, the consecutive phase belts in the same phase winding will be spaced $180°$ apart, and will carry currents that are opposite in time phase. The resultant mmf of the complete (two-layer) phase winding, expressed in ampere turns at a given point on the air-gap periphery, will be obtained by adding to Eq. 3.29 a similar series, except that the sign of each term is reversed and the angle $(x+\pi)$ is substituted for x throughout. That is:

$$(3.30) \quad 2m = m\underline{|x} - m\underline{|x+\pi},$$

since the mmfs of both poles act on the same flux path. Performing this operation on Eq. 3.29, it will be noted that all the odd harmonic terms add directly, since $-\cos[(2k+1)(x+\pi)] = \cos(2k+1)x$; and all the even harmonic terms cancel out, since $\cos[2k(x+\pi)] = \cos 2kx$. Therefore, the mmf wave of phase A, whose current is $\sqrt{2}\,I\cos\omega t$, is:

$$3.31) \quad m_A = \frac{2Mn}{\pi}\left[K_p K_d \cos(x-\omega t) + \frac{1}{3}K_{p3}K_{d3}\cos(3x-\omega t)\right.$$
$$\left. + \frac{1}{5}K_{p5}K_{d5}\cos(5x-\omega t) + \ldots \right]$$
$$+ \frac{2Mn}{\pi}\left[K_p K_d \cos(x+\omega t) + \frac{1}{3}K_{p3}K_{d3}\cos(3x+\omega t)\right.$$
$$\left. + \frac{1}{5}K_{p5}K_{d5}\cos(5x+\omega t) + \ldots \right],$$

where M = peak ampere turns per coil = $\sqrt{2}\,I$ times turns per coil, and n = coils per pole per phase. The cos $(kx - \omega t)$ terms are forward-revolving waves, and the cos $(kx + \omega t)$ terms are backward-revolving.

12 Addition of MMFs Due to Different Phases

In a q-phase winding, with $360°/2q$ phase belts, the time phases of the currents in adjacent phase belts differ by $360°/2q$, in accord with their sequence in the direction of rotation.

For a three-phase winding, with $60°$ phase belts, and voltages in the time sequence A, C, B, Fig. 3.13, the two revolving fields of phase A (Eq. 3.28)

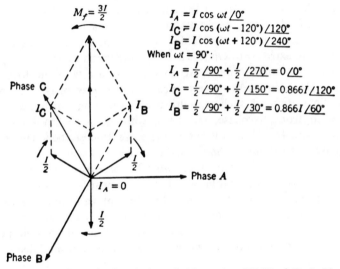

FIG. 3.13. Space Diagram. Resolution of Alternating MMF of Each Phase into Oppositely Revolving Constant-Magnitude Components, Shown at instant when Phase A Current is Zero ($\omega t = 90°$).

will coincide along its center line $\underline{/0}$ at the instant its current is a maximum, when $t = 0$. One-third of a cycle later, when $t = 120°$, each of these fields will have traveled 120 electrical degrees, one forward and the other backward, the former then coinciding with the axis of phase C at $\underline{/120°}$, and the other with phase B at $\underline{/240°}$. But at this moment, phase $\overline{C}$ current is a maximum, so that its forward-revolving mmf component coincides with the forward A mmf, and these two continue to revolve together. The backward mmf of C is $120°$ displaced from that of A, and these mmfs maintain this angle as they continue to revolve. After another third of a cycle, when $t = 240°$, the forward A and C mmfs reach the B axis, at the same moment that phase B current is a maximum. Hence, the forward mmfs of all these phases are directly additive, and together they create a constant-magnitude, sinusoidal, synchronously revolving mmf with a crest value 3/2 the peak value of the alternating field due to one phase alone. The backward-revolving fundamental fields of the three phases are separated by 120°, and their resultant is, therefore, zero, so long as the three phase currents are balanced in both magnitude and phase.

It should be noted that a true six-phase winding ($q = 6$), with three 60° phase belts in each pole, as in Fig. 3.3, is connected with the first and fourth, second and fifth, and third and sixth phase belts in series opposition, so that only three phase currents are actually supplied, and only three leads are required. This doubling of the effective number of phases, by simply reversing the leads of the phase belts located 180° apart, is desirable, because it gives a larger value of K_d (Table 3.1) and, therefore a more effective winding, at no extra cost. On this account, the terms three and six phases are loosely used almost interchangeably, as also are two- and four-phase windings. However, there are certain cases where three-phase, 120° phase-belt windings are used, such as multi-speed motors, in which a winding is reconnected from six-phase, 60° belts, with $2P$ poles, to three-phase, 120° belts, with $4P$ poles. It is better, therefore, to describe a winding as a "60° phase-belt winding", than as either three- or six-phase.

A four-phase winding will have two 90° phase belts per pole, and a similar analysis shows that it will have a forward-revolving constant-magnitude field with a crest value equal to the peak value of one phase, and will have zero backward-revolving fundamental field. A single-phase motor will have equal forward and backward fields, and so will have no tendency to start unless one of the fields is suppressed or modified in some way.

Applying the process of Fig. 3.13 to the kth harmonics, which rotate at $1/k$ times synchronous speed, and again starting with the phase A current at its maximum, when $t = 0$, the forward-revolving component of the kth harmonic of phase A will have traveled only $360/2kq$ mechanical degrees in $1/2q$ cycles, when phase $\bar{B}$ current reaches its maximum. After this moment, the phase A component will continue to revolve in synchronism with the phase $\bar{B}$ component, but $\dfrac{360}{2q}\left(1 - \dfrac{1}{k}\right)$ mechanical degrees behind it. Similarly, these two forward-revolving components will have traveled only $360°/2kq$ farther when phase C current reaches its maximum.

The several components due to successive phase belts, therefore, will be displaced $\dfrac{360}{2q}\left(1 - \dfrac{1}{k}\right)$ mechanical, or $\dfrac{360}{2q}(k-1)$ electrical, degrees in space phase. The resultant of all the phases, for the forward-revolving kth harmonic field, will be proportional to:

$$
(3.32) \quad 1 + 1 \underline{/\tfrac{360}{2q}(k-1)} + 1 \underline{/\tfrac{720}{2q}(k-1)}
$$

$$
+ \ldots \quad + 1 \underline{/\tfrac{360}{2q}(q-1)(k-1)},
$$

and the backward-revolving kth harmonic field will be proportional to:

$$
(3.33) \quad 1 + 1 \underline{/\tfrac{360}{2q}(k+1)} + 1 \underline{/\tfrac{720}{2q}(k+1)}
$$

$$
+ \ldots \quad + 1 \underline{/\tfrac{360}{2q}(q-1)(k+1)}.
$$

As Eqs. 3.32 and 3.33 represent a series of $2q$ phasors equally spaced over 360°, the sum in each case is zero, unless each angle is a multiple of 360°, when the sum is equal to $2q$. Therefore, all the forward-revolving fields cancel out, except those for which $(k-1)2q$ is zero or a whole number, and all the backward-revolving fields vanish except those for which $(k+1)/2q$ is a whole number. Hence, finally, the resultant mmf wave in ampere turns per pole of a 60° phase-belt, three-phase winding with n coils per pole per phase, I rms amperes in each phase, and T turns in series per coil, is, from Eqs. 3.31, 3.32, and 3.33:

$$
(3.34) \quad M_T = \frac{6Mn}{\pi} \Bigg[K_p K_d \cos(x - \omega t) + \frac{1}{5} K_{p5} K_{d5} \cos(5x + \omega t)
$$

$$
+ \frac{1}{7} K_{p7} K_{d7} \cos(7x - \omega t)
$$

$$
+ \frac{1}{11} K_{p11} K_{d11} \cos(11x + \omega t)
$$

$$
+ \frac{1}{13} K_{p13} K_{d13} \cos(13x - \omega t) + \ldots \Bigg],
$$

where $M = \sqrt{2}\, TI =$ peak ampere turns of a single coil.

The coefficient $6Mn/\pi$, more generally, is $2qMn/\pi$, for a winding with q phase belts per pole, and n slots per phase belt.

It will be noted that a balanced three-phase winding produces no triple harmonic fields. Conversely, it is easy to show that, if all three phases are connected in series, as in a Δ-connected winding, a current circulating in the delta will produce triple harmonic fields only. Such a "zero-phase sequence" current, therefore, produces no useful rotor flux linkages nor any useful torque (Sect. 3.15).

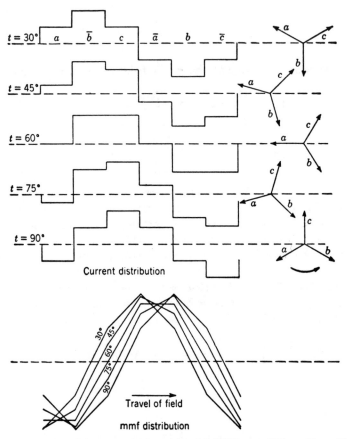

FIG. 3.14. Geometrical Pattern of Revolving MMF Waves of Three-Phase Winding.

Also, it is important to notice that in a single-layer, or a full-pitch, winding with $2q$ phase belts per pair of poles, the $2q-1$ and $2q+1$ harmonics are large, because they most nearly fit the triangles that make the difference between the fundamental sine wave and the actual mmf wave with $2q$ steps. Likewise, the $(S/P)-1$ and $(S/P)+1$ harmonics are large, where S/P = slots per pair of poles, because they most nearly fit the S steps of the mmf wave due to the concentration of the mmf at the slot openings.

Before leaving this subject, it is interesting to note the geometrical patterns assumed by the flux waveform during each cycle of variation. The preceding analysis shows that, at every instant of time, the total air-gap field is composed of identically the same harmonics. However, as the kth harmonic rotates at $1/k$th speed, the several harmonic fields are continually gliding past each other, giving a different shape at each instant to their algebraic sum, or resultant waveform.

In Fig. 3.14, for example, the middle diagram shows the mmf distribution for a three-phase, 60° belt, three-slots-per-pole full-pitch winding, at the instant ($t = 60°$) when phase a current is zero. Phase b current is then -0.866, and phase c current 0.866, their sum being zero. This gives a flat-topped mmf wave. Thirty electrical degrees later ($t = 90°$), the phase currents are -0.50, -0.50, and 1.00, producing a stepped mmf wave that is displaced 30° to the right, and so on. Despite the different wave shapes of the mmf and current distributions, a Fourier's analysis of each successive shape yields precisely the same magnitudes of the various harmonics. If the terms in Eq. 3.34 are summed for successive values of t and plotted along the space axis, the resulting wave will appear as shown at the bottom of Fig. 3.14, if $p = 1$. If $p = 5/6$, the harmonics will be much smaller, and the resultant wave will closely approach a constant-magnitude sine curve.

13 Magnetizing Current

Equation 3.34 gives for the peak of the air-gap fundamental mmf wave, or "armature reaction", of a $2P$-pole, q-phase winding:

$$(3.35) \quad A = \frac{2qMnK_pK_d}{\pi}$$

$$= \frac{\sqrt{2}\,qNIK_pK_d}{\pi P} \text{ ampere turns per pole,}$$

where $\quad N = 2PnT = $ turns in series per phase.

The current loading, Δ, the effective amperes per meter of periphery, can now be defined more exactly:

$$(3.36) \quad \Delta = \frac{2qNIK_pK_d}{\pi D} \text{ amperes per meter}$$

and in terms of Δ, the peak armature mmf is:

$$(3.37) \quad A = \frac{\sqrt{2}\,D\Delta}{2P} \text{ ampere turns per pole.}$$

Combining Eq. 3.35 with Eq. 1.15, the corresponding peak air-gap density of the fundamental flux wave is:

$$B_{g\,max} = \frac{4\sqrt{2}\,qNIK_p K_d}{PG10^7} \text{ webers per sq m,} \quad (3.38)$$

if G is the effective air-gap length in meters; or

$$B_{g\,max} = \frac{1.437 qNIK_p K_d}{PG10^8} \text{ webers per sq in.,}$$

if G is expressed in inches.

In Eqs. 3.35 and 3.38, $q = 3$ for a 60° phase-belt winding, if N is defined on the usual basis as the turns in series between the phase terminals, and P is the number of pairs of poles.

A convenient form of Eq. 3.38 is the corresponding expression in terms of the total flux per pole, ϕ. With sinusoidal air-gap flux distribution:

$$\phi = \frac{2}{\pi} B_{g\,max} \left(\frac{\pi DL}{2P} \right) = \frac{DLB_{g\,max}}{P} \text{ webers,} \quad (3.39)$$

where D and L are the air-gap diameter and core length, respectively.

Substituting Eq. 3.39 in Eq. 3.38, and rearranging:

$$I = \frac{\sqrt{2}\,P^2 G\phi 10^7}{8qNK_p K_d DL} \text{ amperes,} \quad (3.40)$$

if G, D, and L are in meters; or:

$$I = \frac{0.696 P^2 G\phi 10^8}{qNK_p K_d DL} \text{ amperes,}$$

if G, D, and L are in inches. Calculation of the effective air-gap length, G, taking into account the "fringing" of flux at the slot openings, etc., is considered in Sect. 6.9.

Equation 3.40 is the basic design equation for calculating the magnetizing current in a distributed winding, required to produce a desired flux per pole. For given ϕ and N, the magnetizing current varies in proportion to the square of the number of poles, and to the effective air-gap length.

From Eqs. 3.40 and 3.15, the total magnetizing volt-amperes required are:

$$\text{Magnetizing volt-amperes} = qEI = \frac{\pi f G P^2 \phi^2 10^7}{4DL}, \quad (3.41)$$

if G, D, and L are measured in meters.

Equation 3.41 can be independently derived, or verified, by consideration of the magnetic energy that must be stored in the air gap to maintain ϕ.

In a single-phase reactive circuit, with I rms amperes and V rms volts, the instantaneous power is:

(3.42) $$\text{Power} = ei = (\sqrt{2}\,E\sin\omega t)(\sqrt{2}\,I\cos\omega t)$$
$$= EI\sin 2\omega t.$$

Equation 3.42 represents a flow of energy at frequency $2f$, with a maximum transfer in either direction equal to the integral of ei over one-half of the double frequency cycle, or:

(3.43) $$W = \text{max energy of field} = \int_0^{1/4f} EI\sin 4\pi ft\, dt$$
$$= \frac{EI}{2\pi f}.$$

But this is a single-phase supply, which produces an alternating flux in a single space axis. To maintain a polyphase field, equal amounts of energy must be supplied to the two field axes 90° apart, so that the total energy of the polyphase field will remain constant as the field rotates. Hence, the total volt-amperes required to maintain a stored energy W in the q-phase magnetic field are twice those given by Eq. 3.43, or:

(3.44) $$qEI = 4\pi fW.$$

From Eqs. 1.24 and 3.39, for a $2P$-pole field, with a flux of ϕ webers per pole,

(3.45) $$W = \frac{1}{2}\left(\frac{P\phi}{DL}\right)^2\left(\frac{\pi DLG}{8\pi}\right)10^7$$
$$= \frac{GP^2\phi^2 10^7}{16DL} \text{ wsec.}$$

The factor $1/2$ in Eq. 3.45 comes from the fact that the average value of B^2 over the pole pitch is one-half of B^2_{max}. Substituting Eq. 3.45 in Eq. 3.44:

(3.41) $$qEI = \frac{\pi fGP^2\phi^2 10^7}{4DL} \text{ volt-amperes,}$$

which is the same as Eq. 3.41.

Almost the entire energy of the magnetic field is stored in the air gap. Since this energy must be supplied and removed each cycle, the magnetizing volt-amperes, which constitute the energy flow, are nearly proportional to the length of the air gap. The induction motor, in which all the magnetizing current is supplied at full line frequency, must have a relatively short air gap for good performance, as otherwise the current and reactive power required will be excessive.

14 Winding Analysis

No matter how complicated or irregular the arrangement of any winding may be, its performance may be determined by the principles outlined in the preceding section. Every individual coil produces a flux wave like that of Fig. 3.10; every such flux wave may be analyzed into a series of harmonic components, with 2, 4, 6, etc., poles; and each of these harmonic flux waves may be considered as acting independently of all the others, so far as currents, voltages, and fluxes are concerned. The fact that the various harmonics are superposed must be considered in calculating hysteresis losses and magnetic saturation. Also, the torques and magnetic forces produced by interaction of the different harmonic fields must be taken into account, just as they would be if two or more independent windings were superposed. But these interactions between fields present few difficulties, once the individual fields are known.

To analyze a winding, therefore, consider the series-connected coils between any two terminals of a motor or generator winding as consisting of the following series-connected impedances, or voltages (Fig. 3.15):

(*a*) A pure resistance, equal to the actual resistance between terminals.

(*b*) A pure reactance, equal to the leakage reactance of the winding due to all the leakage flux that does not cross the air gap.

(*c*) A transformer, whose magnetizing reactance is equal to the reactance corresponding to the fundamental air-gap flux wave of the winding, and whose secondary impedance is the impedance of the rotor winding for currents with fundamental sine-wave distribution.

(*d*), (*e*), ... A series of smaller transformers, each having a magnetizing reactance corresponding to the air-gap flux wave of a particular harmonic, and a secondary impedance equal to the impedance of the actual rotor winding for currents of the same number of harmonic poles.

This circuit is expanded from that of Fig. 2.4 for the transformer, to include the two principal phase-belt harmonics of a three-phase motor, the 5th and 7th; and the two principal slot harmonics, the mth and nth, as given by Eq. 3.34. Also, the mth and nth slot permeance harmonics, representing the ripples in the fundamental flux wave due to the stator slot openings, are included in the magnetizing current branch. And there are shown in dotted lines the core loss branch and a branch for an additional rotor winding, if any. For each of these stator harmonics, there is shown the corresponding rotor circuit in which currents are induced. The harmonics due to the rotor currents are not shown, as ordinarily these do not induce any stator currents, and therefore they are merely components of X_2. The significance of the R and X values shown, and the use of the circuit for calculating torques and losses will be considered in Chapters 4, 5, and 8.

WINDING ANALYSIS

Methods of calculating the reactances and the rotor impedances are given in Chapters 7 and 8. The point emphasized here is that the performance of the entire winding may be determined by the step-by-step process of first calculating the performance for the fundamental, and then calculating the effects of each harmonic in turn. Often it is sufficient to consider the fundamental field only, the harmonic fields producing only second- or

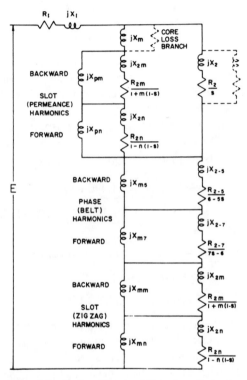

FIG. 3.15. General Equivalent Circuit of Polyphase Induction Motor.

third-order effects. The method is the same that the mathematician follows when he expresses a complicated function in the form of an infinite series, and neglects all but the first two or three terms.

If the student will form this habit of looking at the fundamental and the harmonic fields as separate phenomena, focusing his mind on only one of them at a time, he can gain facility in solving all sorts of machine problems.

The heart of the method lies in the quick and accurate calculation of the effective number of turns in series for each harmonic field. To do this, it is convenient to take the following steps, illustrated for a 120° phase-belt, 2-layer, 3-phase, 8-pole, 4/3-pitch, 24-slot winding:

1. Represent the winding by a sequence of letters, showing the current and location of each coil side (*A* for phase *A*, *B* for phase *B*, etc., and $\bar{A}, \bar{B}$, etc., for the opposite directions of current flow).

Top layer $A\ A\ B\ B\ C\ C\ A\ A\ B\ B\ C\ C\ A\ A\ B\ B\ C\ C$ AA BB CC
Bottom layer $\bar{B}\ \bar{B}\ \bar{C}\ \bar{C}\ \bar{A}\ \bar{A}\ \bar{B}\ \bar{B}\ \bar{C}\ \bar{C}\ \bar{A}\ \bar{A}\ \bar{B}\ \bar{B}\ \bar{C}\ \bar{C}\ \bar{A}\ \bar{A}$ BB CC AA
$\longrightarrow$
8-Pole Field Rotation.

2. Represent each phase by a sequence of phasors, one for each coil side, separated by angles equal to their space displacements in terms of the harmonic under consideration. The center of the first coil side of phase *A* may be taken as the reference angle of 0°. For example, the second harmonic, or sixteen-pole field, of phase *A* (for which 1 slot pitch = $(360 \times 8)/24 = 120°$) is:

$1\underline{/0} + 1\underline{/120} - 1\underline{/480} - 1\underline{/600} + 1\underline{/720} + 1\underline{/840} - 1\underline{/1,200}$

$\qquad - 1\underline{/1,320} + 1\underline{/1,440} + 1\underline{/1,560} - 1\underline{/1,920} - 1\underline{/2,040}$

$\qquad + 1\underline{/2,160} + 1\underline{/2,280} - 1\underline{/2,640} - 1\underline{/2,760},$

which reduces to:

$4\underline{/0} - 4\underline{/240} = 4\sqrt{3}\underline{/30°}$ effective coil sides (phase *A*).

A similar process for phases *B* and *C* gives:

$4\sqrt{3}\underline{/270}$ (phase *B*) and $4\sqrt{3}\underline{/150}$ (phase *C*).

If the number of turns per coil is not uniform, the corresponding turn ratios are used in place of unity in the appropriate terms.

3. Express the resultant forward field as half the phasor sum of the three phase resultants, with each of their space positions displaced forward by the time-phase angle of its own phase current.

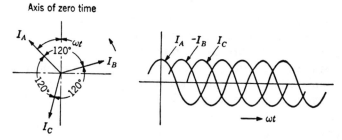

FIG. 3.16. Time Relations of Three-Phase Currents.

WINDING ANALYSIS

Since the A, B, C phase currents as given in Step 1 with forward rotation from left to right have time-phase angles of $0°$, $/\overline{120°}$, $/\overline{240°}$ (Fig. 3.16), we have from Step 2:

16-pole forward field $= 2\sqrt{3}\,I_A\,\underline{/30°-0}$

$+ 2\sqrt{3}\,I_B\,\underline{/270°-120°}$

$+ 2\sqrt{3}\,I_C\,\underline{/150-240°} = 0.$

The sixteen-pole forward field in this case, is, therefore, equal to zero.

4. Express the resultant backward field as half the phasor sum of the three-phase resultants, with each of their space positions displaced backward by the time-phase angle of its own phase current. For this case:

16-pole backward field $= 2\sqrt{3}\,I_A\,\underline{/30°+0}$

$+ 2\sqrt{3}\,I_B\,\underline{/270°+120°} + 2\sqrt{3}\,I_C\,\underline{/150+240°}$

$= 6\sqrt{3}\,I\,\underline{/30°}$, if $I_A = I_B = I_C$.

The same process gives a forward (fundamental) eight-pole field $= 18I\,\underline{/60°}$, and zero backward eight-pole field.

5. The final resultants of the several harmonic fields so obtained, each multiplied by $2M/\pi$ and divided by its harmonic order, will be the coefficients of the successive terms in the complete air-gap mmf wave, referred to the fundamental number of poles, similar to Eq. 3.34. The factor $2M/\pi = 2\sqrt{2}\,TI/\pi$ is the peak ampere turns of the fundamental mmf wave produced by a single coil side with T conductors and I rms amperes per conductor.

For this example, of an 8-pole, 3-phase, 120° phase-belt, 4/3-pitch, 24-slot winding, the complete mmf wave found in this way is:

$$m = \frac{36M}{\pi}\left[\cos(4x-\omega t) + \frac{\sqrt{3}}{6}\cos(8x+\omega t - 90°)\right.$$

$$+ \frac{\sqrt{3}}{12}\cos(16x-\omega t - 90°) + \frac{1}{5}\cos(20x+\omega t)$$

$$+ \frac{1}{7}\cos(28x-\omega t) + \frac{\sqrt{3}}{24}\cos(32x+\omega t - 90°)$$

$$+ \frac{\sqrt{3}}{30}\cos(40x-\omega t + 90°)$$

$$\left. + \frac{1}{11}\cos(44x+\omega t) + \ldots\right] \text{ ampere turns per pole.}$$

In deriving this last equation, the space angle of reference has been shifted to coincide with the center of the fundamental wave, by subtracting $60k°$ from the phase angle of the kth harmonic, as determined in Steps 3 and 4 above. Thus, the space angle for the second harmonic is $30 - 2(60) = -90$.

6. The coefficients for the harmonic fields can be checked, for symmetrical windings, by the usual K_p and K_d formulas, Eqs. 3.16 and 3.17. For this winding, by Eq. 3.17:

$$K_{p_m} = \sin\frac{2m\pi}{3},$$

and, by Eq. 3.16:

$$K_{d_m} = \frac{\sin\dfrac{m\pi}{3}}{2\sin\dfrac{m\pi}{6}}.$$

For the fundamental, $m = 1$, and

$$K_{p_1} K_{d_1} = \frac{(0.866)(0.866)}{2(0.500)} = \frac{3}{4},$$

while for the second harmonic, $m = 2$, and

$$K_{p_2} K_{d_2} = \frac{(0.866)(0.866)}{2(0.866)} = \frac{\sqrt{3}}{4}.$$

The ratio of these is $\sqrt{3}/3$, and this divided by $m = 2$ checks the second harmonic coefficient, $\sqrt{3}/6$, found above.

15 Symmetrical Components

Whenever dissymmetry exists, as when the phase voltages, currents, or impedances are unbalanced, there will be both forward and backward rotating fields of the same frequency and number of poles. For these cases, and especially when external causes of dissymmetry, such as single-phase loads, are to be taken into account, the symmetrical component method is useful.

The method consists in resolving the q actual phase currents and voltages into three separate, symmetrical, systems, called the positive, negative, and zero sequence components. The positive-phase sequence component is a balanced, forward-rotating, q-phase set of currents and voltages, and the negative sequence component is a similar backward-rotating set. The zero sequence component is the single-phase resultant of the three actual phase values, representing the line to neutral current.

SYMMETRICAL COMPONENTS

If there is no neutral connection, or ground fault, the zero sequence current vanishes. Since the zero-phase sequence component of current is the same in each of the q phases, it cannot produce any fundamental flux, and so does not produce any useful torque. It does produce fields with q times the fundamental number of poles, and some losses, but it rarely needs consideration in induction motor problems.

Since the positive-, negative-, and zero-phase sequence components are all balanced with respect to the q phases, calculations for each system can be made independently, and the results superposed (in the absence of magnetic saturation), just as the effects of harmonic fields with different numbers of poles or frequencies are considered independently. A single-line equivalent circuit is made for each component, similar to Fig. 2.4 and the q phases are treated as one for calculation purposes.

In deriving the sequence components, we need to keep the power the same. For example, in changing from q phases to the equivalent single phase of a sequence component, the power becomes q times as great. Therefore, the sequence currents and voltages must be $\sqrt{q}$ times as great as those of one of the q phases. In earlier explanations of the symmetrical component system, the sequence currents and voltages were taken equal to the phase values, so that the power obtained for each sequence component had to be multiplied by q to obtain the true power. While this procedure has been used successfully for many years, it leads to inconsistencies, and is not consistent with the modern view that the flow of energy is what counts.

If a three-phase symmetrical stator winding has unequal (sinusoidal) incoming phase currents:

$$I_A\underline{/\alpha}, \quad I_B\underline{/\beta}, \quad \text{and} \quad I_C\underline{/\gamma},$$

and also has a neutral connection with an outgoing current:

$$I_N\underline{/\delta}$$

flowing in it; all these currents having the same frequency, the zero-phase sequence current is equal to $1/\sqrt{3}$ times the neutral current, and is:

(3.46) $$I_0\underline{/\delta} = \sqrt{(1/3)}I_N\underline{/\delta} = \sqrt{(1/3)}(I_A\underline{/\alpha} + I_B\underline{/\beta} + I_C\underline{/\gamma}),$$

since the phasor sum of all the currents entering the winding must be zero.

Representing the positive-phase sequence current component in phase A by $I_f\underline{/\phi}$, and the negative-phase sequence component in phase A by $I_b\underline{/\theta}$, the balanced positive-phase sequence currents in the three phases, A, B, and C, are (Fig. 3.16):

$$I_f\underline{/\phi-0}, \quad I_f\underline{/\phi-120}, \quad I_f\underline{/\phi-240}.$$

THE ROTATING MAGNETIC FIELD

And the corresponding negative-phase sequence currents are:

$$I_b\underline{/\theta+0}, \quad I_b\underline{/\theta+120}, \quad I_b\underline{/\theta+240}.$$

Thus, the total currents in the three phases are:

(3.47) $\quad \sqrt{3}\,I_A\underline{/\alpha} = I_f\underline{/\phi-0} + I_b\underline{/\theta+0} + I_0\underline{/\delta},$

(3.48) $\quad \sqrt{3}\,I_B\underline{/\beta} = I_f\underline{/\phi-120} + I_b\underline{/\theta+120} + I_0\underline{/\delta},$

(3.49) $\quad \sqrt{3}\,I_C\underline{/\gamma} = I_f\underline{/\phi+120} + I_b\underline{/\theta-120} + I_0\underline{/\delta}.$

Alternatively, the positive-sequence current can be found by advancing I_B 120° and I_C 240°, to bring the I_f components in phase, then adding, and dividing by $\sqrt{3}$. This gives:

(3.50) $\quad I_f\underline{/\phi} = \sqrt{(1/3)}(I_A\underline{/\alpha+0} + I_B\underline{/\beta+120} + I_C\underline{/\gamma+240}).$

The negative-phase sequence current is:

(3.51) $\quad I_b\underline{/\theta} = \sqrt{(1/3)}(I_A\underline{/\alpha+0} + I_B\underline{/\beta+240} + I_C\underline{/\gamma+120}).$

To take a numerical example, suppose that the measured values of currents (amperes or per unit) are:

$$I_A = 1+j0,$$
$$I_B = 0+j1,$$
$$I_C = 0.5+j0.5,$$
$$I_N = 1.5+j1.5.$$

Then the zero-phase sequence current, from Eq. 3.46, is:

$$I_0 = \sqrt{(1/3)}(1.5+j1.5) = 0.866+j0.866,$$

which equals $1/\sqrt{3}$ times the measured neutral current.

The positive-phase sequence current is, from Eq. 3.50:

$$I_f = \sqrt{(1/3)}\,[1+j0+j1\underline{/120}+(0.5+j0.5)\underline{/240}]$$
$$= \sqrt{(1/3)}\,[1+j1(-0.5+j0.866)+(0.5+j0.5)(-0.5-j0.866)]$$
$$= 0.183-j0.683.$$

The negative-phase sequence current is, from Eq. 3.51:

$$I_b = \sqrt{(1/3)}\,[1+j0+j1\underline{/240}+(0.5+j0.5)\underline{/120}]$$
$$= \sqrt{(1/3)}\,[1+j1(-0.5-j0.866)+(0.5+j0.5)(-0.5+j0.866)]$$
$$= 0.683-j0.183.$$

As a check, applying Eqs. 3.47, 3.48, and 3.49:

$$\sqrt{3}\,I_A = 0.183 - j0.683 + 0.683 - j0.183 + 0.866 + j0.866$$
$$= 1.732 + j0,$$

$$\sqrt{3}\,I_B = (0.183 - j0.683)(-0.5 - j0.866) + (0.683 - j0.183)$$
$$(-0.5 + j0.866) + 0.866 + j0.866 = 0 + j1.732,$$

$$\sqrt{3}\,I_C = (0.183 - j0.683)(-0.5 + j0.866) + (0.683 - j0.183)$$
$$(-0.5 - j0.866) + 0.866 + j0.866 = 0.866 + j0.866.$$

Bibliography

Reference to Section

1. "Fractional Pitch Windings for Induction Motors," C. A. Adams, W. K. Cabot, and G. Irving, Jr., *A.I.E.E. Trans.*, Vol. 26, Part II, 1907, pp. 1485–503. — 5
2. "The Rotating Electric Machine in an Electrical Engineering Degree Course," B. A. Adkins, *Electrical Engineering*, Vol. 80, No. 5, May, 1961, p. 347.
3. "Induction Machines," P. L. Alger, C. W. Falls, and A. F. Lukens, Section 7, *Standard Handbook for Electrical Engineers*, Ninth Edition, McGraw-Hill, 1957, pp. 704–48.
4. "Amplitudes of Magnetomotive Force Harmonics for Fractional Slot Windings—I; J. F. Calvert, *A.I.E.E. Trans.*, Vol. 57, 1938, pp. 777–84. — 10
5. *A Study of the Induction Motor*, F. T. Chapman, John Wiley, 1930.
6. *Connecting Induction Motors*, Fourth Edition, A. M. Dudley and J. F. Henderson, McGraw-Hill, 1960. — 14
7. *Electric Machinery*, A. E. Fitzgerald and Charles Kingsley, Jr., Second Edition, McGraw-Hill, 1961.
8. "Irregular Windings in Wound-Rotor Induction Motors," R. E. Hellmund and C. G. Veinott, *A.I.E.E. Trans.*, Vol. 53, 1934, pp. 342–6. — 14
9. *Theory of Alternating Current Machinery*, Second Edition, A. S. Langsdorf, McGraw-Hill, 1955.
10. *Elektrische Maschinen–IV, Die Induktionsmaschinen*, R. Richter, Julius Springer, Berlin, 1936.
11. *Symmetrical Components*, C. F. Wagner and R. D. Evans, McGraw-Hill, 1933. — 15

CONTENTS—CHAPTER 4

The Polyphase Induction Motor

		PAGE
1	Principle of Operation and Utility	107
2	Electromagnetic Structure	108
3	Torque and Slip	109
4	Phasor Diagram	112
5	Circle Diagram	114
6	Starting Performance	116
7	No-load Tests	117
8	Load Tests	122
9	Stray-Load Loss Measurement	123
	Appendix to Chapter 4—Behrend and Tesla	126

4

THE POLYPHASE INDUCTION MOTOR

1 Principle of Operation and Utility

The induction motor, invented by Nikola Tesla in 1886,† is merely an electric transformer whose magnetic circuit is separated by an air gap into two relatively movable portions, one carrying the primary and the other the secondary windings. To permit continuous rotary motion, the secondary core is usually made circular, the primary windings being so arranged as to cause the magnetic flux to cross the air gap radially, and to link the secondary windings with a minimum of magnetic leakage. Alternating current supplied to the primary winding from an electric power system induces an opposing current in the secondary winding when the latter is short-circuited, or closed through an external impedance. Relative motion between the primary and secondary structures is produced by the electromagnetic forces corresponding to the power thus transferred across the air gap by induction.

The essential feature which distinguishes the induction machine from other types of electric motors is that the secondary currents are created solely by induction, as in a transformer, instead of being supplied by a d-c exciter or other external power source, through slip rings or a commutator, as in synchronous and d-c machines.

If the magnetic circuit of a single-phase transformer were cut in two, with the primary and secondary windings on opposite sides of the gap, the only force produced by the currents would be a direct (pulsating) magnetic pull, tending to close the gap. Similarly, a true single-phase induction motor has only a radial pull across its gap when the rotor is stationary, with no tendency to rotate. However, by arranging two or more transformers, fed by out-of-phase currents, around the periphery of a circular core, a rotating instead of merely a pulsating field is produced (Sects. 3.10 and 3.12), giving rise to a tangential component of force, or torque. The transformers may be placed along a straight line, instead of around a circle, so producing linear instead of rotary motion, but in this case the phase rotation must be reversed when the limit of travel is reached in each direction. Thus, each coil of the induction-motor primary winding, Fig. 3.3,

† Trans. A.I.E.E., Vol. 5, 1888, p. 305.

with the opposing coils of the rotor winding, may be considered a transformer, and the whole motor may be thought of as a series of such transformers arranged radially around the periphery.

The polyphase induction machine, having the fewest windings and least insulation, and requiring no commutator, brushes, or slip rings, forms the lowest-cost and most widely used industrial motor. It is estimated that more than forty million of these motors are now in use in American industry, totalling some 120 million horsepower in 1960, and that each normal year's production adds a million motors more. This is exclusive of the single-phase induction motors, chiefly built in fractional-horsepower sizes, which are universally employed for operating electric fans, refrigerators, washing machines, and domestic and farm appliances generally. The American production of these single-phase types is of the order of 20,000,000 annually.

2   Electromagnetic Structure

Figure 1.1 illustrates a typical polyphase induction motor, showing the distributed stator winding in partially closed slots, and the rotor with a "squirrel-cage" winding, cast of aluminum in nearly closed slots, as in Fig. 3.3. A relatively large number of stator slots and a fractional-pitch stator winding with compactly wound ends are used to minimize leakage reactance. A very short radial air gap is required, to reduce the magnetizing current (Eq. 3.40). The rotor is uninsulated, and forms practically a solid block of metal, adapted to the diverse and rugged industrial services in which it is employed.

The rotor bars, solidly connected by conducting end rings, form a closed electric circuit, linking any magnetic flux that crosses the air gap. When polyphase alternating voltages are supplied to the stator winding, the resulting currents establish a rotating field in the air gap, as shown in Chapter 3, and this field cuts the rotor conductors. The secondary voltages, thus induced, produce opposing currents in the rotor winding, which form an approximate mirror image of the stator currents, rotating with them at the synchronous speed of $60f/P$ revolutions per minute. To supply the IR drop of these secondary currents, there must be a component of voltage in time phase with the secondary current, and the secondary current, therefore, must lag in space position behind the rotating air-gap field. A torque is then produced, corresponding to the product of the air-gap field by the secondary current, times the sine of the angle of their space-phase displacement. These relationships are illustrated in Fig. 4.1, which is similar to Fig. 2.1 for the transformer. The phasors I_1 and I_2 represent the primary and secondary currents, drawn along their axes of maximum current density. The phasor difference of these, I_M, is the net current available for producing the air-gap field, called the "magnetizing current". The maximum air-gap density, B_g, is at the center of the coil carrying the maximum magnetizing current, and is 90° displaced from I_M.

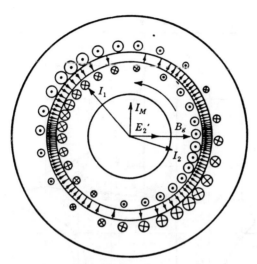

FIG. 4.1. Magnetic Field and Current Distribution in Polyphase Induction Motor.

3 Torque and Slip

At standstill, the secondary current is equal to the air-gap voltage divided by the secondary impedance at line frequency, or:

$$(4.1) \qquad I_2 = \frac{E_2}{Z_2} = \frac{E_2}{R_2 + jX_2},$$

where R_2 is the effective secondary resistance, and X_2 is the secondary leakage reactance at primary frequency.

The speed at which the magnetic field cuts the secondary conductors is the difference between the synchronous speed, $N_s = 60f/P$, and the actual rotor speed, N. The ratio of the speed of the field relative to the rotor to synchronous speed is called the "slip", s:

$$(4.2) \qquad s = \frac{N_s - N}{N_s},$$

or $$N = (1-s)N_s.$$

As the rotor speeds up, with a given air-gap field, the secondary induced

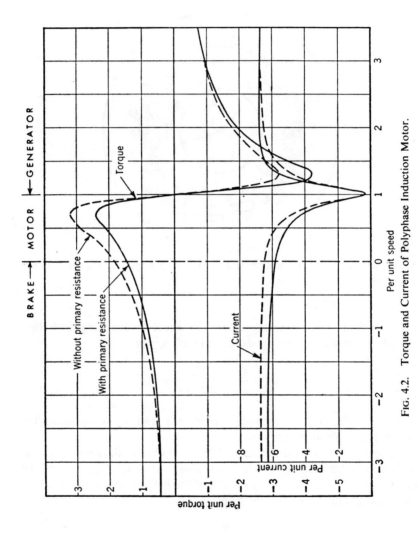

Fig. 4.2. Torque and Current of Polyphase Induction Motor.

TORQUE AND SLIP

voltage and frequency both decrease in proportion to s. Thus, the secondary voltage becomes sE_2, and the secondary impedance $R_2 + jsX_2$, or:

(4.3)
$$I_2 = \frac{sE_2}{R_2 + jsX_2} = \frac{E_2}{\dfrac{R_2}{s} + jX_2}.$$

The only way that the primary is affected by a change in the rotor speed, therefore, is that the secondary resistance as viewed from the primary varies inversely with the slip. The apparent secondary resistance, R_2/s, should be thought of as the sum of the actual resistance, R_2, and the load resistance, $R_2(1-s)/s$, equal to the in-phase motor speed-voltage divided by the secondary current.

In practice, the effective secondary resistance and inductance vary in some degree at different secondary frequencies, because of magnetic saturation and also the varying "skin effect", or current shifting into the lowest impedance paths, at the particular frequency existing at any speed. This may be exaggerated by a deep-bar or double squirrel-cage rotor construction (Sects. 8.4 and 8.6), causing the effective secondary resistance to be appreciably higher, and the leakage inductance to be somewhat lower, at large values of slip.

From Eq. 4.3, as the motor speed increases, and the apparent rotor resistance, R_2/s, increases also, the line current decreases, while the power factor at first rises, reaches a maximum, and then falls. As the motor approaches synchronism, R_2/s is very large, the secondary current becomes very small, and the torque approaches zero. The torque-speed curve of an ideal motor at constant impressed voltage, therefore, has the form shown in Fig. 4.2. This figure is based on calculations for a typical 15-hp, 1,800-rpm, squirrel-cage motor, as given in Sect. 5.2.

If driven above synchronous speed, the slip becomes negative, the torque reverses, and the motor becomes a generator. If driven backward, the torque continues to oppose the motion, and the motor acts as a brake.

If direct current is supplied to the stator, $f = 0$, synchronism occurs at zero speed, and the machine then acts as a brake for both directions of rotation. In this case, the secondary current induced by rotation produces a stationary counter mmf, but no voltage is induced in the stator, and the machine, therefore, operates under constant current. This gives the speed-torque curve a much sharper peak, and maximum torque occurs at a smaller slip than with a-c supply. The term "dynamic braking" is applied to this mode of operation. Figure 4.3 illustrates the braking performance of the same motor as in Fig. 4.2, with a d-c mmf in the stator equal to twice that of full-load alternating current. The peak braking torque is sharply

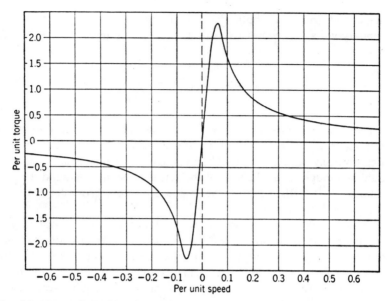

Fig. 4.3. Dynamic Braking Torque of Three-Phase Induction Motor at Twice Rated Current.

limited by magnetic saturation. This subject is considered further in Sect. 5.9.

4 Phasor Diagram

It should be borne in mind that the configuration of electric and magnetic fields portrayed in Fig. 4.1 is revolving synchronously in space, cutting the primary or stator conductors at line frequency, and the secondary at slip frequency. Thus, the phasors represent not only the relative space-phase positions, but also the time phases of the currents, voltages, and fluxes.

The concept of the polyphase induction motor as a transformer, Eq. 4.3 and Fig. 4.1, enables a phasor diagram of the motor currents and voltages to be drawn, Fig. 4.4. With symmetrical phase windings and a balanced polyphase power supply, a single phasor diagram is adequate, the diagrams for the other two phases being identical, but displaced 120° in phase.

In Fig. 4.4, E_1 is the impressed voltage, and I_1 the primary current phasor, lagging behind the voltage by an angle θ_1. The secondary voltage, due to the air-gap field, is $E_2 = -E_1'$, found by vectorial subtraction of the primary resistance drop, $I_1 R_1$, and the primary leakage reactance drop, $jI_1 X_1$, from the impressed voltage. The voltage actually induced in the secondary at slip frequency is $E_2' = sE_2$, and this is entirely consumed in the secondary impedance drop, composed of $I_2 R_2$ and $jI_2 s X_2$. The actual flux linkages in the stator and rotor are found by phasor addition of the mutual and leakage fluxes, as in the Blondel diagram for the transformer, Fig. 2.3. The torque is the product of the secondary current by the air-gap flux, multiplied by the secondary power factor, or sine of the angle of displacement between this current and flux. The power delivered by the stator to the rotor is the product of this torque by synchronous speed.

However, the rotor is slipping past the field at the speed sN_s, so that the power delivered to the motor shaft is only $(1-s)$ times the power coming from the stator. The difference, or s times the power crossing the air gap, called the "slip loss", is consumed in supplying the rotor copper loss.

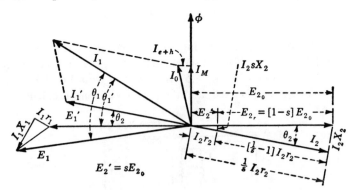

FIG. 4.4. Phasor Diagram of Induction Motor.

Thus, the secondary copper losses are analogous to a friction loss, as if the air-gap field were the band of a slipping clutch grasping the rotor with a radial force proportional to the secondary current. It is important to have a clear visual perception of this division of the power crossing the air gap into the shaft output and the slip loss, as in this way the energy relations during starting, reversing, and generator operation can be easily understood or quickly reasoned out.

To produce the air-gap voltage, E_1', there is required a magnetic flux, ϕ, which lags 90° behind E_1'. A magnetizing current, I_M, is required to produce this flux, which, together with the small energy current, I_{e+h}, required to supply the core losses, must be subtracted from the primary current, I_1, to obtain the primary load current, I_1'. This in turn is equal and opposite to the secondary current, I_2.

When the rotor is stationary, $s=1$, the full voltage is impressed on the primary and induced in the secondary, and a large current flows at low power factor. As the motor accelerates, s decreases, and R_2/s becomes predominant over X_2, so that the current decreases, and the power factor rises to a maximum. Without load, the motor will continue to accelerate until it reaches very nearly synchronous speed, when the secondary current will be just enough to supply a torque equal to the windage and friction, plus the rotational core loss. The primary current will then be practically equal to the magnetizing current, and will have again a very low power factor.

5 Circle Diagram

Figure 4.6 represents the equation:

$$(4.4) \qquad I_1 = \frac{E_1}{\left(R_1 + \dfrac{R_2}{s}\right) + j(X_1 + X_2)},$$

derived from Eq. 4.3 by substituting the primary for the secondary voltage, and the total impedance for the secondary impedance, in accordance with the approximate circuit of Fig. 4.5. It is evident from Eq. 4.4 that, if the impressed voltage, E_1, is fixed, and the primary current is calculated for a

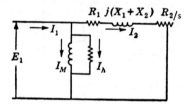

Fig. 4.5. Approximate Equivalent Circuit of Polyphase Induction Motor used as Basis for the Circle Diagram.

series of motor speeds, the locus of the tip of the phasor, I_1, will form an arc of a circle. This circle diagram is very convenient for visualization purposes, but it is based on several approximations. The usual form of the diagram assumes a constant air-gap field at all speeds; i.e., it lumps the primary and secondary impedances on the secondary side of the magnetizing current, as shown in Fig. 4.5. The diagram is the same as that for the transformer, Fig. 2.5, except that $R_2(1-s)/s$ replaces the transformer load resistance, R_L; and X_L is taken to be zero.

The data necessary to construct Fig. 4.6 are the magnitude of the no-load current, ON, and of the blocked-rotor current, OS, and their phase angles with reference to the line voltage, OE. A circle with its center on the line NU at right angles to OE is drawn to pass through N and S.

Each line on the diagram can be measured directly in amperes, but it also represents volt-amperes or power, when multiplied by the phase voltage times the number of phases. The line VS drawn parallel to OE represents the total motor input with blocked rotor, and on the same scale VT represents the corresponding primary I^2R loss. Then ST represents the power input to the rotor at standstill, which, divided by the synchronous speed, gives the starting torque.

At any load point A, OA is the primary current, NA the secondary current, and AF the motor input. The motor output is AB, the torque times synchronous speed is AC, the secondary I^2R loss is BC, the primary I^2R loss is CD, and the no-load copper loss plus core loss is DF. The

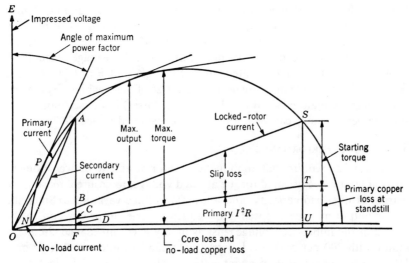

FIG. 4.6. Circle Diagram.

maximum power factor point is P, located by drawing a tangent to the circle from O. The maximum output and maximum torque points are similarly located at Q and R by tangent lines parallel to NS and NT, respectively.

Besides the error previously mentioned due to assuming constant air-gap voltage at all loads, other errors occur in the circle diagram, if there are any variations in reactance and in secondary resistance over the speed range, such as result from changes in magnetic saturation and secondary frequency. The diagram does, however, afford a convenient means of checking the overall performance of a motor with a minimum of test data.

The diameter of the circle is equal to the voltage divided by the standstill reactance, or equal to the blocked-rotor current value on the assumption of zero resistance in both windings. The maximum torque of the motor, measured in kilowatts at synchronous speed, is equal to a little less than the radius of the circle multiplied by the voltage, OE, and by the number of phases.

Since the diameter of the circle is a little greater than the starting current, and the radius is a little greater than the power component of the current at maximum torque, the per unit maximum torque is roughly equal to one-half the per unit starting current. This approximate relation, derived more exactly in Chapter 5, Eq. 5.62, is an extremely important and basic concept.

It means that the starting current of a low-resistance motor is a direct measure of the maximum torque. No true induction motor, without hysteresis losses (with fixed primary voltage and frequency, and no change in the connections or physical arrangement of the windings) can possibly have a maximum output in watts greater than half its starting volt-amperes.

This relation between the maximum output and the leakage reactance is another illustration of the statement in Sect. 2.6 and Eq. 2.24, that the reactance is an inverse measure of the useful size of the machine. By placing hard magnetic materials (with high hysteresis losses) in the secondary, the starting performance can be improved, as considered in Sect. 8.7.

6 Starting Performance

It is usual to require that an induction motor be able to deliver momentarily at least twice its rated torque at rated voltage, to ensure a reasonable margin of overload capacity, and allow for voltage variations. Since the motor output at a given speed and frequency varies in proportion to the square of the impressed voltage (just as in any fixed impedance circuit), a motor with 200 per cent breakdown (maximum) torque will deliver only 162 per cent torque at 90 per cent voltage. Also, as Fig. 4.2 indicates, the motor speed varies widely, or becomes unstable, as the load approaches the maximum torque limit. These considerations, and the desire to provide a safe torque margin in application, have led to the requirement in the American Standards that general-purpose motors shall have at least 200 per cent breakdown torque (Sect. 14.2).

To meet a 200 per cent breakdown-torque guarantee, and allow for manufacturing variations, the motor will be designed for a somewhat greater torque, of the order of 225 per cent or more. In accordance with the circle diagram, the per unit starting current must be not less than twice this, or somewhat more than 500 per cent. If the motor is to develop 150 per cent of full-load torque at starting, the circle diagram indicates that the per unit resistance of the secondary winding must then be of the order of 6 per cent, since $0.06 \times 5^2 = 1.50$. By designing the motor with a lower reactance, and, therefore, greater maximum output and starting current, the same starting torque can be obtained with a lower secondary resistance, and, therefore, with a higher operating efficiency.

Thus, high starting torque is obtainable by sacrifice of efficiency, on the one hand, or by increased starting current, on the other hand. A chief problem of designing and application engineers is to make the best compromise between these factors.

The American Standards for integral-horsepower three-phase induction motors of 15 hp and larger prescribe that the starting current on 220 volts shall not exceed $14\frac{1}{2}$ amp per rated horsepower (Sect. 14.3). Thus, the ratio of the permitted starting current to the rated current (assuming an apparent efficiency of 81 per cent) is:

$$(4.5) \qquad \frac{14.5 \times \sqrt{3} \times 220 \times 0.81}{746} = 6.00.$$

The difference between this value of 6 and the value of somewhat more than 5 necessary to obtain the required breakdown torque allows some freedom to the designer to use preferred dimensions, to employ double squirrel-cage rotors, and to give more or less emphasis to efficiency and full-load slip.

High starting torque could be secured without sacrifice of efficiency, if some way could be found to vary the resistance of the secondary winding from a high value at standstill to a low value at full speed. One way to do this is to design the squirrel cage with deep bars, or with two distinct windings separated by a flux-leakage path, as in Fig. 8.7. In this case, the effective secondary resistance is high at starting, when the rotor frequency is high, and low at full speed, when the reduced rotor frequency permits the current to flow in the high reactance winding (Sect. 8.6).

Another way is to provide an insulated secondary winding, similar to the primary, which is brought out to slip rings, and to connect an external resistance between the slip rings at starting. The resistance may be adjusted to give the desired torque at standstill, decreased as the speed rises, and short-circuited at full speed. This method assures minimum starting current, and permits longer duration of the starting period, or prolonged operation at reduced speeds, without overheating the motor, and so is generally used for large and adjustable-speed motors (Sect. 8.12). A third way is to place hard magnetic materials in the secondary flux leakage paths, making use of the "Saturistor" principle (Sects. 8.7 and 8.13).

7 No-Load Tests

As the circle diagram indicates, the complete performance of a polyphase induction motor can be quite accurately determined once the geometry of the diagram is known. The necessary data can be found by simple running-light and locked-rotor tests. In the running-light test, the motor is run at no load with normal frequency and voltage applied, until the power input becomes constant. On slip-ring motors, the brushes are short-circuited. Readings of amperes and watts are taken at one or more values of impressed voltage, with rated frequency maintained. Accurately balanced phase voltages and a sine-waveform of voltage are necessary for good results. The watts input at rated voltage will be the sum of the friction

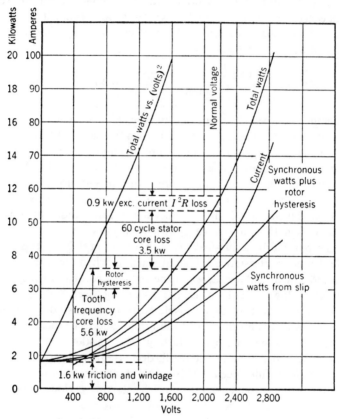

FIG. 4.7. Running-Light Test Data, 16-Pole, 400-hp, 450-rpm, 2,200-Volt Motor.

and windage, core loss, and no-load primary I^2R loss. Subtracting the calculated primary I^2R loss at the temperature of test from the input gives the sum of the friction and windage and core loss. Segregation of the core loss from the windage and friction is not necessary for normal efficiency or other rated voltage performance calculations. However, the segregation can be made by taking amperes and watts input readings, at rated frequency, at different voltages varying from 125 per cent of normal down to about 15 per cent voltage, or the point of minimum current. Plotting the input watts, less primary I^2R, against the square of the voltage, and extrapolating the lower part of the curve in a straight line to intercept the zero-voltage axis determines the friction and windage. Typical data of such a test are shown in Fig. 4.7.

All the high-frequency losses, due to the passage of the rotor teeth past the stator-slot openings, must be supplied by a torque produced by the slip-frequency rotor currents, in exactly the same way that the friction and windage are supplied. Therefore, by measuring the motor slip at no load, the total of high-frequency losses, plus friction and windage, can be calculated from the simple formula:

(4.6) $\qquad$ High-frequency losses $+F$ and $W = \dfrac{qE^2 s}{R_2}$ watts,

where E is the phase voltage, q the number of phases, R_2 the secondary resistance in ohms per phase, and s the slip at no load. By making these calculations over a range of voltage, the total core loss can be segregated into its fundamental-frequency, or transformer loss; and its high-frequency, or pulsation loss, components. For accurate results, E in the above equation should be the air-gap voltage, or line voltage less primary impedance drop (Eq. 5.31). Also, to the right-hand side of Eq. 4.6 should be added the (synchronous) torque due to the rotor hysteresis loss. This hysteresis torque is constant (with a given air-gap flux) at all speeds below synchronism, but reverses as the motor passes through zero slip, and is negative in generator operation. It can be estimated from a knowledge of the rotor flux densities and Epstein test data on the steel in the rotor core and teeth.

The value of the no-load current at rated voltage and frequency fixes the starting point of the circle diagram. The magnetizing reactance may be calculated by the formula:

(4.7) $\qquad X_M = \dfrac{E}{I_0} - X_1.$

The primary leakage, X_1, is determined from the locked-rotor test data (usually taken as half the total reactance at standstill).

For the locked-rotor test, the motor is blocked so it cannot rotate, a reduced voltage of rated frequency is applied to the terminals, and readings of volts, watts, and amperes are taken. Readings should be taken quickly, and the temperature of the windings should be observed before and after the test to minimize errors due to changing resistance values. In machines with closed-slot rotors, or very small air gaps, magnetic saturation of the leakage paths will occur, and it is then desirable to take additional readings

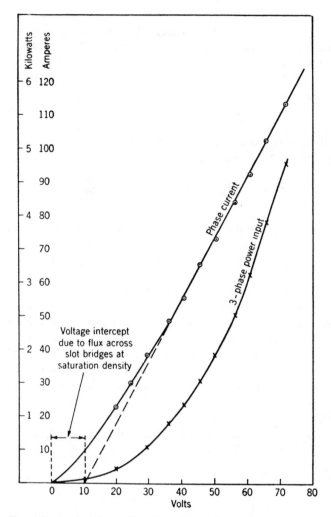

FIG. 4.8. Locked-Rotor Test Data, Three-Phase Induction Motor.

at half or even at full voltage to establish the actual value of starting current (Fig. 4.8).

The motor impedance per phase is determined from the volts, amperes, and watts readings. The total resistance component for a three-phase motor is:

(4.8) $$R = \frac{W}{3I^2} \text{ (ohms per phase } Y\text{)},$$

NO-LOAD TESTS

and the reactance component is:

(4.9) $$X_t = \sqrt{\left(\frac{E^2}{3I^2} - R^2\right)} \text{ (ohms per phase } Y\text{)},$$

where W = watts input, I = line current, and E = voltage between lines.

The test value of reactance, X_t, from Eq. 4.9 is a little smaller than the true value of reactance, because the formula assumes the line current to flow through the primary and secondary impedances in series, whereas actually the magnetizing component of the current flows in the primary winding only. A small correction factor should be applied to X_t, therefore, to obtain the reactance, X, for use in equivalent circuit calculations (Eq. 5.13).

Normally, the primary and secondary leakage reactance values, X_1 and X_2, are assumed equal, each having the value $X/2$, but a correction term, ΔX, should be added to X_2 when the motor is at full speed, to allow for the deep-bar or double squirrel-cage effect (Eq. 4.11).

The primary resistance is measured with direct current, a current about one-quarter of full-load value being preferably used, and readings being taken quickly to avoid errors due to temperature changes during the test. The primary resistance per phase Y is equal to one-half the resistance between any two terminals.

Subtracting the primary resistance at the temperature of test from the resistance component of the total impedance gives the effective secondary resistance at standstill. The starting torque may be calculated from this value by the equation:

(4.10) $$\text{Starting torque\dag} = 9.55KqI^2R_t/N_s \text{ newton meters}$$
$$= 7.04KqI^2R_t/N_s \text{ ft-lb,}$$

where I = amperes starting current per phase at specified voltage; q = number of phases; N_s = synchronous speed in rpm; R_t = resistance component of motor impedance, less primary resistance at temperature of test, in ohms per phase; K = an empirical constant, usually about 0.9, which allows for non-fundamental secondary losses.

In practice, it is usual to measure (with lever arm and scale) the torque produced, in which case Eq. 4.10 provides a useful check on the accuracy of the measurements, and on the value of K.

For a deep-bar or double squirrel-cage motor, the effective secondary

---

\dag 1 watt is $\dfrac{33,000}{746}$ = 44.24 ft-lb per min, whence

$$\text{Torque} = \left(\frac{44.24}{2\pi} = 7.04\right)\left(\frac{\text{watts}}{N_s}\right) \text{ ft-lb.}$$

reactance at speed is materially higher than at standstill, because of the progressive shifting of the secondary current from the low-reactance, high-resistance paths into the low-resistance, high-reactance paths as the secondary, or slip, frequency decreases. Hence, for accurate performance calculations, it is necessary to determine the motor reactance at low secondary frequency. If a low-frequency supply is available, the locked-rotor test may be repeated at 15 cycles, or at most 25 cycles, for a 60-cycle motor. Calculation of the low-frequency reactance by Eq. 4.9, and multiplying this by the ratio of the rated to the test frequency will give the proper value to use in operating performance calculations.

Alternatively, the reactance value at full speed may be obtained by adding an amount ΔX to the reactance determined by full-frequency locked-rotor test. The value of ΔX is approximately (Eq. 8.65, with $m = 1$, and Table 5.4):

$$\Delta X = R_t - R_2, \tag{4.11}$$

where R_2 = secondary resistance at full-load slip, determined by the slip test of the following section.

8 LOAD TESTS

To determine the secondary resistance and slip losses, a current-slip curve should be taken under actual load conditions, with rated voltage and frequency maintained at the motor terminals. Measurements at a few points in the neighborhood of full-load current are usually sufficient, but for slip-ring motors a wider range should be covered, because of the variable resistance of the brushes. The slip is normally too small to be determined by tachometer readings, and should, therefore, be measured with a slipmeter or stroboscope. The slipmeter method makes use of a revolution counter differentially geared to the motor under test and to a small synchronous motor driven from the same power supply at the same synchronous speed. Or, an electric counter may be used to measure the motor speed and also that of a synchronous motor on the same power circuit, and the difference taken. Care must be taken to correct the observed values of slip for the difference between the test temperature and the standard value of 75 C, or the temperature attained in a full-load heat run with an ambient temperature of 25 C.

In practice, the value of current corresponding to an assumed value of R_2/s may be calculated exactly by the equivalent circuit (Sect. 5.2), the corresponding value of s read off the slip-current curve, and the true value of R_2 obtained by multiplying R_2/s by this value of s. Or, R_2 may be approximately determined as follows:

The secondary resistance is equal to

(4.12) $$R_2 = \frac{kEs}{I_1} \text{ ohms per phase, approximately,}$$

where E = terminal voltage per phase, s = ratio of rpm of slip to synchronous speed, and I_1 = observed phase current.

The coefficient k varies over a range of about 1 to 1.2, depending on the motor characteristics and the value of the test load. It may be calculated by Eq. 5.27.

If direct slip measurements are not feasible, the value of R_2 determined by Eq. 4.8 in a low-frequency locked-rotor test may be used. Or, for a wound rotor, the actual resistance between slip rings may be measured, and multiplied by the square of the ratio of primary to secondary volts to obtain the resistance referred to primary. The voltage ratio is obtained by measurement of primary and secondary voltages at standstill with the slip rings open-circuited. Averages of several rotor positions are taken to avoid errors due to possible unbalance.

9 Stray-Load Loss Measurement

Stray-load losses are defined as the excess of the total measured losses above the sum of the friction and windage, core, and copper losses, calculated for the load conditions from the no-load tests described above. These extra losses are chiefly made up of high-frequency core losses in stator and rotor teeth, and rotor copper losses, caused by the induced harmonic currents due to the phase belts and slot openings. While the stray-load losses may be determined by direct input-output tests with a dynamometer or calibrated driving motor, the result is a small difference between two large quantities, and so accuracy is very difficult to obtain. Whenever such tests are made, it is desirable to repeat them with the direction of power flow reversed, as thereby the measurement errors may be substantially canceled out.

A preferred way of making an input-output test when two duplicate motors are available is to couple the two motors together and connect them to sources of electric power supply of slightly different frequencies. One machine is then operated as a motor at normal voltage and frequency, the other as a generator at normal voltage, but with a slightly lower frequency output. Electrical input or output readings are taken for each machine, and the slip values and winding temperatures of each are also measured. The tests are then repeated with the power flow reversed, the former motor becoming a normal-frequency, normal-voltage, generator, and the former generator becoming an above-frequency, normal-voltage motor, the supply frequency of the latter being raised to give the desired load. The locations of all meters, instrument transformers, etc., being unchanged, and the power flow reversed, errors of instrument calibration and ratio are canceled out.

To obtain the overall motor efficiency from this test, the procedure is:

1. The stator I^2R at the test temperature is calculated for each machine, using the observed currents.

2. The motor secondary I^2R loss is:

$$\text{Motor slip} \times (\text{motor input} - \text{stator } I^2R),$$

where slip is expressed as a decimal fraction of the impressed frequency.

3. The generator secondary I^2R loss is:

$$\text{Generator slip} \times (\text{generator output} + \text{stator } I^2R).$$

4. The combined stray-load loss is determined by subtracting from the total measured loss (difference between motor input and generator output) the sum of the stator I^2R losses, secondary I^2R losses, core losses, and friction and windage of the two machines. The core losses and friction and windage are in each case assumed to be the same as for the motor at rated voltage and frequency, the small frequency differences making inappreciable errors in this respect.

5. The motor stray-load loss is determined as the product of the total stray-load loss in the two machines by the ratio of the motor secondary I^2R loss to the sum of the motor and generator secondary I^2R losses, the latter being corrected to the motor temperature. The average of the results obtained with the two directions of power flow is taken as the correct value.

6. The motor efficiency is then equal to the ratio of the motor input less the sum of its stator and secondary I^2R losses (corrected to 25 C ambient temperature plus full-load temperature rise), the core loss, friction and windage, and the stray-load loss, to the motor input.

The reverse-rotation method of measuring stray-load loss is the most useful for larger motors. This requires that the motor be driven at synchronous speed in the reversed direction, against its own torque, while sufficient normal-frequency voltage (15 to 20 per cent) is applied to maintain full-load current in the stator windings. The watts input to the stator, and the mechanical power input required to drive the rotor, are measured, as well as the voltage, current, and speed. Under these conditions, the stray-load loss should be practically the same as in normal full-load running, since both current and speed are normal. There is an error, however, due to magnetic saturation, especially with closed rotor slots (Sect. 9.4), which makes the true loss greater than the measured value.

The measured power required to drive the rotor in this test, W_R, is equal to the stray-load loss plus friction and windage, plus just half the total rotor I^2R losses, as the speed of the magnetic field with respect to the rotor is twice synchronous, and the externally supplied loss is, therefore, equal to the induced rotor loss. The stray-load loss is given by subtracting from the measured stator input, W_s, the stator I^2R loss and the core loss (corrected to the low voltage of the test), and subtracting this result, and the friction and windage loss, from W_R:

Stray-load loss $= W_R - W_s - (F+W) +$ corrected core loss $+ I^2(2R_s - R_{dc})$.

The value of R_s to be used in this equation is the effective a.c. stator resistance, determined by measuring the stator input at rated current and normal frequency, with the rotor removed. R_s will be larger than the d.c. resistance, R_{dc}, due to the eddy current losses (skin effect) in the stator conductors, and in the end structure due to end leakage flux. For small machines, R_s may be nearly equal to R_{dc}, and in this case the rotor removed test may be omitted, when the last term of the equation becomes simply I^2R_{dc}.

Appendix to Chapter 4

BEHREND AND TESLA

In view of its historic importance, and the colorful personalities of the two participants, there is recorded here in full the address given by Mr. B. A. Behrend in New York City on May 18, 1917, on the occasion of the presentation of the Edison Medal to Nikola Tesla. Tesla was the inventor of the polyphase induction motor, and Behrend originated the circle diagram, almost concurrently with Alexander Heyland.

MR. CHAIRMAN, MR. PRESIDENT OF THE AMERICAN INSTITUTE OF ELECTRICAL ENGINEERS, FELLOW MEMBERS, LADIES AND GENTLEMEN:

By an extraordinary coincidence, it is exactly twenty-nine years ago, to the very day and hour, that there stood before this Institute Mr. Nikola Tesla, and he read the following sentences:

"To obtain a rotary effort in these motors was the subject of long thought. In order to secure this result, it was necessary to make such a disposition that while the poles of one element of the motor are shifted by the alternate currents of the source, the poles produced upon the other elements should always be maintained in the proper relation to the former, irrespective of the speed of the motor. Such a condition exists in a continuous current motor; but in a synchronous motor, such as described, this condition is fulfilled only when the speed is normal.

"The object has been attained by placing within the ring a properly subdivided cylindrical iron core wound with several independent coils closed upon themselves. Two coils at right angles are sufficient, but a greater number may be advantageously employed. It results from this disposition that when the poles of the ring are shifted, currents are generated in the closed armature coils. These currents are the most intense at or near the points of the greatest density of the lines of force, and their effect is to produce poles upon the armature at right angles to those of the ring, at least theoretically so; and since this action is entirely independent of the speed—that is, as far as the location of the poles is concerned—a continuous pull is exerted upon the periphery of the armature. In many respects these motors are similar to the continuous current motors. If load is put on, the speed, and also the resistance of the motor, is diminished and more current is made to pass through the emerging coils, thus increasing the effort. Upon the load being taken off, the counter-electromotive force increases and less current passes through the primary or energizing coils. Without any load the speed is very nearly equal to that of the shifting poles of the field magnet.

"It will be found that the rotary effort in these motors fully equals that of the continuous current motors. The effort seems to be greatest when both armature and field magnet are without any projections."

Not since the appearance of Faraday's *Experimental Researches in Electricity* has great experimental truth been voiced so simply and so clearly as this description of Mr. Tesla's great discovery of the generation and utilization of poly-phase alternating currents. He left nothing to be done by those who followed him. His paper contained the skeleton even of the mathematical theory.

Three years later, in 1891, there was given the first great demonstration by Swiss engineers, of the transmission of power at 30,000 volts from Lauffen to Frankfort by means of Mr. Tesla's system. A few years later this was followed by the development of the Cataract Construction Company, under the presidency of our member, Mr. Edward D. Adams, and with the aid of the engineers of the Westinghouse Company. It is interesting to recall here tonight that in Lord Kelvin's report to Mr. Adams, Lord Kelvin recommended the use of direct current for the development of power at Niagara Falls and for its transmission to Buffalo.

The due appreciation or even enumeration of the results of Mr. Tesla's inventions is neither practicable nor desirable at this moment. There is a time for all things. Suffice it to say that, were we to seize and to eliminate from our industrial world the results of Mr. Tesla's work, the wheels of industry would cease to turn, our electric cars and trains would stop, our towns would be dark, our mills would be dead and idle. Yea, so far reaching is this work, that it has become the warp and woof of industry.

The basis for the theory of the operating characteristics of Mr. Tesla's rotating field induction motor, so necessary to its practical development, was laid by the brilliant French savant, Professor André Blondel, and by Professor Kapp of Birmingham. It fell to my lot to complete their work and to coordinate—by means of the simple "circle diagram"—the somewhat mysterious and complex experimental phenomena. As this was done twenty-one years ago, it is particularly pleasing to me, upon the coming of age of this now universally accepted theory—tried out by application to several million horsepower of machines operating in our great industries—to pay my tribute to the inventor of the motor and the system which have made possible the electric transmission of energy. *His* name marks an epoch in the advance of the electrical science. From *that* work has sprung a revolution in the electrical art.

We asked Mr. Tesla to accept this medal. We did not do this for the mere sake of conferring a distinction or of perpetuating a name; for so long as men occupy themselves with our industry, his work will be incorporated in the common thought of our art, and the name of Tesla runs no more risk of oblivion than does that of Faraday, or that of Edison.

Nor indeed does this Institute give this medal as evidence that Mr. Tesla's work has received its official sanction. His work stands in no need of such sanction.

No, Mr. Tesla, we beg you to cherish this medal as a symbol of our gratitude for a new creative thought, the powerful impetus, akin to revolution, which you have given to our art and to our science. You have lived to see the work of your genius established. What shall a man desire more than this? There rings out to us a paraphrase of Pope's lines on Newton:

> Nature and Nature's laws lay hid in night:
> God said, "Let Tesla be," and all was light.

Bibliography

		Reference to Section
1.	"I.E.E.E. Test Procedure for Polyphase Induction Motors and Generators." *I.E.E.E. No. 112A.*	7, 8, 9
2.	"Application of Deceleration Test Methods to the Determination of Induction-Motor Performance," R. W. Ager, *A.I.E.E. Trans.*, Vol. 58, 1939, pp. 72–6.	8
3.	"Induction Motor Core Losses," P. L. Alger and R. Eksergian, *A.I.E.E. Jour.*, Vol. 39, Oct. 1920, pp. 906–20.	7
4.	*The Induction Motor and Other Alternating Current Motors*, B. A. Behrend, McGraw-Hill, 1921.	
5.	*Asynchronous Machines*, R. Langlois-Berthelot, Dunod, Paris, 1934.	
6.	*Theorie der Wechselstrommaschinen*, O. S. Bragstad and R. S. Skancke, Julius Springer, Berlin, 1932.	
7.	*A Study of the Induction Motor*, F. T. Chapman, John Wiley, 1930.	
8.	"A Torque Transducer System for Observing, Recording, and Controlling, A. A. Emmerling, I.E.E.E. Conference Paper No. CP 63-1005, 1963.	9
9.	"Measurement of Stray Load Loss in Polyphase Induction Motors," C. J. Koch, *A.I.E.E. Trans.*, Vol. 51, Sept. 1932.	9
10.	*Theory of Alternating Current Machinery, Second Edition*, A. S. Langsdorf, McGraw-Hill, 1955.	
11.	"Efficiency Tests of Induction Machines," C. C. Leader and F. D. Phillips, *A.I.E.E. Trans.*, Vol. 53, 1934, pp. 1628-32.	
12.	"Reverse-Rotation Test for the Determination of Stray Load Loss in Induction Machines," T. H. Morgan, W. E. Brown, and A. J. Schumer, *A.I.E.E. Trans.*, Vol. 58, 1939, pp. 319-22.	9
13.	"Induction-Motor Characteristics at High Slip," T. H. Morgan, W. E. Brown, and A. J. Schumer, *A.I.E.E. Trans.*, Vol. 59, 1940, pp. 464-8.	8
14.	"Stray-Load-Loss Test on Induction Machines," T. H. Morgan and P. M. Narbutovskih, *A.I.E.E. Trans.*, Vol. 53, 1934, pp. 286-90.	9
15.	"Stray-Load-Loss Tests on Induction Machines"–II, T. H. Morgan and V. Siegfried, *A.I.E.E. Trans.*, Vol. 55, 1936, pp. 493-7.	9
16.	"Induction Motor Locked Saturation Curves," H. M. Norman, *A.I.E.E. Trans.*, Vol. 53, 1934, pp. 536-41.	7
17.	"Die-Cast Rotors for Induction Motors," L. C. Packer, *A.I.E.E. Trans.*, Vol. 68, Part I, 1949, pp. 253-60.	
18.	*Modern Polyphase Induction Motors*, F. Punga and O. Raydt, Translated from the German by H. M. Hobart, Isaac Pitman & Sons, Ltd., London, 1933.	
19.	*Elektrische Maschinen-IV, Die Induktionsmaschinen*, R. Richter, Julius Springer, Berlin, 1936.	
20.	The "Superimposed Frequency Test" for Induction Motors, or S. F. Test, M. P. Romeira, *A.I.E.E. Trans.*, Vol. 67, Part II, 1948, pp. 952-5.	8
21.	"An Analysis of the Induction Machine," H. C. Stanley, *A.I.E.E. Trans.*, Vol. 57, 1938, pp. 751-7.	
22.	"A New System of Alternate Current Motors and Transformers," N. Tesla, *A.I.E.E. Trans.*, Vol. 5, 1888, pp. 309-24.	
23.	"Measurement of Stray-Load Loss in Induction Motors," D. H. Ware, *A.I.E.E. Trans.*, Vol. 64, 1945, pp. 194-7.	9

CONTENTS—CHAPTER 5

Polyphase Induction Motor Performance Calculations

		PAGE
1	The Equivalent Circuit	131
2	Circuit Calculations	133
3	Generalized Circuit Calculations	140
4	Calculations in the Region of Standstill	142
5	Calculations in the Region of Full Load	144
6	Operation as a Generator	148
7	Power-Factor Determination	148
8	Calculations in the Region of Maximum Torque	153
9	Dynamic Braking	156
10	Summary of Formulas	159

5

POLYPHASE INDUCTION MOTOR PERFORMANCE CALCULATIONS

1 The Equivalent Circuit

Although the performance of a polyphase induction motor can be easily visualized by the phasor or circle diagram, it is not convenient to make exact or repetitive calculations by these graphic methods. For this purpose, the equivalent circuit offers a far more convenient and versatile method of analysis. Digital computers have largely replaced the slide rule for making extensive calculations of motor performance, and have permitted allowance for factors formerly omitted or guessed at. However, the equivalent circuit provides a sound basis for computer programming, and a full understanding of the circuit is essential for effective use of computers, as well as for judging the effects of various alternatives in design. The approximate formulas developed in this chapter are helpful and informative in many ways.

The "exact" equivalent circuit, due to Steinmetz, is shown in Fig. 5.1. It is the same as the transformer circuit of Fig. 2.4 except for the replacement of the transformer load impedance by a variable secondary resistance:

$$R_2(1-s)/s,$$

representing the ratio of the in-phase component of the motor speed-voltage to the secondary current (sect. 4.3). The magnetizing reactance, X_M, of the induction motor is relatively much smaller than that of the transformer, because of the large number of ampere turns required to force the flux across the motor air gap.

The impressed voltage on each phase, E_1, causes a current, I_1, to flow in the primary winding. Usually, for three-phase machines, a Y-connected winding is assumed, making E_1 equal to the line voltage divided by $\sqrt{3}$.

A voltage drop occurs in the primary winding, due to its resistance, R_1, and its leakage reactance, X_1. The leakage flux in part links the end turns, and in part crosses the stator slots or tooth tips, without linking the secondary winding. The remaining voltage, E_2, is consumed in the magnetizing reactance; or, to put it in another way, is balanced by the back emf produced by the fundamental sine wave of air-gap flux.

If the rotor is operating at synchronous speed, so that s is zero, the apparent secondary resistance is infinite, there is no rotor current, and the entire primary current is magnetizing, or $I_1 = I_M$. At a speed below synchronism, the air-gap voltage, E_2, produces a secondary current, I_2, which creates a secondary resistance drop, $I_2 R_2$, and a secondary leakage reactance drop equal to $jI_2 X_2$, referred to the primary.

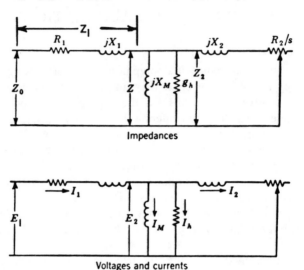

Fig. 5.1. Equivalent Circuit of Polyphase Induction Motor.

Even though the actual secondary voltage, as measured at a fixed position on the rotor, occurs at slip frequency, the rotation makes the full frequency value appear in the equivalent circuit, as viewed from the stator, by Eq. 4.3. Many early inventors thought that high inductance could be inserted in the rotor winding to cut down the current at starting, when the secondary frequency is high, without limiting the current at full speed, when this frequency is low. This assumption is clearly erroneous, as the secondary mmf is always revolving at full synchronous speed with respect to the primary, so that the stator always "sees" the rotor at full frequency.

The remaining voltage in the circuit,

$$\frac{I_2 R_2}{s} - I_2 R_2, \quad \text{or} \quad \frac{I_2 R_2 (1-s)}{s},$$

is the "speed voltage" produced by the rotating air-gap field cutting the stator winding at rotor speed; i.e., it is the in-phase component of the voltage induced by the air-gap field but not consumed in the rotor resistance. This voltage, multiplied by the secondary current, represents the power delivered to the motor shaft, which appears as mechanical output. The output indicated by the circuit must be multiplied by the number of phases to obtain the total output of the motor, and the friction and windage losses must be subtracted from this total to obtain the net output. The torque developed at any speed is obtained by dividing this output by $(1-s)$, and by the appropriate conversion factor (Eq. 3.11).

In practice, the so-called constants of the equivalent circuit, or circuit impedances, vary materially with changes in motor current, speed, voltage, temperature, etc. These variations can be taken care of by introducing appropriately modified values for each condition of operation. For example, it is standard practice to assume the temperature of the windings to be 75 C in making efficiency calculations for Class A and B insulated motors, so the ohmic resistances are calculated at this temperature. Also, it is customary to take a lower value of motor reactance under full voltage starting conditions, when the current is high and the leakage-flux paths are saturated, than at full speed, when the current is low. Likewise, the non-uniform current distribution in the rotor conductors causes the secondary resistance to be higher, and the secondary reactance to be lower, at standstill than they are at full speed. All these variations are recognized in practice by using test or calculated values appropriate for the particular condition.

Instead of varying the circuit constants to take account of all conditions, it may be preferable to make circuit calculations with fixed values of the circuit constants, and then to correct the results, if necessary, by suitable formulas. Convenient charts and correction formulas have been devised that make it easy to determine the effects of any moderate change in the impedance values without recalculating the whole circuit. Or, the calculations can be programmed on a computer.

2 CIRCUIT CALCULATIONS

Definitions of the circuit constants, and a worksheet for circuit calculations, are given in Tables 5.1 and 5.2, respectively.

To explain the procedure, calculations in accordance with Table 5.2 for a typical 3-phase, 220-volt, 15-hp, 1,800-rpm induction motor are given in Table 5.3, and the resulting performance curves are shown in Fig. 5.2. In Fig. 4.2, the complete speed-torque curves of the same motor are shown, calculated with and without primary resistance, bringing out the important influence of this resistance on the breakdown-torque values as motor and as generator.

If the calculations are made in ohms, volts, and amperes, item (9) becomes E/Z_0 instead of $1/Z_0$, and 11, 12, 13, and 14, Table 5.2, are all

Table 5.1

Definitions of Equivalent Circuit Constants

Unless otherwise noted, all quantities except watts, torque, and power output are per phase for two-phase motors and per phase Y for three-phase motors.

E_1 = impressed voltage (volts) = line voltage $\div \sqrt{3}$ for three-phase Y motors.
I_1 = primary current (amperes).
I_2 = secondary current in primary terms (amperes).
I_M = magnetizing current (amperes).
R_1 = primary resistance (ohms).
R_2 = secondary resistance in primary terms (ohms).
R_0 = resistance between primary terminals (ohms).
R_t = apparent secondary resistance determined by locked-rotor test.
$R = R_2/s$ = ratio of secondary resistance to per unit slip.
X_1 = primary leakage reactance (ohms).
X_2 = secondary leakage reactance (ohms).
$X = X_1 + X_2$.
X_0 = reactance at primary terminals (ohms).
X_t = reactance determined by locked-rotor test.
ΔX = decrease in X_2 at standstill below the full speed value, due to presence of eddy currents in secondary conductors (deep-bar effect).
X_M = magnetizing reactance (ohms).
Z_1 = primary impedance (ohms).
Z_2 = secondary impedance in primary terms (ohms).
Z_0 = impedance between primary terminals (ohms).
Z = combined secondary and magnetizing impedance (ohms).
s = per unit slip (expressed as a fraction of synchronous speed).
N = synchronous speed (rpm).
q = number of phases.
f = rated frequency (cycles per second).
f_t = frequency used in locked-rotor test.
T^* = torque (newton-meters)
W = watts input.
W_H = core loss (watts).
W_F = friction and windage (watts).
W_{RL} = running-light watts input.
W_s = stray-load loss (watts).
W_t = watts input at standstill.

$$Y = G - jB = \frac{1}{Z} = \frac{1}{R + jX} = \frac{R}{Z^2} - \frac{jX}{Z^2}.$$

$|Z| = \sqrt{(R^2 + X^2)}$.

Table 5.2

Equivalent Circuit Calculations

Per Unit Terms

s	Slip (negative for generator).
(1) $Z_2 = \dfrac{R_2}{s} + jX_2$	Secondary impedance.
(2) $Y_2 = G_2 - jB_2$	Reciprocal of (1).
(3) $Y_M = G_h - jB_M$	Magnetizing admittance.†
(4) $Y = G - jB$	Sum of (2) and (3).
(5) $Z = R + jX$	Reciprocal of (4).
(6) $Z_1 = R_1 + jX_1$	Primary impedance.
(7) $Z_0 = R_0 + jX_0$	Sum of (5) and (6).
(8) $\|Z_0\| = \sqrt{(R_0^2 + X_0^2)}$	Phase impedance.
(9) $I_1 = 1/Z_0$	Phase current, reciprocal of (8).
(10) Power factor $= R_0/Z_0$	
(11) Input $= I_1^2 R_0$	Input of motor (output of generator).
(12) Gross output $= \dfrac{I_1^2 R G_2 (1-s)}{G}$	Input to shaft times $(1-s)$ (negative for generator).
(13) $W_F(1-s)$	Friction and windage.
(14) Net output	(12) minus (13) (negative input of generator).
(15) Efficiency	(14) divided by (11). For generator, (11) divided by (14).
(16) Torque	(14) divided by $(1-s)$.

Actual Values

(17) Phase amperes	(9) times rated load current.
(18) Rpm speed	$(1-s)$ times synchronous rpm.
(19) Horsepower output	(14) times rated volt-amperes divided by 746.
(20) Foot-pounds torque	5,250 times (19) divided by (18).
Newton-meters torque	7,130 times (19) divided by (18).

† G_h is usually taken as zero, and the core loss watts are then added to the calculated input, and assumed to be constant at all loads.

watts per phase. If the calculations are made in per unit terms, as in Table 5.3, these are per unit values, and give total watts when multiplied by the total rated volt-amperes, or watts per phase when multiplied by rated volt-amperes per phase. For a motor, the rated output is less than the rated volt-amperes input, so that the per unit torque at full load is equal to the product of efficiency times power factor, or 0.82 for Fig. 5.2.

The stray-load loss, W_s, may be determined by test, or given an arbitrary fixed, or "conventional", value. This loss may be subtracted from the calculated output, or it may be included in the calculation by an arbitrary addition to the primary resistance, ΔR_1, where $\Delta R_1 = W_s$ at full load divided by phases times full-load current squared.

136 POLYPHASE INDUCTION MOTOR PERFORMANCE CALCULATIONS

TABLE 5.3

EQUIVALENT CIRCUIT CALCULATIONS FOR SINGLE SQUIRREL-CAGE MOTOR IN ACCORDANCE WITH TABLE 5.2

	Braking	Standstill	Near Maximum Torque	Near Full Load	Generating
s	3.00	1.00	0.30	0.045	−0.045
			Per Unit Values		
(1) Z_2	$0.015+j0.075$	$0.045+j0.075$	$0.150+j0.075$	$1.00\ +j0.075$	$-1.00\ +j0.075$
(2) Y_2	$2.56\ -j12.8$	$5.88\ -j9.80$	$5.34\ -j2.67$	$0.995-j0.075$	$-0.995-j0.075$
(3) Y_M	$0.04\ -j0.33$	$0.04\ -j0.33$	$0.04\ -j0.33$	$0.04\ -j0.333$	$0.04\ -j0.333$
(4) Y	$2.60\ -j13.13$	$5.92\ -j10.1$	$5.38\ -j3.00$	$1.035-j0.408$	$-0.955-j0.408$
(5) Z	$0.015+j0.073$	$0.043+j0.074$	$0.142+j0.079$	$0.836+j0.329$	$-0.884+j0.378$
(6) Z_1	$0.045+j0.075$	$0.045+j0.075$	$0.045+j0.075$	$0.045+j0.075$	$-0.045+j0.075$
(7) Z_0	$0.060+j0.148$	$0.088+j0.149$	$0.187+j0.154$	$0.881+j0.404$	$-0.839+j0.453$
(8) $\|Z_0\|$	0.160	0.173	0.243	0.969	0.954
(9) I_1	6.25	5.78	4.12	1.032	1.050
(10) Power factor	0.372	0.508	0.770	0.908	−0.880
(11) Input	2.32	2.94	3.18	0.940	−0.925
(12) Gross ouput	−1.11	0	1.68	0.820	−1.06
(13) W_F	0.030	0	0.01	0.015	0.015
(14) Net output	−1.14	0	1.67	0.805	−1.075
(15) Efficiency		0	0.525	0.857	0.860
(16) Torque	0.57	1.44	2.39	0.843	−1.03
			Actual Values		
(17) Amperes	236	218	156	39.1	39.7
(18) Rpm	−3,600	0	1,260	1,692	1,908
(19) Horsepower	−220	0	32.3	15.5	−20.7
(20) Newton-meters	43.6	110	183	65.6	−77.3−

CIRCUIT CALCULATIONS

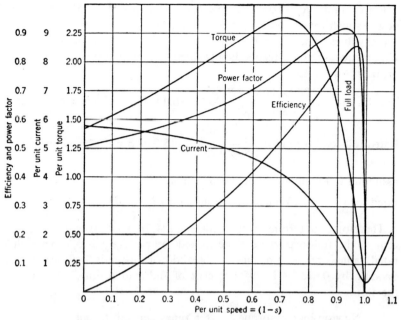

FIG. 5.2. Performance Curves of Typical Polyphase Induction Motor.

The equivalent circuit constants assumed for the motor are:

$R_1 = 0.045$ per unit, including an allowance for stray-load losses, W_s,
$R_2 = 0.045$ per unit, assumed constant at all speeds,
$X_1 = X_2 = 0.075$ per unit, assumed constant at all speeds,
$X_M = 3.0$ per unit, assumed constant at all speeds,
$G_h = 0.040$ per unit, assumed constant at all speeds,
$W_F = 0.015 (1-s)$ per unit, assumed to vary in proportion to the speed.

As shown by the curves, the motor full-load efficiency is 0.858, and the power factor is 0.905, giving a rated current per phase Y (at 15 hp) equal to:

$$I_1 = \frac{15 \times 746}{0.858 \times 0.905 \times 220 \sqrt{3}} = 37.8 \text{ amp.}$$

To convert the per unit impedances into ohms, therefore, they should be multiplied by the unit impedance (Sect. 2.4):

$$\text{Unit impedance} = \frac{220}{37.8 \sqrt{3}} = 3.35 \text{ ohms.}$$

For example, $R_1 = 0.045 \times 3.35 = 0.151$ ohm per phase Y. The per unit power and torque values must be multiplied by the rated input to obtain actual values. The per unit torque calculated for $s = 1$ (at standstill) is 1.44. Unit torque is that corresponding to rated kva input at synchronous speed, or, from Eq. 3.11:

$$\text{Unit torque} = \frac{7{,}040 \times 37.8 \times 220 \sqrt{3}}{1{,}800 \times 1{,}000} = 56.4 \text{ ft-lb} = 76.5 \text{ newton-meters.}$$

$$\text{Rated full-load torque} = \frac{15 \times 5{,}252}{1{,}800(0.955)} = 45.9 \text{ ft-lb} = 62.2 \text{ newton-meters,}$$

assuming full-load slip to be 0.045. Thus,

rated torque = 0.813 p.u., corresponding to $I_1 = 1.00$.

Therefore, the torque at standstill is $1.44 \times 76.5 = 110$ nm, or 177 per cent of the full-load value. The locked-rotor current is 578 per cent of rated value, or 218 amp; and the breakdown torque is 2.40 per unit, or, dividing by 0.813, it is 295 per cent of full-load torque. On Fig. 5.2, therefore:

Newton-meters torque = 76.5 times per unit torque,

Torque/torque at rated output = 1.23 times per unit torque,

Line current in amperes = 37.8 times per unit current,

Speed in rpm = 1,800 times per unit speed.

Tables 14.1 and 14.2 give values, for a standard motor of this rating, of 165 per cent torque and 220 amp with locked rotor, and 200 per cent breakdown torque. In practice, such a motor would normally have a double squirrel cage, instead of the single cage for which the calculations are made, giving somewhat better starting torque and efficiency with a lower slip and less breakdown torque for the same locked-rotor current (Sect. 8.6).

The circuit data may be obtained from the no-load tests briefly described in Sect. 4.7, by the formulas given in Table 5.4.

The procedure in making performance calculations based on test data, with Table 5.2, is first to divide E_1 by the approximate expected value of normal current, obtaining an arbitrary value of R_2/s. With this value and the known circuit constants, calculations are carried through for one slip point, determining the actual value of I. From the test slip-current curve, the true value of s is found, and from this and R_2/s, R_2 is calculated. The power factor, efficiency, torque, etc., are then determined by (10) to (16), Table 5.2. Additional points are calculated with different values of s,

TABLE 5.4

FORMULAS FOR CALCULATING CIRCUIT CONSTANTS FROM TEST DATA FOR THREE-PHASE MOTORS

(5.1) $X_t = \dfrac{f}{f_t}\sqrt{\left[\dfrac{E_1^2}{3I^2} - \left(\dfrac{W}{3I^2}\right)^2\right]}$ (Eqs. 4.8 and 4.9).

(5.2) $X = X_t\left(1 + \dfrac{I_M X_t}{4E_0}\right) + \Delta X$ (see Eq. 5.14).

(5.3) $X_1 = X_2 = 0.5X$ for single squirrel-cage or wound-rotor motors.

(5.4) $\Delta X = R_t - R_2$ (see Eqs. 4.11, 8.64, and 8.65).

(5.5) $W_H + W_F = W_{RL} - 3I_M^2 R_1$.
 W_s from the load tests described in Sect. 4.8 or 4.9 (Usually $W_s = 0.01$ to 0.03 times rated output at full load, and varies as the square of the load current at other loads.)

(5.6) $X_M = \dfrac{E_1}{I_M} - X_1$.

covering the desired range of loads, and the exact characteristics are taken off curves plotted from the calculated results.

It will be noted that no use is made in these calculations of the value of secondary resistance, R_t, as determined from standstill impedance readings. The reason is that the resistance of the secondary at standstill is generally increased materially above its full-speed (slip-frequency) value by the presence of eddy currents and secondary iron losses (Sect. 8.4). If the locked-rotor torque is measured, and impedance readings are taken simultaneously, the secondary power input ought to check the measured torque, expressed in watts at synchronous speed, by the formula:

(5.7) $qI^2 R_t = W_t - qI^2 R_1 = 0.105 NT$, with T in newton meters,

from Eq. 3.11.

Usually, the measured torque is about 10 per cent less than given by Eq. 5.7, since some of the power input at standstill represents non-synchronous harmonics (Eq. 4.10).

If values of torque, current, etc., are desired for considerable overloads, or throughout the accelerating range, the values of R_2 and X should be modified to allow for magnetic saturation and eddy currents (Sect. 8.4). Curves of reactance against current obtained by locked-rotor tests over the desired range of values, and values of R_t and corresponding values of ΔX obtained by locked-rotor tests at different frequencies are desirable for this purpose, especially for closed-slot or double squirrel-cage rotors.

There are many special forms of induction motor that employ deep rotor bars, or multiple squirrel cages, or reactors, or "Saturistors" in the secondary circuits. The equivalent circuits for these motors are derived from that of Fig. 5.1, by locating the appropriate impedances and/or additional branch circuits in the secondary. Examples of such motors and their performance calculations are considered in Chapter 8.

3 Generalized Circuit Calculations

The magnetizing reactance, X_M, is usually ten or more times as great as X, whereas R_1 and R_2 are usually much smaller than X, except for special motors designed for frequent starting service, that is:

$$X_M \gg X > (R_1 \text{ or } R_2).$$

These relationships enable the solution of the equivalent circuit to be expressed in the form of an infinite series, of which only the first few terms normally need be considered.

To facilitate the derivation of formulas and charts from which any desired characteristics of a polyphase motor can be determined, it is convenient to make use of the following symbols:

$$a = \frac{I_M}{I_1} = \text{ratio of no-load current to primary current at assumed load,}$$

$$b = \frac{I_1 X}{E_1} = \text{ratio of apparent leakage reactance drop at assumed load, to impressed primary voltage,}$$

$$ab = \frac{I_M X}{E_1} = \frac{X}{X_M + X_1} = \text{the "leakage factor"}.$$

It is, further, convenient to make the following assumptions, in accord with usual practice:

1. $X_1 = X_2$, or primary reactance equals secondary reactance. This assumption is made for the reasons that the departures from it met with in practice can cause only small errors, that it is difficult to determine X_1 and X_2 separately by test, and that to keep X_1 and X_2 separate in the calculations requires extra work.

2. $I_M = I_1$ at no load, or the running-light power factor is very low. This assumption is made for the reasons that it is closely true for integral-horsepower motors, that the departure from fact is greatest in high-power-factor motors, where the value of I_M is of least importance, and that the approximation greatly simplifies the work, without causing appreciable error.

3. I_M is proportional to air-gap flux, or the magnetizing current is proportional to E_2. The effect of magnetic saturation makes the true value of I_M a little less at high overloads than this indicates. The value of X_M used in calculations should be reduced if the voltage is above normal.

4. The core loss is proportional to E_2^2. This is nearly true at magnetic

densities usual in induction motors. On over-voltage, the exponent is greater than 2, whereas at low voltages it may be a little less than 2.

5. All impressed current and voltage harmonics are neglected, or sinusoidal power supply waveforms are assumed throughout.

6. Stray-load losses are taken into account by increasing the effective primary resistance by an appropriate amount.

To make generalized circuit calculations, it is desirable to express the variables as functions of a small number of dimensionless parameters. There are six independent constants of the circuit, which can be reduced to two principal and two subsidiary constants, by assuming the leakage reactance to be equally divided between primary and secondary ($X_1 = X_2$), and by recognizing that the secondary resistance, R_2, and the slip occur together only as the ratio R_2/s, so that R_2/s can be used as the independent variable. The principal constants are then the magnetizing and leakage reactances, X_M and X, and subsidiary constants are the primary resistance, R_1, and the core loss conductance, G_h. There is, in fact, one other constant, the friction and windage, which must be specified to determine the motor efficiency, but this is taken care of by subtracting it from the output indicated by the circuit, without affecting in any way the circuit calculations.

The problem now is to derive simple means of determining any particular characteristic of the motor by means of the equivalent circuit, for any set of values of the four motor constants. To solve this problem, the four constants will be expressed as voltage or current ratios, and the circuit will be solved for the general case by developing it in the form of an infinite power series. Since the two auxiliary constants have much smaller values than the other two, the series need be carried only to the first- or second-order terms in these parameters, though it should be carried out to include the fourth- or even sixth-order terms of the principal parameters. Although the algebraic expressions resulting from this process are rather extensive, a little familiarity with them leads to their expression in convenient forms, and, if only approximate results are desired, they become very simple.

In order to obtain series converging as rapidly as possible, it is important to arrange the parameters in the order of their normal magnitudes. The magnitudes depend, of course, upon the operating range of the motor under consideration; the magnetizing current being a very small part of the total under starting conditions, for example, although it is a very large proportion of the whole at light loads. For this reason, it is convenient to carry out the solution separately for the three different ranges of operation which are of interest, corresponding to normal operation under load, the region of maximum torque, and the region of starting. In the following sections, the process indicated is carried through for these three conditions in turn.

To summarize, the five important design constants of an induction machine, expressed as dimensionless parameters for circuit calculations, are:

$$a = \frac{I_M}{I} = \text{ratio of no-load magnetizing current to total current at the load under consideration,}$$

$$b = \frac{I_1 X}{E_1} = \text{ratio of total reactance drop to impressed voltage,}$$

$$c = \frac{I_1 R_1}{E_1} = \text{ratio of primary resistance drop to impressed voltage,}$$

$$d = \frac{I_1 R_2}{E_1} = \text{ratio of secondary resistance drop, based on primary current, to impressed voltage,}$$

$$h = \frac{\text{core loss}}{\text{volt-amperes input}} = \text{ratio of core loss watts to volt-amperes input at the load under consideration.}$$

Instead of using d, it is more convenient in most cases to use:

$$k = \frac{d}{s} = \frac{R_2}{sZ_0} = \frac{R}{Z_0} = \text{ratio of apparent per cent secondary resistance drop to per cent slip.}$$

Since it is small, and to simplify the results, h is assumed to be zero in most of the calculations made here. Its effects are given more fully in reference number one.

4 Calculations in the Region of Standstill

Near standstill, and also in reversed rotation, the reactance is much larger than R_2/s (except when R_2 is abnormally high), so that we can develop approximate formulas in the form of a series in powers of R_2/sX. The core loss can be neglected, and the value of a is very small in this region. Using R for convenience in place of R_2/s, the secondary impedance is:

(5.8) $$Z_2 = R + j\frac{X}{2},$$

and the admittance is:

(5.9) $$Y_2 = \frac{4R - j2X}{4R^2 + X^2}.$$

CALCULATIONS IN THE REGION OF STANDSTILL 143

The magnetizing admittance is $-j/X_M$. Adding this to Y_2, and taking the reciprocal,† we have for the apparent secondary impedance:

$$(5.10) \quad Z = \left[R + j\frac{X}{2}\left(1 + \frac{X^2 + 4R^2}{2XX_M}\right)\right]\left[\frac{4R^2 + X^2}{4R^2 + X^2\left(1 + \frac{X^2 + 4R^2}{2XX_M}\right)^2}\right]$$

$$-R\left(1 - \frac{X}{X_M} + \frac{3X^2}{4X_M^2} - \frac{R^2}{X_M^2}\right)$$

$$+ j\frac{X}{2}\left(1 - \frac{X}{2X_M} + \frac{X^2}{4X_M^2} + \frac{2R^2}{XX_M} - \frac{R^2}{X_M^2}\right), \text{ approximately.}$$

Adding the primary impedance, $R_1 + j\frac{X}{2}$, to this, the total impedance is

$$(5.11) \quad Z_0 = R_1 + R\left(1 - \frac{X}{X_M} + \frac{3X^2}{4X_M^2} - \frac{R^2}{X_M^2}\right)$$

$$+ jX\left(1 - \frac{X}{4X_M} + \frac{X^2}{8X_M^2} + \frac{R^2}{XX_M} - \frac{R^2}{2X_M^2}\right).$$

However, from Eq. 4.7:

$$(5.12) \quad X_M = \frac{E}{I_M} - \frac{X}{2} = \frac{X}{ab}\left(1 - \frac{ab}{2}\right),$$

so that, neglecting the R^2 terms, and replacing R by R_2/s, Eq. 5.11 becomes:

$$(5.13) \quad Z_0 = R_1 + \frac{R_2}{s}\left(1 - ab + \frac{a^2b^2}{4}\right)$$

$$+ jX\left(1 - \frac{ab}{4}\right), \text{ approximately.}$$

Therefore, the true value of X to use in the equivalent circuit is large than the test value, X_t, found by Eq. 4.9, by the factor $1 + \frac{ab}{4}$, or:

$$(5.14) \quad X = X_t\left(1 + \frac{ab}{4}\right), \text{ approximately,}$$

as given in Table 5.4.

† In these calculations, the binomial theorem is extensively used:
$$(1 + a)^n = 1 + na + \frac{n(n-1)a^2}{2} + \ldots.$$

Also, the apparent secondary resistance, R_t, approximately measured in the locked-rotor test, is smaller than the true value by the factor $(1-ab)$.

The starting torque is proportional to the square of the current and the apparent secondary resistance, R_t (Eq. 4.10), and so is less than the torque that would be obtained if the magnetizing current were zero, by the factor:

$$(5.15) \qquad \left(1+\frac{ab}{4}\right)^2 (1-ab) = 1-\frac{ab}{2}, \text{ approximately.}$$

5 Calculations in the Region of Full Load

At full load as a motor, s is small, and R_2/s is the largest part of the total motor impedance, Z_0. Hence, we may most conveniently calculate all the impedances of the circuit as products of R_2/s by a power series. Having done this, we may find an expression for:

$$k = \frac{R_2}{sZ_0} = \frac{R}{Z_0}$$

by successive approximations, and, by substituting its value in the other expressions, obtain the ratios of all the impedances to R_2/s. These ratios and quantities derived from them will give the power factor, torque, and other performance factors.

As a prelude to the calculations, it is useful to note the following equalities:

$$(5.16) \qquad X_1 = X_2 = \frac{bR_2}{2ks},$$

$$(5.17) \qquad E_2 = E_0\left(1-\frac{ab}{2}\right), \text{ approximately,}$$

$$(5.18) \qquad B_M = \frac{s}{R_2}\left(\frac{2ak}{2-ab}\right)$$

$$= \frac{s}{R_2}ak\left(1+\frac{ab}{2}+\frac{a^2b^2}{4}\right), \text{ approximately,}$$

$$(5.19) \qquad G_h = \frac{khs}{R_2}(1+ab), \text{ approximately.}$$

The first step is to express the secondary circuit admittance, Y_2, as an infinite series, derived by taking the reciprocal of Z_2:

$$(5.20) \qquad Z_2 = \frac{R_2}{s}\left(1+\frac{jb}{2k}\right),$$

$$(5.21) \qquad Y_2 = \frac{s}{R_2}\left[\left(1-\frac{b^2}{4k^2}+\frac{b^4}{16k^4}\right)-j\left(\frac{b}{2k}-\frac{b^3}{8k^3}+\cdots\right)\right].$$

CALCULATIONS IN THE REGION OF FULL LOAD

Adding the magnetizing admittance, Y_M, to Y_2, and taking the reciprocal, we have their combined impedance, Z:

$$(5.22) \quad Z = \frac{R_2}{s}\bigg[1-(a^2k^2+ab+kh)$$

$$+\bigg(a^4k^4+a^3k^2b+3a^2k^3h$$

$$+\frac{a^2b^2}{4}+2abkh+k^2h^2+\frac{b^2h}{4k}\bigg)+\cdots\bigg]$$

$$+j\bigg[\bigg(ak+\frac{b}{2k}\bigg)$$

$$-\bigg(a^3k^3+a^2bk+\frac{ab^2}{4k}+2ak^2h+bh\bigg)+\cdots\bigg].$$

The addition of the primary impedance, $\dfrac{R_2}{s}\bigg(\dfrac{c}{k}+j\dfrac{b}{2k}\bigg)$, to Z, gives the total motor impedance, Z_0:

$$(5.23) \quad Z_0 = \frac{R_2}{s}\bigg[1-\bigg(a^2k^2+ab+kh-\frac{c}{k}\bigg)+\bigg(a^4k^4+a^3k^2b$$

$$+3a^2k^3h+\frac{a^2b^2}{4}+2abkh+k^2h^2+\frac{b^2h}{4k}\bigg)+\cdots\bigg]$$

$$+j\bigg[\bigg(ak+\frac{b}{k}\bigg)-\bigg(a^3k^3+a^2bk+\frac{ab^2}{4k}+2ak^2h+bh\bigg)+\cdots\bigg].$$

The absolute value of Z_0, found by extracting the square root of the sum of the squares of the resistance and reactance components of Eq. 5.23, is:

$$(5.24) \quad |Z_0| = \bigg(\frac{R_2}{s}\bigg)\bigg[1+\bigg(-\frac{a^2k^2}{2}+\frac{b^2}{2k^2}+\frac{c}{k}-kh\bigg)$$

$$+\bigg(\frac{3a^4k^4}{8}+\frac{3a^2k^3h}{2}-\frac{a^2kc}{2}-\frac{a^2b^2}{4}-\frac{abc}{k}-\frac{ab^3}{4k^2}$$

$$+k^2h^2-\frac{b^2h}{4k}-\frac{b^2c}{2k^3}-\frac{b^4}{8k^4}\bigg)+\cdots\bigg].$$

The complete solution of the equivalent circuit has now been formally obtained, and it remains to reduce the expressions to useful forms. The first objective we shall set is the calculation of the slip, s, for a given primary current, from known values of the circuit constants. Taking the reciprocal of Eq. 5.24, and multiplying through by R_2/s, we find:

$$(5.25) \quad k = \frac{R_2}{sZ_0} = 1 + \left(\frac{a^2k^2}{2} - \frac{b^2}{2k^2} - \frac{c}{k} + kh\right)$$

$$- \left(\frac{a^4k^4}{8} + \frac{a^2k^3h}{2} + \frac{a^2kc}{2} + \frac{a^2b^2}{4} - \frac{abc}{k}\right.$$

$$\frac{ab^3}{4k^2} + \frac{3b^2h}{4k} - \frac{3b^2c}{2k^3} - \frac{3b^4}{8k^4}$$

$$\left. - \frac{c^2}{k^2} + 2ch\right) + \dots$$

As the value of k in the neighborhood of full load is not far from unity, we may readily solve Eq. 5.25 by successive approximations, giving:

$$(5.26) \quad k = 1 + \left(\frac{a^2}{2} - \frac{b^2}{2} + h - c\right) + \left[\frac{3a^4}{8} - \frac{a^2b^2}{4}\right.$$

$$\left. + \frac{ab^3}{4} - \frac{b^4}{8} + ac(b-a) + h\left(a^2 - \frac{b^2}{4}\right) + \dots\right].$$

For practical purposes, the second- and higher-order terms in Eq. 5.26 may be replaced by the single term $2a^2b^2$, which is roughly equal to them for normal motors in the neighborhood of full load, so that we finally obtain:

$$(5.27) \quad k \cong 1 + \frac{a^2 - b^2}{2} + h - c + 2a^2b^2.$$

Equation 5.27 has been tested in comparison with the results of complete circuit calculations over a wide range of motor constants, and it has been found to be accurate within 1 or 2 per cent for usual values of k (1 to 1.2) and to 5 per cent for abnormal values of k (up to 1.4).

This equation is useful in making performance calculations from test results, for there is no suitable means of directly measuring R_2, and a knowledge of this is essential for the start of the equivalent circuit calculations for a definite value of slip. Measurement of R_2 by standstill impedance tests involves large errors due to iron losses and eddy currents, and, on slip-ring motors, its measurement with direct current applied across the rings involves transformer ratio calculations which may lead to errors, as well as errors due to the brush contact voltage drop.

The normal procedure in testing is, therefore, to operate the motor under load, and take simultaneous readings of line current and slip,

CALCULATIONS IN THE REGION OF FULL LOAD

establishing a slip-current curve (Sect. 4.8). This test does not require any torque or wattmeter measurements, and so is conveniently and accurately performed. The usual primary resistance, running-light, and blocked-rotor tests determine the values of c, a, h, and b for any assumed value of primary current, I_1. Substituting values in Eq. 5.27 gives the value of k, which in conjunction with the test s versus I_1 curve gives R_2 from the identity $R_2 = ksZ_0$. A check may be obtained by repeating the calculations for several values of current, and the same value of R_2 should be found in all cases. A single pair of test values of s and I_1 is sufficient, instead of the whole curve. Without using Eq. 5.27, the same answer can be found by guessing the value of k, thence finding R_2, carrying through an equivalent circuit calculation for the known slip, and comparing the value of I_1 thus found with the measured value. By repeating the process with successively closer values of k until agreement is reached, the correct answer can finally be obtained.

Other useful expressions are those for the secondary current, secondary induced voltage, and power factor. The secondary current is readily determined from the identity $I_2 = I_1(|Y_2 Z|)$. $|Y_2|$ is the reciprocal of the absolute value of Z_2, and $|Z|$ is the absolute value of Z from Eq. 5.22. Evaluating these quantities with the aid of Eq. 5.26 and multiplying them gives:

$$(5.28) \quad \frac{I_2}{I_1} = 1 - \left(\frac{a^2}{2} + \frac{ab}{2} + h\right) + \left(-\frac{a^4}{8} + \frac{a^3 b}{4} + \frac{a^2 b^2}{2} + a^2 c + \frac{abh}{2} + \frac{b^2 h}{2} + ch\right),$$

which is approximately equal to:

$$(5.29) \quad \frac{I_2}{I_1} \cong 0.98 - \frac{a^2}{2} - \frac{ab}{2}.$$

This equation is convenient for determining the actual secondary copper loss, and hence the torque and rotor heating, for any slip, when the usual no-load test data are available.

Similarly, the induced secondary voltage, referred to primary, is given by evaluating the identity $E_2 = E_0 |Y_0 Z|$ with the aid of Eqs. 5.22, 5.24, and 5.26, whence:

$$(5.30) \quad \frac{E_2}{E_0} = 1 - \left(\frac{ab}{2} + \frac{3b^2}{8} + c\right) + \left(\frac{3a^2 b^2}{8} + \frac{a^2 c}{2} + \frac{7ab^3}{16} + \frac{b^2 h}{2} + \frac{5b^2 c}{8} - \frac{9b^4}{128} + 2ch\right),$$

which is approximately equal to:

$$\text{(5.31)} \qquad \frac{E_2}{E_0} \cong 1 - \left(\frac{ab}{2} + \frac{3b^2}{8} + c\right) + a^2b^2.$$

Equation 5.31 is convenient for determining the actual secondary flux densities and the air-gap flux under load conditions. At no load, the per unit primary reactance drop is $ab/2$, so that, assuming the core loss to vary as E_2^2, the ratio of actual core loss under load to that at no load is:

$$\text{(5.32)} \qquad \frac{E_2^2}{E_0^2\left(1 - \frac{ab}{2}\right)^2} = (1+ab)\left(\frac{E_2^2}{E_0^2}\right)$$

$$= 1 - \frac{3b^2}{4} - 2c, \text{ approximately.}$$

This equation does not make any allowance for stray-load losses, which are due to the leakage fluxes produced by the load currents. It gives the same value for core loss as found by the formal solution of the equivalent circuit of Fig. 5.1.

6 Operation as a Generator

All the equations 5.16 to 5.32 can be applied to induction generators by simply changing the signs of the h and c terms in the equations, wherever they appear, and remembering that the real power output is negative; i.e., the shaft receives power instead of delivering it. The reactive power remains lagging with respect to the line voltage, whence we say that the induction generator *delivers leading* kva, whereas the motor *receives lagging* kva, both of which mean that the magnetizing volt-amperes are supplied by the power system. The reversal of sign of c gives a little lower power factor and a little higher breakdown torque in generator, than in motor, operation (Fig. 4.2).

7 Power-Factor Determination

The power factor may be calculated from the identity, power factor = $R_0 |Y_0|$, R_0 being given by the resistance component of Eq. 5.23, and $|Y_0|$ by the reciprocal of Eq. 5.24. Carrying out the operations indicated, and substituting Eq. 5.26 in the result to eliminate k, we find:

$$\text{(5.33)} \qquad \text{Power factor (pf)} = 1 - \frac{(a+b)^2}{2} + \left[\left(-\frac{a^4}{8} + \frac{a^3b}{2}\right.\right.$$

$$\left.\left. + \frac{3a^2b^2}{2} + \frac{3ab^3}{4} - \frac{b^4}{8}\right) + bh(a+b)\right.$$

$$\left. + 2ac(a+b)\right] + \cdots.$$

The complete form of Eq. 5.33 is too long for practical use, and no way has been found for reducing it to a simple expression of accuracy better than 2 or 3 per cent. It can be expressed in a variety of approximate forms, however, which are useful in rough calculations, such as:

$$\text{(5.34)} \qquad \text{Power factor} = 1 - \frac{(a+b)^2}{2} + 3a^2b^2.$$

This equation indicates clearly the symmetrical way in which a and b determine the power factor, but it is not accurate enough for most performance calculations.

To have a ready method of finding the power factor for any load, when no-load test data only are available, we must resort to a new expedient. By assuming definite values of the parameters c and h, we can reduce the variables determining the power factor to three: a, b, and R_2/s. This enables a power-factor chart to be plotted, the power factor being given as a function of R_2/s for various values of a and b. Such a chart, shown in Fig. 5.3, can readily be drawn with sufficient accuracy to enable the power factor to be read off to about 0.2 per cent for any conditions. The values read directly will, of course, be in error due to the differences between the actual and assumed values of c and h, but these errors can be corrected for by means of the proper terms of Eq. 5.33. By constructing two or three such charts for particular values of c and h with definite increments between, the true power factor can be determined with sufficient accuracy for most purposes by a single chart reading, and can be found to within less than 1 per cent either by interpolating between two chart readings or by adding a small fraction to the reading from a single chart.

The power-factor chart shown in Fig. 5.3 has proved extremely useful, as it enables rapid and consistent predictions of performance to be made with good accuracy. For establishing results from guarantee tests, however, the complete equivalent circuit calculations (Fig. 5.1) are generally preferable, as they are more exact, and they can be made very quickly with a computer.

Somewhat similar power-factor charts have been disclosed by others, but no adequate methods for locating power-factor load curves and for correcting the chart results to exact values have been given. The chart here described was first devised by H. Maxwell about 1910, and was successively modified by C. Macmillan and others of the author's associates before it was put into its present exact form in 1923.

The detailed procedure in making the chart is substantially as follows, the numerical values given being those used in constructing Fig. 5.3.

1. Arbitrary numerical values that will be convenient in calculation are

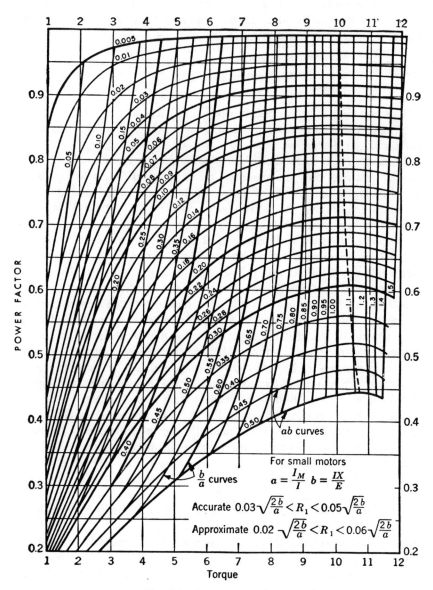

FIG. 5.3. Power-Factor Chart.

assumed for the impressed voltage, E, and the leakage reactance, X, of the circuit in Fig. 5.1.

2. A series of values of the leakage factor, ab, covering the range from 0 to 0.50, is assumed.

3. The primary resistance for any value of ab is determined by arbitrarily assuming a definite value for c, as 0.04, at a load current corresponding to a definite ratio of a to b, as 2. These values fix R_1 for any value of ab, the numbers cited giving:

$$R_1 = 0.04X \sqrt{(2/ab)}.$$

4. The magnetizing reactance for any value of ab is

$$X_M = \frac{X}{2}\left(\frac{2}{ab} - 1\right).$$

5. An arbitrary value, as 0.02, is assumed for core loss-current ratio, h, at a load current corresponding to a definite ratio of a to b, as 2. These values fix G_h for any value of ab, the numbers cited giving:

$$G_h = \frac{0.01\sqrt{(2ab)}}{X(1-ab)}.$$

6. For each value of ab, a circuit is set up with the constants given above, as shown in Fig. 5.4, and calculations of the circuit impedance are made over a range of values of R_2/s from approximately X to ∞.

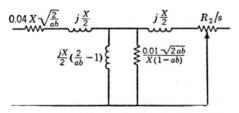

FIG. 5.4. Equivalent Circuit for Calculating Power-Factor Chart.

7. A curve of circuit power factor versus secondary input is plotted for each value of ab, the curves for different values of ab being put on a common scale by putting 10 for the secondary power input for each curve at the load at which $b = a$, or where $Z_0 = X/\sqrt{(ab)}$.

8. On each curve, points corresponding to definite ratios of b to a, as 0.1, 0.2, 0.3, ..., are marked, and points for each value of b/a are connected by other curves.

9. If desired, a correction for normal friction and windage loss may be made by an appropriate shift of the horizontal scale.

The use of the chart may best be illustrated by an example. Suppose no-load tests on a motor have shown that its running-light current is 10 amp, and that its leakage reactance, determined from standstill tests, divided into normal voltage, gives a current of 200 amp; and it is desired to find its power factor at a load current of 30 amp, and its maximum power factor. The value of ab is 10 divided by 200, or 0.05, and the value of b/a at 30 amp input is 0.15 divided by 0.333, or 0.45. The intersection of the two corresponding curves on Fig. 5.3 gives a power factor of 0.902. Following the $ab = 0.05$ curve to its summit gives the maximum power factor as 0.917. The ratio of abscissae at the two points, 6.63 to 10.03, indicates that the torque at maximum power factor is 1.53 times the torque at 30 amp, and this ratio divided by the power-factor ratio indicates that the current at maximum power factor is approximately 45.3 amp. At half the load torque corresponding to 30 amp input, the abscissa is half of 6.63, and the power factor is, therefore, 0.792.

To obtain exact power-factor values, it is necessary to correct the chart values for the errors due to discrepancies between the actual and the chart values of c and h, as indicated above. By referring to Eq. 5.33, and taking out all the terms that involve c, we find that the coefficient of c, or the per cent change in power factor due to 1 per cent change in c, is:

(5.35) $$\left(\frac{\Delta pf}{\Delta c}\right) = 2a(a+b),$$

and this equation gives the correction accurately enough for all practical purposes.

Similarly, the coefficient of h in Eq. 5.33 is:

(5.36) $$\left(\frac{\Delta pf}{\Delta h}\right) = b(a+b).$$

Thus, if the motor used in the example given above had a per unit primary resistance drop at 30 amp of 0.06, and a no-load core loss current of 0.9 amp per phase, the values of c and h at this load, 0.06 and 0.03, respectively, would be in excess of the chart values, and so the true power factor would be higher than that indicated. The chart value of c is 0.04 when $a = ab$, or $0.04 \sqrt{(2b/a)}$ at any other load; hence is 0.038 at 30 amp input, where $b/a = 0.45$. Hence, by Eq. 5.35, the discrepancy in c makes the true power factor greater than the chart value by:

$$(0.060 - 0.038)(2)(0.333)(0.333 + 0.15) = 0.007.$$

Similarly, the chart value of h at 30 amp input is $0.02 \sqrt{(a/2b)}$, or 0.021, and the increase in power factor due to the discrepancy between the motor's and the chart constants is, by Eq. 5.36:

$$(0.030 - 0.021)(0.15)(0.333 + 0.15) = 0.0006.$$

CALCULATIONS IN THE REGION OF MAXIMUM TORQUE   153

The true motor power factor at 30 amp input is, therefore, 0.008 greater than the chart indicates, or is 0.910.

Equation 5.33 can also be used to find the location of maximum power factor. By differentiating Eq. 5.33 with respect to current, and recognizing that the numerical per cent variations of a, b, c, and h for 1 per cent change in current are all the same, but that a and h decrease whereas b and c increase with increase in current, the per cent change in power factor due to 1 per cent change in primary current is found to be:

$$(5.37) \qquad \frac{\Delta \text{pf}}{\frac{\Delta I_1}{I_1}} = (a^2 - b^2) + \frac{1}{2}(a^4 - 2a^3 b + 3ab^3 - b^4)$$

$$- 2ac(a-b) - bh(a-b).$$

Equating this to zero, the point of maximum power factor is found to occur when $b/a = 1 + (ab/4)$ very nearly. The corresponding locus is indicated by the dashed line on Fig. 5.3.

Other relations obtainable from Eq. 5.33 show that the effect on power factor of an increase in frequency at constant line current and voltage is practically the same as the effect of an equal per cent increase in line current at constant voltage and frequency. Also, an increase in voltage with fixed frequency and current has exactly the same effect on power factor as an equal per cent increase in current with voltage and frequency constant, neglecting saturation.

All the results given apply equally well to induction generators and motors, if the signs of the c and h terms are changed wherever they appear. However, the power-factor chart of Fig. 5.3 is drawn for positive values of c and h, so that the corrections to it for the negative values corresponding to generator operation will be relatively large. The power factor so obtained for an induction generator is, of course, leading with respect to the delivered power.

8 CALCULATIONS IN THE REGION OF MAXIMUM TORQUE

Equation 2.23 gives the load resistance at maximum power output of a transformer. In the induction-motor circuit $\frac{R_2(1-s)}{s}$ takes the place of the load impedance $R_L + jX_L$ in the transformer circuit, so that the value of $\frac{R_2(1-s)}{s}$ at maximum motor output is, neglecting the effect of magnetizing current:

$$(5.38) \qquad \frac{R_2(1-s)}{s} = \sqrt{[(R_1 + R_2)^2 + (X_1 + X_2)^2]},$$

whence the maximum output is, from Eq. 2.24:

$$\text{(5.39)} \quad \text{Maximum output} = \frac{qE_0^2}{2\{(R_1+R_2)+\sqrt{[(R_1+R_2)^2+X^2]}\}} \text{ watts,}$$

and it occurs at a value of slip:

$$\text{(5.40)} \quad s \text{ at maximum output} = \frac{R_2}{R_2+\sqrt{[(R_1+R_2)^2+X^2]}}.$$

Maximum torque occurs when the total power input to the secondary (power transferred across the air gap) is a maximum, regardless of its division between secondary copper loss and useful output. It occurs, therefore, when

$$\text{(5.41)} \quad \frac{R_2}{s} = \sqrt{(R_1^2+X^2)}.$$

The maximum torque in synchronous watts is:

$$\text{(5.42)} \quad \text{Maximum torque} = \frac{qE_0^2}{2[R_1+\sqrt{(R_1^2+X^2)}]},$$

occurring at a slip:

$$\text{(5.43)} \quad s \text{ at maximum torque} = \frac{R_2}{\sqrt{(R_1^2+X^2)}} = \frac{R_2}{X}\left(1-\frac{c^2}{2b^2}\right) \text{ approx.}$$

These conventional (transformer) formulas do not allow for the effects of magnetizing current, which lowers the air-gap voltage, E_2, and, therefore, the output, appreciably below the corresponding values for a transformer. To take account of this, we may write, in place of Eq. 5.21:

$$\text{(5.44)} \quad Y_2 = \frac{s}{R_2}\left(\frac{4k^2-j2bk}{4k^2+b^2}\right).$$

Adding Y_M to this, taking the reciprocal, and neglecting terms in h, we find the apparent secondary impedance to be:

$$\text{(5.45)} \quad Z = \frac{R_2}{s}\left\{\left[1-ab+\left(\frac{a^2b^2}{4}-a^2k^2\right)+\ldots\right]\right.$$
$$\left.+j\left[\frac{b}{2k}+\left(ak-\frac{ab^2}{4k}\right)-a^2bk+\ldots\right]\right\}$$

CALCULATIONS IN THE REGION OF MAXIMUM TORQUE

Adding the primary impedance to Z gives the total motor impedance:

(5.46) $\quad Z_0 = (R+R_1)+j(X+X_1)$

$$= \frac{R_2}{s}\left\{\left[1+\left(\frac{c}{k}-ab\right)+\left(\frac{a^2b^2}{4}-a^2k^2\right)+\cdots\right]\right.$$

$$\left.+j\left[\frac{b}{k}+ak-\frac{ab^2}{4k}-a^2bk+\cdots\right]\right\}.$$

The square of the absolute value of Z_0 is found by adding the squares of the real and j components of Eq. 5.46:

(5.47) $\quad |Z_0|^2 = \frac{R_2^2}{s^2}\left[\left(1+\frac{b^2}{k^2}\right)+\left(\frac{2c}{k}-\frac{ab^3}{2k^2}\right)\right.$

$$\left.+\left(\frac{c^2}{k^2}-\frac{2abc}{k}-a^2b^2-a^2k^2-2a^2kc\right)+\cdots\right].$$

From Eq. 5.47 and the identity $R_2 = sk\,|Z_0|$, we find:

(5.48) $\quad (k+c)^2 = 1-b^2+\frac{ab^3}{2}+(2abck+a^2b^2k^2+a^2k^4+2a^2k^3c+\cdots),$

which, when solved by successive approximations, reduces to:

(5.49) $\quad k = \sqrt{(1-b^2)}-c+\frac{ab^3}{4}+\frac{a^2}{2}+abc-\frac{a^2c}{2}-\frac{a^2b^2}{4}+\cdots,$

or, very approximately, to:

(5.50) $\quad k \cong \sqrt{(1-b^2)}-c+\frac{ab^3}{4}+\frac{a^2}{2},$

which corresponds to Eq. 5.27. The torque developed is equal to:

(5.51) $\quad T = qE_2^2 G_2 = \frac{qE_2^2 R_2}{sZ_2^2} = \frac{qI_1^2 R_2 Z^2}{sZ_2^2}$

$$= \frac{qsE_0^2 k^2 Z^2}{R_2 Z_2^2} = \frac{qE_0^2 bkZ^2}{XZ_2^2},$$

which reduces to:

(5.52) $\quad T = \frac{qbE_0^2}{X}\left[\sqrt{(1-b^2)}-\left(\frac{a^2}{2}+ab+c\right)+\left(\frac{3ab^3}{4}+\frac{3a^2b^2}{2}+\frac{a^3b}{2}\right.\right.$

$$\left.\left.+2abc+2a^2c\right)+\cdots\right].$$

156   POLYPHASE INDUCTION MOTOR PERFORMANCE CALCULATIONS

By differentiating Eq. 5.52 with respect to current (remembering that the rates of per cent change of b and c are equal to the rate of per cent change in current, and the rates of per cent change of a and h are numerically the same but are negative), equating to zero, and solving by successive approximations, we find the approximate expression for b at maximum torque:

$$(5.53) \qquad b = 0.707\left(1 + \frac{ab}{32} + \frac{5a^2b^2}{8} - \frac{c\sqrt{2}}{4b} + \frac{ac\sqrt{2}}{2} + \cdots\right).$$

Substituting Eq. 5.53 in 5.52, the value of the maximum torque itself is found to be:

$$(5.54) \qquad T_{\max} = \frac{qE_0^2}{2(R_1 + X)}\left(1 - \frac{5\sqrt{2}\,ab}{8} + \frac{a^2b^2\sqrt{2}}{2} + \cdots\right).$$

This equation is the same as Eq. 5.42, except for the factor

$$\left(1 - \frac{5\sqrt{2}\,ab}{8} + \frac{a^2b^2\sqrt{2}}{2} + \cdots\right),$$

which, therefore, measures the reduction of the maximum torque due to the magnetizing current. By reference to the power-factor chart, Fig. 5.3, it will be seen that normal motors with full-load power factors of 0.90, 0.80, and 0.70 (full load generally occurs when $b = 0.5a$, approximately) have ab values of 0.053, 0.11, and 0.17, respectively, and consequently have maximum torques 5, 9, and 13 per cent below the values indicated by the transformer formula, Eq. 5.42, which neglects the magnetizing current.

From Eqs. 5.49 and 5.53, the value of k at maximum torque is:

$$(5.55) \qquad k = 0.707\left[1 + \frac{(4\sqrt{2}-1)ab}{32} + \frac{(6\sqrt{2}-5)a^2b^2}{8} + \cdots\right],$$

and from this the slip at maximum torque is found to be:

$$s = \frac{IR_2}{kE_0} = \frac{bR_2}{kX} = \frac{R_2}{X}\left[1 - \frac{(2\sqrt{2}-1)ab}{32} + \frac{(5-3\sqrt{2})a^2b^2}{4} + \cdots\right]$$

or, approximately:

$$(5.56) \qquad s = \frac{R_2}{X}(1 - 0.06ab)$$

(compare with Eq. 5.43).

9 Dynamic Braking

As mentioned in Sect. 4.3, supplying direct current to the stator of an induction motor makes synchronism occur at zero speed, and converts the

DYNAMIC BRAKING

motor into a brake that opposes rotation in either direction, but has zero torque at standstill (Fig. 4.3).

Under these conditions, the stator field does not rotate, the secondary induced currents produce no counter voltage, and the stator current remains constant at all speeds.

The braking torque-speed curve is very peaked if the rotor has a fixed resistance, as explained in Sect. 4.3. With a Saturistor in the secondary circuit, the torque-speed curve can be flattened out, as shown in Fig. 5.5, enabling much smoother braking to be obtained without excessive peak torque. In the figure, the braking torque curves with the rotor short-circuited and with a Saturistor are nearly the same for 20 amperes stator current, since this is below the pickup current for the Saturistor. At 50 amperes, however, the Saturistor gives more than twice as much torque at full speed with the same peak torque.

The performance of a motor during braking can be calculated by assuming the primary current to be fixed, instead of the primary voltage, as in Table 5.5. For example, assume that a direct current equivalent to twice rated alternating current is supplied to the stator of the same motor as before, and that the slip (speed) is 0.045. From the next to last column of Table 5.3, we have:

TABLE 5.5

DYNAMIC BRAKING PERFORMANCE

	Without Saturation	With Saturation
s	0.045	0.045
$Z_2 = R_2 + jX_2$	$1.00 + j0.075$	$1.00 + j0.075$
$Y_2 = G_2 - jB_2$	$0.995 - j0.075$	$0.995 - j0.075$
$Y_M = G_h - jB_M$	$0.040 - j0.333$	$0.040 - j0.85$
$Y = G - jB$	$1.035 - j0.408$	$1.035 - j0.925$
$Z = R + jX$	$0.836 + j0.329$	$0.535 + j0.48$
$|Z|$	0.899	0.72
I_1 (impressed)	2.00	2.00
E	1.80	1.44
$T = \dfrac{I_1^2 R G_2}{G}$	3.21	2.06

The first column of this table is calculated with the same magnetizing impedance as on alternating current, but the result shows an induced voltage, E, or per unit flux, equal to 180 per cent of rated voltage, and a per unit torque of 3.21. In a normal motor, magnetic saturation at any such density as this will greatly increase the magnetizing current, making B_M increase from 0.333 to 0.85 (say). With this corrected value of B_M, E is reduced to 1.44, and the torque to 2.06. The torque curve is calculated in

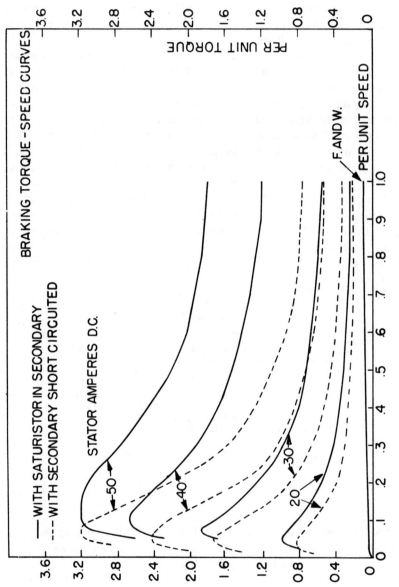

Fig. 5.5. Braking Torque-Speed Curves for Motor with Saturistor.

this manner, adjusting B_M at each value of slip to correspond to the magnetizing current required to produce the calculated value of E (Fig. 4.3).

Following the same procedure as in Sect. 2.5, the watts output of a transformer at constant primary current, neglecting core loss, is the real part of:

$$\frac{I_1^2(R_2+jX_2)jX_M}{R_2+jX_2+jX_M}, \text{ or is } \frac{I_1^2 R_2 X_M^2}{R_2^2+(X_2+X_M)^2}.$$

This is a maximum when $R_2 = X_2 + X_M$; or, putting R_2/s for R_2, the motor slip at maximum braking torque is:

$$s = \frac{R_2}{X_2+X_M}.$$

Comparing this with Eq. 5.43, it is evident that the braking torque (constant-current) curve peaks at a very much smaller slip than the motor torque (constant-voltage) curve. The reduction of X_M due to saturation moves the peak torque to progressively higher values of slip as the stator direct current is increased.

10 Summary of Formulas

For convenience, the foregoing formulas may be expressed in terms of the apparent value of leakage reactance, X_t, found in the locked-rotor test (Eq. 4.8). They are summarized here:

The true value of equivalent circuit reactance is:

(5.14) $$X = \left(1+\frac{ab}{4}\right)X_t \text{ ohms.}$$

The standstill current is, from Eq. 5.13:

(5.57) $$I_s = \frac{E_0}{\sqrt{\{[R_1+(1-ab)R_2]^2+X_t^2\}}} \text{ amp.}$$

If E, R, and X are per unit values, this equation gives the per unit value of I_s, or the ratio of current at standstill to rated load current.

The standstill torque is, from Eqs. 4.10 and 5.13:

(5.58) $$T_s = \frac{9.55 K K_E q(1-ab)I_s^2 R_2}{N_s} \text{ newton-meters*},$$

where K is an empirical constant of the order of 0.9, which allows for non-fundamental secondary losses, and K_E is a factor greater than one to allow for the "deep-bar effect" in rotor conductors.

*The constant becomes 7.04 to give foot-pounds.

160   POLYPHASE INDUCTION MOTOR PERFORMANCE CALCULATIONS

The relation between K_E, R_2, and the apparent secondary resistance, R_t, determined by impedance test, is:

$$R_t = K_E(1-ab)R_2.$$

The maximum torque is, from Eqs. 5.42 and 5.54:

(5.59) $$T_m = \frac{q(1-1.1ab)E_0^2}{2[R_1+\sqrt{(R_1^2+X_t^2)}]} \text{ synchronous watts.}$$

Dividing this equation by the number of phases, q, gives the motor torque in watts per phase. If E, R, and X in Eq. 5.59 are per unit values, this equation (without the factor q) gives the per unit maximum torque directly.

The slip at maximum torque is, from Eqs. 5.43, 5.56, and 5.14:

(5.60) $$S_{mT} = \frac{R_2(1-0.3ab)}{\sqrt{(R_1^2+X_t^2)}} \text{ numeric.}$$

The maximum output is:

(5.61) $$W_m = \frac{q(1-1.1ab)E_0^2}{2\{R_1+R_2+\sqrt{[(R_1+R_2)^2+X_t^2]}\}} \text{ synchronous watts.}$$

The same comments apply as under Eq. 5.59.

The ratio of maximum torque in synchronous watts to volt-amperes at locked rotor, obtained by dividing T_m from Eq. 5.59 by qE_0I_s from Eq. 5.57, is:

(5.62) $$\frac{T_m}{I_s} = \frac{(1-1.1ab)}{2}\left[1-\frac{R_1}{X_t}+\frac{R_2}{2X_t^2}(R_2+2R_1)\right] < 0.5 \text{ numeric,}$$

approximately, which is always less than 0.5, as pointed out in Sect. 4.5. Equation 5.62 is also true in per unit terms, if rated input volt-amperes are taken as unit power. If locked-rotor current is expressed in per unit of rated load *input* current, and torque expressed in per unit of rated load *output*, the apparent efficiency enters as a factor in Eq. 5.62, and on this basis the *per unit* maximum torque is closely equal to, or may exceed, 0.5 of the per unit locked-rotor current.

The power factor is given by Fig. 5.3, or, roughly:

(5.34) $$\text{pf} = 1 - \frac{(a+b)^2}{2} + 3a^2b^2.$$

Bibliography

Reference to Section

1. "Induction Motor Performance Calculations," P. L. Alger, *A.I.E.E. Trans.*, Vol. 49, July 1930, pp. 1055–66.
2. "Induction Machines," P. L. Alger, C. W. Falls, and A. F. Lukens, Section 7, *Standard Handbook for Electrical Engineers*, Ninth Edition, McGraw-Hill, 1957, pp. 704–48.
3. "Polyphase Induction Motors," W. J. Branson, *A.I.E.E. Trans.*, Vol. 49, Jan. 1930, pp. 319–28.
4. "Stopping Time and Energy Loss of A-C Motors with D-C Braking," O. I. Butler, *A.I.E.E. Trans.*, Vol. 76, Part III, 1957, pp. 285–90.
5. "The Limitations of Induction Generators in Constant Frequency Aircraft Systems." E. Erdelyi, E. E. Kolatorowicz, and W. R. Miller, *A.I.E.E. Trans.*, Vol. 77, 1958, pp. 348-51.
6. *Electric Machinery*, A. E. Fitzgerald and Charles Kingsley, Jr., Second Edition, McGraw-Hill, 1961.
7. "The Squirrel-Cage Induction Generator," H. M. Hobart and E. Knowlton, *A.I.E.E. Trans.*, Vol. 31, Part II, 1912, pp. 1721-47.   6
8. "Electric Braking of Induction Motors," H. C. Specht, *A.I.E.E. Trans.*, Vol. 31, Part I, 1912, pp. 627-40.   9
9. "Alternate Current Machinery—Induction Alternators," W. Stanley and G. Faccioli, *A.I.E.E. Trans.*, Vol. 24, 1905, pp. 851-72.   6
10. "The Alternating Current Induction Motor," C. P. Steinmetz, *A.I.E.E. Trans.*, Vol. 14, 1897, pp. 185-217.
11. "Performance Calculations on Induction Motors," C. G. Veinott, *A.I.E.E. Trans.*, Vol. 51, Sept. 1932, pp. 743-54.

CONTENTS—CHAPTER 6

Design of Induction Machines

		PAGE
1	The Designing Process	165
2	The Frame Structure	166
3	Standard Frame Dimensions	168
4	Ventilation and Temperature Rise	171
5	Shafts and Bearings	172
6	Insulation	174
7	The Magnetic Core	177
8	The Windings	179
9	Magnetizing Current Calculation	181
10	Loss Calculations	188

6

DESIGN OF INDUCTION MACHINES

1 THE DESIGNING PROCESS

Under modern conditions of high motor production and intense competition, the designer must plan his line of machines as a whole, selecting basic features and preferred sizes that will simplify manufacture and require the least variety of special tools. The National Electrical Manufacturers' Association Standards of mounting dimensions and the American Standard C50 for motor performance provide a framework (Sect. 14.2) that is very helpful in defining the performance requirements and the varieties of motor needed. Broadly, the purposes of these Standards, in the public interest, are to eliminate misunderstandings between the manufacturer and the user, and to assist the user in securing the correct product for his particular need. More exactly, N.E.M.A. Standards define products, processes, or procedures with reference to such things as nomenclature, construction, dimensions, safety, performance, and tests.

The designer's problem, in the light of these standards, is to select materials, proportions, manufacturing methods, styling, and margins of liberality that will give his line a distinction of its own. Skill and judgment are required to do this, and at the same time to satisfy the full range of application requirements and to secure satisfactory costs.

The six major elements the designer must consider are the type of frame, the ventilation, the bearing and shaft proportions, the insulation system, the magnetic core dimensions, and the winding. These are put in this order advisedly, since the general principles of design must be determined first, and held as uniform as possible throughout, to save cost and facilitate use of common parts, distinctive styling, and quick shipments. Of course, with increasing horsepower ratings and corresponding increases in core dimensions, dimensional theory (Sect. 2.13) dictates changes in the proportions and in various design features. The designer must also decide where such "breaks" in the line should occur, and how far it is advisable to go before another break is made.

In American practice, two size ranges of motors are given separate consideration: fractional-horsepower motors, built in frame sizes smaller than 1 hp at 1,800 rpm; and integral-horsepower motors, which include all larger sizes. Each of these two size ranges is divided into three types, designated as general-purpose, definite-purpose, and special-purpose, depending on their adaptability to different uses.

2 The Frame Structure

The mechanical structures, or frames and end shields, which support and enclose the core and windings, serve three distinct purposes. First, they transmit the torque to the motor supports, and so are designed to withstand twisting forces and shocks. Second, they serve as a ventilating housing, or means of guiding the cooling medium into effective channels.

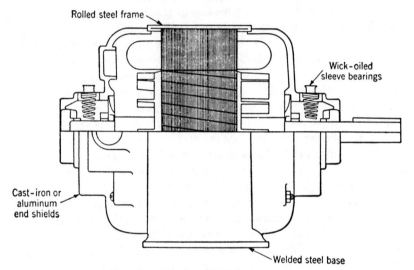

Fig. 6.1. Fractional-Horsepower Motor.

Third, they shield live and moving motor parts from human contact and from injury caused by falling objects or weather exposure.

A great variety of designs is employed to meet these requirements, and to adapt machines to particular service conditions. The two basic types are the drip-proof frames with open ventilation, and the totally enclosed fan-cooled machines. Two other frame types, the open and the splash-proof designs, are now used in special circumstances only. The skill of designers in making frames that thoroughly protect the motor windings from injury, with free ventilation comparable with that of the old wide-open frames, has contributed greatly to the pleasing appearance, greater reliability, and more versatile performance characteristic of modern motors. Figure 6.1 shows a typical fractional-horsepower motor with rolled steel shell, Figs. 1.1 and 6.2 drip-proof motors with cast-iron frames, and Fig. 6.3 a larger size with fabricated steel frame.

THE FRAME STRUCTURE   167

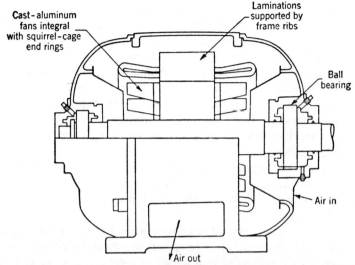

Fig. 6.2. Drip-Proof Motor with Cast-Iron Frame and Double End Ventilation.

The guiding objectives of frame designs are to:

1. Make the standard drip-proof motor frame with open ventilation in such a way that it costs no more to produce, and gives no higher temperature rise, than a wide-open frame. If these results are achieved, the wide-open frame serves no useful purpose, and becomes obsolete.

2. Make the splash-proof motor frame with as much additional protection against entry of water or spray as can be secured without incurring more than 10° additional temperature rise (by resistance) at full load above

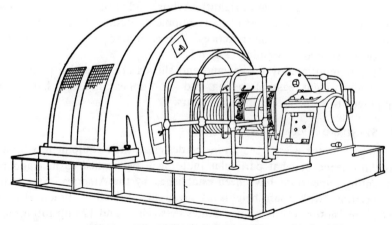

Fig. 6.3. Large Mill-Type Wound-Rotor Induction Motor with Steel-Plate Frame.

that of the protected-type frame, and with a minimum of extra cost. If this can be done at a cost warranted by improved reliability in service, and below the cost of the totally enclosed fan-cooled motor, the splash-proof motor has an economic place; otherwise no.

3. Make the totally enclosed fan-cooled motor frame to exclude completely dirt as well as moisture, with ventilation to give not more than 10° additional temperature rise above that of the drip-proof frame, and with a minimum cost. If the mounting dimensions and cost of this design could be made the same as those of the splash-proof motor, it would supersede the latter.

Totally enclosed frames without fan cooling have a limited field of use in small motor sizes, and explosion-proof designs are needed for some applications, but so far as possible such types are adapted from other standard frame parts.

Cast iron as a frame material has the advantages of greater resistance to corrosion, and easier adaptation to special shapes and pleasing contours. It also permits strengthening vital points by local increases in wall thickness, as at corners, without adding materially to the total weight or cost. Welded steel frames have the advantages of lightness and ready modification, or repair in the field. The wide use of both types in competitive designs is an illustration of the continuing rivalry between the arts of casting and fabricating metals. The fabricated frame is most easily made in the form of a rolled shell, with feet welded on (Fig. 6.1), and is well adapted to axial ventilation. The cast frame with integral feet is readily provided with ventilating openings in the lower half (Fig. 6.2), and is, therefore, especially suitable for double end, or radial, ventilation. Which design is preferable depends on many factors, such as the importance of corrosion resistance, weight, rigidity, styling, and temperature limits; and also on the economic factors of tool equipment, production quantities, and associated manufacturing practices. Large motor frames are almost always fabricated by welding (Fig. 6.3).

Cast-aluminum-alloy frames and end shields are attractive because of their light weight, ease of casting, and good appearance, and have been employed to some extent in recent years.

3 Standard Frame Dimensions

To give motor users freedom in the design of their equipment, without foreknowledge of what particular motor will be applied to it, standard mounting dimensions have been established by the American Standards Association, with shaft heights as given in Table 6.1. Practically all induction motors in the United States between $\frac{1}{8}$- and 125-hp ratings are built in one or another of these standard "frame sizes".

STANDARD FRAME DIMENSIONS

TABLE 6.1

POLYPHASE INDUCTION MOTORS IN STANDARD N.E.M.A. MOTOR FRAMES

	Shaft Height, Inches from Base to Center	Diameter of Shaft Extension, Inches	Usual Horsepower Rating (1969) For 60 Cycles, Class B Insulation, 80° C Rise Open Type, 1.15 Service Factor			
			\ \ \ \ \ \ \ \ \ \ \ Synchronous Speed, RPM			
			3600	1800	1200	900
143T	$3\frac{1}{2}$	$\frac{7}{8}$	$1\frac{1}{2}$	1	$\frac{3}{4}$	$\frac{1}{2}$
145T	$3\frac{1}{2}$	$\frac{7}{8}$	2, 3	$1\frac{1}{2}$, 2	1	$\frac{3}{4}$
182T, 184T	$4\frac{1}{2}$	$1\frac{1}{8}$	5, $7\frac{1}{2}$	3, 5	$1\frac{1}{2}$, 2	1, $1\frac{1}{2}$
213T, 215T	$5\frac{1}{4}$	$1\frac{3}{8}$	10, 15	$7\frac{1}{2}$, 10	3, 5	2, 3
254T, 256T	$6\frac{1}{4}$	$1\frac{5}{8}$	20, 25	15, 20	$7\frac{1}{2}$, 10	5, $7\frac{1}{2}$
284T, 286T	7	$1\frac{7}{8}$	30, 40	25, 30	15, 20	10, 15
324T, 326T	8	$2\frac{1}{8}$	50, 60	40, 50	25, 30	20, 25
364T, 365T	9	$2\frac{3}{8}$	75, 100	60, 75	40, 50	30, 40
404T, 405T	10	$2\frac{7}{8}$	125, 150	100, 125	60, 75	50, 60
444T, 445T	11	$3\frac{3}{8}$	200, 250*	150, 200	100, 125	75, 100

* This rating has 1.0 service factor.

Ratings have not been formally assigned to frames in the fractional-horsepower sizes. Ratings in the integral-horsepower frames have been increased from time to time over past years (Table 1.1).

The frame number for a fractional-horsepower motor is equal to 16 times the height from base to center of shaft. Only one frame number for each diameter of these motors is required, since the same base and foot dimensions are used for all lengths of frame in the same diameter. For an integral-horsepower motor, the first two digits of the frame number are equal to 4 times the shaft height, while the third digit indicates the length of frame, or axial distance between the motor feet.

The ratio of successive shaft heights, and, therefore, of frame diameters, averages about 1.14, the ratio being a little greater for the small sizes, and a little smaller for the larger sizes. Two frame lengths are usually recognized for each height. Since the output obtainable from a homologous series of frames varies roughly as the 4.5 power of the diameter (Sect. 3.4), the ratio of successive horsepower ratings obtainable in the standard dimensions of Table 6.1 should be about $1.14^{2.25}$, or 1.33. This agrees well with the average step between standard horsepower ratings, also shown in Table 6.1. There is, therefore, a good alignment between standard frames and ratings, assuming that the space available within the frame dimension is utilized to the same degree in all sizes.

The frame used for a four-pole, 1,750-rpm motor of a given rating is normally used also for the next larger rating of two-pole motors, and for successively smaller ratings of six-pole and lower-speed motors. Each motor designer selects his own values for the diameter and length of the magnetic core in a given frame, as only the mounting dimensions are standardized. A good deal of judgment is called for to select the slot sizes, core heights, and other elements that will fit all these requirements with the least number of parts and the least excess of material over that required to meet the specifications.

As indicated in Table 1,1, steady technical progress in better magnetic steels, better insulation, and improved design and manufacturing techniques has made it possible to increase the rating of each frame size one step about every ten years. The industry standards are, therefore, in a continual state of development. Changes are not made at precisely the same time by all manufacturers. For this reason, and because special motors, with higher temperature rise or special torque requirements, are not fully standardized, the horsepower and frame-size relations of Table 6.1 are not "frozen".

The problem of maintaining free competition among manufacturers, and steady technical progress in the more efficient use of material, and of yet preserving interchangeability between all makes of motors, poses some very interesting questions of standards and engineering policy. The logic behind the accepted plan of periodically decreasing the motor dimensions, one step at a time, as technical progress permits, is indicated by a consideration of two possible alternatives.

If the horsepower rating of each frame size were permanently "frozen", the manufacturers could take advantage of technical progress either by giving greater margins of performance for the same rating, keeping the same amount of active material; or by shrinking the active material and saving in cost, keeping the same performance margins.

The first alternative would result in continually higher maximum outputs, higher starting currents, and lower temperature rises for the same horsepower rating. Economics would then force the users to employ the motors at continually higher loads with respect to the rating—i.e., this would result in more and more overloading. In effect, a motor rated 1 hp would then be used as a 1.5-hp motor, and so on. All overcurrent protective devices and supply circuits would then have to be adjusted to suit. If these devices were given new ratings to describe their performance truthfully, they would not match the motor nameplates. If they were simply overloaded to match the motor, all ratings would in the end become indefinite. The confusion this would lead to can be imagined. It is as if every grocery store decided to change its weighing scales now and then, to give more weight to the pound, and the competing stores changed theirs to give still a little more.

The second alternative leads to motors with continually smaller "insides", leaving more and more vacant space inside the mounting dimensions. The economic importance of space saving to the user makes this unattractive in the long run. A third alternative, of having no standard frames, was given up long ago, because of the great value to users of having known and interchangeable mounting dimensions.

The use of the design letters A to F to describe standard electrical characteristics makes it possible to secure interchangeability in performance as well as in mechanical dimensions.

4 Ventilation and Temperature Rise

The temperature of a motor must be limited because (a) organic insulations and lubricants deteriorate chemically (as by oxidation) at a rate that doubles for each 8 to 12 C increase in temperature; (b) the resistivity of copper increases linearly with temperature; and (c) the differing thermal expansions of iron, copper, and other materials give rise to mechanical stresses and displacements that cause progressive deterioration. On the basis of (a), motor insulations are defined in four major classes: A, B, F, and H, for which limiting hot-spot temperatures of 105 C, 130 C, 155 C, and 180 C in continuous service are recognized.† On the basis of (b), (Eq. 1.10), the resistance of copper increases in the ratios of 1 to 1.24 to 1.33 to 1.51, as the temperature is increased from 40 to 105 to 130 to 180 C. Thus, a motor designed for the limiting temperature of Class H insulation is handicapped by 22 per cent greater copper resistivity than one similarly designed for Class A insulation.

American Standards assume an ambient temperature of not over 40 C, and allow a temperature rise by resistance of 80 C at rated load, or 90 C at the service factor rating, for a general-purpose motor with Class B insulation. 105 C rise is allowed for Class F insulation, and 125 C for Class H, that are used in some special motors. The change from Class A insulation, with 50 C rise, that was standard for many years, to Class B was made about 1966, and this accounts for the marked jump in rating of a given frame that occurred at that time, as shown in Table 1.1.

Good design practice requires that the rise in temperature of the outgoing over the incoming ventilating air be of the order of one-quarter the total rise or less. Since 1 kw of loss raises the temperature of 100 cu ft of air 18 C per min, a Class B insulated motor with 90 per cent full-load efficiency requires an air supply of about 10 cu ft per min per rated hp, for normal ventilation. To make the air effective in cooling, the designer must provide fans, guide vanes, and air passages over the coils and core that will

†I.E.E.E. Standard No. 1.

ensure as uniform and as high a velocity of air flow over all the exposed surfaces as practicable. Skill is required to accomplish this without excessive windage loss or audible noise, and with a low-cost reliable construction. The problem is especially difficult in the larger totally enclosed fan-cooled motors (Fig. 6.4), where two independent air flows must be established, one inside the frame to carry the heat from the active material to the frame and end shield surfaces, and another outside the

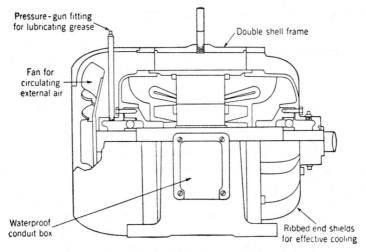

FIG. 6.4. Totally Enclosed Fan-Cooled Motor.

frame to hold down the temperatures of these surfaces. Water cooling coils are sometimes employed to save space and reduce the "inside ambient" temperature, and hydrogen, nitrogen, or carbon dioxide in place of air cooling may be used on large machines with substantially gas-tight frames and bearing seals.

5 SHAFTS AND BEARINGS

Relatively large-diameter shafts and close spacing of bearings are desirable in induction motors, because the small air gaps that must be used lead to large and rapidly varying radial magnetic forces between rotor and stator. With some combinations of rotor and stator slots, and especially when the air gap is slightly eccentric, these forces are unsymmetrical and pulsate at tooth frequency, so that resonant vibration of the shaft may occur during the starting period, unless the critical frequency of the shaft is safely above the tooth frequency. The diameter of the shaft extension must be large enough to limit the shaft bending caused by belt pull or overhung weight. Excessive shaft deflection will cause wear at the outboard end of a sleeve bearing, and may lead to fatigue failure of the shaft itself as a result of the flexure repeated each revolution.

Both ball and sleeve bearings are widely used, the intense rivalry between them providing an urge for continual improvement of both. The ball bearing has an inherent capacity for thrust loads, allowing mounting in any position, and an accurately centered shaft position, which make it preferable for many applications (Fig. 6.5). The sleeve bearing is better adapted for

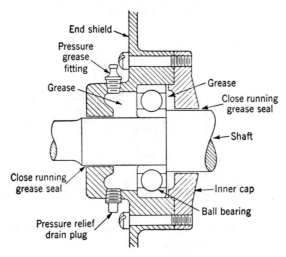

FIG. 6.5. Ball-Bearing Housing.

high values of load times speed, tends to be quieter, and usually has a longer life (Fig. 6.6).

Grease lubrication is generally employed for ball bearings, because grease is much easier to retain in place without leakage than oil. The soap content of the grease serves as a convenient oil retainer. With modern greases containing an antioxidant, tightly housed bearings, and good operating conditions, relubrication is required only at long intervals, sometimes several years, depending on size and type of service. Modern sleeve-bearing designs have careful provisions for preventing oil leakage and for by-passing air flow that might carry away oil vapor, so that they also can operate for long periods without reoiling. Grooving is provided in sleeve-bearing linings to assure a supply of oil being drawn into the loaded area of the bearing at the start of rotation in either direction, and oil rings, or disks, or waste packings are provided to insure a continuous flow at speed. On very large, or high-speed, shafts, oil pumps are required

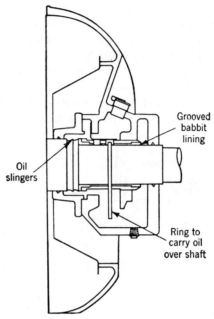

Fig. 6.6. Sleeve Bearing.

to assure a sufficient flow of oil to carry away the heat of the bearing losses without excessive temperature rise.

6 INSULATION

The problems in designing the insulation system are fourfold: to establish desired levels of insulation to ground, and between turns and phases of the winding; to select reliable, economical, and readily applicable insulating materials; and to apply these so that they will be uniform in quality and will have long service life.

American Standards require that the electrical insulation withstand a high-potential test to ground of $2E$ plus 1,000 volts, for a period of one minute, where E is the rated circuit voltage. Somewhat lower requirements are made for very small and low-voltage machines. It is also customary to give motors a high-frequency, or surge-voltage, test of some kind, to prove the adequacy of the insulation between turns. Usually this is done by applying $2E$ plus 1,000 volts across the motor terminals, at a frequency high enough (1,000 cycles more or less) to hold the test current down to a low value.

When voltage is applied instantaneously, as in switching, a wave of current enters the winding, travels through to the opposite terminal, is reflected, and returns to its starting point (Sect. 2.14). The steady-state current is finally built up from successive reflections of this sort. Of course, the impressed voltage has no way of knowing in advance whether the winding is open or closed, so that this building-up process must occur whenever a switch is closed. In a motor, which may have a total winding length between terminals of 500 ft or so, the whole process is over in a few microseconds, as the wave velocity through the winding is of the order of 200 to 500 ft per microsecond. During the process, however, the voltage is not uniformly distributed over the windings, the turns adjacent to the line terminals having to bear momentarily 10 to 100 or more times their normal share.

For this reason, there is an advantage in using a surge-voltage test, which inherently concentrates the voltage on the end turns, rather than a simple high-frequency test. Form-wound preinsulated coils are generally given a separate turn-insulation test before assembly, permitting the application of a much higher voltage between turns of the coil than is possible after the winding is fully assembled (without exceeding the line-to-ground test voltage).

Synthetic resins, such as Formvar (Class A) or Alkanex (Class B), are generally used to cover individual conductors of coil windings. The film has a thickness of only a few mils, and is easily able to withstand 500 volts or more, even after considerable abuse by pounding or stretching. Such factors as roughness of the copper wire, eccentricity of the film with the wire, mechanical tension in the film leading to cracking under suddenly applied heat, crazing, and solvent attack must all be considered, and margins allowed for them in the finished winding. Cotton, silk, paper, or synthetic film wrappings of somewhat greater thickness than the resin films are used in special cases, particularly for larger conductors with Class A insulation. For Class B insulation, synthetic films such as Alkanex are used for small wires, and asbestos or glass fibers for larger sizes, with thorough varnish impregnation.

In the early days insulation materials were assigned to temperature classes by name only, and it was assumed that the service life would be adequate with the designated temperatures. Now, however, a great many kinds of synthetic insulations are available, and the number is growing every day. The only way to determine the temperature life of these new materials is by test. Therefore, the A.I.E.E. has developed a life test procedure, consisting in repeated exposures to high temperatures, moisture, and voltage stress, and it is expected that all new insulating materials will be classified by these procedures before being accepted for use. Thus, the formal definition of a Class F insulation system is one that has been shown by experience or tests to have a life at 155 C equivalent to that of a system composed of Class A insulation materials at 105 C.

For ground insulation, multiple layers of tape, made of paper, varnished cloth and various combinations of these with synthetic resins, such as Mylar, are employed in Class A windings. Mica tape, with either paper, synthetic film, or glass-fiber backing, is employed for Class B windings; and mica with silicone varnish for Class H windings.

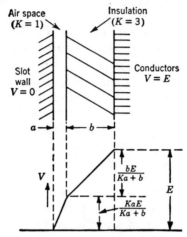

FIG. 6.7. Effect of Dielectric Constant, K, on Voltage Distribution between Conductors and Slot Wall.

The voltage stress in air at which breakdown, with corona, first appears is about 80 volts per mil (2,000 volts/mm). Corona is the result of ionization of the air occurring near the peak of the voltage wave, and is manifested by visible light and ozone formation. Since the nascent oxygen produced by corona in air rapidly deteriorates organic materials, including varnishes, it is necessary to design the insulation so that corona will not occur at normal voltages, or else to provide a compact mica structure to withstand it. If the dielectric constant of the insulation is K, the dielectric stress in any air film that is present between an insulated conductor and the slot wall will be K times the stress in the insulation itself (Fig. 6.7). As K is about 3 for usual insulating materials, freedom from corona requires that the average voltage stress across the insulation thickness should not exceed about 25 volts per mil in service. Another factor to be allowed for is the surface conductivity, or "creepage", resulting from dirt and moisture, which may carry the ground potential well out on the coil ends. To protect against this, and to minimize vibration of small coils, the windings are impregnated with insulating varnish.

7 THE MAGNETIC CORE

The magnetic flux in both stator and rotor is carried by alloy-steel laminations, usually containing 1 to 3 per cent silicon. The lamination thickness may be 0.014 in. for large machines, and 0.019 or 0.025 for small motors, the choice representing a compromise between the cost and the eddy-current loss, which latter increases as the square of the thickness. Economy requires use of the highest flux density compatible with moderate core losses and magnetizing ampere turns, making a steel of high magnetic quality desirable. The laminations are, therefore, annealed at high temperature (about 800 C, preferably after the slots are punched), and are insulated with core plate enamel. In the smaller-sized motors and with controlled atmosphere anneal, however, the oxide scale on the laminations may provide sufficient insulation.

Even though the fundamental flux in the rotor alternates at the very low frequency of slip during normal operation, it is nevertheless nearly as important to have high-quality, thin laminations in the rotor as the stator. This is true because, with the small air gap of an induction motor, the mmf harmonics due to the phase belts and slot openings induce high-frequency voltages in the squirrel cage bars, and the consequent I^2R losses may be large. If thick laminations are used, the eddy currents in the steel oppose the mmf and so materially reduce the leakage reactance of the bars at the harmonic frequencies, so allowing greater currents to flow, with consequent higher stray losses under load (Sect. 9.4).

The actual dimensions of the core and teeth are selected to carry the required amount of magnetic flux, found by Eq. 3.15, with a maximum density in the core of the order of 1.4 webers per square meter, and a somewhat higher peak value in the teeth. When the number of poles is greater than four, it is often desirable to make the core section back of the teeth deeper than required by the flux-density limit, in order to provide greater stiffness, with freedom from radial vibrations and noise.

Motors smaller than 2 or 3 ft in diameter employ one-piece stator core laminations, the center circles being used for the rotor. For larger motors, to avoid wasting the steel from the center of the rotor and from the outside corners of the stator, the cores are made up of segments assembled in ring form with butt and lap joints, as in transformers. These are held together by axial key bars fitting into dovetailed slots in the outer rim of the core, or they may be welded at the back of the core. In this case, however, better insulation is required between laminations, as the core voltage acts across one instead of two inter-laminar spaces. The peripheral length of one segment, usually between 1 and 2 ft, is chosen to give the most economical balance between the cost of dies, the cost of assembly, and the amount of scrap left over in cutting the laminations from steel strips.

It is desirable to choose the total number of segments in such a way as to provide an equal number of core joints in the core-flux paths of alternating poles. For, if the flux entering the stator core from every north pole encounters a core joint if it turns clockwise, and no joint if it turns counter-clockwise, the different reluctances will give rise to a net difference between the core fluxes in the two directions of flow, and this net diffcrence will link the shaft. This net flux will create an alternating voltage between the two ends of the shaft, giving rise to shaft currents, with consequent risk of

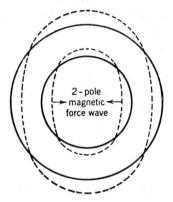

FIG. 6.8. Four-Node Vibration of Stator Core.

damage to the bearings, unless the bearings are insulated from the end shields or pedestals.

The assembled core may be considered a ring of steel, loaded with the distributed weight of the teeth and coils. It has several frequencies of radial vibration, corresponding to two, four, and more nodes (Fig. 6.8 and Sect. 10.5). As the varying magnetic forces impressed on it also have many frequencies, the designer should take care not to allow these to coincide (Sect. 10.2). Large fabricated frames with considerable areas of unsupported side plates also have many frequencies of resonant vibration that may be excited if the high-frequency magnetic forces are of appreciable magnitude. These considerations limit the choice of slot numbers to a narrow range.

Partly closed slots are used in the smaller motors because they increase the effective gap area, giving a lower magnetizing current for a given size of motor, and allowing the use of fewer turns in the winding, with lower resistance. The width of opening is chosen to permit easy assembly of the coils, and to avoid magnetic saturation of the tooth tips at high currents. Larger motors, wound for voltages greater than 600, are normally made with open slots to permit the use of form-wound coils insulated before assembly.

Half-closed slots, with formed coils in two halves side by side, are sometimes advantageous in the intermediate sizes. Tunnel windings, with the wires threaded through completely closed slots, have been used, but at a considerable sacrifice of insulation space factor and cost of assembly. These windings also increase the reactance under running conditions. Another undesirable effect is magnetic saturation of the slot bridge at high currents, which reduces the reactance under starting conditions, giving a high ratio of starting current to maximum output.

8 The Windings

Armature windings of induction machines are designed to give low-leakage reactance and as nearly sinusoidal a current distribution around the periphery as possible. Otherwise, with the relatively small length of air gap between stator and rotor, the extra leakage fluxes produced would cause reduced output and high losses. In consequence, two-layer fractional-pitch windings are normally used, with a relatively large integral number of slots per pole, usually nine or twelve. It is also usual to connect the windings of large motors with two or more circuits in parallel in each phase. The same voltage being impressed on each circuit insures equality of the magnetic fluxes linking them, preserving a balanced magnetic pull, despite inequalities of air gap.

The number of turns in series in the primary winding is fixed by the required voltage, and by the desired value of magnetic flux per pole, in accordance with Eq. 1.21. The conductor size is chosen to fit in the desired slot, and large enough to carry the required load current, usually with a current density at full load of the order of 500 amp per sq cm, more or less (in copper). The permissible current density is limited by the requirement that, if the motor is stalled at standstill, the winding temperature must not exceed a permissible value before the overload relays trip the motor off the line (usually after about 15 seconds). The radial depth of one conductor must be small, of the order of 0.4 cm or less, to minimize eddy-current losses (Sect. 8.4). The balance between magnetic flux density in the core, conductor size, and number of turns is adjusted to hold the maximum output, efficiency, power factor, and heating to the desired values. The large number of dimensions that can be independently selected gives a great deal of latitude to the designer, and this latitude in turn enables him to use a limited variety of core dimensions, slot numbers, and wire sizes to cover a wide range of applications. The choice of proportions is not critical, as departures of one variable from normal can be largely compensated for by adjusting others.

When the numbers of turns in series per phase, N, and of stator slots, S, are decided for a $2P$-pole winding, there still remains the choice of winding pitch and the number of parallel circuits, c. Normally, the pitch of a three-phase winding is made nearly $\frac{5}{6}$ (except that two-pole windings have shorter pitch to reduce the length of the end turns), in order to make the mmf distribution as sinusoidal as possible. To obtain a regular three-phase winding, the number of slots, S, must be a multiple of 3 and also of $2P$. Exact circuit balance requires that both $2P/c$ and $S/3c$ be integers. The value of c is then chosen to make the number of turns in each coil, $3cN/S$, a convenient whole number. Fractional values of the slots per phase belt, $S/6P$, are permissible, but, to keep the phases balanced, the denominator of the residual fraction should not be a multiple of 3 (for three-phase windings). These irregular windings are usually employed in synchronous machines to improve the voltage waveform, and to avoid standstill locking of motors. The resulting irregularities between phase belts, however, are a source of extra reactance, noise, and iron losses, when the air gap is very small, and so they are seldom used in induction machines. A great variety of special winding arrangements has been developed. One of the most useful of these is the pole-changing winding for multispeed operation (Sect. 8.11).

The selection of the slot width to fit the desired coil size with a given flux per pole provides an illustration of the choices the designer is continually required to make. With a fixed air-gap diameter and number of slots, the slot pitch is fixed, and any increase in slot width decreases the tooth width. This in turn raises the flux density in the teeth, requiring an increase in core length to bring the density back to normal. The core density is lowered by this, so the outside diameter of the core can be reduced to match, with only a small net change in the weight of laminations required. The wider slot requires an increase in the end length of coil, as well as a lengthened core. For, as Fig. 6.9 indicates, with a given clearance between coil sides, c, maintained for ventilation purposes, and a given slot pitch, $w+t$, the angle of the coil, α, must be steeper if the slot is widened. Thus, a narrow slot helps materially to reduce the amount of copper required for the winding, and the overall motor length. However, the narrower slot, with a fixed insulation thickness, requires that the slot be deepened more than proportionately, to maintain the same copper section. The slot reactance goes up and the end reactance goes down, when the slot is made narrower and deeper (and the coil ends shorter), so that here again the net effect is small for a considerable range of variation.

As pointed out in Sect. 4.2, small induction motors commonly have permanently short-circuited rotor windings, composed of one bar in each slot solidly connected to end rings. These "squirrel cages" are usually cast in one piece, of aluminum, but larger motors (200 hp or more) have copper or alloy bars brazed or welded to copper end rings.

However, three-phase insulated rotor windings similar to those in the stator are employed when it is desired to vary the secondary resistance to control the speed, or to reduce the starting current. These windings are brought out to slip rings, which are connected to external resistors through carbon or metal-graphite brushes. Since the rotor always has partially

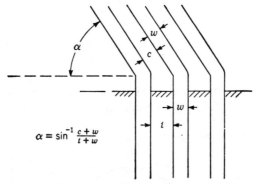

FIG. 6.9. Effect of Coil End Clearance on Coil End Length.

closed slots to provide maximum air-gap area, and as it is difficult to provide high-voltage rotor insulation that will be adequate in the presence of high centrifugal forces, brush dust, etc., the secondary is normally wound for a lower voltage than the primary. The larger motors generally have bar windings, with only one turn per coil, one end of each conductor being bent into shape and connected to the adjacent coil after being pushed through the semi-closed slot. Deep solid conductors can be used because, with the low frequency of rotor current in normal operation, eddy-current losses will not be increased.

9 Magnetizing Current Calculation

Equation 3.40 gives for the magnetizing current, in rms amperes per phase:

$$(6.1) \qquad I_M = \frac{E}{X_M} = \frac{0.696 k_i P^2 G \phi \, 10^8}{q K_p K_d DLN}$$

if G, D, and L are expressed in inches. P = pairs of poles, N = turns in series, and q = number of phases. In this equation, G is the effective length of radial air gap (greater than the actual gap, g), and k_i is a factor greater than unity introduced to allow for the mmf consumed in the iron (Eq. 6.9). To apply this to a practical motor, therefore, we must first determine the ratio of effective to total air gap, as influenced by the slot openings, and, second, calculate the ampere turns consumed in the yoke and teeth of the stator and rotor laminations.

Dr. F. W. Carter was the first (in 1899) to calculate mathematically the magnitude of flux fringing in an open slot facing a smooth pole face. He solved Laplace's equation governing the flux distribution in an air space

$$\frac{d^2V}{dx^2} + \frac{d^2V}{dy^2} = 0,$$

where V is the magnetic potential at any point, by using the Schwarzian method of transformation of coordinates to map the slot and pole-face contours of the real axis in a second plane. The formulas he derived in this way, for an infinite depth of slot and infinite tooth width, have been extended by C. F. Green to include finite tooth widths and overhung

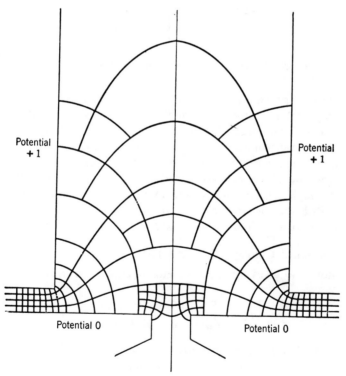

FIG. 6.10. Magnetic Flux and Potential Lines for Open Slot and Air Gap.

teeth. R. W. Wieseman has also graphically derived curves for the ratio of effective to actual air gap with open slots of various proportions, by flux plotting; drawing the potential and flux lines at right angles, and so adjusting the "tubes" of flux that they form curvilinear squares with the equipotential lines (Fig. 6.10). His curves check Carter's analytical results very closely.

Figure 6.11 shows the magnitude of the drop in flux density at the slot center as a function of the ratio of slot width to air gap. Figure 6.12 gives

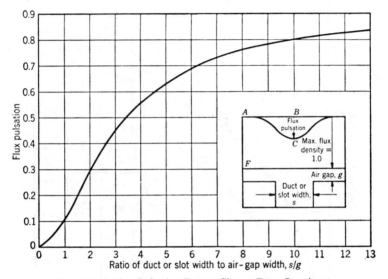

FIG. 6.11. Flux Pulsation Due to Slot or Duct Openings.

the corresponding air-gap fringing coefficient for open slots, in a convenient form for design use, showing Carter's original values (top curve) and as modified by Green (next three curves).

For a slot opening 6 times as wide as the air gap, for example, and an equal tooth width, the figure shows that the flux fringing into the slot is equal to that entering a tooth of a width 2.66 times the gap length, or 0.444 times the slot width. Thus, the effective width of tooth is $t+2.66g$, and this, divided by the slot pitch, $t+s = 2t$, gives the ratio of the effective to total air-gap area, 0.722. Or, the effective air-gap length is:

$$(6.2) \qquad G = \frac{g(t+s)}{t+fg}.$$

With overhung slots, the fringing flux entering the center of the slot has a longer distance to go to reach the sides of the slot beyond the tooth overhang than with open slots. On this account, the ordinates of Fig. 6.12 should be reduced about 10 per cent in calculating the effective air gap with partially closed slots. If there are slot openings on both sides of the air gap, the effective gap length is found by multiplying the actual gap by the product of the primary and secondary fringing factors, each calculated separately, from Fig. 6.12.

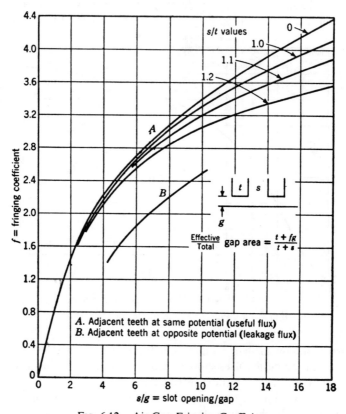

FIG. 6.12. Air-Gap Fringing Coefficients.

Finally, it should be noted that the fringing when adjacent teeth are at potentials $+1$ and -1, as for reactance flux, is considerably less than when adjacent teeth are at the same potential, as for the no-load flux. In the former case, the flux density is zero at the center of the slot, whereas in the latter case, it does not fall to zero (Fig. 6.11). The lowest curve in Fig. 6.12, calculated by A. A. Bennett, gives the fringing for this condition.

Since the flux density in the teeth varies in proportion to the air-gap flux density, magnetic saturation will be reached by the teeth at the center of the flux wave before it is reached elsewhere. The magnetizing current drawn from the line is nearly sinusoidal, however, and the impressed mmf wave preserves its sinusoidal space distribution at all currents. Therefore, the air-gap flux wave has a flat top, the flattening becoming more pronounced as the flux is increased, Fig. 6.13. The flux density in the core back of the teeth is a maximum at 90 electrical degrees away from the peak of the air-gap flux-density wave, since the core flux is the space integral of the air-gap flux. Hence, the peripheral distribution of core ampere turns is 90° displaced from that for the teeth. A saturated core, therefore, tends

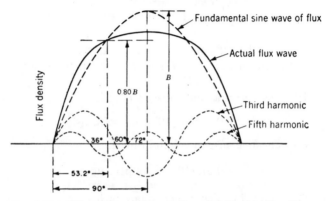

FIG. 6.13. Effect of Saturation on Shape of Air-Gap Flux Wave.

to result in a peaked air-gap flux wave, offsetting in part the flattening effect of saturation in the teeth.

The flux distribution curve retains its fixed shape as the field revolves, so that the "saturation harmonics" in the air-gap flux wave revolve at the same speed as the fundamental, and produce rotor voltages at multiples of the slip frequency. At high values of slip, they induce rotor currents which damp them out quite completely, restoring the sinusoidal flux wave shape. And the decreased air-gap flux and saturation at increasing values of load current make these harmonics of no importance with respect to breakdown torque or starting performance. Only the fundamental of the air-gap flux wave need be considered, therefore, for our present purposes.

A simple way to calculate the extra magnetizing current required to supply the ampere turns consumed in the core and teeth is to prorate the ampere turns required for a single element of flux that crosses the air gap at a point 53° from the point of zero density, or where the air-gap density is 0.80 of the maximum on a sine-wave basis. This is very near the point at which the actual air-gap flux-density wave and its fundamental intersect, since the third harmonic falls to zero at 60°, the fifth harmonic is negative between 36 and 72°, and the seventh harmonic is zero at 51.4°, Fig. 6.13. In calculating the flux density at the 53° point, therefore, we are determining a true point on the fundamental wave, at 0.80 of its peak value, to a close approximation. The exact value is 53.2°, if the third harmonic has 2.84 times the amplitude of the fifth, and all higher harmonics are neglected.

It is convenient to make all ampere-turn calculations on the basis of the average density over the effective pole area. Instead of assuming a longer gap, as in Eq. 6.2, the reduced pole area is used. The average gap density is:

$$(6.3) \qquad B_g = \frac{2P\phi}{\pi DL}\left(\frac{t+s}{t+fg}\right) \text{ webers per square meter,}$$

where P is the number of pairs of poles, ϕ the flux per pole in webers, D the air-gap diameter in meters, L the net core length in meters, and f is the fringing factor from Fig. 6.12.

By Eq. 1.15, the corresponding air-gap mmf is:

$$(6.4) \qquad M_g = 0.795 G B_g\, 10^6 \text{ ampere turns,}$$

where G is the effective air-gap length in meters,

$$= 0.313 G B_g\, 10^8 \text{ ampere turns,}$$

if G is in inches and B in webers per square inch.

The average density in the teeth over the pole pitch is:

$$(6.5) \qquad B_t = \frac{2P\phi}{SLw_t} \text{ webers per square inch,}$$

where S is the number of slots, if L is in inches and w_t is the width of each tooth in inches, usually taken as $\frac{1}{3}(w_{t\text{max}} + 2w_{t\text{min}})$. The tooth mmf corresponding to B_t may be taken from the lower curve of Fig. 6.14. This curve was derived by first reading H corresponding to $0.4\pi B_t$ from a normal d-c saturation curve of the lamination steel, and plotting $1/0.4\pi$ times this value of H against B_t. The 0.4π coefficient comes from first multiplying by $\pi/2$ to change from the average to the maximum of the sine wave, and then multiplying by 0.8 to get the ordinate at the 53.2° point, where the actual wave is assumed to coincide with the fundamental. Allowance is also made for the additional flux flowing radially in the slot and air-duct spaces, which becomes appreciable when B_t exceeds about 0.7×10^{-3} webers per sq in. The total ampere turns are then:

$$(6.6) \qquad M_t = H_t \text{ (from Fig. 6.14) } d_t \text{ ampere turns,}$$

where d_t is the radial length of tooth in inches.

MAGNETIZING CURRENT CALCULATION

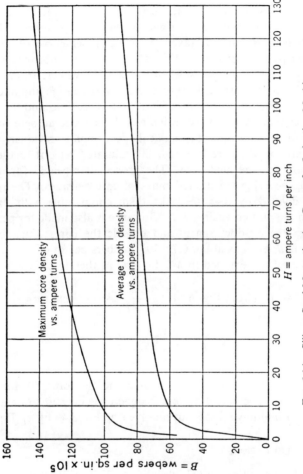

FIG. 6.14. Silicon Steel Magnetization Curves for Induction Machines.

The maximum flux density in the core back of the teeth, averaged over the core depth, is:

(6.7) $$B_c = \frac{\phi}{2d_c L} \text{ webers per sq in.,}$$

where d_c is the radial depth of the core in inches. The flux of each pole divides into two parts on its entrance from the teeth into the core, half going to the left and half to the right, to the adjacent poles. The mmf corresponding to B_c may be taken from the upper curve of Fig. 6.14. This curve was derived by first reading values of H from a normal d-c saturation curve of the lamination steel, for successive values, from $\theta = 0$ to $\theta = 90°$, of $B_c \cos \theta$, plotting these on a uniform scale of θ, and integrating the area of the resulting curve from $\theta = 0°$ to $\theta = 53.2°$, to find the average ordinate over that range. This ordinate is the average ampere turns per inch length of path, for flux starting at the 53.2° point previously selected and going around the core (through the saturated region) to the corresponding point in the next pole. The process is repeated for different values of B_c, covering the desired range of core densities. The resulting values are divided by 0.4π to make them comparable with M_g from Eq. 6.4, and are plotted against B_c. Allowance is also made for peripheral flux through the air-duct spaces and outside the core, which becomes appreciable if B_c exceeds about 0.7×10^{-3} webers per sq in.

The core mmf to be added to M_g, Eq. 6.4, is, therefore:

(6.8) $$M_c = H_c \text{ (from Fig. 6.14)} \left[\frac{\pi(D + 2d_t)}{4P}\right] \text{ ampere turns.}$$

The ratio

(6.9) $$k_i = 1 + \frac{M_t + M_c}{M_g}$$

given by Eqs. 6.4, 6.6, and 6.8 may be called the "saturation factor" of the magnetic circuit. The actual no-load magnetizing current to be expected is obtained by multiplying the current given by Eq. 3.40 by k_i, Eq. 6.1.

10 Loss Calculations

The power losses of an induction motor consist of five elements:

1. The primary copper loss.
2. The slip loss.
3. The core loss.
4. The stray-load losses.
5. The friction and windage loss.

1. *Primary Copper Loss.* To calculate the primary resistance, it is necessary only to include a factor for the number of phases in Eq. 1.11, giving:

(6.10) $$R_1 = \frac{1.65 q N_1^2 L_{t1}}{10^6 C_1} \text{ ohms per phase at } 75\,C,$$

where N_1 = turns in series per phase,
L_{t1} = mean length of turn in inches,
q = number of phases, and
C_1 = total cross section of copper (all phases) in all the primary slots, in square inches.

The coefficient 1.65 of Eq. 6.10 is twice the 0.827 value in Eq. 1.11, because in the induction motor the total cross section of copper, C_1, includes both the going and returning sides of each coil, whereas C in the transformer formula, Eq. 1.11, is the copper section of one side of the winding only. It becomes 4.19 if all dimensions are in centimeters.

2. *Slip Loss.* The secondary resistance, referred to primary, is calculated by a similar formula, using the secondary copper section, C_2, and length of turn, L_{t2}, and multiplying the value of N_1^2 by the ratio of the squares of $K_p K_d$ for primary and secondary. This latter factor enters because any change in K_p or K_d gives a reciprocal change in the number of amperes in each conductor required to produce the same air-gap mmf. The actual number of turns in the secondary winding does not enter the equation. For example, doubling the secondary turns in series per phase, with the same copper section, will quadruple the ohmic resistance measured across the secondary terminals, but will halve the current required for the same mmf, leaving the $I^2 R$ the same. Hence, for a wound-rotor motor,

(6.11) $$R_2 = \frac{1.65 q K_{p1}^2 K_{d1}^2 N_1^2 L_{t2}}{10^6 C_2 K_{p2}^2 K_{d2}^2} \text{ ohms per phase at } 75\,C,$$

referred to primary, with dimensions in inches.

For a squirrel-cage winding, with all the bars solidly connected to end rings, it is convenient to consider the end rings as if they were extensions of the bar length; and the bar currents to be sinusoidally distributed around the periphery, in $2P$ poles. The maximum current in each end ring is the integral of the bar currents over one-half pole pitch. Or, as there are $S_2/4P$ bars in a half pole pitch; and the average of a sine wave is $2/\pi$ times the maximum, the ratio of maximum end-ring current to maximum bar current is:

(6.12) $$\frac{I_{E\,\text{max}}}{I_{B\,\text{max}}} = \frac{2 S_2}{4 P \pi} = \frac{S_2}{2 P \pi}.$$

Both end-ring and bar currents vary sinusoidally around the periphery. Hence, the ratio of end-ring loss to bar loss is equal to the ratio of the squares of their current densities multiplied by the ratio of the (two) end-ring volume to the volume of all the bars. That is:

$$(6.13) \quad \frac{R_E}{R_B} = \left(\frac{S_2}{2P\pi}\right)^2 \left(\frac{C_2}{C_E S_2}\right)^2 \left(\frac{2\pi D_R C_E}{C_2 L_B}\right)$$

$$= \frac{C_2 D_R}{2\pi C_E L_B P^2},$$

where D_R = diameter of end rings,
C_2 = total cross section of all the bars,
L_B = length of each bar, and
C_E = cross section of one end ring.

Substituting Eq. 6.13 in Eq. 6.11, remembering that $K_{p2}K_{d2} = 1$ (for a squirrel cage in the absence of skew), and that $2L_B$ corresponds to the length of one secondary turn, the resistance of a copper squirrel-cage winding referred to the primary is:

$$(6.14) \quad R_2 = \frac{1.65 q K_{p1}^2 K_{d1}^2 N_1^2}{10^6} \left(\frac{2L_B}{C_2} + \frac{D_R}{\pi C_E P^2}\right) \text{ ohms at 75 C,}$$

all dimensions being measured in inches. The coefficient becomes 4.19, if centimeters are used.

More exactly, if the number of bars per pole, $S_2/2P$, is 3 or less, the second term in Eq. 6.14 should be multiplied by:

$$\left[\frac{P\pi}{S_2 \sin\left(\frac{P\pi}{S_2}\right)}\right]^2,$$

since the end-ring current distribution will then be "lumped" instead of sinusoidal. This becomes infinite when $S_2 = P$, as a one-bar-per-pair-of-poles winding is effectively open-circuited; and one if S_2 is large.

The coefficient 1.65 applies to copper. As the resistivity of pure aluminum is 1.62 times that of copper, the coefficient should be 2.67 for aluminum, or proportionately higher for other metals with higher resistivities. In practice, the aluminum used in cast windings contains small amounts of silicon and iron, making its resistance 20 per cent or more higher, bringing the coefficient to approximately 3.3 for inch units, or 8.4 for centimeter units.

In deriving Eqs. 6.11 and 6.14, the rms current has been assumed to be

LOSS CALCULATIONS

the same in every secondary conductor, a true condition for a balanced polyphase motor, free of harmonic fields. When one stator phase of a q-phase squirrel-cage motor is excited alone, however, the secondary resistance at standstill is found to be not R_2, as given by Eq. 6.14, but $2/q$ times this value (fundamental field only). For, if a stationary, alternating, sinusoidally distributed mmf of A peak ampere turns and $2P$ poles is applied to a squirrel-cage winding having R bars, the induced currents will vary in proportion to the sine of the electrical space angle, θ, between each bar and the mmf axis. The contribution of each pair of bars to the fundamental mmf will then be proportional to the sine of the space angle times the current, or to $\sin^2 \theta$, and we shall have for the peak value of mmf:

(6.15) $$A = \frac{4\sqrt{2}}{\pi}\left(\frac{R}{4P}\right)(I_2) \quad (\tfrac{1}{2} = \text{avg value of } \sin^2 \theta),$$

or $$I_2 = \frac{\sqrt{2}\pi P A}{R},$$

where I_2 = rms current in the outermost bars ($\theta = \pi/2$). The peak mmf produced by the current of a single stator phase is:

(6.16) $$A = \frac{2\sqrt{2}\,K_{p1}\,K_{d1}\,N_1\,I_1}{\pi P},$$

of which half represents a forward- and half a backward-revolving field. Combining Eqs. 6.15 and 6.16:

(6.17) $$I_2 = 4K_{p1}\,K_{d1}\,N_1\,I_1/R.$$

The average copper loss per bar is one-half that in the outside bars, since the average value of $\sin^2 \theta = \tfrac{1}{2}$, making the total copper loss in the R squirrel-cage bars:

(6.18) $$I_1^2 R_2 = \left(\frac{R}{2}\right)(I_2)^2 \left(\frac{0.827 R L_B}{10^6 C_2}\right).$$

From Eqs. 6.17 and 6.18, the bar resistance is:

(6.19) $$R_2 = 2\left(\frac{1.65 K_{p1}^2\,K_{d1}^2\,N_1^2}{10^6}\right)\left(\frac{2L_B}{C_2}\right) \text{ for one phase alone,}$$

which is $2/q$ times the value of R_2 given by Eq. 6.14. The expected single-phase $I^2 R$ loss at standstill is, therefore, $2/q^2$ times the corresponding loss on three-phase.

This result may be checked by considering that the change from single-phase to q-phase excitation multiplies the forward-field mmf by q, and reduces the backward field to zero. Hence, the single-phase $I^2 R$ loss in the squirrel cage should be multiplied by the factor $q^2 \times \frac{1}{2}$ to obtain the total loss on three-phase.

For example, the secondary resistance of one leg (60° winding) of a three-phase motor is $2/q = 2/3$ of the three-phase value given by Eq. 6.14, considering the fundamental field only. With two legs in series, however (120° winding), the distribution factor is reduced by the factor $\sqrt{3}/2$, so that the effective turns are $\sqrt{3}$ times as great, and the secondary resistance becomes $3 \times \frac{2}{3} = 2$ times the normal three-phase value. Therefore, the same resistance per phase at standstill is obtained by test on a three-phase motor, for either three-phase or single-phase line-to-line connection, but a lower value is obtained with a line-to-neutral connection. In practice, the presence of harmonic fields, of greater magnitude as the phase-belt width is narrowed, causes the actually measured single-phase resistance to be higher than given by the $2/q$ factor, by an amount that increases as q becomes larger, and is dependent on the coil pitch (Sect. 7.5 and Table 7.2).

The slip loss, or secondary I^2R, can be calculated from Eq. 6.11 or 6.14, by multiplying R_2 by qI_2^2, where I_2 is the secondary current referred to primary, as determined by equivalent circuit calculations (Sect. 5.2) or by the approximate formula, Eq. 5.29.

3. *Core Loss.* The stator core loss is calculated by multiplying the volume of core and teeth by empirical watt loss coefficients, depending on the maximum flux density. A medium grade of 0.025 in. thick silicon sheet steel, well annealed, has a transformer core loss equal to, very roughly:

(6.20) $\qquad W_c = 0.9 B_{max}^2 \, 10^6$ watts per cu in. at 60 cycles,

where B_{max} = maximum flux density in webers per square inch.

The coefficient may be less than 0.6 for the best grades of 0.014 sheet steel, or more than 1.2 for the lower grades in 0.025 thickness; and, of course, it depends markedly on the anneal, sheet insulation, freedom from burrs, stresses, etc. If B_{max} exceeds 10^{-3} webers per sq in., the loss rises very rapidly, because of magnetic leakage caused by saturation, and additional eddy-current losses. The total stator core loss is calculated from Eqs. 6.5, 6.7, and 6.20, remembering that B_t in Eq. 6.5 must be multiplied by $\pi/2$ to correct it to B_{max}.

As shown in Chapter 3, the space and time waveforms of the primary mmf in an induction machine are nearly sinusoidal—because harmonics in the line currents are prevented by the winding distribution and pitch factors, and any that exist are opposed by induced secondary currents. Therefore, as shown in Fig. 1.10 (*b*), at high densities the initial rate of rise, dB/dt, is much greater than that of a sine wave with the same peak density (in contrast to the conditions in a transformer). Consequently, when the flux density rises above the knee of the saturation curve, the effective frequency of the voltages induced in the core laminations rises rapidly. The combined increases in frequency and flux density cause the eddy-current losses to rise much faster than as the square of the density—the exponent may be 6 or more at densities above 10^{-3} webers per square inch.

P. D. Agarwal was the first to give this explanation of the much higher core losses that are observed in both synchronous and induction machines than in transformers operating at comparable high densities.

In addition, there are high-frequency pulsation and "pole face" losses in the teeth, whose total values may be 0.5 or more times the stator core loss, depending on the slot/gap and S_1/S_2 ratios, and many other details of design and manufacture (Fig. 4.7). These can be estimated by equivalent circuit calculations, taking account of the slot-frequency harmonics, as discussed in Section 9.3. Here again there is a challenge to the designer to secure minimum core losses with low cost and uniform, predictable performance.

4. *Stray-Load Losses.* The stray-load losses are additional core and eddy-current losses, caused by the increase in air-gap leakage fluxes with load, and by the high-frequency pulsations of these fluxes. A large part of the stray-load loss is due to induced harmonic currents in the secondary winding, which can be calculated by the extended equivalent circuit, Fig. 3.15, considered in Sect. 9.4. In general, both open-circuit core loss and load losses are low when the ratio of secondary slots to primary slots is near unity, and they increase rapidly as this ratio is increased. If the ratio is made less than unity, the load losses tend to rise also, but less rapidly than when the ratio is increased.

The basis for this may be seen easily by consideration of the per cent flux pulsation that will occur in a single rotor tooth of three different widths. If the rotor-tooth face is much wider than the primary tooth pitch, the total flux entering it will vary only slightly as it moves past the stator. If the rotor-tooth face is equal to the stator-tooth pitch, there will be no pulsation in the rotor-tooth flux, since it always includes the flux from one stator tooth and one slot. If the rotor-tooth face is much narrower than the stator-tooth pitch, its flux will pulsate considerably, as it will first face a slot plus a small part of a tooth, and then face a tooth only. The flux pulsation will penetrate the rotor winding, inducing counter currents, which will in turn produce reflected pulsations of flux in the stator teeth. These pulsations will be greatest in a motor with considerably more secondary than primary slots, and without skew, and will produce ripples in the voltage at the stator terminals (Sect. 10.7). Also, the flux pulsations produce high-frequency magnetic forces that may produce objectionable noise (Sect. 10.1). The number of nodes of these forces is roughly proportional to the difference between the numbers of stator and rotor slots, so that high noise is likely to accompany low losses. By skewing the slots and taking other design precautions, these factors can be kept under control, but it is a fine art to do this and to secure simultaneously low losses, low reactance, quietness, and uniform torque. In practice, the core loss may be 1 to 4 per cent, and the stray-load loss may be between $\frac{1}{2}$ and 3 per cent of the full-load input, or higher in poorly designed motors.

5. *Friction and Windage.* The friction and windage loss is usually empirically determined, although it is possible to calculate with fair precision the oil-film losses in high-speed bearings and also the ventilation losses. A typical value for the total friction and windage of a 10-hp, 1,800-rpm motor is 1.5 per cent of the full-load output, this percentage figure increasing roughly as the square root of the speed, and decreasing slowly as the rated output becomes larger. The losses in ball and sleeve bearings are not greatly different, averaging perhaps $\frac{1}{3}$ per cent of the rated output. The ventilation loss necessary to move a volume of air (at 40 C at sea-level pressure) of Q cubic feet per minute is:

$$(6.21) \qquad W_v = \frac{1.1 Q}{10^4} \left(\frac{V}{4,000} \right)^2 \text{ kw,}$$

where V is the delivered velocity of the air in feet per minute. This formula is nothing more than the familiar expression, kinetic energy $= \frac{1}{2} M V^2$, the weight of air being taken as 0.067 lb per cu ft. The ratio $V/4,000$ is used in Eq. 6.21 because 4,000 fpm corresponds to a pressure of 1 in. of water, a unit frequently employed in ventilation calculations. It is conservative to assume that V will not exceed the peripheral speed of the rotor. The net efficiency of motor fans is generally low, of the order of 20 to 40 per cent, so that the total ventilation loss is several times as large as Eq. 6.21 indicates. Since, as indicated in Sect. 6.4, the normal air supply is only about 15 cfm per rated hp, and as the delivered air velocity is seldom greater than 4,000 fpm in moderate-sized machines, the ventilation loss is roughly:

$$(6.22) \qquad 15 \times 1.1 \times 10^{-4} = 0.0017 \text{ kw per rated hp,}$$

and even with 20 per cent fan efficiency, the total loss need not exceed about 1 per cent.

BIBLIOGRAPHY

Reference to Section

1. "Impulse Testing of Rotating A-C Machines," *A.I.E.E. Trans.*, Vol. 79, Part III, 1960, pp. 182–8.   6
2. "Survey of Induction Motor Protection," *A.I.E.E. Trans.*, Vol. 79, Part III, 1960, pp. 188–92.   6

BIBLIOGRAPHY

Reference to Section

3. "The Design of Induction Motors with Special Reference to Magnetic Leakage," C. A. Adams, *A.I.E.E. Trans.*, Vol. 24, 1905, pp. 649–84. — 1
4. "Eddy-Current Losses in Solid and Laminated Iron," P. D. Agarwal, *A.I.E.E. Trans.*, Vol. 78, Part I, 1959, pp. 169–79. — 10
5. "Stray-Load Losses in Polyphase Induction Machines," P. L. Alger, G. Angst, and E. John Davies, *A.I.E.E. Trans.*, Vol. 78, Part IIIA, June 1959, pp. 349–55. — 10
6. "Induction Motor Core Losses," P. L. Alger and R. Eksergian, *A.I.E.E. Jour.*, Vol. 39, Oct. 1920, pp. 906–20. — 10
7. "Shaft Currents in Electric Machines," P. L. Alger and H. W. Samson, *A.I.E.E. Trans.*, Vol. 43, 1924, pp. 235–44. — 5
8. *Two-Dimensional Fields in Electrical Engineering*, L. V. Bewley, The Macmillan Co., 1948. — 9
9. "Intersheet Eddy-Current Loss in Laminated Cores," L. V. Bewley and H. Poritsky, *A.I.E.E. Trans.*, Vol. 56, 1937, pp. 344–6. — 7
10. "Voltage Oscillations in Armature Windings under Lightning Impulses—I; E. W. Boehne, *A.I.E.E. Trans.*, Vol. 49, Oct. 1930, pp. 1587–1607. — 8
11. "The Magnetic Field of the Dynamo-Electric Machine," F. W. Carter, *The Jour. of the I.E.E.*, Vol. 64, 1926, pp. 1115–38. — 9
12. *Conformal Transformations in Electrical Engineering*, W. J. Gibbs, Chapman and Hall, London, 1958. — 9
13. "Forces in Machine End Windings," Dean Harrington, *A.I.E.E. Trans.*, Vol. 71, Part III, 1952, pp. 849–58. — 8
14. "The Predetermination of the Performance of Induction Motors," D. B. Hoseason, *The Jour. of the I.E.E.*, Vol. 63, 1925, pp. 280–6. — 1
15. "The Cooling of Electrical Machines," D. B. Hoseason, *The Jour. of the I.E.E.*, Vol. 69, 1931, pp. 121–43. — 4
16. "Application of Computers to Solution of Induction Motor Thermal Circuits," A. E. Johnson, *A.I.E.E. Trans.*, Vol. 75, Part III, 1956, pp. 1543–6. — 4
17. "Magnetic Flux Distribution in Annular Steel Laminae," A. E. Kennelly and P. L. Alger, *A.I.E.E. Trans.*, Vol. 36, 1917, pp. 1113–31. — 7
18. *Design of Electrical Apparatus*, Third Edition, J. H. Kuhlmann and N. F. Tsang, John Wiley, 1950. — 1
19. "Some Aspects of Electric-Motor Design—Polyphase Induction Design to Meet Fixed Specifications," T. C. Lloyd, *A.I.E.E. Trans.*, Vol. 63, 1944, pp. 14–20. — 1
20. "Dimensional Studies of Lightweight Motors for Aircraft," W. J. Morrill, *A.I.E.E. Trans.*, Vol. 63, 1944, pp. 698–701. — 1
21. "Measurement of Temperature in General-Purpose Squirrel-Cage Induction Motors," C. P. Potter, *A.I.E.E. Trans.*, Vol. 58, 1939, pp. 468–472. — 4
22. "Effects of Axial Slits on the Performance of Induction Machines with Solid Iron Rotors," P. K. Rajagopalan and V. B. Murty, *I.E.E.E. Trans.*, 1969, Paper No. 69 TP5-PWR.
23. *The Performance and Design of Alternating Current Machines*, Third Edition, M. G. Say, Isaac Pitman & Sons, Ltd., London, 1958.
24. "Temperature-Aging Tests on Class-A-Insulated Fractional-Horsepower Motor Stators," J. A. Scott and B. H. Thompson, *A.I.E.E. Trans.*, Vol. 61, 1942, pp. 499–501. — 4
25. "Tooth Pulsation in Rotating Machines," T. Spooner, *A.I.E.E. Trans.*, Vol. 43, 1924, pp. 252–61. — 10
26. "Squirrel-Cage Induction-Motor Core Losses," T. Spooner, *A.I.E.E. Trans.*, Vol. 44, 1925, pp. 155–60. — 10

		Reference to Section
27.	"No-Load Copper Eddy-Current Losses," T. Spooner, *A.I.E.E. Trans.*, Vol. 45, 1926, pp. 231–8.	10
28.	"No-Load Induction Motor Core Losses," T. Spooner and C. W. Kincaid, *A.I.E.E. Trans.*, Vol. 48, April 1929, pp. 645–54.	10
29.	"Determination of Temperature Rise of Induction Motors," E. R. Summers, *A.I.E.E. Trans.*, Vol. 58, 1939, pp. 459–67.	4
30.	"Shorter Sleeve Bearings Designed for Widest Range of Conditions," E. R. Summers, *Product Engineering*, Vol. 12, Oct. 1941, pp. 530–6.	5
31.	"Synthesis of Induction Motor Designs on Digital Computer," C. G. Veinott, *A.I.E.E. Trans.*, Vol. 71, Part III, 1952, pp. 849–58.	1
32.	"Induction Machinery Design Revolutionized by the Digital Computer," C. G. Veinott, *A.I.E.E. Trans.*, Vol. 75, Part III, 1956, pp. 1509–15.	1
33.	"Electrical Insulation—How Temperature Classification Methods Can Be Used," H. P. Walker, I.E.E.E. Trans. Supplement, 1963, pp. 859-869.	9
34.	*Electromagnetic Fields*, Vol. 1, *Theory and Applications*, E. Weber, John Wiley, 1950.	9
35.	"Graphical Determination of Magnetic Fields," R. W. Wieseman, *A.I.E.E. Trans.*, Vol. 46, 1927, pp. 141–8.	9

CONTENTS—CHAPTER 7

REACTANCE CALCULATIONS

		PAGE
1	Elements of Reactance	199
2	The Magnetizing Reactance	201
3	The Primary Slot Reactance	201
4	The Zero-Phase Sequence Reactance	206
5	The Secondary Slot Reactance	208
6	The Differential Reactance	209
7	The Zigzag Reactance	216
8	Overlap Method of Calculating Zigzag Reactance	222
9	Effect of Skew on Reactance	227
10	The Belt-Leakage Reactance	228
11	Peripheral Air-Gap Leakage	232
12	Coil End Leakage of Armature Windings	233
	Appendix A—Sine-Wave Linkages of an Open-Slot Stator	244
	Appendix B—The Flux Distribution and the Associated Energy Flow in an Annular Core	246

7
REACTANCE CALCULATIONS

1 ELEMENTS OF REACTANCE

The equivalent circuit, Fig. 5.1, shows three distinct elements of the motor reactance, X_M, X_1, and X_2. From a physicist's viewpoint, $X_M + X_1$ is the total reactance of the primary winding, acting alone; and $X_M + X_2$ is the total reactance of the secondary winding (in primary terms). From the designer's viewpoint, X_M is the reactance due to the fundamental sine wave, or useful, air-gap flux; and X_1 and X_2 are the leakage reactances due to all other elements of flux produced by the respective windings.

It is desirable to have X_M as large, and X_1 and X_2 as small, as possible to obtain good motor performance. As pointed out in Sect. 2.3 and as shown by Eqs. 2.24 and 5.39, the total leakage reactance, $X_1 + X_2$, is a nearly exact inverse measure of the maximum torque of the motor. Thus, in practice, the designer selects the motor windings and dimensions to secure a definite value of $X_1 + X_2$, and then makes X_M as large as he can.

The leakage reactance is divided, for convenience, into eight distinct components:

    (a) The primary slot reactance.
    (b) The secondary slot reactance.
    (c) The zigzag-leakage reactance.
    (d) The reactance due to skew.
    (e) The belt-leakage reactance.
    (f) The coil end leakage reactance.
    (g) The incremental reactance.
    (h) The peripheral leakage.

(c), (d), and (e) together constitute the air-gap leakage, or differential leakage reactance, due to harmonics of the air-gap field. (g) is the additional reactance that exists at full speed, above that at standstill, due to decreased magnetic saturation of the tooth tips, and to the redistribution of secondary current that occurs as the slip frequency decreases. (h) is the flux that leaks from pole to pole in the air-gap space without entering the rotor. This is negligible in usual induction machines, but may be appreciable in machines with a diaphragm between stator and rotor, or when the gap is large.

Under all conditions, when full voltage is impressed, the equivalent flux (given by Eq. 3.15) must link the stator winding (neglecting the *IR* drop). At no load, when the stator current is very small, and the rotor current is practically zero, substantially all this flux crosses the air gap and follows the normal magnetic circuit through the stator and rotor cores and teeth (line 4, Fig. 7.1). When the load increases, the increasing rotor current produces a back mmf on the air gap, "damming up" the flow of flux across the air gap. This momentarily decreases the total flux and the

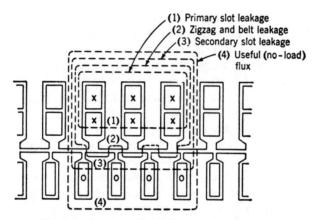

Fig. 7.1. Leakage-Flux Paths of Induction Motor.

induced voltage, allowing more current to flow from the line, which in turn restores the flux to normal, after the rotor has settled back to a slightly lower speed. The load current in the stator, and the opposing rotor current, combine to produce a peripheral flow of flux in the leakage paths across slots and tooth tips, between the two windings (lines 1, 2, and 3, Fig. 7.1). This leakage flux is diverted from the radial flow across the air gap, making the ratio of flux that gets as far as the rotor core to the total stator core flux smaller and smaller as the secondary current increases.

The changing picture of flux distribution as load comes on may be visualized as the rearrangement of a fixed total flux, more and more of which passes peripherally across and between the stator and rotor windings as the load increases, instead of going through into the rotor core. This picture is especially useful to keep in mind when considering the phenomena of sudden short circuit (Sect. 2.9). Although the several elements of leakage flux are actually elements of the main flux diverted from its normal path, the principle of superposition enables us to consider them as independent quantities, if we neglect saturation.

2 The Magnetizing Reactance

X_M is given by dividing I_M from Eq. 6.1 into the no-load voltage per phase (Eq. 3.15):

(7.1) $$X_M = \frac{E}{I_M} = \frac{2.51 q f N^2 K_p^2 K_d^2 DL}{k_i GP^2 10^8} \text{ ohms per phase,}$$

with dimensions in centimeters. The coefficient becomes 6.38 with dimensions in inches.

Since the air-gap flux is reduced under starting conditions (Eq. 5.31), the factor k_i (Eq. 6.9) should be correspondingly reduced or taken as unity for such calculations.

X_M varies inversely as P^2. That is, with all dimensions fixed, and a fixed number of turns in series, the magnetizing reactance is quartered if the number of poles is doubled. For this reason, the air-gap length must be made as small as possible for low-speed motors with large numbers of poles. High-speed motors, with few poles and high peripheral speeds, may have larger air gaps, cutting down the flux pulsations and stray losses without incurring excessive magnetizing current.

The output equation, Eq. 3.9, shows that the output increases roughly in proportion to the speed, or inversely as the poles for fixed dimensions and frequency. Since the magnetizing current increases as the square of the poles, the ratio of magnetizing to full-load current rapidly increases for motors of lower speeds. For induction motors of less than about $\frac{1}{10}$-hp output per pole, the no-load current is nearly equal to that at full load. Such motors have very low power factor.

These relations explain the American practice of using higher-speed induction motors almost exclusively. Low speeds are obtained by employing belts, reduction gears, or d-c motors, or, in the larger sizes, direct-connected synchronous motors.

3 The Primary Slot Reactance

A similar procedure to that in Sect. 1.8 shows that the primary slot reactance is equal to the product:

$2\pi f \times$ slots per phase
$\times$ (series conductors per slot)$^2 \times$ embedded length of slot
$\times$ slot permeance ratio $\times (4\pi \times 10^{-9})$,

or

(7.2) $$X_{1s} = (2\pi f)\left(\frac{S_1}{q}\right)\left(\frac{2qN_1}{S_1}\right)^2 (L)(P_{s1})(12.57)(10^{-9})$$
$$= \frac{3.16 f q L N_1^2 P_{s1} 10^{-7}}{S_1} \text{ ohms per phase,}$$

with dimensions in centimeters, where N_1 is the number of series-connected turns per phase, P_{s1} is the effective ratio of depth to width of slot, and L is the slot length expressed in centimeters. The coefficient becomes 8.02 with dimensions in inches.

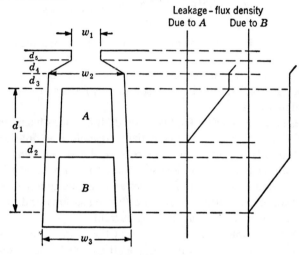

FIG. 7.2. Slot Leakage.

From Fig. 7.2, the slot flux linkages are made up of four parts:

(a) The flux crossing the opening, and linking all the conductors in the slot. This is measured by the ratio of depth to width of opening, or d_5/w_1.

(b) The flux crossing the tapered part, or neck, of the slot, also linking all the conductors in the slot. This is measured by the integral of the depth over width ratio, or:

$$-\int_0^{d_4} \frac{d_4\,dx}{(d_4-x)w_2+xw_1} = \frac{-d_4}{w_2-w_1} \ln\left[(d_4-x)w_2+xw_1\right]_0^{d_4}$$

$$= \frac{d_4}{w_2-w_1} \ln \frac{w_2}{w_1} = \frac{2d_4}{w_2+w_1} \text{ approximately,}$$

since $\ln(1+x) = x - \frac{x^2}{2} + \ldots = \frac{x}{1+x/2} + \ldots$, if x is small.

(c) The flux crossing the slot above the top of the coil, linking all the conductors in the slot, or d_3/w_2.

(d) The flux crossing the body of the slot and linking only a part of the conductors (Sect. 1.8). For simplicity in dealing with the tapered slot of

Fig. 7.2, let $w_3 = w_2(1+\varepsilon)$. The slot width at height x from the bottom of the slot will be:

$$(d_1-x)w_3 + xw_2 = w_2 d_1\left[1 + \frac{(d_1-x)\varepsilon}{d_1}\right].$$

The fraction of the total slot current below x, which creates, and is linked by, the flux at this point, is:

$$\frac{x[(2d_1-x)w_3 + xw_2]}{d_1^2(w_2+w_3)} = \frac{x\left[1+\frac{(2d_1-x)\varepsilon}{2d_1}\right]}{d_1\left(1+\frac{\varepsilon}{2}\right)}$$

$$= \frac{x}{d_1}\left[1+\frac{(d_1-x)\varepsilon}{2d_1}\right] \text{ approximately.}$$

The permeance ratio corresponding to the total linkages is, therefore:

(7.3) $$\int_0^{d_1} \frac{x^2\left[1+\frac{(d_1-x)\varepsilon}{2d_1}\right]^2 dx}{d_1^2 w_2\left[1+\frac{(d_1-x)\varepsilon}{d_1}\right]} = \int_0^{d_1} \frac{x^2\, dx}{d_1^2 w_2} = \frac{d_1}{3w_2} \text{ approximately.}$$

Thus, if terms of the order of ε^2 and higher are neglected, these linkages of a tapered slot are just the same as those of a slot of uniform width w_2.

The sum of the four parts gives for the total slot permeance ratio (Eq. 7.2):

(7.4) $$P_{s1} = \left(\frac{d_5}{w_1} + \frac{2d_4}{w_1+w_2} + \frac{d_3}{w_2}\right) + \frac{d_1}{3w_2}.$$

If d_2 is large, a corrective term, $d_2/12w_2$, should be subtracted from Eq. 7.4, since the flux crossing between coil sides is produced by, and links, only half the conductors. These linkages are, therefore, proportional to $d_2/4$, instead of $d_2/3$.

Substituting Eq. 7.4 in Eq. 7.2 gives the primary slot reactance for a full-pitch winding, in which the top and bottom coil sides carry identical, in-phase currents.

If the winding has a fractional pitch, the currents in the A and B coil sides are out of phase by the amount of the electrical angle between phases, θ. In a balanced fractional-pitch winding there are always in each phase equal numbers of (lower) coil sides in the same slots with currents $\theta°$ behind in time phase, and of other (upper) coil sides in the same slots with currents $\theta°$ ahead in time phase (Fig. 3.6). As, by definition, the mutual inductance between an upper and a lower coil side in the same slot is the same for each of them, the voltages lagging $\theta°$ behind that are induced in the B conductors will be exactly equal to the $\theta°$ ahead voltages induced in the A conductors. Hence, the out-of-phase components of these voltages will cancel, and only the in-phase components of induced voltage need be considered. The mutual inductance between upper and lower coil sides in any slot carrying currents $\theta°$ out of phase must, therefore, be multiplied by $\cos \theta$ to obtain the effective inductance.

Referring again to Fig. 7.2, the total slot reactance is made up of three parts: the self-reactance of A; the mutual reactance of A and B (entering twice); and the self-reactance of B. Since A and B are effectively in series, the total reactance may be represented as the sum of the total reactances of A and B, or:

(7.5) $$X = X_A + 2X_{AB} + X_B.$$

The slot permeance ratio for the self-reactance of the A conductor is, taking $w_2 = w_3 = w$:

(7.6) $$P_A = \frac{d_3}{w} + \frac{d_1 - d_2}{6w} \text{ numeric.}$$

The corresponding permeance ratio for the mutual reactance of A and B, including the effect of pitch, is:

(7.7) $$P_{AB} = \left(\frac{d_3}{w} + \frac{d_1 - d_2}{4w}\right) \cos \theta.$$

The coefficient $\frac{1}{4}$ of the second term of Eq. 7.7 may be derived in two ways. The current in B produces a uniform flux density crossing the slot throughout the height of A. The voltage induced by B in an element of A, therefore, varies linearly from the top to the bottom of A, and its total is the same as if one-half of all the flux that is produced by B and crosses A linked all of A. Hence, one-half the height of A, or $(d_1 - d_2)/4$ is effective in producing mutual inductance. On the other hand, a current in A produces a flux crossing the slot that varies linearly from zero at the bottom of A to full value (Fig. 7.2) at the top of A. Hence, the total flux produced by A, in its conductor height, which also links B, is the same as if the full mmf acted across half the height of A. This is the same result as before.

The slot permeance ratio for the B conductor is, similarly:

(7.8) $$P_B = \frac{d_3}{w} + \frac{4(d_1 - d_2)}{6w} + \frac{d_2}{w}.$$

THE PRIMARY SLOT REACTANCE

Substituting these values in Eq. 7.5, and dividing by 4 to refer the answer to a single conductor per slot, in accord with Eq. 7.3:

$$P_{s1} = \frac{d_3}{4w}(1+2\cos\theta+1) + \frac{d_1-d_2}{4w}\left(\frac{1}{6}+\frac{\cos\theta}{2}+\frac{4}{6}\right) + \frac{d_2}{4w}$$

$$= \frac{d_3}{2w}(1+\cos\theta) + \frac{d_1-d_2}{4w}\left(\frac{5+3\cos\theta}{6}\right) + \frac{d_2}{4w},$$

or, putting $K_s = \dfrac{1+\cos\theta}{2}$:

(7.9) $$P_{s1} = \frac{K_s}{w}\left(d_3+\frac{d_1}{3}\right) + \frac{d_1}{12w}(1-K_s) - \frac{d_2}{4w}\left(K_s-\frac{2}{3}\right).$$

The bracketed expression of Eq. 7.4 is used in place of d_3/w in Eq. 7.9 in the case of partially closed slots.

If all the slots were alike, as in a full-pitch, or a three-phase, $\tfrac{2}{3}$-pitch, winding, a single value of K_s could be used in Eq. 7.9. In the general case, however, there are at least two kinds of slots, carrying coil sides of different phases. It is necessary, therefore, to multiply the distinct values of K_s for the different kinds of slots by the proportion of each, and add these, to find the effective value of K_s for the entire winding. Evidently, for a three-phase winding, the effective value of K_s will vary linearly with pitch between the limiting values of one for a full-pitch winding, 0.75 for a $\tfrac{2}{3}$-pitch winding, 0.25 for a $\tfrac{1}{3}$-pitch winding, and 0 for a zero-pitch winding. Table 7.1 gives these data more completely.

TABLE 7.1

PROPORTIONS OF SLOTS CARRYING COIL SIDES FROM DIFFERENT PHASES FOR A BALANCED THREE-PHASE WINDING

Phase Difference between Currents in Top and Bottom Coil Sides	Winding Pitch, p	Ratio to Total Number of Slots
$\theta° = 0°$	$1 \leq p \leq \tfrac{4}{3}$	$4 - 3p$
$(\cos\theta = 1)$	$\tfrac{2}{3} \leq p \leq 1$	$3p - 2$
	$0 \leq p \leq \tfrac{2}{3}$	0
$\theta = 60°$	$1 \leq p \leq \tfrac{4}{3}$	$3p - 3$
$(\cos\theta = 0.50)$	$\tfrac{2}{3} \leq p \leq 1$	$3 - 3p$
	$\tfrac{1}{3} \leq p \leq \tfrac{2}{3}$	$3p - 1$
	$0 \leq p \leq \tfrac{1}{3}$	0
$\theta = 120°$	$\tfrac{2}{3} \leq p \leq \tfrac{4}{3}$	0
$(\cos\theta = -0.50)$	$\tfrac{1}{3} \leq p \leq \tfrac{2}{3}$	$2 - 3p$
	$0 \leq p \leq \tfrac{1}{3}$	$3p$
$\theta = 180°$	$\tfrac{1}{3} \leq p \leq \tfrac{4}{3}$	0
$(\cos\theta = -1)$	$0 \leq p \leq \tfrac{1}{3}$	$1 - 3p$

REACTANCE CALCULATIONS

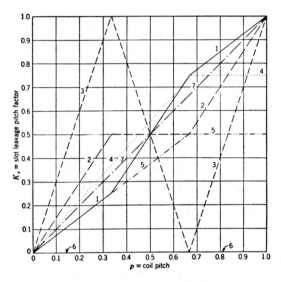

Fig. 7.3. Slot Leakage Pitch Factor.

3 phase, 60° phase belts
1. 3-phase or line-to-line
2. Line-to-neutral
3. Zero-phase sequence

3-phase, 120° phase belts
4. 3-phase or line-to-line
5. Line-to-neutral
6. Zero-phase sequence

2-phase, 90° phase belts
7. All connections

The effective values of K_s for the entire winding are given in Fig. 7.3 for all usual types of polyphase windings.

4 The Zero-Phase Sequence Reactance

A three-phase winding operated on single-phase, with line to line connection, has the same value of K_s as on three-phase. For, on single-phase, half of the slots that normally carry currents out of phase will carry equal in-phase currents, I and I, and the other half will carry current in only one coil side, due to the idle phase. Thus, the average value of $\cos \theta$ is $(1+0)/2 = 0.50$, which is the same as $\cos 60° = 0.50$, its value on three-phase.

When the current flows from line to neutral in a single phase, there is no current in the other halves of slots normally carrying out-of-phase currents, making $\cos \theta = 0$ and $K_s = 0.5$ in Eq. 7.9 for all these slots. From Table 7.1, the average value of K_s then becomes:

Line to neutral slot pitch factor

$K_s = \frac{1}{2}(5-3p)$ if $1 \leq p \leq \frac{4}{3}$,

$K_s = \frac{1}{2}(3p-1)$ if $\frac{2}{3} \leq p \leq 1$,

$K_s = \frac{1}{2}$ if $\frac{1}{3} \leq p \leq \frac{2}{3}$,

$K_s = \frac{3p}{2}$ if $0 \leq p \leq \frac{1}{3}$.

The zero-phase sequence reactance (Sect. 3.15), which applies when the currents in all three phases are equal and in time phase, is found by making $\cos \theta = 1$ for coil sides in slots carrying the same phase current, and those carrying currents that are usually 120° out of phase; and $\cos \theta = -1$ for coil sides in slots carrying currents normally 60° out of phase; since with this connection alternate phase belts carry equal but opposite currents. For this case, therefore, from Table 7.1:

Zero-phase sequence slot pitch factor

$K_s = 4-3p$ if $1 \leq p \leq \frac{4}{3}$,

$K_s = 3p-2$ if $\frac{2}{3} \leq p \leq 1$,

$K_s = 2-3p$ if $\frac{1}{3} \leq p \leq \frac{2}{3}$,

$K_s = 3p$ if $0 \leq p \leq \frac{1}{3}$.

These latter values are plotted in Fig. 7.3, curve 3. The line-to-neutral, three-phase, and zero-phase sequence reactances are related by a simple equation.

For, by Sect. 3.15, if

$$I_A = I, I_B = 0, I_C = 0,$$

the sequence currents are $I_f = I_b = I_0 = I/\sqrt{3}$, and the reactive line-to-neutral voltage of phase A is:

$$E_A = \frac{1}{\sqrt{3}}(E_f + E_b + E_0) = \frac{I}{3}(X_f + X_b + X_0) = X_{LN}.$$

At standstill, the forward- and backward-field reactances are each equal to the three-phase reactance, so that:

(7.10) $\qquad X_{LN} = \frac{1}{3}(2X_{3\text{ phase}} + X_{0\text{ phase}}).$

The three-phase values of K_s may be found from the single-phase values just given and this equation, or, alternatively, by reference to Fig. 7.3:

Three-phase slot pitch factor

$$K_s = \tfrac{1}{4}(7-3p), \qquad 1 \leq p \leq \tfrac{4}{3},$$
$$K_s = \tfrac{1}{4}(3p+1), \qquad \tfrac{2}{3} \leq p \leq 1,$$
$$K_s = \tfrac{1}{4}(6p-1), \qquad \tfrac{1}{3} \leq p \leq \tfrac{2}{3},$$
$$K_s = \frac{3p}{4}, \qquad 0 \leq p \leq \tfrac{1}{3}.$$

It is useful also to recognize that the mutual reactance between phases A and B of a regular three-phase winding, which are spaced 120° apart in space, is:

$$X_{AB} = -\tfrac{1}{2}(\tfrac{2}{3} X_{3\,\text{phase}}) + \tfrac{1}{3} X_{0\,\text{phase}}.$$

The factor outside the first-term parentheses is the cosine of the angle of displacement, or $\cos 120 = -\tfrac{1}{2}$. The term in parentheses is $\tfrac{2}{3} X_{3\,\text{phase}}$, because the sum of the backward- and forward-field voltages produced by a single phase is proportional to $I(\overset{F}{\tfrac{1}{2}} + \overset{B}{\tfrac{1}{2}})$, and is $\tfrac{2}{3}$ that due to balanced three-phase currents:

$$I(\overset{F}{\tfrac{3}{2}} + \overset{B}{0}).$$

The coefficient of the second term is $\tfrac{1}{3}$, because the zero-phase sequence mmf produced by a single phase is one-third that due to three phases carrying equal in-phase currents, and the zero-phase sequence voltage is the same in every phase.

5 The Secondary Slot Reactance

The same procedure is followed in calculating the secondary slot reactance. To express this in primary terms, it is necessary to multiply the value of P_{s2} for the secondary given by Eq. 7.9 by the ratio of the squares of the primary and secondary pitch and distribution factors, and by the ratio of primary to secondary slots.

This relation follows since, for equal and opposite mmfs in stator and rotor, the product of number of slots times effective amperes per slot must be the same on both sides; and the actual amperes per slot are the effective amperes divided by $K_p K_d$ for each winding.

Thus, the total slot reactance of primary and secondary is, from Eq. 7.2:

(7.11) $\quad X_s = 3.16 f q L N_1^2 \, 10^{-7} \left(\dfrac{P_{s1}}{S_1} + \dfrac{K_{p1}^2 K_{d1}^2 P_{s2}}{K_{p2}^2 K_{d2}^2 S_2} \right)$ ohms per phase,

with dimensions in centimeters, where K_p and K_d are given by Eqs. 3.12 and

3.14, P_{s1} and P_{s2} by Eq. 7.10, and K_s by Fig. 7.3. For a squirrel-cage winding, K_{p2} and K_{d2} are unity, and P_{s2} is given by Eq. 7.4. For inch units, the coefficient is 8.02.

6 THE DIFFERENTIAL REACTANCE

As shown by Eq. 3.34, the air-gap field includes, besides the fundamental sine wave, a series of harmonic waves with multiples of the fundamental number of poles, revolving at submultiples of the synchronous speed. All these induce normal-frequency voltages in the winding that produced them, and, therefore, add to the winding reactance. The voltages they generate in the opposing winding when the motor is at speed are not at slip frequency, and, consequently, they produce no useful result. The total reactance due to all these harmonic fields of both stator and rotor is called the differential reactance, or the air-gap leakage. For convenience, the differential reactance is considered as the sum of two elements, the zigzag and the belt leakage.

The zigzag leakage is that due to all the air-gap harmonics that would be produced if the winding had one slot per pole per phase; i.e., if each slot formed a complete phase belt and each slot carried the same rms current, equally spaced apart in time as well as space phase.

The belt leakage is the additional reactance due to the actual phase belts, which may be several slots wide, and, in fractional slot windings, may have varying widths. The belt-leakage harmonics have orders $2kq \pm 1$, whereas the zigzag-leakage harmonics are of orders $2ks \pm 1$, where $q =$ phase belts per pole, $s =$ slots per pole, and k is any integer.

This arbitrary distinction is drawn between the zigzag- and belt-leakage harmonics, because (1) the secondary induced voltages due to the former must be nearly open-circuited for good squirrel-cage motor performance, whereas the voltages due to the latter are substantially short-circuited in the secondary; and because (2) the former are independent of winding pitch but vary with the number of slots, and the latter are nearly independent of the number of slots, but vary with the pitch.

Nevertheless, it is worth while to develop a procedure for calculating the total differential leakage (first used by Chapman) before taking up its two elements separately. Assume a q-phase winding with s nearly closed slots per pole, and a coil pitch of s-b teeth, or b teeth short of 100 per cent. Figure 7.4 (center) shows the stepped rectangular air-gap flux wave produced by phase A of such a winding, and (above) the overlapping waves produced by phases A and B, all on the basis of negligible slot openings. The self-linkages of phase A alone are the sum of the linkages due to the center teeth carrying full flux, and linking all the turns; and of the pairs of outer teeth carrying successively less flux and linking fewer

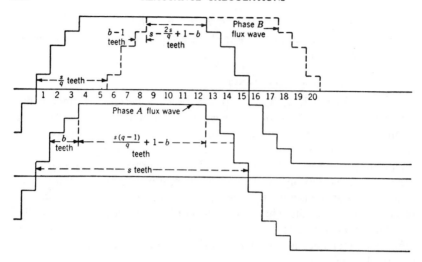

FIG. 7.4. Air-Gap Flux Linkages of Polyphase Winding at the Instant when $I_A = -I_B$; $I_C = 0$.

turns. There are $\dfrac{s(q-1)}{q}+1-b$ central teeth (4 to 12). Taking as unity the self-linkages of a single full-pitch coil, the linkages the center teeth produce are:

$$\frac{1}{s}\left[s\frac{(q-1)}{q}+1-b\right].$$

The next two teeth (3 and 13) outside the central ones are linked by $(s-q)/s$ of the total A phase turns, and have $2/s$ of the total permeance. Their linkages are, therefore:

$$\frac{2}{s^3}(s-q)^2.$$

Each of the successive pairs of outer teeth is linked by q/s less turns, until the bth pair is reached. The next outer pair of teeth is linked by $(s-2q-bq)/s$ of the turns, contributing the linkages:

$$\frac{2(s-2q-bq)^2}{s^3}.$$

THE DIFFERENTIAL REACTANCE

Each of the succeeding outer pairs of teeth has $2q/s$ fewer turns, until the outermost pair is reached. Hence, the total self-linkages of phase A are:

$$L_{AA} = \frac{1}{s^3}\left[\frac{s^3(q-1)}{q}+s^2(1-b)+2(s-q)^2+2(s-2q)^2+\ldots\right.$$
$$\left.+2(s-bq)^2+2(s-2q-bq)^2+2(s-4q-bq)^2+\ldots\right].$$

As the sum of the squares of the first n integers is

$$\frac{n(n+1)(2n+1)}{6},$$

this reduces to:

(7.12) $\qquad L_{AA} = 1 - \frac{2}{3q} + \frac{q}{3s^3}(2s - 3b^2s - bq + b^3q), \quad 0 \leq b \leq \frac{s}{q},$

for a winding pitch between 1 and $(q-1)/q$.

The linkages produced by phase A with phase B are similarly found, except that, as the currents in the two phases are displaced π/q degrees in time, and as symmetry makes the out-of-phase components of the linkages cancel with the corresponding components due to the other phases, the effective current in phase A must be reduced by the factor $\cos(\pi/q)$.

The number of teeth (9 to 12, Fig. 7.4) that is linked by all of phase B, as well as all of phase A, is $s - \frac{2s}{q} + 1 - b$, and these create linkages with B equal to:

$$\frac{1}{s}\cos\frac{\pi}{q}\left(s - \frac{2s}{q} + 1 - b\right).$$

Of the remaining teeth common to A and B, there are $b-1$ that carry full phase A flux, but are linked by successively fewer coils of the B phase (tooth 8, Fig. 7.4). These contribute linkages:

$$\frac{1}{s^2}\cos\frac{\pi}{q}\{(s-q)+(s-2q)+\ldots+[s-(b-1)q]\}.$$

There are also $b-1$ pairs of teeth (13, 18), one of each pair carrying positive and the other negative flux, the former being linked by all, and the latter by successively fewer of the coils of phase B. Each pair is linked by successively fewer of the coils of phase A. Hence, these teeth contribute linkages:

$$\frac{1}{s^3}\cos\frac{\pi}{q}\{q[s-(b-1)q]+2q[s-(b-2)q]+\ldots+(b-1)q(s-q)\}.$$

The remaining $(s/q)+1-b$ pairs of teeth carry equal and opposite phase A fluxes, which link the same number of turns of phase B, and, therefore, contribute no linkages. Hence, the total linkages of phase B due to phase A (and vice versa) are:

$$\frac{1}{s^3}\cos\frac{\pi}{q}\left\{s^2\left(s-\frac{2s}{q}+1-b\right)\right.$$

$$+s[s-q+s-2q+\ldots+s-(b-1)q]+q[s-(b-1)q]$$

$$\left.+2q[s-(b-2)q]+\ldots+(b-1)q(s-q)\right\},$$

which reduces to:

(7.13) $\quad L_{AB} = \cos\frac{\pi}{q}\left[\left(\frac{q-2}{q}\right)-\frac{bq^2}{6s^3}(b^2-1)\right], \quad 0 \leq b \leq \frac{s}{q}.$

The bracketed expression in Eq. 7.13 is to be taken as zero, if it is negative.

The corresponding expressions for values of b between s/q and $2s/q$, corresponding to coil pitches between one and two phase belts short of full pitch are, if q is greater than 2:

(7.14) $\quad L_{AA} = 1 - \frac{1}{3q} - \frac{1}{s^2}\left(bs-\frac{q}{3}\right), \quad \frac{s}{q} \leq b \leq \frac{2s}{q};$

(7.15) $\quad L_{AB} = \cos\frac{\pi}{q}\left(\frac{1}{2q}+\frac{q}{2s^2}+\frac{3b}{2s}-\frac{3b^2q}{2s^2}+\frac{b^3q^2}{3s^3}-\frac{bq^2}{3s^3}\right), \quad \frac{s}{q} \leq b \leq \frac{2s}{q}.$

And the value of L_{AA} for coil pitches greater than zero by not more than one phase-belt width is:

(7.16) $\quad L_{AA} = \frac{q(s-b)}{3s^3}[q+(s-b)(3s-qs+bq)], \quad s-\frac{s}{q} \leq b \leq s.$

The total inductance of one phase is obtained by adding the self- and effective mutual inductances. For a three-phase, 60° phase-belt winding, $q = 3$, $\cos(\pi/q) = \frac{1}{2}$, we have:

$$L_t = L_{AA}+L_{AB}+L_{AC} = L_{AA}+2L_{AB},$$

or

(7.17) $\quad L_{60} = \frac{10}{9}+\frac{1}{2s^3}(4s-6b^2s-3b+3b^3), \quad 0 \leq b \leq \frac{s}{3};$

(7.18) $\quad = \frac{19}{18}+\frac{1}{2s^3}(5s+bs^2-6b-9b^2s+6b^3), \quad \frac{s}{3} \leq b \leq \frac{2s}{3}.$

THE DIFFERENTIAL REACTANCE

Equations 7.17 and 7.18 are identical for $\frac{2}{3}$ pitch, or $b = s/3$.

For a two-phase winding, L_{AB} is zero, and Eqs. 7.12 and 7.16 are identical, giving for the total linkages:

$$(7.19) \quad L_{90} = \frac{2}{3} + \frac{2}{3s^3}(2s - 3b^2 s - 2b + 2b^3), \quad 0 \leq b \leq s.$$

Table 7.2 compares the total air-gap reactance of different types of winding, as given by these formulas.

TABLE 7.2

TOTAL AIR-GAP REACTANCE IN PER UNIT TERMS

Coil Pitch	60° Belts 3 Phase, and 1 Phase Line to Line	60° Belts 1 Phase, Line to Neutral	90° Belts 2 Phase
	Eqs. 7.17, 7.18	Eqs. 7.12, 7.14	Eq. 7.19
1	$\frac{10}{9} + \frac{2}{s^2}$	$\frac{7}{9} + \frac{2}{s^2}$	$\frac{2}{3} + \frac{4}{3s^2}$
$\frac{5}{6}$	$\frac{149}{144} + \frac{7}{4s^2}$	$\frac{17}{24} + \frac{3}{2s^2}$	$\frac{50}{81} + \frac{10}{9s^2}$
$\frac{3}{4}$	$\frac{1091}{1152} + \frac{13}{8s^2}$	$\frac{367}{576} + \frac{5}{4s^2}$	$\frac{9}{16} + \frac{1}{s^2}$
$\frac{2}{3}$	$\frac{5}{6} + \frac{3}{2s^2}$	$\frac{5}{9} + \frac{1}{s^2}$	$\frac{40}{81} + \frac{8}{9s^2}$
$\frac{1}{2}$	$\frac{5}{9} + \frac{1}{s^2}$	$\frac{7}{18} + \frac{1}{s^2}$	$\frac{1}{3} + \frac{2}{3s^2}$
$\frac{1}{3}$	$\frac{5}{18} + \frac{1}{2s^2}$	$\frac{2}{9} + \frac{1}{s^2}$	$\frac{14}{81} + \frac{4}{9s^2}$
$\frac{1}{4}$	......	......	$\frac{5}{48} + \frac{1}{3s^2}$

For a single-phase winding formed by using two legs of a 60° belt, three-phase winding, the total inductance is found by substituting 1 for $\cos(\pi/q)$ in Eq. 7.13 or Eq. 7.15, and adding this to Eq. 7.12 or Eq. 7.14, which gives identically Eqs. 7.17 and 7.18. Thus, the per unit line-to-line reactance of a three-phase, 60° belt winding is identical with its per unit three-phase reactance. For a single-phase winding consisting of one leg of the three-phase winding only (line-to-neutral connection), the total reactance is given by Eq. 7.12 or 7.14, as shown in the third column of Table 7.2.

The ratio of the useful (fundamental) flux to the total generated by a single full-pitch coil (Fig. 7.5 (a) and Eq. 3.26) is $8/\pi^2$. Multiplying this by the squares of the pitch and distribution factors (Eqs. 3.12 and 3.14), and

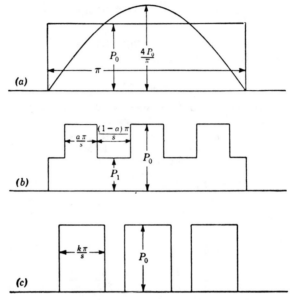

FIG. 7.5. Alternative Forms of Flux Wave due to Full-Pitch Coil.

by the ratio $q/2$ of the total revolving field to the field due to one phase alone, gives the per unit inductance due to the useful flux. Or:

$$(7.20) \qquad L_s = \frac{4q^3}{\pi^2} \frac{\sin^2 \frac{\pi}{2q} \sin^2 \frac{\pi(s-b)}{2s}}{s^2 \sin^2 \frac{\pi}{2s}}.$$

Dividing Eq. 7.20 into the value of the total air-gap inductance, and subtracting unity from the result gives the sought-for value of the total per unit differential leakage of the winding. This gives, for a two-phase, 90° belt winding:

$$(7.21) \qquad \frac{X_{D90}}{X_M} = \frac{\pi^2(s-b)(s^2+bs-2b^2+2)\sin^2 \frac{\pi}{2s}}{24s \sin^2 \frac{\pi(s-b)}{2s}} - 1, \quad 0 \leq b \leq s.$$

For a three-phase, 60° belt winding:

$$\text{(7.22)} \quad \frac{X_{D60}}{X_M} = \frac{\pi^2(20s^3 + 36s - 54b^2s - 27b + 27b^3)\sin^2\frac{\pi}{2s}}{486s\sin^2\frac{\pi(s-b)}{2s}} - 1, \quad 0 \leq b \leq \frac{s}{3};$$

$$\text{(7.23)} \quad = \frac{\pi^2(19s^3 + 45s + 9bs^2 - 54b - 81b^2s + 54b^3)\sin^2\frac{\pi}{2s}}{486s\sin^2\frac{\pi(s-b)}{2s}} - 1,$$

$$\frac{s}{3} \leq b \leq \frac{2s}{3}.$$

All these expressions are exact if the slot openings are extremely narrow, but are on the high side when the slot openings are wide and s is small, because of the difference between the fringing coefficients for useful and leakage flux (Fig. 6.12).

When s is very large, the zigzag leakage approaches zero, and the differential leakage becomes equal to the belt leakage alone. The belt leakage of a full-pitch, two-phase, 90° phase-belt winding, therefore, is given by developing Eq. 7.21 in powers of $1/s$, and putting $b = 0$:

$$\text{(7.24)} \quad \frac{X_{D90}}{X_M} = \frac{\pi^2(s^2+2)}{24}\sin^2\frac{\pi}{2s} - 1, \quad b = 0,$$

$$= \frac{\pi^4}{96} - 1 + \frac{\pi^4}{48s^2}\left(0.412 - \frac{0.687}{s^2} + \ldots\right), \quad b = 0, s > 8.$$

And, for a three-phase, full-pitch, 60° phase-belt winding, from Eq. 7.22:

$$\text{(7.25)} \quad \frac{X_{D60}}{X_M} = \frac{2\pi^2(5s^2+9)}{243}\sin^2\frac{\pi}{2s} - 1, \quad b = 0,$$

$$= \frac{5\pi^4}{486} - 1 + \frac{\pi^4}{54s^2}\left(0.543 - \frac{0.672}{s^2} + \ldots\right), \quad b = 0, s > 8.$$

Numerical values of these per unit differential reactances for two- and three-phase windings are given in Tables 7.3 and 7.4. The limiting values, when s is very large, are checked by the independently derived belt-leakage formulas developed in Sect. 7.10.

Table 7.3
Per Unit Differential Reactance of 90° Phase-Belt Windings, Eq. 7.21

Pitch Deficiency b	$s/2$ = Slots per Pole per Phase					
	1	2	3	4	5	6
0	0.2337	0.0840	0.0468	0.0330	0.0264	0.0229
1	0.2337	0.0583	0.0334	0.0251	0.0213	0.0193
2		0.0840	0.0285	0.0177	0.0147	0.0138
3		0.0840	0.0468	0.0189	0.0116	0.0096
4			0.1019	0.0330	0.0148	0.0089
5			0.2337	0.0649	0.0265	0.0130
6				0.1222	0.0468	0.0229
7				0.2337	0.0839	0.0398
8					0.1382	0.0650

Table 7.4
Per Unit Differential Reactance of 60° Phase-Belt Windings, Eqs. 7.22 and 7.23

Pitch Deficiency b	$s/3$ = Slots per Pole per Phase					
	1	2	3	4	5	6
0	0.0966	0.0284	0.0140	0.0089	0.0064	0.0052
1	0.0966	0.0235	0.0115	0.0074	0.0055	0.0045
2	0.0966	0.0284	0.0111	0.0063	0.0044	0.0035
3		0.0284	0.0140	0.0069	0.0041	0.0030
4		0.0284	0.0143	0.0089	0.0050	0.0031
5			0.0137	0.0092	0.0064	0.0040
6			0.0140	0.0089	0.0068	0.0052
7				0.0083	0.0067	0.0056
8				0.0089	0.0062	0.0055
9					0.0057	0.0052

7 The Zigzag Reactance

By making $q = s$ and $b = 0$ in Eqs. 7.12 and 7.13, the total inductance of one phase of a full-pitch winding with one slot per pole per phase (as a squirrel cage, for example) is found to be:

$$(7.26) \quad L_t = L_{AA} + L_{AB} + L_{AC} + \ldots$$

$$= 1 + \frac{s-2}{s}\cos\frac{\pi}{s} + \frac{s-4}{s}\cos\frac{2\pi}{s} + \ldots + \frac{2}{s}\cos\frac{(s-1)\pi}{s}$$

$$= 1 - \frac{2}{s}\left[\cos\frac{\pi}{s} + 2\cos\frac{2\pi}{s} + 3\cos\frac{3\pi}{s} + \ldots \right.$$

$$\left. + (s-1)\cos\frac{(s-1)\pi}{s}\right] \text{ numeric.}$$

To evaluate this expression, we note that it is equal to $1 - \dfrac{2dT}{sd\theta}$, where $\theta = \pi/s$, and:

(7.27) $$T = \sin\theta + \sin 2\theta + \sin 3\theta + \ldots + \sin(s-1)\theta.$$

By substituting $\sin n\theta = \dfrac{-j(e^{jn\theta} - e^{-jn\theta})}{2}$, performing the differentiation indicated, and summing the resultant exponential series by means of the identity:

(7.28) $$\frac{r(r^{s-1} - 1)}{r - 1} = r + r^2 + r^3 + \ldots + r^{s-1},$$

we find, finally, that:

(7.29) $$L_t = \frac{1}{s}\csc^2\frac{\pi}{2s},$$

which gives the total air-gap reactance of a one-slot-per-pole-per-phase winding with nearly closed slots, in per unit terms. This is the sum of the magnetizing and zigzag-leakage reactances, the belt leakage being zero in this case.

Since s is usually 6 or more, $\csc \pi/2s$ may be replaced by the series:

(7.30) $$\csc\frac{\pi}{2s} = \frac{2s}{\pi}\left[1 + \frac{1}{6}\left(\frac{\pi}{2s}\right)^2 + \frac{7}{360}\left(\frac{\pi}{2s}\right)^4 + \ldots\right],$$

and Eq. 7.29 becomes:

(7.31) $$L_t = \frac{4s}{\pi^2}\left(1 + \frac{\pi^2}{12s^2} + \ldots\right).$$

The fundamental sine wave, Fig. 7.5 (a), has an area $8/\pi^2$ times that of the reactangle itself. And the sum of the sine-wave linkages due to all the s phases acting conjointly is $s/2$ times those of a single phase by itself (Eq. 3.34). The total fundamental sine-wave linkages of one phase are, therefore:

(7.32) $$L_s = \left(\frac{8}{\pi^2}\right)\left(\frac{s}{2}\right) = \frac{4s}{\pi^2}.$$

The same result is obtained by putting $b = 0$ and $q = s$ in Eq. 7.20.

Dividing $L_t - L_s$ by L_s, the per unit zigzag leakage for this case is found to be:

(7.33) $$\frac{L_t - L_s}{L_s} = \frac{\pi^2}{12s^2} \text{ approximately.}$$

Carrying through the same procedure for the secondary winding with s_2 slots per pole, and adding the primary and secondary values, the total zigzag reactance of the motor is found to be:

$$(7.34) \qquad X_Z = \frac{\pi^2 X_M}{12}\left(\frac{1}{s_1^2}+\frac{1}{s_2^2}\right) \text{ ohms per phase,}$$

where X_M is the magnetizing reactance, Eq. 7.1, due to the fundamental sine-wave flux only. Or,

$$(7.35) \qquad X_Z = \frac{\pi^2 X_M}{12}\left(\frac{4P^2}{S_1^2}+\frac{4P^2}{S_2^2}\right) \text{ ohms per phase,}$$

where P is the number of pairs of poles, and S_1 and S_2 are the total numbers of slots in primary and secondary, respectively.

Equation 7.34 gives a close approximation to the zigzag reactance for the case of closed slots, or a smooth air-gap surface.

If the slots are open, however, the flux wave of a single coil has the notched shape of Fig. 7.5 (b). The total flux produced by the full-pitch coil will now be k times that for the closed-slot case, where:

$$(7.36) \qquad k = a + \frac{P_1}{P_0}(1-a), \quad \text{numeric} < 1,$$

if a = ratio of tooth width to slot pitch, and

$$\frac{P_1}{P_0} = \frac{\text{avg flux density over slot face}}{\text{avg flux density over tooth face}}.$$

k is also equal to g/G, the ratio of actual to effective air gap, Fig. 6.12.

The total linkages of one phase are still given by Eq. 7.31 except for the multiplier k, so that:

$$(7.37) \qquad L_t = \frac{4ks}{\pi^2}\left(1+\frac{\pi^2}{12s^2}+\ldots\right) \text{ numeric.}$$

The fundamental sine-wave linkages of the notched flux wave of Fig. 7.5 (b) are:

$$(7.38) \qquad L_s = \frac{4ks}{\pi^2}\left[1+\frac{\pi^2}{24ks^2}(a)(1-k)(1+a)\right],$$

as derived in Appendix A of this chapter (Eq. 7.3A).

We must remember at this point that L_s is the per unit mutual reactance between primary and secondary windings, and that we must know the total reactance of the secondary winding as well as of the primary before the zigzag leakage can be found:

$$(7.39) \qquad X_Z = X_M \left(\frac{L_{t1} + L_{t2} - 2L_s}{L_s} \right).$$

This is similar to Eq. 7.5, except that the primary and secondary currents are opposing, making the mutual inductance term negative, whereas in the slot reactance case the upper and lower coil side currents are in the same direction, making the mutual term positive.

For a smooth air gap, with a large number of secondary slots, $L_{t2} = L_s$ (by making s very large in Eq. 7.31), and so Eq. 7.39 reduces to Eq. 7.33. But in the open-slot case, the total air-gap linkages, L_{t2}, produced by the opposing secondary winding (which only "sees" the average permeance as the rotor moves by the stator) remain given by Eq. 7.37. Hence, from Eqs. 7.37, 7.38, and 7.39, for the open-slot case of Fig. 7.5 (b):

$$(7.40) \qquad X_Z = X_M \left[\frac{\pi^2}{12 s_1^2} - \frac{\pi^2 a(1-k)(1+a)}{12 k s_1^2} + \frac{\pi^2}{12 s_2^2} \right]$$
$$= \frac{\pi^2 X_M}{12} \left\{ \frac{1}{k s_1^2} [k - a(1+a)(1-k)] + \frac{1}{s_2^2} \right\},$$

which reduces to Eq. 7.34, if $k = 1$.

It is worth while to pause here, and consider more carefully the separate values of X_Z for the primary and secondary in this case. The value of X_Z for the primary only is merely the difference between L_t and L_s, or, by Eqs. 7.37 and 7.38:

$$(7.41) \qquad X_{Z1} = X_M \left(\frac{L_{t1} - L_s}{L_s} \right)$$
$$= \frac{\pi^2 X_M}{12 s_1^2} \left[1 - \frac{a(1+a)(1-k)}{2k} \right] \text{ ohms per phase.}$$

On the same basis, the secondary zigzag leakage is:

$$(7.42) \qquad X_{Z2} = X_M \left(\frac{L_{t2} - L_s}{L_s} \right)$$
$$= \frac{\pi^2 X_M}{12} \left[\frac{1}{s_2^2} - \frac{a(1+a)(1-k)}{2 k s_1^2} \right] \text{ ohms per phase,}$$

and the sum of Eqs. 7.41 and 7.42 gives the correct value, Eq. 7.40. The value of X_{Z2} in Eq. 7.42 is negative when s_2 is large.

That is to say, when a secondary winding with a large number of closed slots faces an open-slot primary, the secondary mmf produces a fundamental sine wave indicated by Eq. 7.37, and also produces harmonics of order $2s_1 - 1$ and $2s_1 + 1$, exactly similar to the primary tooth harmonics created by the primary winding itself (Eq. 3.34). These secondary harmonics create fundamental-frequency voltages in the primary, and, therefore, are true elements of useful mutual flux, even though they are distinct from the fundamental flux wave. In this way, we have the seeming paradox of a mutual flux greater than the fundamental sine wave produced by the secondary, giving a negative value to the secondary differential leakage (Eq. 7.42), if that is arbitrarily defined with reference to the fundamental sine wave.

This illustrates the necessity for considering the differential leakage as the difference between primary and secondary fields, rather than as the sum of primary and secondary parts, obtained independently by reference to the fundamental sine wave.

It is clear that the waveform of Fig. 7.5 (b) is only one of many shapes that may be taken by the air-gap field of a single coil, depending on the slot and tooth configurations and the assumptions made about the fringing flux paths.

Actually, the flux at the outer edge of the full-pitch coil must slope down to a zero value, instead of remaining at the P_1 value, as incorrectly shown in Fig. 7.5 (b). That is, the end-slot fringing is in accord with the lower curve on Fig. 6.12, while in the center slot of the coil fringing is in accord with the upper curve, the in-between slots having intermediate values. If this tapering of the stator flux wave is to be allowed for, however, it must also be allowed for in calculating the flux of the rotor. The rotor is moving by the stationary slot openings, and thus sees only the average permeance, but it is still true that the fringing coefficient has, on the average, a greater value at the center of the rotor coil than at the edges. To go into this problem would take more space than we can allow here, and it must suffice to say that the differential leakage given by Eq. 7.40 is on the high side of the true value.

A simpler form of permeance variation is shown in Fig. 7.5 (c), in which all the permeance is lumped in the teeth, which are widened to allow for fringing. In this case, putting $P_1 = 0$ in Eq. 7.36:

(7.43) $\quad k = a =$ ratio of tooth width to a slot pitch,

and L_t is still given by Eq. 7.37.

By substituting $k = a$ in Eq. 7.38, the per unit fundamental sine-wave linkages are found to be:

$$(7.44) \quad L_s = \frac{4ks}{\pi^2}\left(1 + \frac{\pi^2(1-a)^2}{24s^2}\right).$$

A similar expression holds for the secondary, if it, too, has open slots.

Hence, the zigzag leakage for Fig. 7.5 (c) is, by Eq. 7.39:

$$(7.45) \qquad X_Z = \frac{\pi^2 X_M}{12}\left(\frac{a_1^2}{s_1^2} + \frac{a_2^2}{s_2^2}\right).$$

Comparing this with Eqs. 7.40 and 7.34, it is seen that lumping the permeance as in Fig. 7.5 (c) markedly reduces the differential leakage at low values of k, as compared with the closed-slot case, Fig. 7.5 (a).

As noted in Sect. 6.9, the value of k, the ratio of effective to actual air gap, in the case of slot openings on both sides of the air gap, is:

$$(7.46) \qquad k = a_1 a_2 = \frac{g^2}{G_1 G_2},$$

where $a_1 = g/G_1$ and $a_2 = g/G_2$ are found from Fig. 6.12, considering the primary and secondary slot openings separately.

Reviewing these methods of calculating the zigzag reactance, given by Eqs. 7.34, 7.40, and 7.45, the formula:

$$(7.47) \qquad X_Z = \frac{\pi^2 X_M}{12}\left[\frac{(6a_1-1)F_{sc}}{5s_1^2} + \frac{(6a_2-1)}{5s_2^2}\right]$$

is suggested as a good value to use. Figure 7.6 charts the (bracketed term in s_1) coefficients of Eqs. 7.34, 7.40, 7.45, and 7.47, against k (or g/G), showing the range of variation caused by the different assumptions.

The factor F_{sc} in Eq. 7.47 (less than unity) allows for induced secondary currents, such as occur in a squirrel-cage winding. These induced currents "damp out" the flux pulsations, or prevent their penetration through the squirrel-cage winding, and thus serve to reduce the zigzag reactance (Sect. 9.3). As the induced currents are greatly reduced by skewing, and they are very small in a phase-wound secondary, it is usually sufficient to take F_{sc} as 1. There is no corresponding factor for the secondary term of the zigzag reactance, since the pitch and distribution factors for voltages induced in the primary winding by the secondary tooth harmonics are normally low, making the induced currents of this type very small. In special cases, as when the rotor-tooth harmonics coincide with important harmonics of the primary winding, these currents may need to be considered.

Equation 7.47 has been derived entirely on the basis of a one-slot-per-pole-per-phase winding. When the winding has several slots per phase belt, the mmf across each slot opening varies in proportion to K_s (Fig. 7.3), and not in proportion to $K_p^2 K_d^2$, as the value of X_M does. Hence, Eq. 7.47 should be multiplied by $K_s/K_p^2 K_d^2$, as developed in the next section, to be strictly correct. This ratio does not differ greatly from unity, however.

222 REACTANCE CALCULATIONS

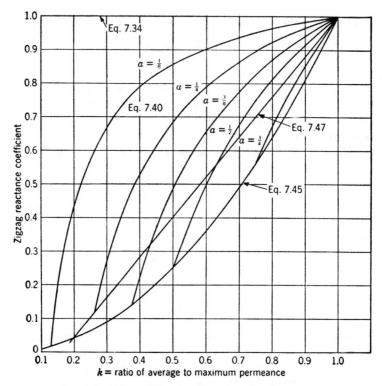

FIG. 7.6. Effect of Slot Openings on Zigzag Reactance.

8 OVERLAP METHOD OF CALCULATING ZIGZAG REACTANCE

Equation 7.45 can be derived in an entirely different way, using the "tooth overlap" method of analysis, as developed by C. A. Adams, E. Arnold, L. H. A. Carr, and, more recently, H. R. West. In this method, the zigzag-leakage flux is calculated by applying the mmf of rotor and stator windings to the zigzag peripheral path across overlapping tooth tips around the air gap (Fig. 7.1).

Consider a symmetrical polyphase machine with full-pitch windings in stator and rotor, the rotor being in such a position that its winding axis coincides with that of the stator. In this position, the belt leakage is zero, and the total air-gap leakage is zigzag reactance. Figure 7.7 (a) shows this condition. Equal and opposing ampere turns are assumed in rotor and stator. The number of slots per pole is assumed to be very large, and stator tooth 1 is taken to be at the midpoint of a phase belt; so that the current is the same in all the slots for an indefinite distance on both sides of it. Between stator tooth 1 and the opposite rotor tooth 1, there will be no mmf

OVERLAP METHOD OF CALCULATING ZIGZAG REACTANCE 223

drop, and we may take their common center line as a reference axis. The mmf drop between any other two opposing rotor and stator teeth in the figure will then be proportional to the algebraic sum of all the ampere

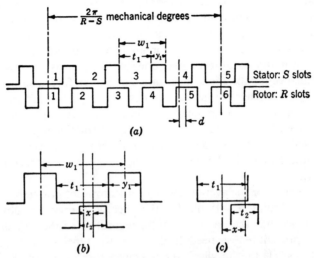

FIG. 7.7. Air Gap with Open Slots.

turns between the two teeth in question and the reference center line of the first teeth. For instance, between stator tooth 4 and rotor tooth 5, the mmf drop is:

$$4\pi n_1 i_1 \left(3 - \frac{4w_2}{w_1}\right),$$

where $n_1 i_1 =$ ampere conductors per stator slot, and

$$\frac{w_2}{w_1} = \frac{\text{rotor slot pitch}}{\text{stator slot pitch}}.$$

The distance between the center lines of stator tooth 4 and rotor tooth 5 is:

$$d = (3w_1 - 4w_2) \quad \text{or} \quad \left(3 - \frac{4w_2}{w_1}\right) w_1.$$

Substituting this in the expression for the mmf drop, we get:

$$4\pi n_1 i_1 \frac{d}{w_1}.$$

That is, the magnetic potential difference between any two opposing teeth is equal to the magnetic potential difference per stator slot times the ratio of the distance between the two tooth center lines to one stator slot pitch.

This simple rule makes it possible to calculate the flux distribution as completely and accurately as our ability to allow properly for fringing will permit.

By means of this rule, we can determine the air-gap flux density between any two opposing teeth, and hence the magnetic energy; and this in turn gives a measure of the reactance. The next step is to derive an expression for the average energy per tooth for all relative positions of a pair of opposing rotor and stator teeth, over one slot pitch. This gives the average zigzag reactance for the conditions stated, and, therefore, also the average effective zigzag-leakage permeance per slot. Having a formula for effective zigzag permeance for the assumed conditions, we may use this same value of permeance in calculating zigzag reactance of fractional-pitch windings.

For other, non-coaxial, positions of the rotor, belt leakage appears. This is superposed on the zigzag leakage and may be accounted for separately.

An expression for the average energy per stator tooth can most easily be obtained by finding a formula for the average energy per slot of that member which has the narrower teeth, regardless of the slot pitch. By using this procedure, we need study only two types of slot and tooth combinations: (1) that in which the narrower tooth is narrower than the slot in the opposite member, and (2) that in which the narrower tooth is wider than the slot in the opposite member. This will cover all possible slot and tooth combinations. In fact, only one of these combinations need be analyzed here, since the same analysis applied to the other yields identically the same formula.

In Fig. 7.7 (*b*) is represented a slot and tooth combination of the first type; i.e., the rotor teeth are narrower than either the stator teeth or the stator slots.

As the rotor tooth moves through one stator slot pitch, there are three regions to be considered: (1) Fig. 7.7 (*b*), where the rotor tooth is entirely opposite the stator tooth; (2) Fig. 7.7 (*c*), where the rotor tooth is partly opposite the stator tooth and partly opposite the stator slot; and (3) where the rotor tooth is entirely opposite the stator slot. We must integrate through these regions separately in order to get the average energy.

Region 1

$$0 \leq x \leq \frac{t_1 - t_2}{2}.$$

Let the ampere turns per rotor slot be $n_2 i_2$. Then, the potential difference between the two teeth is, according to the rule stated previously:

(7.48) $$v = 4\pi n_2 i_2 \frac{x}{w_2}.$$

OVERLAP METHOD OF CALCULATING ZIGZAG REACTANCE   225

The magnetic energy in the permeance P between the teeth is:

$$W_a = \frac{v^2 P}{8\pi}. \tag{7.49}$$

Substituting for v and P their values of $4\pi n_2 i_2 (x/w_2)$ and t_2/g, we have:

$$W_a = 2\pi (n_2 i_2)^2 \frac{x^2 t_2}{g w_2^2} \tag{7.50}$$

for the energy as a function of the rotor-tooth position in region 1.

Region 2

$$\frac{t_1 - t_2}{2} \leq x \leq \frac{t_1 + t_2}{2}.$$

In this region, the rotor tooth is partly opposite the stator tooth and partly opposite the slot. The permeance is:

$$P = \frac{1}{g}\left(\frac{t_1 + t_2}{2} - x\right), \tag{7.51}$$

so that the energy is:

$$W_b = 2\pi (n_2 i_2)^2 \frac{x^2}{g w_2^2}\left(\frac{t_1 + t_2}{2} - x\right). \tag{7.52}$$

Region 3

$$\frac{t_1 + t_2}{2} \leq x \leq \frac{1}{2} w_1.$$

Here the rotor tooth is entirely opposite the stator slot, so that it carries no flux, and, therefore, $W_c = 0$.

The average energy of the rotor-tooth flux through one stator slot pitch is:

$$W_2 = \frac{2}{w_1}\left[\int_0^{(t_1-t_2)/2} W_a \, dx + \int_{(t_1-t_2)/2}^{(t_1+t_2)/2} W_b \, dx + \int_{(t_1+t_2)/2}^{w_1/2} W_c \, dx\right] \tag{7.53}$$

$$= \frac{\pi (n_2 i_2)^2}{6 g w_2^2 w_1} t_1 t_2 (t_1^2 + t_2^2).$$

The energy per stator tooth will be:

$$W_1 = \frac{w_1}{w_2} W_2, \tag{7.54}$$

and since $n_2 i_2 = n_1 i_1 (w_2/w_1)$, the average energy per stator tooth is:

$$W_1 = \frac{4\pi (n_1 i_1)^2}{24 g w_1^2 w_2}[t_1 t_2 (t_1^2 + t_2^2)], \tag{7.55}$$

and, therefore, the effective permeance per stator slot is, from Eq. 7.49:

$$P_z = \frac{t_1 t_2 (t_1^2 + t_2^2)}{12 g w_1^2 w_2}. \tag{7.56}$$

The value of P_z is an equivalent depth over width ratio that, if substituted for P_{s1} in Eq. 7.2, will give the total zigzag reactance of primary and secondary in ohms for a full-pitch winding. The factor K_s (Fig. 7.3) must be included for fractional-pitch windings. Letting:

$$\frac{t_1}{w_1} = a_1, \quad \frac{t_2}{w_2} = a_2, \quad \text{and} \quad k = a_1 a_2$$

(see Eq. 7.46), P_z (Eq. 7.56) becomes:

$$P_z = \frac{K_s k (a_1^2 w_1^2 + a_2^2 w_2^2)}{12 g w_1} \quad \text{numeric.} \tag{7.57}$$

Substituting this for P_{s1} in Eq. 7.2:

$$X_Z = (3.16 f q L N_1^2 \, 10^{-7}) \left[\frac{k K_s (a_1^2 w_1^2 + a_2^2 w_2^2)}{12 g S_1 w_1} \right] \text{ohms,} \tag{7.58}$$

with dimensions in centimeters.

Substituting the value of X_M from Eq. 7.1, while taking the saturation factor, k_i, as unity, and recognizing that $g = kG$ and $S_1 = 2 P s_1$:

$$X_Z = \frac{\pi^2 X_M K_s}{12 K_{p1}^2 K_{d1}^2} \left(\frac{a_1^2}{s_1^2} + \frac{a_2^2}{s_2^2} \right). \tag{7.59}$$

The mmf across the slot opening that creates the zigzag flux is the same as that acting on the upper part of the slot, which we have already seen has an effective pitch factor equal to K_s. In the winding we are considering, also, $K_d = 1$, as there is only one slot per pole per phase. Hence, K_s is very nearly equal to $K_{p1}^2 K_{d1}^2$, and Eq. 7.59 may be written:

$$X_Z = \frac{\pi^2 X_M}{12} \left(\frac{a_1^2}{s_1^2} + \frac{a_2^2}{s_2^2} \right), \tag{7.45}$$

which is the same as Eq. 7.45 derived by the difference method. The method as here developed is inexact, because the fringing is assumed to be abrupt, as in Fig. 7.5 (c). There is also the question of whether to use the upper or lower curve of Fig. 6.12, or some intermediate value in calculating a_1 and a_2. Hence, the mean formula (Eq. 7.47), as shown in Fig. 7.6, is recommended for practical use.

9 Effect of Skew on Reactance

If the slots of either rotor or stator are skewed, or spiraled, the variations in flux density, magnetic pull, and torque due to the slot openings will be displaced in time phase over the core length, resulting in more uniform torque, less noise, and better voltage waveform. However, skew increases the reactance, because the voltage induced by the fundamental sine wave of flux is displaced in successive elements of the conductor length.

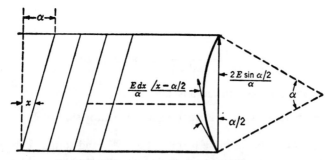

Fig. 7.8. Effect of Skew on Induced Voltage.

In Fig. 7.8, α represents the total electrical angle of skew, and x the electrical angle at any point along the conductor, measured from one end. The element of voltage induced in a short length of the conductor at point x is $\dfrac{E\,dx}{\alpha}\underline{/x}$, where E is the total voltage induced in a straight conductor. The net voltage across the conductor is the sum of the projections of the elementary voltages on the midpoint value, or

$$\text{Net } E = \int_0^\alpha \frac{E}{\alpha}\cos\left(x - \frac{\alpha}{2}\right)dx = \frac{2E\sin\alpha/2}{\alpha}.$$

Consequently, the resultant voltage in each secondary conductor is decreased by the factor

$$\frac{2\sin\alpha/2}{\alpha},$$

representing the ratio of the chord to the arc. Nevertheless the voltage induced in each primary conductor remains at full value. The ratio of mutual to total air-gap reactance is thus reduced by the above ratio, and this gives an additional component of leakage inductance expressed by Eq. 2.6.

Since L_1 and L_2 have the same values as previously, but the old value of M for straight slots becomes $M' = \dfrac{2M\sin\alpha/2}{\alpha}$, the added reactance is due to the difference in the M^2 term, or:

$$L_{\text{spiral}} = \frac{L_1 L_2 - M'^2}{L_2} = \frac{L_1 L_2 - M^2}{L_2} + \frac{M^2 - M'^2}{L_2},$$

and

(7.60) $$X_{\text{skew}} = X_M\left(1 - \frac{4\sin^2\dfrac{\alpha}{2}}{\alpha^2}\right)$$

$$= \frac{\alpha^2 X_M}{12} = \frac{\pi^2 \sigma^2 X_M}{12 s_1^2} \text{ approximately,}$$

where α = angle of skew in radians, and

σ = angle of skew as a fraction of one stator slot pitch.

Equation 7.60 has the same form as Eq. 7.47, and so may conveniently be added as an extra term to the formula for zigzag reactance. By symmetry, half is on the stator and half on the rotor side of the magnetizing current branch of the circuit. As pointed out previously, spiraling also largely eliminates parasitic currents induced in a squirrel-cage winding by tooth flux pulsations, and so has the further effect of increasing the factor F_{sc} in Eq. 7.47, thus increasing the reactance even more than indicated by Eq. 7.60.

However, the skew reactance is obtained by an increase in radial flux density at the ends of the core length, over the density in the center of the core, in proportion to the changing phasor difference of primary and secondary current densities at these positions (Fig. 9.10). This increase is sharply limited by magnetic saturation of the rotor and stator teeth. In practice, therefore, spiraling adds very little to the standstill reactance at full voltage, but increases materially the reactance at low currents.

10 THE BELT-LEAKAGE REACTANCE

Consider now a polyphase winding with a very large number of slots per pole, so that the zigzag reactance (Eq. 7.47) is practically zero. Each coil of the winding will then produce a smooth rectangular flux waveform (Fig. 7.5 (a)), and there will be no slot harmonics. However, all the harmonics of a rectangular wave will be present (Eq. 3.34), the magnitude of the mth harmonic being proportional to $K_{pm}K_{dm}$, except that, as shown in Sect. 3.12, all triple harmonics are eliminated in a symmetrical three-phase winding.

THE BELT-LEAKAGE REACTANCE

All the harmonic fields rotate at lower speeds than the fundamental, and so will produce parasitic, non-useful, secondary voltages when the motor is at speed. They are all elements of leakage reactance. They create a continually changing shape of the flux wave as it revolves (Fig. 3.14).

The voltage induced in the primary winding by the mth primary winding harmonic is, from Eq. 3.15:

(7.61) $$E_m = 4.44 f N K_{pm} K_{dm} \phi_M \text{ volts.}$$

From Eq. 3.34, the ratio of the maximum flux density of the mth harmonic to the fundamental sine-wave density is:

(7.62) $$\frac{B_{gm}}{B_g} = \frac{K_{pm} K_{dm}}{m K_{p1} K_{d1}} \text{ numeric,}$$

and the ratio of the mth harmonic flux per pole to the fundamental flux will be:

(7.63) $$\frac{\phi_m}{\phi} = \frac{K_{pm} K_{dm}}{m^2 K_{p1} K_{d1}} \text{ numeric,}$$

since the harmonic pole pitch is one-mth that of the fundamental. Substituting Eq. 7.63 in Eq. 7.61, and dividing by the fundamental voltage, the per unit leakage reactance due to the mth harmonic is:

(7.64) $$\frac{E_m}{E} = \frac{X_{Bm}}{X_M} = \frac{K_{pm}^2 K_{dm}^2}{m^2 K_{p1}^2 K_{d1}^2}.$$

Noting, also, that the winding we are considering is assumed to have a very large number of slots per pole, we may make n very large in Eq. 3.16, deriving:

(7.65) $$K_{dm} = \frac{2q}{\pi m} \sin \frac{m\pi}{2q}.$$

For a three-phase, balanced, 60° phase-belt winding, $q = 3$, and $\sin(m\pi/2q)$ has the values 0.500, -0.500, 0.500, -0.500, 0.500, ... for the fundamental, 5th, 7th, 11th, 13th, etc., harmonics; and unity for all the triple harmonics. (As stated above, the triple harmonics of the three phases cancel each other; but when current flows in a single phase alone, line to neutral, the triple harmonics remain and are important.) We may, therefore, replace Eq. 7.64 for this winding with a very large number of slots by:

(7.66) $$\frac{X_{Bm}}{X_M} = \frac{K_{pm}^2}{m^4 K_{p1}^2}.$$

This result may be visualized by noting from Eq. 7.1 that the magnetizing reactance of a normal $2P$-pole winding varies inversely as P^2, and that another P^2 factor enters in this case, because the winding is only one-mth as effective in producing an mth harmonic flux and in generating an mth harmonic voltage as it is for the fundamental.

The total belt-leakage reactance is merely the sum of all the reactances due to these phase-belt harmonics of the infinite slot winding, Fig. 3.15. It is obtained by calculating the sum of all the values of Eq. 7.66 for successive values of m. This process has been carried through for 60°, 90°, and 120° phase-belt windings, for a full range of values of winding pitch, with the results shown in Fig. 7.9. For full pitch, these three values may be summed exactly. Making $K_{pm} = K_{p1} = 1$ in Eq. 7.66:

$$(7.67) \quad \frac{X_B}{X_M} = \frac{1}{3^4} + \frac{1}{5^4} + \frac{1}{7^4} + \frac{1}{9^4} + \cdots$$

$$= \frac{\pi^4}{96} - 1 = 0.0147 \text{ for two-phase, } 90° \text{ phase belts.}$$

$$(7.68) \quad \frac{X_B}{X_M} = \frac{1}{5^4} + \frac{1}{7^4} + \frac{1}{11^4} + \frac{1}{13^4} + \cdots$$

$$= \frac{5\pi^4}{486} - 1 = 0.00214 \text{ for three-phase, } 60° \text{ or } 120° \text{ phase belts.}$$

These equations check with Eqs. 7.24 and 7.25.

Although X_B is the same at full pitch for 120° as for 60° phase belts, the 120° reactance increases rapidly as the pitch departs from unity, since the even harmonics then become large, while the 60° reactance decreases.

Equation 7.68 states that, if a motor has a ratio of magnetizing to total leakage reactance (Sect. 5.3, value of $1/ab$) of 25, giving a full-load power factor around 91 or 92 per cent, and has a full-pitch winding, its belt leakage reactance will be 0.00214×25, or 5 per cent of the total leakage reactance. This appears small, but it represents a serious handicap to the motor performance, because of the induced secondary currents, losses, and parasitic torques produced. If the three-phase winding pitch is reduced to about 80 per cent, Fig. 7.9 indicates that the belt-leakage reactance will be reduced to 0.00024, little more than a tenth of the full-pitch value.

The reactance of a phase-wound secondary is also given by Fig. 7.9, the total for both windings being the sum of the primary and secondary values. A squirrel-cage winding has no phase belts, and its belt leakage is, therefore, zero.

Although it is true that the exact belt leakage for a winding with a finite number of slots per pole is somewhat more than given by Fig. 7.9, because of the approximation of Eq. 7.65, the difference is unimportant. The effect of open slots on the belt leakage is small, since the important phase-belt harmonics span several slot openings. As shown in Sect. 7.6, the presence of slot openings materially reduces the zigzag leakage (Fig. 7.6). This is another reason for considering the zigzag and belt leakages independently.

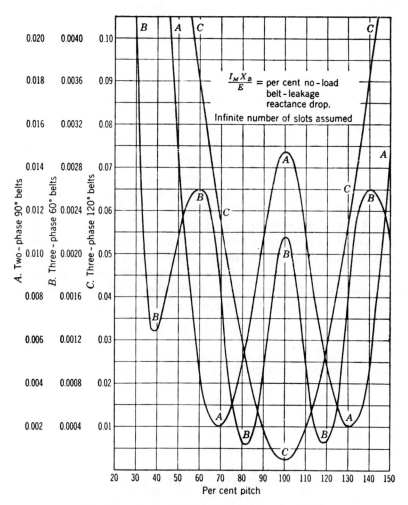

Fig. 7.9. Chart of Values of $I_M X_B/E$ against Pitch.

If the number of slots per pole is fractional, making successive phase belts of different width, or if other irregularities are introduced, the belt leakage will be increased accordingly. The analysis of such irregular windings, and the selection of the most favorable sequence of slot numbers in successive phase belts, form a very interesting technical problem of too great a scope to consider here. It can be carried through by the method outlined in Section 3.14.

The belt-leakage harmonics of each winding induce voltages in the opposing winding, causing opposing circulating currents, which reduce the effective reactance. A squirrel-cage rotor thus reduces the belt-leakage reactance of the stator nearly to zero. For this reason, X_B is neglected in squirrel-cage motor reactance calculations.

11 Peripheral Air-Gap Leakage

So far we have assumed that all the air-gap flux produced by the stator mmf crosses the air gap in radial paths. We have allowed for the peripherally directed leakage flux crossing the slots, and that bridging the slot openings by passing across the air gap and back. There is also a certain amount of flux flowing in a peripheral direction entirely within the air gap, because of the mmf acting on this narrow air path from pole to pole. Evidently this component will become rapidly larger if the air gap is increased. If the rotor is entirely removed, all the flux within the stator bore will be of this air leakage character.

In Appendix B, equations are derived for the flux distribution in the annular air gap, or in the air core if the rotor is removed. This analysis shows that the ratio of the fundamental component only of the air-gap peripheral leakage from pole to pole (flux that never reaches the rotor surface) to the fundamental sine-wave flux is (Eq. 7.31B):

$$(7.69) \qquad \frac{X_p}{X_M} = \frac{2P^2 g^2}{D^2}.$$

This is completely negligible for normal induction machines, which rarely have an air gap to pole pitch ratio $(2Pg/\pi D)$ larger than 0.01, but it may be quite appreciable for high-speed synchronous or d-c machines having large air gaps.

From Eq. 7.69 and Eq. 7.66, the air-gap peripheral leakage component of the mth harmonic (belt-leakage) flux is:

$$(7.70) \qquad \frac{X_{pm}}{X_M} = \frac{2m^2 P^2 g^2}{D^2}\left(\frac{K_{pm}^2}{m^4 K_{p1}^2}\right) = \frac{2P^2 g^2 K_{pm}^2}{m^2 D^2 K_{p1}^2}.$$

For a full-pitch, three-phase, 60° phase-belt winding, the sum of all the harmonic components of peripheral leakage is:

$$(7.71) \qquad \frac{X_{pB}}{X_M} = \frac{2P^2 g^2}{D^2}\left(\frac{1}{5^2}+\frac{1}{7^2}+\frac{1}{11^2}+\frac{1}{13^2}+\cdots\right)$$
$$= \frac{2P^2 g^2}{D^2}\left(\frac{\pi^2}{9}-1\right) = 0.096\left(\frac{2P^2 g^2}{D^2}\right).$$

The total three-phase air-gap peripheral leakage, therefore, lies between 100 and 110 per cent of the value given by Eq. 7.69, depending on the winding pitch, and for two-phase it lies between 100 and 124 per cent.

12 Coil End Leakage of Armature Windings

The usual diamond-shaped pattern of the stator end windings, Fig. 7.10, together with the varying configurations of the end rings of a squirrel-cage rotor, or a field winding, and the locations of adjacent magnetic parts, give

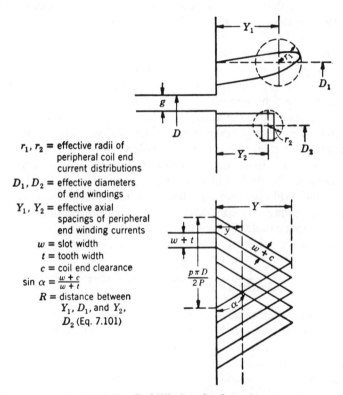

r_1, r_2 = effective radii of peripheral coil end current distributions

D_1, D_2 = effective diameters of end windings

Y_1, Y_2 = effective axial spacings of peripheral end winding currents

w = slot width
t = tooth width
c = coil end clearance
$\sin \alpha = \frac{w+c}{w+t}$
R = distance between Y_1, D_1, and Y_2, D_2 (Eq. 7.101)

FIG. 7.10. End Winding Configuration.

a complicated three-dimensional pattern of end leakage flux that is very difficult to determine exactly. However, the principle of superposition permits the total field to be resolved into components, each of which can be clearly visualized and calculated approximately. By this plan, the effects of arbitrary changes in design on each element may be judged, and the total end reactance estimated with good accuracy, for a wide range of proportions.

All the end leakage flux lines are assumed to flow in radial planes at the end of the armature core, making the core end a plane of symmetry. In other words, the calculations are made as if the armature core were removed, and the winding projections on the two ends were brought together, with all the flux paths entirely in air. This assumption is in part justified by the symmetry of the windings, and in part by the damping effect of eddy currents in the plane of the laminations, which oppose any flux entering the stator core in an axial direction. In the rotor or field, however, where the current is either direct or of low frequency, axially directed end leakage flux can enter the core unopposed, so that a corrective term to allow for this should be added to the secondary reactance as calculated by the above method. A corrective term for the stator may also be useful, especially as an index of load losses in the end structure.

On this basis, the flux flowing in the leakage paths linking the end windings will be considered as the sum of two entirely independent elements, one produced by the axially directed components, and the other by the peripherally directed components of the end winding current. Even though these two elements of leakage flux have portions of their paths in common, they may be treated entirely independently. This may be seen by consideration of the flux linkages of two parallel conductors in air, an example which will also provide a foundation for this analysis.

A current flowing in a conductor A, far removed from any other conductor, produces circular lines of flux around itself, the flux density at any radius R from the conductor being (from Eq. 1.17):

$$(1.22) \qquad B_R = \frac{4\pi I}{2\pi R 10^7} = \frac{2I}{R 10^7} \text{ webers per square meter,}$$

if I is given in amperes and R in meters, as shown in Fig. 1.5. If, now, a similar, parallel conductor, B, is brought near A, the resultant field at any point in the surrounding space will be equal to the phasor sum of the two circular fields calculated for the two conductors independently. Although the resultant field has a complex geometric pattern, the linkages can be simply and accurately calculated by summing the fields of the two conductors, as if they existed independently. Thus, the total inductance of a transmission loop with outgoing and ingoing wires, each of radius r and separated by a distance R, from Fig. 1.5, is:

$$L = 2 \times 2 \times 10^{-7} \left(\int_0^r \frac{x^3}{r^4} dx + \int_r^\infty \frac{1}{x} dx - \int_0^\infty \frac{dx}{R+x} \right)$$

$$= 4 \times 10^{-7} \left[\frac{1}{4} + \int_r^R \frac{1}{x} dx - \int_0^R \frac{dx}{R+x} + \int_R^\infty \left(\frac{1}{x} - \frac{1}{R+x} \right) dx \right];$$

or

(7.72) $$L = 4 \times 10^{-7}\left(\frac{1}{4} + \ln\frac{R}{r}\right) \text{ henries per meter of line}$$

$$= 6.44 \times 10^{-4}\left(\frac{1}{4} + 2.303 \log\frac{R}{r}\right) \text{ henries per mile of line.}$$

This gives a reactance for the loop, outgoing and return, equal to $2\pi f$ times Eq. 7.72, or:

(7.73) $$X = 2.51 f 10^{-6}\left(\frac{1}{4} + 2.303 \log\frac{R}{r}\right) \text{ ohms per meter.}$$

The first term in the brackets represents the internal inductance, due to flux within the conductor itself. This internal flux increases the reactance for current flowing in the interior of the wire, as compared with the surface current paths, so that in large wires, or at high frequencies, the current is concentrated in a surface layer of the conductor. This "skin effect" reduces the inductance and increases the resistance.

Rayleigh and Niven have shown that the self-inductance of a one-turn coil of circular cross section, of diameter D, and cross-section radius r, is:

(7.74) $$L = 2\pi D\left[\left(1 + \frac{r^2}{2D^2}\right)\ln\frac{4D}{r} + \frac{r^2}{6D^2} - 1.75\right]10^{-7} \text{ henries,}$$

approximately, if D and r are expressed in meters; and Clerk Maxwell showed that the mutual inductance of two similar coaxial coils, of diameters D_1 and D_2 meters, and with a slant spacing between them (Fig. 7.10) equal to R, is:

(7.75) $$M = 2\pi\sqrt{(D_1 D_2)}\left[\left(1 + \frac{3R^2}{4D_1 D_2}\right)\ln\frac{4\sqrt{(D_1 D_2)}}{R}\right.$$
$$\left. - \left(2 + \frac{R^2}{4D_1 D_2}\right)\right]10^{-7} \text{ henries, approximately.}$$

For our purposes, the terms in r^2 and R^2 in these equations will be neglected, as r and R are small in comparison with D. Hence, the inductance of two identical one-turn coaxial coils connected in series opposition is, from Eqs. 7.74 and 7.75:

(7.76) $$2(L - M) = 4\pi D\left(\ln\frac{R}{r} + \frac{1}{4}\right)10^{-7} \text{ henries,}$$

with dimensions in meters.

Multiplying Eq. 7.72 by the periphery of one coil, πD, gives Eq. 7.76 also, showing that the two large-diameter coils can be treated as if they formed a transmission loop.

For the diamond-shaped coil ends of Fig. 7.10, at any point distant $\bar{y}$ from the bend of the coil, the amperes per meter of periphery in the outgoing coil side may be represented by:

$$\Delta \cos\left[\theta - \frac{p\pi}{2}\left(1-\frac{y}{Y}\right)\right],$$

and the current in the returning coil side is:

$$\Delta \cos\left[\theta + \frac{p\pi}{2}\left(1-\frac{y}{Y}\right)\right],$$

where Δ is the maximum amperes per meter of periphery in one layer of the winding of diameter D_1, as it leaves the core. If A = maximum ampere turns of armature mmf per pole for the entire winding (Eq. 3.35):

(7.77) $$\Delta = \frac{PA}{D_1 K_p} = \frac{2qNK_d I}{\pi D_1 \sqrt{2}}.$$

When the end turns bend at an angle α, the spacing between coil side centers is decreased in the ratio of $\sin \alpha$ to 1 (Fig. 6.9), so the value of Δ in the end winding (taken at right angles to the conductors in either layer) is $(1/\sin \alpha)$ times the value given by Eq. 7.77.

In the above, P is the number of pairs of poles, D_1 is the mean diameter of the stator end windings, N is the turns in series per phase, and K_p and K_d are the normal pitch and distribution factors of the winding. Y is the axial length of end winding projection on each end (Fig. 7.10):

(7.78) $$Y = \frac{p\pi D_1}{4P}\tan\alpha.$$

The total end winding current in the axial direction is obtained by multiplying each current by $\sin \alpha$, to obtain the axial component, and then subtracting the returning from the outgoing current. The $(1/\sin \alpha)$ factor in Δ mentioned above cancels the $\sin \alpha$ factor, so that the resultant axial current in the end windings is:

(7.79) $$I_a = \Delta\left\{\cos\left[\theta - \frac{p\pi}{2}\left(1-\frac{y}{Y}\right)\right] - \cos\left[\theta + \frac{p\pi}{2}\left(1-\frac{y}{Y}\right)\right]\right\}$$

$$= \frac{2PA}{D_1 K_p}\sin\theta\sin\frac{p\pi}{2}\left(1-\frac{y}{Y}\right) \text{ amperes per meter}.$$

The integral of I_a between $\theta = 0$ and $\theta = \pi/2$ is:

$$(7.80) \quad A_y = \int_0^{\pi/2} I_a \left(\frac{D_1}{2P}\right) d\theta = 2\left(\frac{PA}{D_1 K_p}\right)\left(\frac{D_1}{2P}\right) \cos\theta \sin\left[\frac{p\pi}{2}\left(1 - \frac{y}{Y}\right)\right]$$

$$= \frac{A}{K_p} \cos\theta \sin\left[\frac{p\pi}{2}\left(1 - \frac{y}{Y}\right)\right].$$

If Y is 0, this becomes simply A, as $K_p = \sin(p\pi/2)$ by Eq. 3.12; and if $y = Y$, it is zero.

That is, the resultant axial current in the end windings may be regarded as a sheet of current whose intensity, given by Eq. 7.79, varies sinusoidally around the periphery (with θ) and axially (with y), and whose total amount in one-half pole pitch at the bend of the coil is the peak armature mmf.

Assuming the flux produced by this axial current sheet to flow entirely in radial planes, the total linkages produced may be found by first calculating the flux produced in a small axial distance dy, and then integrating from 0 to Y.

As shown in Appendix B (Eq. 7.25B), the total flux produced in air by a $2P$-pole mmf of A peak ampere turns per pole, distributed sinusoidally around a cylinder of any diameter, is:

$$(7.81) \quad \phi_y = 4\pi (10^{-7}) A_y \text{ webers per meter of core length.}$$

Since A_y, at a distance y from the bend of the coil, is given by Eq. 7.80, and the linkages produced will also be proportional to the pitch factor times the distribution factor at each point, the total voltage due to the axially directed currents, at the two ends of the core, will be (with $\theta = 0$), from Eqs. 1.21, 7.80, and 7.81:

$$V_{ea} = 2 \times 4.44 f N \phi_{max} K_d \sin\frac{p\pi}{2}(1 - y/Y)$$

$$= 8.88 f N \left(\frac{4\pi A K_d}{K_p 10^7}\right) \int_0^Y \sin^2\frac{p\pi}{2}(1 - y/Y) \, dy$$

$$= \frac{1.116 f A N K_d}{10^5 K_p} \left[\frac{y}{2} + \frac{Y \sin p\pi(1 - y/Y)}{2p\pi}\right]_0^Y$$

$$= \frac{0.558 f A Y N K_d}{10^5 K_p}\left(\frac{p\pi - \sin p\pi}{p\pi}\right),$$

$$(7.82) \quad V_{ea} = \frac{0.439 f A N D_1 K_d \tan\alpha}{10^5 K_p P}\left(\frac{p\pi - \sin p\pi}{\pi}\right) \text{ volts per phase,}$$

with D expressed in meters.

Since $A = \dfrac{\sqrt{2}\,qNK_p K_d I}{\pi P}$, the corresponding reactance is:

(7.83) $\quad X_{ea} = \dfrac{0.197 fq N^2 D_1 K_d^2 \tan\alpha}{P^2 10^5}\left(\dfrac{p\pi - \sin p\pi}{\pi}\right)$ ohms per phase.

The coefficient becomes 0.501×10^{-7} if D is expressed in inches.

Before Eq. 7.83 is used to calculate end leakage reactance of an actual machine, three corrections must be made in it to allow, respectively, for the mutual inductance with the rotor end windings, for the presence of magnetic retaining rings or of circulating currents in the binding bands, if any, and for the effects of slot and tooth harmonics of the end leakage flux.

As shown in Appendix B, Eq. 7.34B, the fraction of the radial flux per pole produced by a sinusoidal winding of diameter D_1, which links an inner winding at diameter D_2, assuming all the flux paths entirely in air, is:

(7.84) $\quad\quad\quad\quad \phi_2/\phi_1 = (D_2/D_1)^P.$

If the flux paths are entirely in air except that there is a magnetic retaining ring of outer diameter D_3 and inner diameter D_4 between the windings, the mutual inductance is zero, while the self-inductance of the outer winding at diameter D_1 is increased by the factor, from Eq. 7.39B:

(7.85) $\quad\quad\quad \dfrac{\phi_{\text{mag ring}}}{\phi_{\text{air}}} = 1 + \left(\dfrac{D_3}{D_1}\right)^{2P},$

and the self-inductance of the inner winding at diameter D_2 is increased by the factor:

(7.86) $\quad\quad\quad \dfrac{\phi_{\text{mag ring}}}{\phi_{\text{air}}} = 1 + \left(\dfrac{D_2}{D_4}\right)^{2P}.$

Hence, if the rotor end windings are similar to those in the stator, and are located at a diameter D_2, Eq. 7.83 must be multiplied by $1 - (D_2/D_1)^P$ to obtain the net leakage flux. If the windings are dissimilar, judgment should be employed to choose a factor that reasonably represents the ratio of leakage to total radial flux. If magnetic retaining rings are used, of outer diameter D_3, Eq. 7.83 must be multiplied by the factor $1 + (D_3/D_1)^{2P}$, for the stator. In the case of rotor end windings of diameter D_2, Eq. 7.83 must be multiplied by the same factor $1 - (D_2/D_1)^P$ to take account of the mutual flux, or by a factor $1 + (D_2/D_4)^{2P}$ if a magnetic retaining ring of inner diameter D_4 is used.

Since Eq. 7.25B shows that the radial flux per pole in the end windings is the same for the same peak ampere turns per pole, regardless of the number of poles, the effective air gap for the mth harmonic flux in the end windings is only $1/m$th the air gap for the fundamental. Thus, the belt leakage in the end windings will be given by Eq. 7.66, except that $1/m^3$ must be substituted for $1/m^4$. For a two-phase, full-pitch winding, this gives:

(7.87) $$\frac{X_B}{X_{ea}} = \frac{1}{3^3} + \frac{1}{5^3} + \frac{1}{7^3} + \ldots = 0.051,$$

and for a three-phase full-pitch winding, it is:

(7.88) $$\frac{X_B}{X_{ea}} = \frac{1}{5^3} + \frac{1}{7^3} + \frac{1}{11^3} + \frac{1}{13^3} + \ldots = 0.013.$$

If the winding pitch is less than unity, the (pitch factor)^2 for the mth harmonics, averaged from $y = 0$ to $y = Y$, will be:

(7.89) $$\int_0^Y \sin^2 \frac{m p \pi}{2} \left(1 - \frac{y}{Y}\right) dy = \frac{Y}{2}\left(\frac{m p \pi - \sin m p \pi}{m p \pi}\right),$$

and the ratio of this to the fundamental value, from Eq. 7.82, is:

(7.90) $$\frac{m p \pi - \sin m p \pi}{m(p \pi - \sin p \pi)}.$$

This increases with m, and also becomes larger as p decreases, but can hardly ever exceed 3. The product of Eq. 7.90 by Eq. 7.88 is, therefore, <0.04, and this belt-leakage term can be neglected.

To calculate the slot harmonic components of the radial end leakage flux, we must recognize that for them also the air gap is reduced in proportion to the harmonic order by Eq. 7.25B. As a first approximation, therefore, the value of per unit zigzag reactance for the stator, Eq. 7.33, should be multiplied by the average value of m for the tooth harmonics, or by $2s_1$. However, Eq. 7.20B shows that the radial flux density of the high-order harmonic fields in an air core decreases as the $(P-1)$ power of the radius on each side of the exciting winding. As the mmf is located in a coil side of depth equal to one-half the slot depth, and as the effective air gap for the $2s$th harmonic is $D_1/4Ps_1$ by Eq. 7.24B, we may assume that the actual air gap for these harmonics is closely represented by the sum of this value and half the slot depth. Therefore, the tooth harmonic component of the radial end winding reactance may be estimated by multiplying X_{ea} by:

(7.91) $$\frac{\frac{\pi^2}{12s_1^2}(2s_1)}{1 + \frac{2Ps_1 d_s}{D_1}} = \frac{1.65}{s_1\left(1 + \frac{S_1 d_s}{D_1}\right)},$$

where s_1 = slots per pole, S_1 = total slots, d_s = depth of slot, and D_1 = average diameter of the end winding, for the stator. This value is still high, because the coil spacing in the bent part of the end windings is reduced, and the mmf is distributed over the conductor width, instead of being concentrated at the slot edges, both of which reduce the amplitudes of the slot harmonic mmfs. On this account, and because the second term in the denominator of Eq. 7.91 will always be considerably greater than 1, we may use as a final value for the ratio of slot harmonic to total radial and leakage reactance of the stator winding about two-thirds of Eq. 7.91, or:

$$\text{(7.92)} \qquad \frac{D_1}{s_1 S_1 d_s} = \frac{D_1}{2Ps_1^2 d_s}.$$

The numerical value of Eq. 7.92 is rarely more than 0.03, so it may be neglected.

The final expression for the net end flux leakage reactance, due to axial currents, will, therefore, be given by multiplying Eq. 7.83 by $[1-(D_2/D_1)^P]$ from Eq. 7.84, or, for magnetic end rings, by $[1+(D_3/D_1)^{2P}]$ from Eq. 7.85. However, Eq. 7.84 applies only when the inner winding end turns are exactly opposite the stator end turns, as in a slip-ring motor. In practice, the rotor winding end projections, or end rings of a squirrel cage, are usually much shorter than the stator end windings. For this reason, it is necessary to modify Eq. 7.84 by using an equivalent (smaller) value of D_2, determined by experience. For present purposes, $0.8D_2$ will be used in place of D_2.

This gives for the end leakage reactance produced by axial end currents:

$$\text{(7.93)} \qquad X_{ea} = \frac{2qfD_1 N^2 K_d^2 \tan \alpha}{P^2 10^6}$$

$$\times \left(\frac{p\pi - \sin p\pi}{\pi}\right)\left[1 - \left(\frac{0.8D_2}{D_1}\right)^P\right]^\dagger \text{ ohms per phase,}$$

if D is expressed in meters.

The peripherally directed component of total end winding current at y, θ, is the sum of the ingoing and outgoing currents multiplied by $\csc \alpha$ as before, and by $\cos \alpha$:

$$\text{(7.94)} \quad I_p = \Delta \left\{ \cos\left[\theta - \frac{p\pi}{2}\left(1 - \frac{y}{Y}\right)\right] + \cos\left[\theta + \frac{p\pi}{2}\left(1 - \frac{y}{Y}\right)\right] \right\} \cot \alpha$$

$$= 2\Delta \cot \alpha \cos \theta \cos\left[\frac{p\pi}{2}\left(1 - \frac{y}{Y}\right)\right].$$

---

† If the machine has magnetic retaining rings of outer diameter D_3, the factor in brackets is to be replaced by $[1 + (D_3/D_1)^{2P}]$, and an additional expression similar to Eq. 7.93 must be added to give the separate rotor end reactance.

COIL END LEAKAGE OF ARMATURE WINDINGS   241

The integral of I_p from 0 to Y gives the total peripheral current:

(7.95) $$I_{ep} = 2\Delta \cot \alpha \cos \theta \int_0^Y \cos \frac{p\pi}{2}\left(1 - \frac{y}{Y}\right) dy$$

$$= \frac{4Y}{p\pi} \Delta \cot \alpha \cos \theta \sin \frac{p\pi}{2} = A \cos \theta \text{ amperes.}$$

Since this varies sinusoidally around the periphery (with θ), its rms value is:

(7.96) $$I_{ep} = \frac{A}{\sqrt{2}} \text{ rms amperes.}$$

The "center of gravity" of this peripheral current is located at a value of y determined by finding the integral of the product of y times the current and dividing this by the total current. This is:

$$Y_1 = \frac{1}{I_{ep}} \int_0^Y y I_p \, dy = \frac{p\pi}{2YK_p} \int_0^Y y \cos\left[\frac{p\pi}{2}\left(1 - \frac{y}{Y}\right)\right] dy$$

$$= \frac{p\pi}{2YK_p}\left\{-\frac{2Yy}{p\pi}\sin\left[\frac{p\pi}{2}\left(1-\frac{y}{Y}\right)\right] + \frac{4Y^2}{p^2\pi^2}\cos\left[\frac{p\pi}{2}\left(1-\frac{y}{Y}\right)\right]\right\}_0^Y$$

$$= \frac{p\pi}{2YK_p}\left[\frac{4Y^2}{p^2\pi^2}\left(1 - \cos\frac{p\pi}{2}\right)\right] = \frac{2Y}{p\pi}\left(\frac{1-\cos\frac{p\pi}{2}}{\sin\frac{p\pi}{2}}\right) = \frac{2Y}{p\pi}\tan\frac{p\pi}{4};$$

or

(7.97) $$Y_1 = \frac{D_1}{2P}\tan \alpha \tan \frac{p\pi}{4} = \frac{\pi p D_1}{8P}\left(1 + \frac{p^2}{5}\right)\tan \alpha, \text{ approximately.}$$

We may, therefore, consider the peripheral end leakage flux of the stator winding as due to a circular coil of diameter equal to the mean diameter of the end windings, located $(D/2P)\tan \alpha \tan(p\pi/4)$ meters out from the stator core, and carrying a current of $A/\sqrt{2}$ rms amperes. The inductance of such a coil is given by Eq. 7.74. Thus, the total reactive volt-amperes due to the peripheral component of the stator end winding currents is, including both ends, from Eqs. 7.74 and 7.96:

(7.98) $$2I_a^2 X_{ep} = (2\pi f)(2)(2\pi D_1 10^{-7})\left(\frac{A}{\sqrt{2}}\right)^2 \left(\ln \frac{4D_1}{r} - 1.75\right)$$

$$= 0.395 f D_1 A^2 10^{-5}\left(\ln \frac{4D_1}{r} - 1.75\right),$$

with D in meters.

The corresponding value of ohms per phase for the stator is obtained by dividing Eq. 7.98 by the number of phases, q, and by the square of the rated current per phase, I_1^2. Substituting also $\sqrt{2}\, qN_1 K_p K_d I/\pi P$ for A, we obtain:

$$(7.99) \quad \text{Total } X_{ep1} = \frac{0.80 q f N_1^2 K_{p1}^2 K_{d1}^2}{P^2 10^6}$$

$$\times \left(D_1 \ln \frac{4D_1}{r_1} - 1.75 D_1 \right) \text{ ohms per phase.}$$

If the secondary, or rotor, end winding currents flow in a similar ring of diameter D_2, and the two mmfs are assumed to be equal and opposite, the mutual reactance between primary and secondary end windings will be the same as Eq. 7.99 except that the expression in brackets will become:

$$\left[\sqrt{(D_1 D_2)} \ln \frac{4\sqrt{(D_1 D_2)}}{R} - 2\sqrt{(D_1 D_2)} \right]$$

by analogy between Eqs. 7.74 and 7.75. We may assume without appreciable error that $\sqrt{(D_1 D_2)}$ equals D, the air-gap diameter, for usual machines. Noting also that $\ln x = 2.30 \log x$, and transforming the second term of each bracket into an equivalent multiplier in the first expression, we may subtract the mutual from the total, obtaining finally the value of peripheral stator end leakage reactance:

$$(7.100) \quad \text{Net } X_{ep1} = \frac{1.84 q f N_1^2 K_{p1}^2 K_{d1}^2}{P^2 10^6}$$

$$\times \left(D_1 \log \frac{0.695 D_1}{r_1} - D \log \frac{0.541 D}{R} \right) \text{ ohms per phase,}$$

with dimensions in meters.

Here, r_1 is the mean radius of the imaginary circular coil in which the peripheral stator end winding current flows, and R is the perpendicular slant distance between this coil and the corresponding secondary coil.

It is suggested that the value of r_1 be taken as one-half the depth of stator slot, and the corresponding value of r_2 for the secondary as either one-half the depth of slot, or half the depth of field coil. The value of R is:

$$(7.101) \quad R = \sqrt{[0.25(D_1 - D_2)^2 + (Y_1 - Y_2)^2]},$$

where Y_1 is given by Eq. 7.97, and Y_2 is the corresponding axial distance from the core to the center of the peripheral current in the rotor end winding.

COIL END LEAKAGE OF ARMATURE WINDINGS

The total end leakage reactance of the stator (primary) is found by adding Eqs. 7.93 and 7.100, taking K_d^2 as 0.92:

$$(7.102) \quad X_e = \frac{1.84 q f N^2 D_1}{10^6 P^2} \left\{ \tan \alpha \left(\frac{p\pi - \sin p\pi}{\pi} \right) \left[1 - \left(\frac{0.8 D_2}{D_1} \right)^P \right] \right. $$
$$\left. + 0.93 K_{p1}^2 \left(\log \frac{1.4 D_1}{d_{s1}} - \frac{D}{D_1} \log \frac{0.54 D}{R} \right) \right\},$$

with D expressed in meters. The coefficient becomes 4.6×10^{-8} if D is in inches.

The rotor end leakage reactance may be found by similar calculations. Usually, however, it is sufficient to assume the per unit rotor and stator end leakages to be the same, making the total twice that given by Eq. 7.102. The importance of these reactance formulas is that they show the effects of changes in dimensions or proportions of the machine, rather than give any exact values. In practice, it is customary to obtain test data on machines more or less similar to a proposed new design, and find the coefficient needed to adjust the calculated value to match the test data. Then this coefficient is applied to the calculations for the new design.

In conclusion, a very rough but convenient formula for the total end leakage reactance is obtained by rounding off the first term of Eq. 7.102 and multiplying it by 4:

$$(7.103) \quad X_e = \frac{7 f q N^2 D}{P^2 10^6} (p - 0.3) \text{ ohms per phase,}$$

with dimensions in meters. The coefficient 7 becomes 0.18 if the dimensions are in inches. This reactance should be divided equally between the stator and rotor branches of the equivalent circuit.

Appendix A to Chapter 7

SINE-WAVE LINKAGES OF AN OPEN-SLOT STATOR

The permeance of Fig. 7.5 (b) is represented by the discontinuous equation:

$$(7.1A) \quad P = P_1 \Big]_0^{(1-a)\pi/2s} + P_0 \Big]_{(1-a)\pi/2s}^{(1+a)\pi/2s} + P_1 \Big]_{(1+a)\pi/2s}^{(3-a)\pi/2s}$$

$$+ P_0 \Big]_{(3-a)\pi/2s}^{(3+a)\pi/2s} + \ldots + P_1 \Big]_{\pi - [(1-a)\pi/2s]}^{\pi}.$$

The amplitude of the fundamental sine wave of flux produced by a full-pitch coil linking this permeance is:

$$F = \frac{2}{\pi}\int_0^\pi P \sin x \, dx = -\frac{2P_1 \cos x}{\pi}\Big]_0^{(1-a)\pi/2s} - \frac{2P_0}{\pi}\cos x \Big]_{(1-a)\pi/2s}^{(1+a)\pi/2s} - \ldots$$

$$= \frac{2(P_0 - P_1)}{\pi}\left[\cos\frac{(1-a)\pi}{2s} - \cos\frac{(1+a)\pi}{2s} + \cos\frac{(3-a)\pi}{2s}\right.$$

$$\left. - \cos\frac{(3+a)\pi}{2s} + \ldots\right] + \frac{4P_1}{\pi}$$

$$= \frac{4P_1}{\pi} + \frac{4(P_0 - P_1)}{\pi}\sin\frac{a\pi}{2s}\left[\sin\frac{\pi}{2s} + \sin\frac{3\pi}{2s} + \ldots + \sin\frac{(2s-1)\pi}{2s}\right].$$

The expression in brackets is the diameter of a regular polygon with $2s$ sides, each of unit length, giving:

$$(7.2A) \qquad F = \frac{4P_1}{\pi} + \frac{4(P_0 - P_1)}{\pi}\sin\frac{a\pi}{2s}\csc\frac{\pi}{2s}.$$

If $a = 1$, which is the closed-slot case, Fig. 7.5 (a), this reduces to $4P_0/\pi$.

Since unit linkages are equal to P_0, the area of a sine wave is $2/\pi$ times the maximum, and the combined linkages due to the s phases are $s/2$ times the self-linkages of a single phase, the per unit sine-wave linkages for the open-slot permeance, Fig. 7.5 (b) are:

$$L_s = \frac{sF}{\pi P_0} = \frac{4s}{\pi^2}\left(\frac{P_1}{P_0} + \frac{P_0 - P_1}{P_0}\sin\frac{a\pi}{2s}\csc\frac{\pi}{2s}\right).$$

Substituting $P_1 = \dfrac{(k-a)P_0}{1-a}$ from Eq. 7.36, and expanding in powers of $1/s$, we have, finally:

(7.3A) $$L_s = \frac{4ks}{\pi^2}\left\{1 + \frac{\pi^2}{24ks^2}[a(1-k)(1+a)] + \ldots\right\}.$$

Appendix B to Chapter 7

THE FLUX DISTRIBUTION AND THE ASSOCIATED ENERGY FLOW IN AN ANNULAR CORE

1B General Equations

It is useful to know the flux distribution that exists in the annular core back of the teeth of a stator or rotor lamination. The equations that apply to this problem apply also to the magnetic fields in air that are produced within the stator bore by the stator currents when the rotor is removed, or by the rotor currents when the stator is removed. Knowledge of these

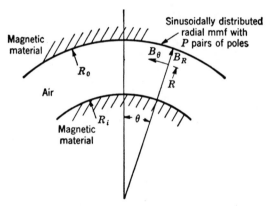

Fig. 7.11. Flux Distribution in Annular Air Gap Space.

fields is especially useful in calculating the leakage reactance of end windings.

It is assumed that a sinusoidally distributed mmf with P pairs of poles and A peak ampere turns per pole exists around the outer periphery of an annular air gap of indefinite axial length, and with inner (rotor) and outer (stator) radii of R_i and R_o inches, respectively (Fig. 7.11). The stator and rotor cores are both assumed to have infinite permeability throughout.

Since the lines of induction are continuous, the total flux leaving any area must equal the flux entering it. Letting B_R and B_θ represent the flux

APPENDIX B

densities in the radial (outward) and peripheral (counterclockwise) directions, respectively:

$$\left(B_\theta + \frac{\partial B_\theta}{\partial \theta}\partial\theta\right)\partial R - B_\theta\, \partial R + B_R(R+\partial R)\,\partial\theta$$
$$+ \frac{\partial B_R}{\partial R}\partial R(R\,\partial\theta) - B_R(R\,\partial\theta) = 0,$$

or:

(7.1B) $$\frac{\partial B_\theta}{\partial \theta} + B_R + R\frac{\partial B_R}{\partial R} = 0.$$

Also, as there is no mmf source in the air space, the line integral of mmf around the periphery of any area is zero, giving:

$$B_R\,\partial R + \left(B_\theta + \frac{\partial B_\theta}{\partial R}\partial R\right)(R\,\partial\theta) + B_\theta\,\partial R\,\partial\theta$$
$$-\left(BR + \frac{\partial B_R}{\partial \theta}\partial\theta\right)\partial R - B_\theta(R\,\partial\theta) = 0,$$

or:

(7.2B) $$\frac{R\,\partial B_\theta}{\partial R} + B_\theta - \frac{\partial B_R}{\partial \theta} = 0.$$

Differentiating Eqs. 7.1B and 7.2B with respect to R and θ:

(7.3B) $$\frac{\partial^2 B_\theta}{\partial R\,\partial\theta} + 2\frac{\partial B_R}{\partial R} + R\frac{\partial^2 B_R}{\partial R^2} = 0,$$

(7.4B) $$\frac{\partial^2 B_\theta}{\partial\theta^2} + \frac{\partial B_R}{\partial\theta} + R\frac{\partial^2 B_R}{\partial R\,\partial\theta} = 0,$$

(7.5B) $$\frac{\partial^2 B_R}{\partial R\,\partial\theta} - 2\frac{\partial B_\theta}{\partial R} - R\frac{\partial^2 B_\theta}{\partial R^2} = 0,$$

(7.6B) $$\frac{\partial^2 B_R}{\partial\theta^2} - R\frac{\partial^2 B_\theta}{\partial R\,\partial\theta} - \frac{\partial B_\theta}{\partial\theta} = 0.$$

Eliminating $\frac{\partial^2 B_\theta}{\partial R\,\partial\theta}$ from Eqs. 7.3B and 7.6B, and $\frac{\partial^2 B_R}{\partial R\,\partial\theta}$ from Eqs. 7.4B and 7.5B:

(7.7B) $$\frac{1}{R^2}\frac{\partial^2 B_R}{\partial\theta^2} + \frac{\partial^2 B_R}{\partial R^2} + \frac{2}{R}\frac{\partial B_R}{\partial R} - \frac{1}{R^2}\frac{\partial B_\theta}{\partial\theta} = 0,$$

and:

(7.8B) $$\frac{1}{R^2}\frac{\partial^2 B_\theta}{\partial\theta^2} + \frac{\partial^2 B_\theta}{\partial R^2} + \frac{2}{R}\frac{\partial B_\theta}{\partial R} + \frac{1}{R^2}\frac{\partial B_R}{\partial\theta} = 0.$$

Eliminating $\partial B_\theta/\partial\theta$ from Eqs. 7.1B and 7.7B and $\partial B_R/\partial\theta$ from Eqs. 7.2B and 7.8B:

(7.9B) $$\frac{1}{R^2}\frac{\partial^2 B_R}{\partial\theta^2}+\frac{\partial^2 B_R}{\partial R^2}+\frac{3\,\partial B_R}{R\,\partial R}+\frac{B_R}{R^2}=0,$$

and

(7.10B) $$\frac{1}{R^2}\frac{\partial^2 B_\theta}{\partial\theta^2}+\frac{\partial^2 B_\theta}{\partial R^2}+\frac{3\,\partial B_\theta}{R\,\partial R}+\frac{B_\theta}{R^2}=0.$$

B_R and B_θ have exactly the same form, therefore, and are given by the solution of Eq. 7.9B or 7.10B:

(7.11B) $$B_R = (C_1 R^{n-1} + D_1 R^{-n-1})\sin n(\theta - \theta_1),$$

(7.12B) $$B_\theta = (C_2 R^{n-1} + D_2 R^{-n-1})\sin n(\theta - \theta_2).$$

At the outer (stator) radius, $R = R_0$, and the radial mmf is given by:

(7.13B) $$M_0 = A\cos P\theta \text{ ampere turns per pole.}$$

The peripheral component of flux density at this radius, B_θ, must equal the peripheral rate of change of mmf, multiplied by the permeability, $\mu[4\pi(10^{-7})$ in meter units]. Hence, from Eqs. 7.12B and 7.13B:

$$(C_2 R_0^{n-1} + D_2 R_0^{-n-1})\sin n(\theta - \theta_2) = \frac{\mu\,\partial M_0}{R_0\,\partial\theta} = \frac{-\mu AP \sin P\theta}{R_0},$$

giving:

(7.14B) $$C_2 R_0^{n-1} + D_2 R_0^{-n-1} = \frac{-\mu AP}{R_0},$$

(7.15B) $$n = P,$$

and

(7.16B) $$\theta_2 = 0.$$

Also, the flux enters the rotor surface in a radial direction at every point, since the rotor permeability is assumed infinite. Therefore, from Eq. 7.12B:

$$C_2 R_i^{n-1} + D_2 R_i^{-n-1} = 0,$$

giving:

(7.17B) $$D_2 = -C_2 R_i^{2n}.$$

Substituting these relations in Eq. 7.12B:

(7.18B) $$B_\theta = \frac{\mu APR_0^P}{(R_0^{2P} - R_i^{2P})}(R_i^{2P}R^{-P-1} - R^{P-1})\sin P\theta \text{ webers per sq m.}$$

Substituting Eq. 7.18B in Eq. 7.2B:

(7.19B) $$\frac{\partial B_R}{\partial \theta} = \frac{\mu A P R_0^P \sin P\theta}{(R_0^{2P} - R_i^{2P})} \{R_i^{2P} R^{-P-1} - R^{P-1}$$
$$- R[R_i^{2P}(P+1)R^{-P-2} + (P-1)R^{P-2}]\}$$
$$= \frac{\mu A P R_0^P \sin P\theta}{(R_0^{2P} - R_i^{2P})} (-PR_i^{2P} R^{-P-1} - PR^{P-1}),$$

giving, by integration:

(7.20B) $$B_R = \frac{\mu A P R_0^P \cos P\theta}{(R_0^{2P} - R_i^{2P})} (R_i^{2P} R^{-P-1} + R^{P-1}) \text{ webers per sq m.}$$

Equations 7.18B and 7.20B give the peripheral and radial flux densities at every point in the air gap, both of these being sinusoidally distributed around the periphery in accord with the applied mmf of Eq. 7.13B.

2B Flux Equations

The total flux entering the stator, per unit length of core, is, from Eq. 7.20B:

(7.21B) $$\phi_0 = \left[\int_{-\pi/2P}^{\pi/2P} RB_R \, \partial\theta \right]_{R=R_0}$$
$$= \frac{2\mu A (R_0^{2P} + R_i^{2P})}{(R_0^{2P} - R_i^{2P})} \text{ webers per meter of core length.}$$

If the sinusoidal mmf of Eq. 7.13B were impressed across a uniform air gap, of g meters, the total flux would be:

(7.22B) $$\phi_0 = \int_{-\pi/2P}^{\pi/2P} \frac{\mu M_0 R_0 \, d\theta}{g} = \int_{-\pi/2P}^{\pi/2P} \frac{\mu A R_0 \cos P\theta \, d\theta}{g} = \frac{2\mu A R_0}{Pg}.$$

And, if R_i is zero, corresponding to the rotor being removed, Eq. 7.21B gives:

(7.23B) $$\phi_0 = 2\mu A = 8\pi (10^{-7}) A \text{ webers per meter of core length}$$
$$= 6.38(10^{-8}) A \text{ webers per in. of core length.}$$

Hence, the ratio of total flux produced in an air core to that with a uniform small air gap is:

(7.24B) $$\frac{\text{Air-core flux}}{\text{Air-gap flux}} = (2\mu A)\left(\frac{Pg}{2\mu A R_0}\right) = \frac{Pg}{R_0}.$$

In other words, removing the rotor converts the effective air-gap length from g meters to R_0/P meters. The flux per pole with an air core, from Eq. 7.23B, is the same for the same peak ampere turns per pole, regardless of the number of poles, and regardless of the core radius. From Eqs. 7.21B and 7.35B also, the flux produced by the stator with rotor removed ($R_i = 0$) is the same as that produced by the rotor with the stator removed ($R_0 = \infty$) with the same applied mmf. Accordingly, the flux per pole produced by a sinusoidal mmf, distributed around a cylinder with air both inside and outside, is just half that given by Eq. 7.23B, or:

(7.25B) $\quad \phi_0 = 4\pi(10^{-7})A$ webers per meter of core length.

If we substitute:

(7.26B) $$y = \log_e \frac{R_0}{R_i}, \quad \text{or} \quad e^y = \frac{R_0}{R_i},$$

Eq. 7.21B becomes:

(7.27B) $$\phi_0 = 2\mu A \left(\frac{e^{Py} + e^{-Py}}{e^{Py} - e^{-Py}} \right)$$

$$= 2\mu A \coth Py.$$

If $R_i = R_0(1 - g/R_0)$, and g is small:

$$e^{-y} = 1 - \frac{g}{R_0},$$

$$y = \frac{g}{R_0}\left(1 + \frac{g}{2R_0} + \frac{g^2}{3R_0^2} + \cdots\right),$$

and $\quad \coth Py = \dfrac{1}{Py}\left(1 + \dfrac{P^2 y^2}{3} - \dfrac{P^4 y^4}{45} + \cdots\right),$

so that Eq. 7.27B becomes:

(7.28B) $\quad \phi_0 = \dfrac{2\mu A R_0}{Pg}\left[1 - \dfrac{g}{2R_0} + \left(\dfrac{4P^2-1}{12}\right)\dfrac{g^2}{R_0^2} + \left(\dfrac{4P^2-1}{24}\right)\dfrac{g^3}{R_0^3} + \cdots\right].$

The flux leaving the rotor is found by making $R = R_i$ in Eq. 7.20B, and integrating:

(7.29B) $$\phi_i = \left. \int_{-\pi/2P}^{\pi/2P} R B_R \, d\theta \right]_{R=R_i}$$

$$= \frac{4\mu A R_0^P R_i^P}{(R_0^{2P} - R_i^{2P})}.$$

APPENDIX B

The flux which "leaks" peripherally around the stator in the air-gap space, without entering the rotor, is the difference between Eq. 7.21B and Eq. 7.29B. Expressed as a fraction of ϕ_0, this is:

(7.30B) $$\frac{\phi_0 - \phi_i}{\phi_0} = 1 - \frac{2R_0^P R_i^P}{R_0^{2P} + R_i^{2P}}.$$

If $R_i = R_0 - g$, this becomes:

(7.31B) $$\frac{\phi_0 - \phi_i}{\phi_0} = \frac{P^2 g^2}{2R_0^2}\left(1 + \frac{g}{R_0} + \cdots\right).$$

In other words, the peripheral air-gap leakage reactance, due to fundamental sine-wave flux that leaves the stator, but never reaches the rotor, is $P^2 g^2 / 2R_0^2$ times the magnetizing reactance. This is ordinarily very small, but may become appreciable in high-speed machines that are designed with large air gaps (Sect. 7.11).

From Eq. 7.20B, the flux reaching an inner radius R, between R_0 and R_i, is:

(7.32B) $$\phi_R = \int_{-\pi/2}^{\pi/2} R B_R \, d\theta$$
$$= \frac{2\mu A R_0^P (R_i^{2P} R^{-P} + R^P)}{R_0^{2P} - R_i^{2P}}.$$

Hence, from Eqs. 7.21B and 7.32B:

(7.33B) $$\frac{\phi_R}{\phi_0} = \frac{R_0^P (R_i^{2P} R^{-P} + R^P)}{R_0^{2P} + R_i^{2P}}.$$

This assumes a magnetic structure of infinite permeability at all radii greater than R_0, or less than R_i. Equation 7.33B is the ratio of the mutual to the total inductance of the winding at R_0 with respect to another winding at radius R, with an outer magnetic ring at radius R_0 and an inner magnetic ring at radius R_i.

If $R_i = 0$, this becomes:

(7.34B) $$\left.\frac{\phi_R}{\phi_0}\right]_{R_i=0} = \left(\frac{R}{R_0}\right)^P.$$

If the mmf is located at R_i and the material outside R_0 as well as that inside R_i has infinite permeability, by analogy with Eq. 7.32B:

(7.35B) $$\phi_R = \frac{2\mu A R_i^P}{R_0^{2P} - R_i^{2P}}(R_0^{2P} R^{-P} + R^P),$$

(7.36B) $$\phi_i = \frac{2\mu A(R_0^{2P}+R_i^{2P})}{R_0^{2P}-R_i^{2P}},$$

(7.37B) $$\frac{\phi_R}{\phi_i} = \frac{R_i^P(R_0^{2P}R^{-P}+R^P)}{R_0^{2P}+R_i^{2P}},$$

and if $R_0 = \infty$:

(7.38B) $$\left.\frac{\phi_R}{\phi_i}\right]_{R_0=\infty} = \left(\frac{R_i}{R}\right)^P.$$

The reluctances of the flux paths inside and outside the radius of applied mmf are in series, so that the flux produced for a given mmf is proportional to the sum of the reciprocals of these values. The above relations may, therefore, be generalized to the following:

If the sinusoidal mmf of A peak ampere turns per pole is located at a radius R_a, and if fully magnetic rings are located both outside the winding at a radius R_0 and inside the winding at a radius R_i, the total flux produced will be:

(7.39B) $$\phi_A = \frac{2\mu A}{\dfrac{R_0^{2P}-R_a^{2P}}{R_0^{2P}+R_a^{2P}} + \dfrac{R_a^{2P}-R_i^{2P}}{R_a^{2P}+R_i^{2P}}}$$

$$= \frac{\mu A(R_a^{2P}+R_i^{2P})(R_a^{2P}+R_0^{2P})}{R_a^{2P}(R_0^{2P}-R_i^{2P})}.$$

If $R_a = R_0$, this reduces to Eq. 7.21B; if $R_a = R_i$ it reduces to Eq. 7.36B.

The ratio of flux reaching any radius $R > R_a$ to the total flux at R_a is given by placing $R_i = R_a$ in Eq. 7.37B, and for any radius $R < R_a$ by placing $R_0 = R_a$ in Eq. 7.33B, since the flux distribution on either side of the winding is not affected by the conditions on the other side.

By a study of Eqs. 7.18B and 7.20B, many other interesting relationships can be derived that are useful in calculating core losses and magnetic forces, as well as reactances.

3B Energy Flow

If we place $R = R_0$, $(P\theta - \pi/2 - \omega t)$ for $P\theta$, and $G = R_0 - R_i$, Eqs. 7.20B and 7.18B become, with $\mu = 4\pi 10^{-7}$:

(7.40B) $$B_R = (4\pi PAR^P/10^7)$$
$$\times \left[\frac{(R-G)^{2P}(R-g)^{-P-1}+(R-g)^{P-1}}{R^{2P}-(R-G)^{2P}}\right]\sin(P\theta-\omega t)$$

and

(7.41B) $B_\theta = -(4\pi PAR^P/10^7)$
$$\times \left[\frac{(R-G)^{2P}(R-g)^{-P-1}-(R-g)^{P-1}}{R^{2P}-(R-G)^{2P}}\right]\cos(P\theta-\omega t),$$

where A is the peak sine-wave mmf at the stator bore:

(7.42B) $A = \sqrt{2}R\Delta/P$ ampere turns per pole (Δ from Eq. 3.36).

Developing these in infinite series, and neglecting all terms smaller than $(G/R)^2$, these are:

(7.43B) $\qquad B_R = B_1(1-G/2R)(1+g/R)\sin(P\theta-\omega t),$

(7.44B) $\qquad B_\theta = (PB_1/R)(G-g)(1+3g/2R)\cos(P\theta-\omega t),$

where B_1 is the peak flux density due to the stator current, or:

(7.45B) $\qquad B_1 = 4\pi A \times 10^{-7}/G$ webers per square meter.

The factor $(1-G/2R)$ adjusts B_R to the average air-gap radius, and $(1+g/R)$ allows for the varying air-gap area at different radii. The peripheral flux density at the outer radius is simply equal to the change in radial mmf over a short peripheral distance, divided by that distance, or:

(7.46B) $\qquad\qquad B_\theta = -(G/R)dB_R/d\theta,$

when $g = 0$.

If the rotor permeability is infinite, and there are no rotor currents, $B_\theta = 0$ at the rotor surface, as shown by the factor $G-g$ in its equation. Similarly, if there is a rotor mmf, the component of B_θ due to it is zero at the stator surface. Under load, there will be an additional load current in the stator, equal to C times the magnetizing current and displaced by an angle $P\alpha = (90-P\alpha_1)$ from the magnetizing current; there will also be an equal and opposite load current in the rotor. These will create additional flux densities in the air gap (B_{RC} and $B_{\theta C}$), found by subtracting their radial and adding their peripheral components:

7.47B) $B_{RC} = -[CB_1\cos(P\theta-P\alpha_1-\omega t)]$
$$\times[(1-G/2R)(1+g/R)-(1+G/2R)(1-g/R)]$$
$$= (CGB_1/R)(1-2g/G)\cos(P\theta-P\alpha_1-\omega t)$$

webers per square meter.

The signs of the correction terms, $G/2R$ and g/R, are opposite for the stator and rotor load currents because they are located on opposite sides of the air gap. This additional radial flux due to the load currents is

CG/R times the no-load flux on each side of the air gap, with opposite signs, and is zero midway across the gap. It is too small to be considered in performance calculations.

(7.48B) $\quad B_{\theta C} = (PB_1 G/R)[(1-g/G)(1+3g/2R)+(g/G)(1-3g/2R)]$
$$\times \sin(P\theta - P\alpha_1 - \omega t)$$
$$= (CPB_1 G/R)[1+(3g/2R)(1-2g/G)]\sin(P\theta - P\alpha_1 - \omega t).$$

This peripheral flux density, when added to the radial density, causes the flux to enter both the stator and the rotor at an angle different from 90°, thus causing a peripheral force, or torque.

To calculate the torque, which is a measure of the energy flow across the air gap, we make use of the Poynting vector theorem, which states that the flow of electromagnetic energy through any surface is equal to the integral over that surface of the Poynting vector, $\mathbf{E} \times \mathbf{H}$, where $\mathbf{E}$ and $\mathbf{H}$ are the electric- and magnetic-field intensities in the plane of the surface, and the energy flow is at right angles to the surface. The electric-field intensity in the air gap at the stator bore, due to the radial flux, is in the axial direction and is equal to:

(7.49B) $\quad\quad\quad\quad \mathbf{E}_L = vB_R \mathbf{l}_L$ volts per meter,

where v is the peripheral speed of the magnetic field in meters per second, equal to $2\pi\, Rf/P$. The magnetic-field intensity in the radial direction is:

(7.50B) $\quad\quad \mathbf{H}_R = B_R \mathbf{l}_R \times 10^7/4\pi$ ampere-turns per meter,

where $\mathbf{l}_L$ and $\mathbf{l}_R$ are unit vectors in the axial and radial directions, respectively. This is in time phase with $\mathbf{E}_L$, so that the energy flow in the peripheral direction at the stator bore is:

(7.51B) $\quad Q_\theta = \mathbf{E}_L \times \mathbf{H}_R = vB_R^2 \times 10^7/4\pi$ watts per square meter
$$= (vB_1^2 \times 10^7/8\pi)[1-\cos 2(P\theta - \omega t)]$$
$$+ \text{(double-frequency and second-order terms)}.$$

There is a peripheral flow of magnetic energy at synchronous speed, equal at each instant to twice the energy stored in the radial field, multiplied by the speed. To find the total energy being carried around, we multiply the average value of Q_θ by the radial cross-section of the air gap, LG, and by the time of one revolution, P/f, giving:

(7.52B) $\quad$ Energy of the rotating field $= (LGP/f)(\pi Df/P)(B_1^2 \times 10^7/8\pi)$
$$= DLGB_1^2 \times 10^7/8 \text{ watt-seconds}.$$

The total energy in the radial field is half that corresponding to the peak flux density, or:

(7.53B) Radial field energy $= (1/2)(\pi DLG)(B_1^2 \times 10^7/8\pi)$
$$= DLGB_1^2 \times 10^7/16 \text{ watt-seconds.}$$

The peripheral magnetic field, B_θ, gives rise to another energy flow, in the radial direction, whose value is:

(7.54B) $Q_R = \mathbf{E}_L \times \mathbf{H}_\theta = vB_R B_\theta \times 10^7/4\pi$ watts per square meter
$$= (vPB_1^2 \times 10^7/8\pi R)(G-g)\sin 2(P\theta - \omega t)$$
$$+ (vCPGB_1^2 \times 10^7/8\pi R)\cos P\alpha_1$$
$$+ \text{(double-frequency and second-order terms).}$$

The first term of Eq. 7.54B represents the flow of energy to and fro between the power system and the radial field. The total energy delivered and returned each quarter cycle, or $4f$ times per second, is found by integrating Q_R from $t = 0$ to $t = \frac{1}{4}f = \pi/2\omega$ to obtain the maximum in time, then integrating over the area of a half pole pitch, $\pi RL/2P$, to obtain the energy per half pole, and multiplying by $4P$. This is:

(7.55B) Radial energy flow
$$= (vPB_1^2 G \times 10^7/8\pi R) \int_{-\pi/4P}^{\pi/4P} 4PRL\, d\theta \int_0^{\pi/2\omega} \sin 2(P\theta - \omega t)\, dt$$
$$= DLGB_1^2 \times 10^7/4\pi \text{ watt-seconds.}$$

The second term of Eq. 7.54B is the real power transferred across the air gap. Integrating this over the entire air-gap area, we find the total power delivered to the shaft is:

(7.56B) Shaft power $= \int_0^{2\pi} Q_R RL\, d\theta = (2\pi RL)(vCPGB_1^2 \times 10^7/8\pi R)\sin P\alpha$
$$= \pi^2 \sqrt{2}\, D^2 LB_1 C\Delta N \sin P\alpha/120 \text{ watts}$$
$$= qCIE \sin P\alpha \text{ watts,}$$

which checks with Eq. 3.9 since $v = \pi DN/60$, and $C\Delta$ is the load current density equivalent to Δ_1 in the earlier equation.

These relations show that the energy flowing radially into the air gap from the stator is proportional to the product of the radial and peripheral flux densities at the stator surface. It is enlightening to trace ways by which the energies of the various components of the field are supplied, and the interchange of torques and energy between the different phases of a polyphase machine like those in the single-phase machine considered earlier, but there is no space for this analysis here.

Bibliography

Reference to Section

1. "The Leakage Reactance of Induction Motors," C. A. Adams, *Trans. Int. Elec. Congress*, St. Louis, 1904, Vol. 1, 1905, pp. 706–24.
2. "Saturation Factors for Leakage Reactance of Induction Motors," P. D. Agarwal and P. L. Alger, *A.I.E.E. Trans.*, Vol. 79, Part III, 1960, pp. 1037–42. — 7
3. "The Calculation of the Armature Reactance of Synchronous Machines," P. L. Alger, *A.I.E.E. Trans.*, Vol. 47, April 1928, pp. 493–512.
4. "Electromechanical Energy Conversion," Philip L. Alger and Edward Erdelyi, *Electro-Technology*, Sept. 1961. — App. B
5. "Torque-Energy Relations in Induction Machines," P. L. Alger and W. R. Oney, *A.I.E.E. Trans.*, Vol. 73, Part III, 1954, pp. 259–64. — App. B
6. "The Air Gap Reactance of Polyphase Machines," P. L. Alger and H. R. West, *A.I.E.E. Trans.*, Vol. 66, 1947, pp. 1331–43. — 6, 7
7. "Saturation Factors for Leakage Reactance of Induction Motors with Skewed Rotors," G. Angst, *I.E.E.E. Trans.*, Power Apparatus and Systems, Oct. 1963, pp. 716–25. — 7, 9
8. "An Experimental Study of Induction Motor End Turn Reactance," E. C. Barnes, *A.I.E.E. Trans.*, Vol. 70, Part III, 1951, pp. 671–7. — 12
9. "Zig-Zag Leakage," L. H. A. Carr, *The Electrician*, London, Vol. 87, July 15, 1921, pp. 76–7. — 6, 7
10. "Saturation Effect of Leakage Reactance," S. S. L. Chang and T. C. Lloyd, *A.I.E.E. Trans.*, Vol. 68, Part II, pp. 1144–7. — 6, 7
11. *A Study of the Induction Motor*, F. T. Chapman, John Wiley, 1930. — 6, 7
12. "Some Remarks on the Energy Flow in Rotating Electric Machines," F. Dahlgren, *Trans. of the Royal Institute of Technology*, Stockholm, Sweden, No. 38, 1950. — App. B
13. "Synchronous Machines," Part I, Appendix B, R. E. Doherty and C. A. Nickle, *A.I.E.E. Trans.*, Vol. 45, 1926, pp. 912–26. — App. B
14. "Doubly-Linked Leakage Flux in Induction Motors," L. A. Dreyfus, *E.U.M.*, Vol. 37, 1919, pp. 149–51. — 6, 7
15. "Applications of Electromagnetic Field Theory to Induction Machines," Edward C. Guilford and Robert M. Saunders, *A.I.E.E.* Paper No. CP 58-201. — App. B
16. "Zigzag Leakage of Induction Motors," R. E. Hellmund, *A.I.E.E. Trans.*, Vol. 26, Part II, 1907, pp. 1505–24. — 6, 7
17. "Theory of End Winding Leakage Reactance," V. N. Honsinger, *A.I.E.E. Trans.*, Vol. 78, Part III, 1959, pp. 426–31. — 12
18. "Differential Leakage with Respect to the Fundamental Wave and to the Harmonics," M. M. Liwschitz, *A.I.E.E. Trans.*, Vol. 63, 1944, pp. 1139–1150. — 6, 7
19. "Differential Leakage of a Fractional-Slot Winding," M. M. Liwschitz, *A.I.E.E. Trans.*, Vol. 65, 1946, pp. 314–20. — 6, 7
20. "Effect of Skew on Induction Motor Magnetic Fields," C. E. Linkous, *A.I.E.E. Trans.*, Vol. 74, Part III, 1955, pp. 760–5. — 9
21. "Reactances of Squirrel-Cage Induction Motors," T. C. Lloyd, V. F. Giusti, and S. S. L. Chang, *A.I.E.E. Trans.*, Vol. 66, 1947, pp. 1349–55.
22. "Die Differenzstreuung der Asynchronen Maschinen," Erik Morath, *Elektrotechnik und Maschinenbau*, Heft 13, 1964.
23. "The Reactances of Synchronous Machines," R. H. Park and B. L. Robertson, *A.I.E.E. Trans.*, Vol. 47, April 1928, pp. 514–35.

	Reference to Section

24. "Formulae and Tables for the Calculation of Mutual and Self-Inductance," E. B. Rosa and F. W. Grover, *Bull. of the U.S. Bureau of Standards*, Vol. 8, 1912, pp. 1–237. — 12
25. "Distribution of Induction and Core Losses in Armatures," R. Rudenberg, *E.T.Z.*, Vol. 27, 1906, pp. 109–14. — App. B
26. "Electromechanical Energy Conversion in Double Cylindrical Structures," R. M. Saunders, A.I.E.E. Conference Paper, 1962, No. CP-62-135.
27. *Fundamentals of Electric Waves*, Second Edition, by H. H. Skilling, John Wiley, 1948. — App. B
28. "End Component of Armature Leakage Reactance of Round Rotor Generators," R. T. Smith, *A.I.E.E. Trans.*, Vol. 77, Part III, 1958, pp. 636-45 — 12

CONTENTS—CHAPTER 8

SPEED-TORQUE-CURRENT RELATIONS

		PAGE
1	Methods of Speed and Torque Control	261
2	Energy Losses and Heating during Acceleration	262
3	High-Impedance Rotor	263
4	Deep-Bar Rotor	265
5	Idle-Bar Rotors	272
6	Double Squirrel-Cage Rotor	277
7	The Series Hysteresis Motor (Idle Alnico Bars)	283
8	Reduced Voltage Starting	286
9	Part-Winding Starting	287
10	Split-Winding Starting	296
11	Pole Changing and Concatenation	302
12	Wound Rotor with Rheostatic Control	306
13	Wound Rotor with Saturistor	309
14	Speed Control with Auxiliary Machines or Thyristors	314

8

SPEED-TORQUE-CURRENT RELATIONS

1 METHODS OF SPEED AND TORQUE CONTROL

The polyphase squirrel-cage induction motor converts a-c power into mechanical power with ideal simplicity and reliability. One simply couples the motor to the load shaft, connects the three motor leads to the terminals of a three-phase power supply, and closes the switch. The motor comes up to speed and turns the drive shaft at nearly constant speed over a wide range of load (Fig. 5.2). It will continue to do so year in and year out, in heat and cold, though covered with dirt and completely neglected. A very little care for the bearings, and keeping the insulation dry, are all the motor asks for a lifetime of service.

Its principal shortcoming is its inability to run efficiently at reduced speed (with a fixed line frequency), since all the power represented by the slip times the output torque is dissipated in rotor heating. A second shortcoming is the relatively high starting current required to ensure adequate running efficiency and breakdown torque (Sect. 4.6) as well as good starting torque. The recent development of silicon rectifiers has opened the way to wide range stepless speed control of squirrel-cage motors, by variable frequency power supply from static frequency changers. This may lead to much wider use of squirrel-cage motors for vehicles and automated industrial drives.

When frequent starting, high starting torque, low starting current, or reduced speed operation is required, various special designs or control schemes can be employed, such as:

(a) High-impedance rotor (Sect. 8.3).
(b) Deep-bar rotor (Sect. 8.4).
(c) Idle-bar rotor (Sect. 8.5).
(d) Double squirrel cage (Sect. 8.6).
(e) Series hysteresis rotor (idle Alnico bars) (Sect. 8.7).
(f) Reduced voltage starting (Sect. 8.8).
(g) Part-winding starting (Sect. 8.9).
(h) Pole changing and concatenation (Sect. 8.11).
(i) Wound rotor with rheostatic control (Sect. 8.12).
(j) Wound rotor with Saturistor (Sect. 8.13).
(k) Speed control with auxiliary machines or Thyristors (Sect. 8.14).

2 Energy Losses and Heating During Acceleration

In calculating the starting performance of induction motors, it is convenient to remember that the specific heat of copper is 3.45 watt seconds per °C per cu cm, or 388 wsec per °C per kg. The temperature rise of the winding, therefore, is equal to 2.6 times the I^2R loss, expressed in kw sec per kg, if all the heat is stored in the copper. A current density of 5000 amperes per sq cm raises the temperature of copper at 75 C by 15 C per sec*, with no heat loss. In practice, this value is nearly correct for bare copper in air, for heating periods of a few seconds. If the copper is surrounded by insulation, by oil, or by iron, the temperature rise will be progressively lower, depending on the intimacy of contact, the relative volumes of material, and the elapsed time.

Another relation useful in calculating starting performance of high-inertia loads is:

> The total heat generated in the rotor windings of an induction motor in bringing the load from rest to full speed is equal to the kinetic energy supplied to the rotating parts, if friction and load torques are neglected.

For, if N_s is the synchronous speed, and T is the motor torque at a slip s, the power delivered to the load is:

$$P = T(1-s)N_s, \tag{8.1}$$

and the power being dissipated in the rotor winding is:

$$H = sTN_s. \tag{8.2}$$

Since, in this case, all the torque is used in accelerating the load,

$$T = \frac{Wk^2}{g}\frac{dN}{dt} = \frac{Wk^2}{g}N_s\frac{d(1-s)}{dt},$$

so that

$$T\,dt = -N_s\frac{Wk^2}{g}\,ds, \tag{8.3}$$

where W = rotor weight in kg, k = radius of gyration in meters, and $g = 9.81$ meters per sec per sec = acceleration due to gravity.

---

*The corresponding figure for aluminum of 60% conductivity is 54 C per sec. Aluminum, copper, and steel weigh 0.0027, 0.0089, and 0.0079 kg per cu cm, respectively; and their heat capacities are 590, 388, and 503 wsec per °C per kg.

The total power supplied to the load in accelerating it to the speed $N_s(1-s)$ is, from Eqs. 8.1 and 8.3:

(8.4) $$\int_{s=1}^{s=s} P\,dt = -\int_1^s (1-s)N_s^2 \frac{Wk^2}{g}\,ds = \frac{N_s^2 Wk^2(1-s)^2}{2g}.$$

This is the kinetic energy of the rotor at speed $(1-s)N_s$, equal to one-half the moment of inertia times the square of the speed of rotation.

The total power dissipated in the rotor winding in coming up to this speed is:

(8.5) $$\int_{s=1}^{s=s} H\,dt = -\int_1^s sN_s^2 \frac{Wk^2}{g}\,ds = \frac{N_s^2 Wk^2}{2g}(1-s^2).$$

If the motor comes up to full speed, $s = 0$, when Eqs. 8.5 and 8.4 become identical.

That is, whatever the shape of the speed-torque curve, every kilowatt-second of kinetic energy stored in the rotating parts requires an equal number of kilowatt-seconds to be dissipated in secondary copper loss, when an induction motor accelerates a load from rest to synchronous speed.

3 HIGH-IMPEDANCE ROTOR

By simply increasing the resistance of the secondary winding, the speed-torque curve can be altered to any one of the forms shown in Fig. 8.1. The uppermost curve represents a normal single squirrel-cage motor of about 25 hp at 1,800 rpm. The other curves represent wound-rotor motors with 10, 20, 30, etc., per cent external secondary resistance, 100 per cent giving full-load torque at standstill. In all cases, the breakdown torque is the same, but it occurs at different slip values, proportional to the secondary resistance (Eq. 5.62).

A high-resistance squirrel cage limits the starting current, and so reduces the I^2R loss in the stator winding during the starting period, at the same time that it increases the starting torque per ampere. The motor is, therefore, able to start more frequently, and to start higher-inertia loads. To limit rotor heating due to the greater I^2R loss in the squirrel cage, it may be desirable to use brass or other alloy bars of higher resistivity, and, therefore, greater volume than copper or aluminum for a given resistance, so increasing the heat-storage capacity, and decreasing the temperature rise during the brief starting period. It may be desirable also to provide means for increased heat dissipation at speed, as by extended bars, or high-resistance end rings with convolutions to provide increased surface area, and extra fans. The best design for a given application will depend on the relative importance of the high efficiency and constancy of speed associated with low resistance, and the improved starting performance given by high resistance.

Many ingenious methods of decreasing the secondary resistance, as the motor speeds up, have been used, as by employing an insulated motor winding closed through stacks of carbon disks, whose resistance is decreased under compression by centrifugal force; but, under modern conditions, the higher costs of such designs are not justified by the benefits obtained.

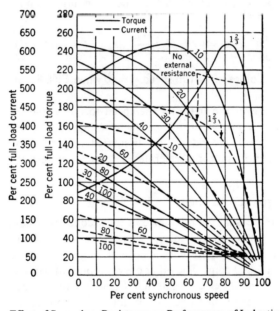

FIG. 8.1. Effect of Secondary Resistance on Performance of Induction Motor.

If the secondary reactance is increased, leaving the resistance the same, the starting current will be reduced, but the starting torque will be reduced too, in proportion to the square of the current. Also, the breakdown torque will be reduced, in inverse proportion to the total reactance. As mentioned in Sect. 4.6, the per unit breakdown torque is a little less than half the reciprocal of the per unit reactance, so that the starting current can not be reduced below about 500 per cent of full-load value by this means, without lowering the breakdown torque below the standard value of 200 per cent, for low-slip motors (except as described in Sect. 8.7).

Nearly all these special methods of increasing the starting impedance have been superseded in American practice by one form or another of the deep-bar or double squirrel-cage motor, which converts part of the running reactance into resistance at starting. With the advent of modern high-energy magnetic materials, it has become possible to use the hysteresis loss in idle Alnico bars to provide greater starting torque per ampere. This new scheme, considered in Sect. 8.7, promises to be widely used in the future.

4 Deep-Bar Rotor

The ideal motor should have a varying secondary resistance, large at standstill, and decreasing as the speed rises. Boucherot was the first to accomplish this by means of the "deep-bar effect", or variation of resistance with frequency.

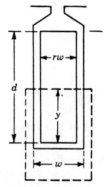

Fig. 8.2. Deep Rotor Bar.

If the squirrel-cage bar is made very deep and narrow (Fig. 8.2), the current will be crowded up toward the top of the bar at full frequency, increasing the effective resistance; while at full speed, when the slip frequency is low, the current will be uniformly distributed, giving low resistance. This is true because the bottom filament in the bar is linked by all the slot-leakage flux, whereas the top filament is linked only by the external flux. This additional reactance of the bottom filament causes its current to be smaller and more lagging in time phase than the current in the top filament.

Let:
  d = depth of bar in cm,
  w = width of slot in cm,
  r = ratio of bar width to slot width,
  f = frequency, in cycles per second,
  ρ = resistivity of bar in ohm-cm ($= 2.10 \times 10^{-6}$ for copper at 75 C),

266   SPEED-TORQUE-CURRENT RELATIONS

$\sqrt{2}\,I\sin 2\pi ft =$ total current in the bar, amperes,
$y =$ distance up from the bottom of the bar in cm,
$B =$ rms flux density across the slot at height y, in webers per square cm,
$\Delta =$ rms current density at height y, in amperes per sq cm,
$R_{dc} = \rho/wd$ ohms per cm of axial length,

$$\alpha = 2\pi\sqrt{\left(\frac{rf}{\rho 10^9}\right)}$$

$$= 1.06\sqrt{\left(\frac{rf}{60}\right)}, \text{ for copper at 75 C.}$$

The rms flux density across the slot at height y is, from Eq. 1.13:

(8.6) $$B = \frac{4\pi r}{10^9}\int_0^y \Delta\,dy.$$

The rms flux linking the conductor filament at height y is:

(8.7) $$\phi = \int_y^d B\,dy.$$

The rms voltage induced by this flux is:

(8.8) $$E = -j2\pi f\phi = -j2\pi f\int_y^d B\,dy.$$

The current density at height y is the net voltage (impressed minus induced) divided by the resistivity:

$$\Delta = \frac{V - E}{\rho},$$

giving:

(8.9) $$\frac{d\Delta}{dy} = -\frac{dE}{\rho\,dy} = \frac{j2\pi fB}{\rho}.$$

From Eqs. 8.6 and 8.9:

(8.10) $$\frac{d^2\Delta}{dy^2} = \frac{j2\pi f}{\rho 10^9}(4\pi r\Delta),$$

$$= 2j\alpha^2\Delta,$$

and

(8.11) $$\frac{d^2B}{dy^2} = \frac{4\pi r}{10^9}\left(\frac{j2\pi fB}{\rho}\right) = 2j\alpha^2 B.$$

The general solution of Eq. 8.11 is:

(8.12) $$B = P\cosh(1+j)\alpha y + Q\sinh(1+j)\alpha y.$$

Since $B = 0$ when $y = 0$, P must be zero.
When $y = d$, at the top of the slot, $B = 4\pi I/(w\,10^9)$, whence

$$Q = \frac{4\pi I 10^{-9}}{w\sinh(1+j)\alpha d},$$

and

(8.13) $$B = \frac{4\pi I}{10^9 w}\left[\frac{\sinh(1+j)\alpha y}{\sinh(1+j)\alpha d}\right] \text{ webers per sq cm.}$$

The solution of Eq. 8.10 is:

(8.14) $$\Delta = M\cosh(1+j)\alpha y + N\sinh(1+j)\alpha y.$$

From Eqs. 8.9 and 8.13, $d\Delta/dy = 0$, when $y = 0$, so that N must be zero; and, from Eqs. 8.9 and 8.14:

$$M(1+j)\alpha = \frac{4\pi I 10^{-9}}{w\sinh(1+j)\alpha d}\left(\frac{j2\pi f}{\rho}\right),$$

so that:

(8.15) $$\Delta = \frac{(1+j)\alpha I}{rw}\left[\frac{\cosh(1+j)\alpha y}{\sinh(1+j)\alpha d}\right].$$

Alternatively, Δ can be found by substituting Eq. 8.13 in Eq. 8.9 and integrating.

The current density given by Eq. 8.15 may be thought of as the superposition on the uniform (average) current density of a circulating current flowing additively along the top of the bar and subtractively along the bottom of the bar, in such a way as to oppose the cross slot-leakage flux. It is possible to build up a formula for the circulating current by successive approximations, assuming for the first step that the circulating current is limited solely by resistance, i.e., is too small to reduce appreciably the cross slot flux. For the second step, the cross slot flux that would be produced by the circulating current acting alone is calculated, and the corresponding additional amount of circulating current is determined. This process leads to an infinite series of terms, from which the additional I^2R loss, reduced reactance, etc., can be obtained.

A general solution for the effects of the eddy currents on I^2R loss and reactance can be obtained more directly, however, by calculating the total voltage drop along the bar, as the sum of the voltages due to IR drop and to the linkages produced by the cross slot flux. The real component of this voltage will give the effective resistance of the bar on alternating current, and the j component will give the effective reactance.

The IR drop along the bar at any height y is:

$$V_r = \Delta\rho \text{ volts per cm,}$$

and the average drop over the entire bar is, from Eq. 8.15:

$$V_r = \frac{1}{d}\int_0^d \Delta\,dy$$

$$= \frac{(1+j)\alpha\rho I}{rwd}\int_0^d \frac{\cosh(1+j)\alpha y\,dy}{\sinh(1+j)\alpha d}$$

(8.16)
$$= \frac{\rho I \sinh(1+j)\alpha d}{rwd \sinh(1+j)\alpha d} = \frac{\rho I}{rwd} = IR_{dc},$$

as would be expected.

The total rms flux crossing the slot above the height y, and, therefore linking the current below y, is, from Eqs. 8.7 and 8.13:

(8.17) $\quad \phi = \int_y^d B\,dy = \frac{4\pi I 10^{-9}}{(1+j)\alpha w}\left[\frac{\cosh(1+j)\alpha d - \cosh(1+j)\alpha y}{\sinh(1+j)\alpha d}\right].$

The total voltage applied to the bar at the height y is:

(8.18) $\quad\quad V = \rho\Delta + j2\pi f\phi.$

Substituting Eqs. 8.15 and 8.17 in this, we find that $\rho\Delta = -j2\pi f$ (the second term of Eq. 8.17), so that these cancel, and we have:

(8.19) $\quad\quad V = \dfrac{j2\pi f(4\pi I 10^{-9})\cosh(1+j)\alpha d}{(1+j)\alpha w \sinh(1+j)\alpha d}$

or

(8.20) $\quad\quad V = IR_{dc}\left[\dfrac{(1+j)\alpha d \cosh(1+j)\alpha d}{\sinh(1+j)\alpha d}\right].$

The real portion of Eq. 8.20, representing the loss component of the voltage, is:

(8.21) $\quad\quad V_{real} = IR_{dc}\left[\dfrac{\alpha d(\sinh 2\alpha d + \sin 2\alpha d)}{\cosh 2\alpha d - \cos 2\alpha d}\right].$

If αd is small, this becomes:

(8.22) $\quad V_{real} = IR_{dc}\left[1 + \dfrac{4(\alpha d)^4}{45} - \dfrac{16}{4{,}725}(\alpha d)^8 + \ldots\right], \quad \alpha d < 1.5.$

If αd is large, this becomes:

(8.23) $$V_{\text{real}} = IR_{dc}(\alpha d), \quad \alpha d > 2.$$

The j part of Eq. 8.20, representing the reactive component of voltage, is:

(8.24) $$V_{\text{reactive}} = jIR_{dc}\left[\frac{\alpha d(\sinh 2\alpha d - \sin 2\alpha d)}{\cosh 2\alpha d - \cos 2\alpha d}\right]$$

If αd is small, this becomes:

(8.25) $$V_{\text{reactive}} = jIR_{dc}\left(\frac{2\alpha^2 d^2}{3}\right)\left[1 - \frac{8\alpha^4 d^4}{315} + \frac{32(\alpha d)^8}{31{,}185} + \cdots\right]$$
$$= j\frac{2fId}{3w10^7}\left[1 - \frac{8\alpha^4 d^4}{315} + \frac{32(\alpha d)^8}{31{,}185} + \cdots\right], \quad \alpha d < 1.5.$$

If αd is large, it becomes:

(8.26) $$V_{\text{reactive}} = jIR_{dc}(\alpha d) = \frac{j2fdI}{3w10^7}\left(\frac{3}{2\alpha d}\right), \quad \alpha d > 2,$$

so that the resistance and reactance approach equality, as the frequency (or bar depth) increases. This is an example of a general law of "skin effect", that the redistribution of current in a conductor forced by eddy currents causes R to approach equality with X asymptotically as the frequency increases. These relations are shown in Fig. 8.3.

The increase in starting resistance is obtained by the presence of the reactive voltage that forces the current into a higher-resistance path when the frequency is high. When the motor speeds up, and the secondary frequency decreases, this voltage declines, and the current assumes its normal path of lower resistance and higher inductance.

A simple criterion of performance is the ratio of the extra reactance incurred at full speed to the extra resistance obtained at starting. That is, if a motor is designed for a given starting current, which practically means a given starting reactance, the deep-bar effect will increase the resistance at starting by an amount ΔR,† and will increase the full-speed reactance by ΔX.† The lower the ratio $\Delta X/\Delta R$, the better the motor performance will be.

Let:

R_0 = maximum permissible bar resistance in primary terms at low slip, to obtain desired full-load speed,

X_0 = maximum permissible bar reactance in primary terms at low slip, to obtain desired maximum torque,

$K = X_0/R_0$.

† Δ here means a small increment, and has no connection with the Δ used previously for current density.

270 SPEED-TORQUE-CURRENT RELATIONS

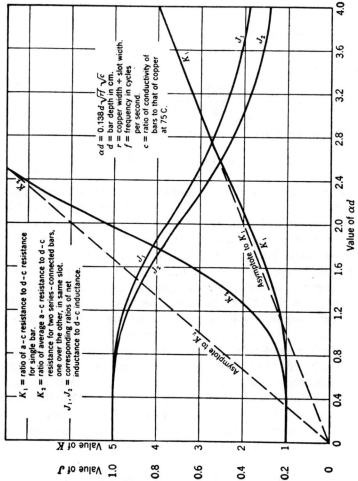

FIG. 8.3. "Deep-Bar" Factors for Copper Bars in Deep Slots.

DEEP-BAR ROTOR

Then, the increase in resistance at starting will be, from Eqs. 8.22 and 8.23:

(8.27) $$\Delta R = (\alpha d - 1)R_0, \text{ if } \alpha d > 2,$$
$$= \frac{4}{45}(\alpha d)^4 R_0, \text{ if } \alpha d < 1.5;$$

and the increase in reactance at full speed over the starting value will be, from Eqs. 8.25 and 8.26:

(8.28) $$\Delta X = \left(1 - \frac{3}{2\alpha d}\right)X_0 = K\left(1 - \frac{3}{2\alpha d}\right)R_0, \text{ if } \alpha d > 2,$$
$$= \frac{8(\alpha d)^4}{315}X_0 = \frac{8K(\alpha d)^4}{315}R_0, \text{ if } \alpha d < 1.5.$$

If αd is large, the bar resistance and reactance are equal at starting, or:

$$\alpha d R_0 = \frac{3X_0}{2\alpha d} = \frac{3KR_0}{2\alpha d},$$

giving:

(8.29) $$\alpha d = \sqrt{\left(\frac{3K}{2}\right)}, \text{ or } K = \frac{2}{3}\alpha^2 d^2.$$

Equation 8.29 is also valid if αd is small, since in this case, from Eq. 8.25:

$$K = \frac{X_0}{R_0} = \frac{2\alpha^2 d^2}{3}.$$

For αd large, from Eqs. 8.27 and 8.28:

(8.30) $$\frac{\Delta X}{\Delta R} = \frac{K\left(1 - \frac{3}{2\alpha d}\right)}{\alpha d - 1} = \frac{\sqrt{(6K)(K-1)} - K}{3K - 2}, \quad K > 2.7.$$

This ratio becomes larger as K increases. Therefore, the extra reactance is less effective in creating extra starting resistance, as the depth of bar is increased.

For αd small:

(8.31) $$\frac{\Delta X}{\Delta R} = \frac{8(45)K(\alpha d)^4}{315(4)(\alpha d)^4} = \frac{2K}{7} = \frac{4(\alpha d)^2}{21}, \quad K < 1.5.$$

Here, also, the benefit of increased starting resistance must be paid for by a more than proportionate increase in running reactance as the bar depth is increased.

Nevertheless, for all values of K below 3.7, Eq. 8.30 shows that ΔR is more than ΔX. This favorable range of design gives a ratio of starting to running resistance, from Eqs. 8.23 and 8.29, of $\alpha d = 2.35$ or less.

In practice, the simple deep-bar design is of limited usefulness, because the large ratio of bar depth to width requires more space and is mechanically less strong than desirable. Hence, modified forms, such as L bars or inverted T bars, are widely used in preference. For all these, formulas similar to Eqs. 8.20 to 8.26 can be worked out by similar methods.

It is important to note that any deep-bar rotor will have a much higher resistance at double frequency, corresponding to backward rotation, than it does at standstill. Hence, for single-phase motors, which have a large backward-revolving field, or for any motor with large harmonic fields, a deep-bar design gives largely increased rotor losses. Its usefulness is, therefore, restricted to balanced polyphase motors.

5 Idle-Bar Rotors

A simple form of low-starting-current, moderate-torque motor is obtained by placing open-circuited conductors, or "idle bars", in the slot space above a depressed squirrel cage, as shown in Fig. 8.4. The width of

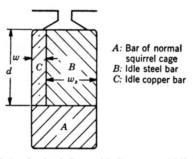

Fig. 8.4. Squirrel Cage with Superposed Idle Bars.

leakage slot, or, in the figure, the width of the idle copper bar, is made such as to give a reactance value at full speed, X_0, corresponding to the desired value of breakdown torque. The remainder of the normal slot width may be filled with a steel bar.

At standstill, the full-frequency leakage flux, due to the current in the squirrel-cage bar, passes across both the idle bars, inducing opposing eddy currents, which in turn create $I^2 R$ losses, and also a counter mmf opposing the passage of the flux. Thus, the effective resistance of the squirrel cage is increased, and its reactance is reduced during the starting period. The performance of this design may be calculated by considering the idle bars to form a short-circuited tertiary winding, which is linked by the slot-leakage flux of the squirrel cage, as in Fig. 8.5.

Assuming the leakage-flux paths through the idle bars to be directly across the slot, the separate copper and steel bars may be replaced by a fictitious single idle bar having the same depth, d; the same width as the

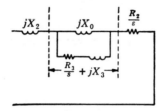

FIG. 8.5. Equivalent Circuit of Idle-Bar Squirrel Cage.

air gap, w; and a conductance equal to that of the two actual bars in parallel. In the same nomenclature as in the analysis of the deep-bar effect, the ratio of the effective bar width, if made of copper, to the effective slot width is:

$$r = 1 + \frac{w_s}{8w},$$

since the resistivity of ordinary cold-rolled steel is about 8 times that of copper. Using this value of r, the value of α for the composite idle bar is:

(8.32) $$\alpha = 1.06 \sqrt{\left(\frac{rf}{60}\right)} \text{ for copper at 75 C.}$$

On this basis, the eddy currents and flux densities in the idle bar will be governed by Eqs. 8.12 and 8.14, provided only that the adjusted value of r is employed.

To find the coefficients of these equations, we note that $B = 4\pi I/w 10^9$, when $y = 0$ and also when $y = d$, if I is the rms current in the squirrel-cage bar in the same slot below the idle bar. Substituting in Eq. 8.12, we obtain:

(8.33) $$P = \frac{4\pi}{w 10^9},$$

and

(8.34) $$Q = \frac{-.\,4\pi I}{w 10^9} \frac{[\cosh(1+j)\alpha d - 1]}{\sinh(1+j)\alpha d},$$

whence the rms flux density at any height y is:

$$(8.35) \quad B = \frac{4\pi I}{w10^9} \left(\frac{\begin{cases} \sinh(1+j)\alpha d \cosh(1+j)\alpha y \\ -[\cosh(1+j)\alpha d - 1]\sinh(1+j)\alpha y \end{cases}}{\sinh(1+j)\alpha d} \right)$$

$$= \frac{4\pi I}{w10^9} \frac{\cosh(1+j)\alpha(y-d/2)}{\cosh(1+j)\alpha d/2}.$$

The total current in the idle bar is zero, so that:

$$(8.36) \quad \int_0^d \Delta \, dy = 0,$$

giving, from Eq. 8.14:

$$(8.37) \quad M \sinh(1+j)\alpha d + N[\cosh(1+j)\alpha d - 1] = 0.$$

Also, from Eqs. 8.9, 8.35, and 8.14, the value of $d\Delta/dy$ when $y = 0$ is:

$$(8.38) \quad \left. \frac{d\Delta}{dy} \right]_{y=0} = \frac{j2\pi f}{\rho} \left(\frac{-4\pi I}{w10^9} \right) = \frac{2j\alpha^2 I}{rw} = (1+j)\alpha N,$$

giving:

$$(8.39) \quad N = \frac{(1+j)}{rw} \alpha I.$$

Substituting Eqs. 8.37 and 8.39 in 8.14, the equation for the current density is obtained:

$$(8.40) \quad \Delta = \frac{(1+j)\alpha I}{rw} \left(\frac{\begin{cases} -[\cosh(1+j)\alpha d - 1]\cosh(1+j)\alpha y \\ +\sinh(1+j)\alpha d \sinh(1+j)\alpha y \end{cases}}{\sinh(1+j)\alpha d} \right)$$

$$= \frac{(1+j)\alpha I \sinh(1+j)\alpha(y-d/2)}{rw \cosh(1+j)\alpha d/2}.$$

The total voltage induced in the squirrel-cage bar below the idle bar is:

$$(8.41) \quad E = -j2\pi f \int_0^d B \, dy \text{ volts per cm.}$$

From Eqs. 8.35 and 8.41:

$$E = \frac{-j2\pi f(4\pi)I}{w10^9(1+j)\alpha} \left\{ \frac{\sinh^2(1+j)\alpha d - [\cosh(1+j)\alpha d - 1]^2}{\sinh(1+j)\alpha d} \right\}$$

or:

$$(8.42) \quad \frac{-E}{I} = Z_3 = 2(1+j)\alpha dR \left[\frac{\cosh(1+j)\alpha d - 1}{\sinh(1+j)\alpha d} \right]$$

$$= 2(1+j)\alpha dR \tanh(1+j)\alpha d/2 \text{ ohms,}$$

IDLE-BAR ROTORS

where R is the d-c resistance of the composite idle bar, and Z_3 is the impedance in the squirrel-cage bar circuit, Fig. 8.5, due to the flux crossing the idle bar.

Equation 8.42 may be conveniently resolved into its real and imaginary components, when αd is small, by letting $x = (1+j)\alpha d$, and expanding in powers of x:

$$Z_3 = 2Rx\left(\frac{\cosh x - 1}{\sinh x}\right) = 2Rx\tanh x/2$$

$$= Rx^2 \frac{1 + \dfrac{x^2}{12} + \dfrac{x^4}{360} + \dfrac{x^6}{20{,}160} + \cdots}{1 + \dfrac{x^2}{6} + \dfrac{x^4}{120} + \dfrac{x^6}{5{,}040} + \cdots}$$

or:

(8.43) $$Z_3 = Rx^2\left(1 - \frac{x^2}{12} + \frac{x^4}{120} - \frac{17x^6}{20{,}160} + \frac{x^8}{11{,}700}\right).$$

Since $x^2 = 2j\alpha^2 d^2$, this gives:

(8.44) $$Z_3 = R\left\{\left[\frac{(\alpha d)^4}{3} - \frac{17(\alpha d)^8}{1{,}260} + \cdots\right]\right.$$

$$\left. + j\left[2(\alpha d)^2 - \frac{(\alpha d)^6}{15} + \frac{(\alpha d)^{10}}{365} - \cdots\right]\right\},$$

whence the added resistance due to the idle bar is:

(8.45) $$R_3 = R\left[\frac{(\alpha d)^4}{3} - \frac{17(\alpha d)^8}{1{,}260} + \cdots\right], \quad \alpha d < 1.5;$$

and the effective reactance due to flux crossing the idle bars is:

(8.46) $$X_3 = 2(\alpha d)^2 R\left[1 - \frac{(\alpha d)^4}{30} + \frac{(\alpha d)^8}{730} - \cdots\right], \quad \alpha d < 1.5,$$

$$= X_0\left[1 - \frac{(\alpha d)^4}{30} + \frac{(\alpha d)^8}{730} - \cdots\right].$$

The quantity $2(\alpha d)^2 R$ is equal to $\dfrac{2fd}{w10^7}$, which is the reactance, X_0, due to the flux crossing the idle bars when unopposed by eddy currents. The factor in brackets in Eq. 8.46 is, therefore, the reduction factor due to the induced idle-bar currents.

When αd is large, Eq. 8.42 reduces to:

(8.47) $$Z_3 = 2\alpha dR(1+j), \quad \alpha d > 2,$$

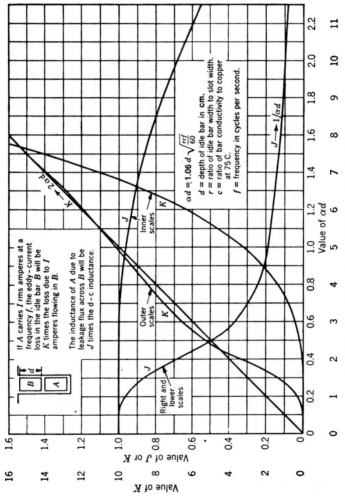

FIG. 8.6. Impedance Ratios for Open-Circuited Bars.

showing that the resistance and reactance due to the idle bars become equal when αd is large.

The general expressions for the idle-bar resistance and reactance ratios are:

$$(8.48) \quad \frac{R_3}{R} = \frac{2\alpha d\left[(\sinh 2\alpha d + \sin 2\alpha d) - 2(\sin \alpha d \cosh \alpha d + \sinh \alpha d \cos \alpha d)\right]}{\cosh 2\alpha d - \cos 2\alpha d},$$

and

$$(8.49) \quad \frac{X_3}{X_0} = \frac{\left[(\sinh 2\alpha d - \sin 2\alpha d) + 2(\sin \alpha d \cosh \alpha d - \sinh \alpha d \cos \alpha d)\right]}{\alpha d(\cosh 2\alpha d - \cos 2\alpha d)}.$$

The relations are shown in Fig. 8.6.

6 Double Squirrel-Cage Rotor

The arrangement of Fig. 8.7, with a top bar, B, of high resistance, and a bottom bar, A, of low resistance, separated by a leakage slot, provides a compact and effective design for securing any desired resistance-reactance ratio. The leakage-slot width, w, must be made wide enough to avoid saturation ($B_{max} < 16{,}000$ webers/sq cm, approx.), when the bottom-bar current is a maximum, as otherwise the reactance will be reduced at starting when it is needed, but will be present at speed when it is not wanted. That is, from Eq. 1.16, if I_A is the maximum current in the bottom bar, the slot width should not be less than

$$w \geqq 8 I_A \, 10^{-5} \text{ cm.}$$

Let:

R_A = bottom-bar resistance, in ohms,

R_B = top-bar resistance, in ohms,

$X + X_A$ = bottom-bar reactance at full frequency, in ohms,

X = mutual reactance of A and B at full frequency, in ohms,

R_0 = parallel resistance of A and B at full speed,

$X + X_0$ = parallel reactance of A and B, at low slip, referred to full frequency.

For good design, X and the self-reactance of B are made as small as possible, so that as much as possible of the total reactance is concentrated in X_A, where it is effective in forcing the desired current distribution at starting.

The slot-leakage flux due to current in A, which crosses the B bar, induces a reactance voltage in B (corresponding to $d_B/2w_B$) only half as great as the voltage it induces in A (corresponding to d_B/w_B). That is, the mutual reactance of A and B, X_2 in Fig. 8.8, includes a term proportional to $d_B/2w_B$ as well as all the flux crossing the slot above the top bar (Sect. 7.3). And the self-reactance of A, X_A, includes another $d_B/2w_B$ term, besides all the flux crossing the slot below the bar. The sum of these terms gives the correct total reactance of A when there is no current in B.

As the self-reactance of the top bar, due to flux crossing itself when there is no current in A, is proportional to $d_B/3w_B$, the B branch of the circuit in Fig. 8.8 should include a term $-(d_B/6w_B)$, so that the sum of this and the

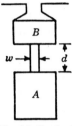

FIG. 8.7. Double Squirrel-Cage Bars, Showing Leakage Slot.

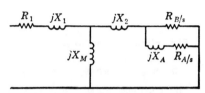

FIG. 8.8. Equivalent Circuit of Double Squirrel-Cage Motor.

$d_B/2w_B$ component of the mutual reactance, X_2, will give the correct self-reactance of B. In practice, however, this very small negative term can usually be neglected.

If the A and B bars have separate end rings, as required for (large) motors with brazed rotor windings, the B end rings will be nearer to the stator end windings than the A end rings, and so the mutual reactance of the stator with the B cage will be higher than that with the A cage by an amount X'. As shown in reference 15, X' must be added to X_A in the A branch of the circuit to give correct results in this case. X' is zero for the common end ring construction normally used for cast rotor windings. The very small negative reactance in the B bar branch will be the same in either case.

The running resistance of the circuit at low slip is:

$$(8.50) \qquad R_0 = \frac{1}{\frac{1}{R_A}+\frac{1}{R_B}} = \frac{R_A R_B}{R_A+R_B}.$$

The running reactance referred to primary (under zero-frequency conditions, when the current divides inversely as the resistances) is:

$$(8.51) \qquad X+X_0 = X+X_A\left(\frac{I_A}{I_A+I_B}\right)^2 = X+X_A\left(\frac{R_B}{R_A+R_B}\right)^2.$$

The starting impedance, at a slip s, is:

$$8.52) \qquad \frac{R_s}{s}+j(X+X_s) = \frac{1}{\frac{s}{R_B}+\frac{s}{R_A+jsX_A}}+jX = \frac{R_B(R_A+jsX_A)}{s(R_A+R_B+jsX_A)}+jX.$$

The starting resistance, from Eq. 8.52, is:

(8.53) $$R_s = \frac{R_B(R_A^2 + R_A R_B + s^2 X_A^2)}{(R_A + R_B)^2 + s^2 X_A^2}.$$

The starting reactance is, from Eq. 8.52:

(8.54) $$X + X_s = X + \frac{R_B^2 X_A}{(R_A + R_B)^2 + s^2 X_A^2}.$$

The increment in starting resistance is, from Eqs. 8.50 and 8.53:

(8.55) $$\Delta R = R_s - R_0 = \frac{R_B^2 s^2 X_A^2}{(R_A + R_B)[(R_A + R_B)^2 + s^2 X_A^2]}.$$

The increment in running reactance is, from Eqs. 8.51 and 8.54:

(8.56) $$\Delta X = X_0 - X_s = \frac{s^2 R_B^2 X_A^3}{(R_A + R_B)^2 [(R_A + R_B)^2 + s^2 X_A^2]}.$$

The best performance will be secured when $\Delta X/\Delta R$ is a minimum, as this gives the least additional running reactance for a given increment of starting resistance. However, from Eqs. 8.55 and 8.56, at all values of slip:

$$\frac{\Delta X}{\Delta R} = \frac{X_A}{R_A + R_B},$$

and, since, from Eqs. 8.50 and 8.51:

(8.57) $$R_B = \frac{R_A R_0}{R_A - R_0},$$

(8.58) $$X_A = \frac{X_0 (R_A + R_B)^2}{R_B^2} = \frac{R_A^2 X_0}{R_0^2},$$

we obtain:

(8.59) $$\frac{\Delta X}{\Delta R} = \frac{X_0}{R_0^2}(R_A - R_0).$$

Equation 8.59 shows that there is no minimum point for $\Delta X/\Delta R$, and that this ratio is independently adjustable by choice of the value of R_A. We may define R_A, arbitrarily, in terms of an independent parameter, m, to be:

(8.60) $$R_A = R_0 \left(1 + \frac{R_0}{m X_0}\right).$$

On this basis, from Eqs. 8.57 and 8.58:

$$R_B = mX_0\left(1 + \frac{R_0}{mX_0}\right), \tag{8.61}$$

$$X_A = X_0\left(1 + \frac{R_0}{mX_0}\right)^2, \tag{8.62}$$

and from Eqs. 8.59 and 8.61:

$$m = \frac{\Delta R}{\Delta X} = \frac{R_B - R_0}{X_0}. \tag{8.63}$$

From Eqs. 8.53, 8.56, 8.60, 8.61, and 8.62, at a slip s:

$$\Delta R_s = R_s - R_0 = \frac{ms^2 X_0}{m^2 + s^2}, \tag{8.64}$$

and

$$\Delta X_s = \frac{s^2 X_0}{m^2 + s^2}. \tag{8.65}$$

A large value of m gives a low increment in running over starting reactance, and, therefore, a relatively low starting current for a given breakdown torque, but it tends to give a "saddlebacked" torque curve, with a low torque at midspeed. Also, a large m requires a very high resistance in the top bar, which may lead to heating difficulties when the rotor is stalled. A small value of m gives a much higher torque in the middle-speed range, but also gives a considerably higher starting current.

A good criterion to follow is to make the starting resistance a maximum for given R_0 and X_0, as this corresponds to the highest starting torque per ampere. As Eq. 8.64 indicates, this occurs at $s = 1$, when $m = 1$. For this case, $\Delta R = \Delta X = X_0/2$, which provides a simple measure of the extra reactance necessary to obtain maximum starting resistance. (See Eq. 4.10 and Table 5.4.)

To summarize, the general procedure in designing any kind of deep-bar or double squirrel-cage motor is first to choose the value of secondary running resistance, R_0, so as to obtain the desired full-load slip. Then, the required starting resistance, $R_0 + \Delta R$, is chosen so as to obtain the needed starting torque with the permissible starting current. Next, the value of m (or of αd for a deep-bar design) is chosen to obtain the desired shape of torque curve, and from this the value of ΔX is determined, by Eq. 8.63, and the value of X_0, by Eq. 8.65. Finally, the stator winding and motor dimensions are chosen to give a value of X such that the running reactance, $X + X_0$, will give the desired breakdown torque. As soon as m, R_0, and X_0

are fixed, the dimensions of the individual bars and the leakage slot in the rotor are given by Eqs. 8.60, 8.61, and 8.62.

Figures 8.9 and 8.10 show idealized speed-current and speed-torque curves for a motor having $R_0 = 0.02$, $X_0 = 0.10$, and $X_1 = 0.10$, the values of primary resistance and magnetizing current being taken as zero for convenience. Curves for double squirrel-cage designs with $m = 0.25, 0.50$,

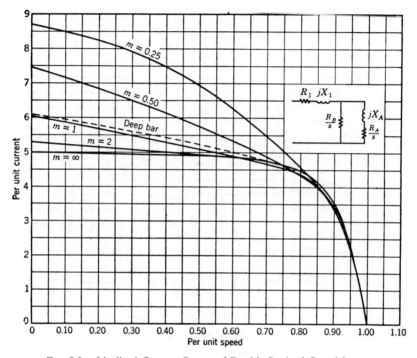

FIG. 8.9. Idealized Current Curves of Double Squirrel-Cage Motors.

1.00, 2.00, and ∞ are shown, and also one curve for a deep-bar rotor with the same R_0 and X_0, giving $K = 5$ and $\alpha d = 2.74$.

The wide range of characteristics obtainable by selecting m for the double squirrel-cage case can be roughly matched by shaping the bars in an L or inverted T form, or by using idle copper or steel bars in the slot space above a single squirrel-cage winding with deep slots. The designer has, therefore, a good deal of scope for originality in providing a variety of performance curves, while using a minimum number of standard parts, as required for low-cost manufacture.

With cast aluminum squirrel-cage windings, now generally used for motors up to 200 hp or more, many different shapes and arrangements of bars are used, limited only by the designer's fancy, the economics of die making, and the casting process. A commonly used construction has upper and lower bars, with a leakage slot between, that is also filled with aluminum, Fig. 8.11. A general method of performance calculation is desirable, therefore, that can be used for any bar configuration.

For this purpose, an equivalent circuit of the ladder type is used, Fig. 8.12. The conductor in each slot is divided into radial sections, each with a depth

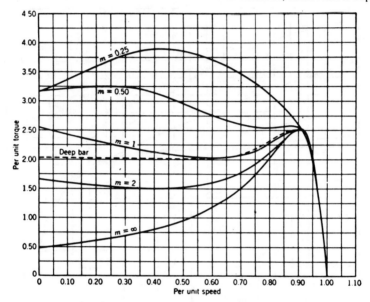

FIG. 8.10. Idealized Torque Curves of Double Squirrel-Cage Motors.

small enough to have a skin effect ratio, Fig. 8.3, nearly unity. Each bar section is represented by a vertical branch in the circuit. The self-reactance of each section is calculated as the second term of Eq. 7.11, in which the slot permeance factor, P_{s2}, has the form $d_n/3w_n$ for the nth section, from Eq. 7.4.

The impedance in each vertical branch of the circuit is the resistance/s of that bar section, minus j times half its reactance, as found above. In the upper horizontal branch of the circuit 3/2 of the section reactance is located on each side of the vertical branch, as shown in Fig. 8.12. It will be found that the R and X values for a deep bar, as given by Fig. 8.3, can be accurately checked by this ladder circuit, if the bar is divided into a sufficient number of sections.

Instead of using many bar sections, it is appropriate to use only the bar sections that have distinct dimensions, as A, B, and C in Fig. 8.11, and then to apply complex number multipliers to all the R and X values of the circuit, to take account of the induced eddy currents. Approximate formulas for these multipliers can be worked out, similar to Eqs. 8.22 and

8.26, and in this way quite accurate impedance values can be calculated for all kinds of bar arrangements, using only a few circuit branches.

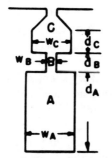

FIG. 8.11. Triple Squirrel Cage.

7 THE SERIES HYSTERESIS MOTOR (IDLE ALNICO BARS)

By using idle bars made of Alnico instead of steel, Fig. 8.4, a large improvement in starting torque per ampere can be obtained for a given value of full-load slip. For, as shown in Sect. 1.10, large hys-

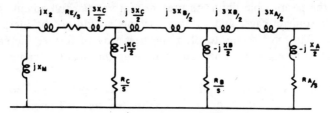

FIG. 8.12. Ladder Circuit for Multiple Squirrel Cage.

teresis losses will be created by the slot leakage flux in the Alnico, and these losses are proportional to the frequency. Thus, in the idle bar equivalent circuit, Fig. 8.5, a series resistance, $R_h s/s$, or simply R_h, is introduced in series with the reactance, X_m, which we shall now call X_h, due to the flux through the bars. At the optimum point, the hysteresis loss at 60 cycles in Alnico 5 is about 300 watts per cu. in. Since the available space in a typical cast aluminum squirrel-cage rotor that can be used for the Alnico bars is about 1 cu. in. per hp (without reducing the flux or current capacity), the Alnico bars will provide about 300/746 or 0.40 per unit torque on an output basis.

The extra leakage flux across the Alnico bars will reduce the maximum torque as well as the starting current. The design procedure, therefore, is to choose the desired maximum torque, and make the Alnico bar depth and width to give the corresponding reactance. The effect on motor performance is quite similar to that of using the double squirrel cage, whose advent permitted the use of smaller stator slots, wider stator teeth, higher magnetic flux, fewer stator winding turns, and, therefore, higher torque and greater output from a given frame. If these changes had been made keeping the single squirrel cage, the starting current would have been much too high, so the extra reactance of the double cage was used to hold down the starting current, without reducing the maximum torque too far. But, the extra reactance due to the double cage is about twice as high at full speed as at standstill (Eq. 8.65), while the extra reactance of the Alnico bars is the same at all speeds (for a given current) except for a very slight reduction at standstill due to eddy currents. In fact, the extra reactance of the Alnico is very much smaller at currents below the pickup value, and therefore its effect on full-load power factor is less than that of the double cage.

The starting current of a standard motor must be limited to about 6 times rated current, because the stator winding must not get too hot if the motor is stalled (as by accidental single phasing), for the time needed for the protective relay to open. Standard overload relays take about 15 seconds to open at 6 times current, and the usual stator current density at full load is of the order of 600 amperes per sq cm, so that the stalled temperature rise reaches about 120 C before the relay opens, which is near the permissible limit. As present motors take full advantage of the double squirrel cage, and starting currents should not be increased, the Alnico bar rotor seems to offer the best way of further increasing the output from a given frame size, continuing the steady advance shown in Fig. 1.1.

Alnico has a resistivity about the same as 3 per cent silicon steel, or about 5 times that of ordinary cold-rolled steel. Therefore, the eddy currents in the Alnico will be much smaller than those in idle steel bars. However, the eddy current losses may be quite appreciable, especially in motors of 50 hp or larger and when operating as a brake in the reverse direction of rotation.

Widely different results can be obtained by making the leakage slot beside the Alnico of different widths, and by filling this space with copper, aluminum, or steel, or leaving it empty. As a practical matter, it is essential to firmly lock the Alnico bars in place, and also to provide intimate contact with the laminations; to prevent them from "shivering" under the high-frequency forces and ultimately coming out in pieces, and to dissipate their losses without excessive temperature rise. For these reasons, Alnico bars are preferably used with cast aluminum rotors.

Fig. 8.13 shows speed-torque-current curves for two 15-hp motors with cast aluminum squirrel-cages and different arrangements of idle Alnico bars. Design A had Alnico 5 bars, and the leakage slots beside the idle bars were filled with aluminum that formed extensions of the current carrying conductors, giving a pronounced deep-bar effect. Since the rotor slots were not skewed, the slot harmonic currents flowing in the rotor bars made a large change in torque between 0.1 and -0.1 per unit speed.

Design B had Alnico 8 bars, with no current in the leakage slot space, and therefore a higher slot reactance with less harmonic torque, and lower locked rotor current. Both of these motors had nearly 200 per cent breakdown torque, and only 2 per cent slip, with locked rotor currents of 4.7 and 4.1 per unit, respectively, values which could not be attained by any single or multiple squirrel-cage construction.

Fig. 8.14 shows the equivalent circuit for a single squirrel-cage rotor with idle Alnico bars cast in place in the slots. The impedance due to the flux through the Alnico is $R_h + jX_h$, whose values are dependent on the

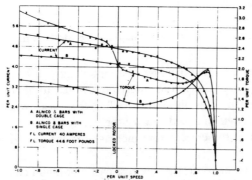

FIG. 8.13. Speed-Torque-Current Curves for Squirrel-Cage Motor with Embedded Alnico Bars.

current, but are substantially independent of frequency (and therefore of rotor speed). If the current exceeds about 150 per cent of the "pick-up" value, Fig. 1.15, R_h falls rapidly, so that it is desirable to make the leakage slot width beside the Alnico bar, Fig. 8.4, wide enough to make the pick-up current not much smaller than the locked rotor current.

If a reduced horsepower rating is accepted, using an oversized frame, space in the rotor slots can be made available for 2 or even 3 cubic inches of Alnico per horsepower. In this way the Saturistor effect can be made much greater, and nearly constant speed and torque can be obtained all the way from full speed backward to full speed forward, with a ratio of per unit current to per unit torque of the order of 1.5 to 2. With a series reactance or an adjustable line voltage, such a motor provides a means of obtaining wide range speed control for fans, pumps, and other drives whose torque varies as the first or higher power of the speed. The slip losses in the rotor limit the permissible rating and/or duration of the low-speed operation, so that wound rotor motors with external Saturistors are required for most integral horsepower ratings with speed control.

The most immediate use for the Alnico bar motor seems to be for fan or pump drive, where it is desired to provide wide range stepless speed control, with the load torque varying as the square of the speed. The usual practice has been to provide high-resistance squirrel cage motors with brazed rotor bars for this service, with voltage control to vary the speed. To limit the current at half speed to less than the full-load value, as

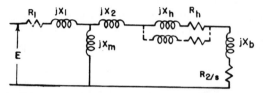

FIG. 8.14. Equivalent Circuit of Single Squirrel-Cage Motor with Idle Alnico Bars.

necessary to prevent undue heating, the motor must have a full-load slip of 10% or more, and the voltage must be lowered to about ⅓ of rated to obtain half speed. This gives a low efficiency. Also, when SCRs with adjustable firing angle are used to vary the voltage (Fig. 12.8), there are large harmonics in the current that increase the losses and heating very considerably.

With an Alnico bar rotor, the low cost cast aluminum construction and low-resistance bars are used, giving a full-load slip of only about 5% with correspondingly better efficiency over the entire speed range. Figures 8.15 and 8.16 show tested torque and current versus speed curves for a 1½-hp motor at various voltages. The half-speed voltage is about 40%, and the line current remains nearly constant over the full range, though it is somewhat higher than the rated current of the high-resistance motor, due to the lower power factor. Other tests have shown that, when SCR instead of Variac voltage control is used, the additional currents and heating due to harmonics in the voltage wave (Fig. 12.9) are small, as the high reactance of the Alnico requires a higher voltage, with smaller harmonics for a given speed, and reduces the current for a given harmonic voltage.

8 REDUCED VOLTAGE STARTING

An autotransformer, or "compensator", is sometimes used to reduce the voltage applied to a large motor at starting, in order to reduce the starting current. If the compensator supplies 50 per cent of rated voltage, the motor starting current will be 50 per cent of the full value, or perhaps 300 per cent of rated current. However, the current on the line side of the compensator, at full voltage, will be only 50 per cent of this, or 150 per cent of rated value, since the kva are equal on both sides of a transformer

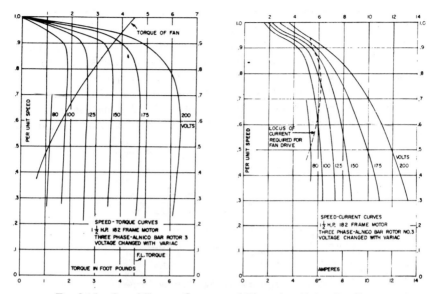

FIG. 8.15. Speed-Torque Curves on 3-Phase for Alnico Bar Rotor.

FIG. 8.16. Speed-Current Curves on 3-Phase for Alnico Bar Rotor.

($I_1 V_1 = I_2 V_2$, neglecting the exciting current). A compensator that reduces the motor voltage to k times normal reduces the starting kva to k^2 times the full voltage value. It is thus far more effective in reducing the line current than a reactor or resistor in series with the line side of the motor.

However, in order to transfer from the starting to the running position, Fig. 8.17, the switch must be opened and then closed on the line side. Immediately after the circuit is opened, the magnetic flux in the motor starts to decrease, sustained only by the induced current in the closed rotor winding. Also, the motor speed decreases, because of the unopposed load and friction and windage torques. When the circuit is again closed, the remaining voltage across the motor terminals is reduced, and is no longer in phase opposition to the line voltage. Therefore, the peak current on reclosing will be as high as, or may be much higher than, the normal starting current at full voltage. If reclosing occurs just when the motor has fallen back one pole, the current may be so high that the torque will damage the shaft or coupling.

The plain compensator method of starting does not reduce the peak starting current; it merely reduces the time duration of the high current. Instead of lasting for the many seconds that are required for the motor to come from rest to full speed, the high current lasts for a few cycles only.

To reduce the reclosing current surge, a transition impedance is sometimes connected between the starting and motor terminals, which maintains continuity of the current during changeover.

An alternative method of reduced voltage starting is to design a three-phase motor for Δ connection, and to connect it in Y for starting. The Y impedance is three times the Δ impedance, giving a starting current only one-third of the full voltage value. The per unit voltage applied to each phase is $1/\sqrt{3}$, or 57 per cent, so that the starting torque and kva, as well as the line current, are one-third of normal. As Fig. 8.16 indicates, this method requires three double-throw switches, and it has the same disadvantage of a high reclosing current that the compensator method has.

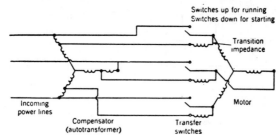

FIG. 8.17. Reduced Voltage Starting Connections.

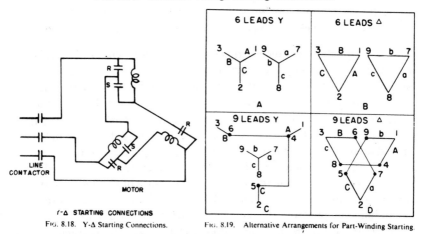

FIG. 8.18. Y-Δ Starting Connections.

FIG. 8.19. Alternative Arrangements for Part-Winding Starting.

For these reasons, and because the one-third value of starting torque is usually insufficient, the Y-Δ method is little used in the United States.

9 Part-Winding Starting

The lowest cost method of reduced current starting for standard induction motors consists in applying voltage to only a part of the winding at first, and connecting in the rest of the winding after a brief interval. Usually, the motor winding is designed with two parallel circuits in each phase, and the voltage is applied initially to only one circuit of each phase, giving "half-winding starting". Whether or no the motor comes up to speed on the half winding, the other half is closed in after a few seconds delay. This is most suitable for the low-voltage connection of dual-voltage motors, which have nine leads brought out. Figure 8.19 shows four alternative winding arrangements, and Fig. 8.20 shows three alternative lead connections that can be used.

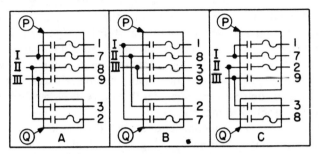

FIG. 8.20. Alternative Lead Connections for Two-Step Starting.

Table 8.1 gives the combinations in which these are used for two-step starting, with two contactors, P having 4 poles and Q having 2 poles, all of the same current rating, equal to half the full winding current. By adding a third, single-pole, contactor, three-step starting can be obtained, with superior performance.

TABLE 8.1

ALTERNATIVE PART-WINDING STARTING SCHEMES

Scheme	Two-Step Starting			
	6 Leads		9 Leads	
	Winding	Connection	Winding	Connection
1/2 Y	Fig. 8.17A	Fig. 8.18A	Fig. 8.17C	Fig. 8.18A
1/2 Δ	Fig. 8.17B	Fig. 8.18A	Fig. 8.17D	Fig. 8.18C
2/3 Y	Fig. 8.17A	Fig. 8.18B	Fig. 8.17C	Fig. 8.18B
2/3 Δ	Fig. 8.17B	Fig. 8.18B	Fig. 8.17D	Fig. 8.18A

The most commonly used schemes are the Y and Δ windings, C and D in Fig. 8.19, with connections A and C, respectively, in Fig. 8.20. For these the contactors can have 3 poles each, but the 4 and 2 pole arrangement is preferred, to permit the 2/3 winding scheme to be used with the same starting panel. To avoid unbalanced air-gap forces on the part winding, due to harmonic fields with pole numbers differing by 2 from the fundamental, it is always desirable to connect alternate poles (all north poles) in one half winding. This ensures balanced air-gap forces, but it gives a large second harmonic air-gap field, which produces a severe torque dip at half speed. On the half winding, therefore, the motor draws about 60 per cent of full winding current on the first step, with about 50 per cent of full torque, but it usually will not accelerate above half speed. When the second contactor is closed, there will be a second current surge, nearly equal to the full winding starting current. For this reason, and because the current density in the active half winding at start is 120 per cent or more of normal full-winding starting value, the heating and end winding forces are fully as severe as with full-voltage starting on the complete winding. Thus, the only reason for using these half winding schemes is to comply with the rules of some power companies that currents larger than a certain value must be applied in increments.

Another feasible starting connection, called double-delta starting, is to employ a two circuit Δ winding, as in Fig. 8.19 D and to connect terminals 1, 2, and 3 only to the lines on the first step. This places the two circuits of each phase in series with 120° space phase displacement, and so reduces the initial current to only about 40 per cent of normal. The currents in all phases are balanced, so that the winding forces and the rate of temperature rise on the first step are much less than for the other part-winding schemes. The locked rotor torque is also low, about 30 per cent of normal, and there is a severe second harmonic torque dip requiring the second contactor to be closed when the motor is at half speed, so giving nearly as great a current inrush as with full winding starting.

The second harmonic torque dip can be greatly reduced by making the winding with a long coil pitch, nearly 100 per cent, but this reduces the half winding reactance, and brings the initial current up to some 80 per cent of full value for the half windings, or 60 per cent for the double delta.

One answer to these difficulties is to use the 2/3 winding scheme connection B (Fig. 8.20) for Y, or A for Δ. This gives almost the same average current and torque on the first step as the half winding, but the torque dip is greatly reduced, so that the motor will come up to full speed quickly on the first step. Then, when the second contactor is closed, there is no additional current surge. And, the timing of the second contactor is no longer critical. The only difference between the 1/2 and 2/3 winding schemes, aside from the different lead connections, is the requirement that the P contactor have 4 poles and the Q only 2, instead of each having 3 poles. As indicated in the table, the 4- and 2-pole contactors are also satisfactory for the half winding schemes, but not for the double delta. On the first step in the 2/3 Y case, 4 of the 6 half phases carry approximately equal currents; while in the 2/3 Δ case, 2 half-phase belts carry equal currents and the other 4 half-phase belts carry about half as much current, but the two connections give nearly the same performance.

The performance of any of these part-winding schemes can be calculated with the equivalent circuit of Fig. 8.21 (derived in Sect. 13.3). This shows separate circuits for the forward and backward fields of each half winding, each with a distinct voltage applied. Each half winding has its own self impedance, Z_1, and also a series reactance, Z_H, due to the air-gap field harmonics that it makes, which are not present when both half windings carry equal currents. This appears in the circuit as a self-reactance of $2Z_H$

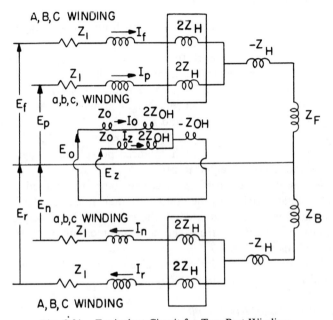

FIG. 8.21. Equivalent Circuit for Two-Part Winding.

and a mutual reactance of $-Z_H$, so that the circuit gives zero for this reactance component when the half windings carry equal in-phase currents, and $4Z_H$ when they carry equal but opposing currents. The rotor impedance is $Z_F = R_2/s + jX_x$ for the forward field, and $Z_B = R_2/(2-s) + jX_2$ for the backward field.

The circuit constants are:

Z_0 = zero phase sequence impedance per phase of half stator winding with normal winding turns.

Z_1 = three-phase self-impedance per phase of half stator winding with normal winding turns, excluding harmonic fluxes not present with full winding.

Z_F, Z_B = forward and backward three-phase mutual impedances of the half windings, including rotor and parallel magnetizing circuit with normal winding turns.

Z_H = three-phase impedance of half stator winding per phase, considering only harmonic fluxes not present with full winding. Since the phases are usually unbalanced with respect to these harmonic fields, an average value of Z_H must be determined.

Z_{0H} = corresponding zero phase sequence impedance of half winding due to extra harmonics.

The values of these impedances may be found by calculation, or by tests.

In the circuit of Fig. 8.21, the forward, reverse, and zero phase sequence voltages applied to the terminals of the A, B, C half windings are designated as E_f, E_r, and E_o, respectively. Likewise, the positive, negative, and zero phase sequence voltages applied to the a, b, c half winding, are designated by E_p, E_n, and E_z, respectively. Under full winding conditions, with the two half windings in parallel, $E_f = E_p$; $E_r = E_n$, and $E_o = E_z$. Normally, of course, with balanced polyphase voltages applied, E_r, E_n, E_o, and E_z are all equal to zero.

To calculate the motor performance, two steps are required. First, the sequence voltages applied to the motor terminals must be calculated from the circuit arrangement; and, second, the motor currents, torques, and so forth, must be calculated from the equivalent circuit with the proper sequence voltages applied to it.

In the calculations and derivations given, for simplicity, the rotor is considered to be at standstill, i.e., $Z_F = Z_B$.

Solution of the Equivalent Circuit

The (balanced) line voltages are taken to be:

(8.66) $\quad E_{\text{I II}} = E\underline{/0}, \quad E_{\text{II III}} = E\underline{/240}, \quad E_{\text{III I}} = E\underline{/120}.$

Considering the general circuit of Fig. 8.21, if the sequence voltages are known,† we have:

(8.67) $\quad E_f = I_f(Z_1 + 2Z_H - Z_H + Z_F) + I_p(Z_F - Z_H),$

or $\quad E_f = I_f(Z_1 + Z_H + Z_F) + I_p(Z_F - Z_H).$

(8.68) $\quad E_p = I_p(Z_1 + Z_H + Z_F) + I_f(Z_F - Z_H).$

(8.69) $\quad E_r = I_r(Z_1 + Z_H + Z_B) + I_n(Z_B - Z_H).$

(8.70) $\quad E_n = I_n(Z_1 + Z_H + Z_B) + I_r(Z_B - Z_H).$

† The subscripts f, r, 0, p, n, and z are used to indicate forward, backward, and zero, and positive, negative, and zero sequence components, respectively.

PART-WINDING STARTING

Let:

(8.71) $\quad P = Z_1 + Z_H + Z_F,$ \hspace{2em} (8.72) $\quad Q = Z_F - Z_H,$

(8.73) $\quad R = Z_1 + Z_H + Z_B,$ \hspace{2em} (8.74) $\quad S = Z_B - Z_H.$

P, Q, R, and S may be expressed as ohms, or as per unit impedance values.

Substituting in Eqs. 8.67 to 8.70:

(8.75) $\quad E_f = PI_f + QI_p,$ \hspace{2em} (8.76) $\quad E_p = PI_p + QI_f,$

(8.77) $\quad E_r = RI_r + SI_n,$ \hspace{2em} (8.78) $\quad E_n = RI_n + SI_r.$

Let:

(8.79) $$\alpha = \frac{1}{P+Q} = \frac{1}{Z_1 + 2Z_F}.$$

(8.80) $$\beta = \frac{Q}{P^2 - Q^2} = \frac{\alpha Q}{P - Q} = \frac{Z_F - Z_H}{(Z_1 + 2Z_F)(Z_1 + 2Z_H)}.$$

Then, at standstill, when $Z_F = Z_B$, so that $R = P$ and $S = Q$, we can solve Eqs. 8.75 to 8.78 simultaneously, giving:

(8.81) $$I_f = \frac{E_f}{P+Q} + \frac{Q(E_f - E_b)}{P^2 - Q^2} = \alpha E_f + \beta(E_f - E_p).$$

(8.82) $\quad I_p = \alpha E_p + \beta(E_p - E_f).$

(8.83) $\quad I_r = \alpha E_r + \beta(E_r - E_n).$

(8.84) $\quad I_n = \alpha E_n + \beta(E_n - E_r).$

We have, also:

(8.85) $\quad E_0 = I_0(Z_0 + Z_{0H}) - I_z Z_{0H},$

(8.86) $\quad E_z = I_z(Z_0 + Z_{0H}) - I_0 Z_{0H},$

whence:

(8.87) $$I_0 = \frac{E_0 Z_0 + (E_0 + E_z)Z_{0H}}{Z_0(Z_0 + 2Z_0 Z_{0H})}.$$

(8.88) $$I_z = \frac{E_z Z_0 + (E_z + E_0)Z_{0H}}{Z_0(Z_0 + 2Z_0 Z_{0H})}.$$

The phase currents are found by combining the sequence components:

(8.89) $\sqrt{3}I_A = I_f + I_r + I_0.$

(8.90) $\sqrt{3}I_B = I_f\underline{/240} + I_r\underline{/120} + I_0 = -0.50(I_f + I_r) - j0.866(I_f - I_r) + I_0.$

(8.91) $\sqrt{3}I_C = I_f\underline{/120} + I_r\underline{/240} + I_0 = -0.50(I_f + I_r) + j0.866(I_f - I_r) + I_0.$

(8.92) $\sqrt{3}I_a = I_p + I_n + I_z.$

(8.93) $\sqrt{3}I_b = -0.50(I_p + I_n) - j0.866(I_p - I_n) + I_z.$

(8.94) $\sqrt{3}I_c = -0.50(I_p + I_n) + j0.866(I_p - I_n) + I_z.$

If R_F and R_B represent the secondary resistance components of Z_F and Z_B, respectively, the motor torque is given by the equation:

(8.95) $T = (I_f + I_p)^2 R_F - (I_r + I_n)^2 R_B,$ or, at standstill:

(8.96) $T = [(I_f + I_p)^2 - (I_r + I_n)^2] R_F$ synchronous watts (all phases).

To change this into foot pounds, it should be multiplied by $7.04/N$, where N is the synchronous speed of the motor in rpm.

When the sequence voltages are known, the currents and torque may be found by direct substitution in the foregoing equations. To find these voltages, we have the following relations (Sect. 3.15):

(8.97) $E_f = \frac{1}{\sqrt{3}}(E_A + E_B\underline{/240} + E_C\underline{/120}).$

(8.98) $E_r = \frac{1}{\sqrt{3}}(E_A + E_B\underline{/120} + E_C\underline{/240}).$

(8.99) $E_p = \frac{1}{\sqrt{3}}(E_a + E_b\underline{/240} + E_c\underline{/120}).$

(8.100) $E_n = \frac{1}{\sqrt{3}}(E_a + E_b\underline{/120} + E_c\underline{/240}).$

(8.101) $E_0 = \frac{1}{\sqrt{3}}(E_A + E_B + E_C).$

(8.102) $E_z = \frac{1}{\sqrt{3}}(E_a + E_b + E_c).$

From Eq. 8.66 and the known lead connections, the equations enable the sequence voltages to be written down directly, or expressed in terms of one or more unknown voltages. In the latter case, the connection diagram fixes enough additional relations between the phase currents to provide as many equations as there are unknowns.

PART-WINDING STARTING

Example:

As a simple example, we may consider the full winding with delta connection. Then, $E_f = E_p = \sqrt{3}\,E$, $E_r = E_n = E_0 = E_z = 0$, and, from Eqs. 8.81 and 8.82:

$$I_f = I_p = \alpha\sqrt{3}\,E.$$

The phase current for the full winding is, from Eqs. 8.89 and 8.92:

(8.103) $$I_A + I_a = (I_f + I_p)/\sqrt{3} = 2\alpha E,$$

and the line current is $\sqrt{3}\,(I_A + I_a)$ or $2\alpha\sqrt{3}\,E$.

With a Y winding, keeping E as the line voltage, the phase voltage is $\dfrac{E}{\sqrt{3}}$, and the current on full winding will be $\dfrac{2\alpha E}{\sqrt{3}}$. ($\alpha$ is three times as great for a Y winding as for a Δ winding for the same line voltage and the same motor rating.)

The torque in synchronous watts for the three phases is given by Eq. 8.96, or is:

$$12\alpha^2 E^2 R_F \text{ synchronous watts.}$$

Proceeding in this way, the ratios of expected locked rotor torques and line currents, with various part-winding schemes, to the corresponding values with the full winding at the same impressed voltages, are found to be:

1/2Δ and 1/2 Y Current Ratio:

$$\frac{(\alpha + 2\beta)}{2(\alpha + B)} \text{ all currents the same.}$$

Torque Ratio:

$$\frac{(\alpha + 2\beta)^2}{4(\alpha + \beta)^2}$$

2/3 Y Current Ratios:

$$\frac{\sqrt{3}\,(\alpha + 2\beta)(2\alpha\,\underline{|210} + \sqrt{3}\,\beta\,\underline{|240})}{2(2\alpha + 3\beta)(2\alpha + \beta)}, \quad \frac{\sqrt{3}\,(\alpha + 2\beta)(2\alpha\,\underline{|150} + \sqrt{3}\,\beta\,\underline{|120})}{2(2\alpha + 3\beta)(2\alpha + \beta)},$$

and

$$\frac{3(\alpha + 2\beta)}{2(2\alpha + 3\beta)}.$$

Torque Ratio:

$$\frac{3(\alpha + 2\beta)^2}{4(2\alpha + 3\beta)(2\alpha + \beta)}.$$

2/3Δ *Current Ratios*:

$$\frac{\sqrt{3}(\alpha+\beta)[4\alpha+(7+j\sqrt{3})\beta]}{4(2\alpha+3\beta)(2\alpha+\beta)}, \quad \frac{\sqrt{3}(\alpha+2\beta)(2\alpha\underline{/30}+\beta\sqrt{3})}{2(2\alpha+3\beta)(2\alpha+\beta)},$$

and

$$\frac{\sqrt{3}[4\sqrt{3}\alpha^2\underline{/330}+\alpha\beta(15-j5\sqrt{3})+\beta^2(7-j3\sqrt{3})]}{4(2\alpha+3\beta)(2\alpha+\beta)}.$$

Torque Ratio:

$$\frac{[4\alpha^2+11\alpha\beta+(6+\underline{/300})\beta^2]^2-(2\alpha^2+2\alpha\beta\underline{/60}+\beta^2)^2}{4(2\alpha+3\beta)^2(2\alpha+\beta)^2}.$$

10 Split-Winding Starting

Another practical method for securing reduced starting current, with nearly normal torque per ampere, is to provide two or more parallel circuits in each phase of the winding, and to connect the circuits of one (or two) phases in series for starting, changing to parallel for full-speed operation. This will give unbalanced currents during the starting period, but other motors already on the line will act as phase balancers to supply the negative-phase sequence currents, and the effect on the power system

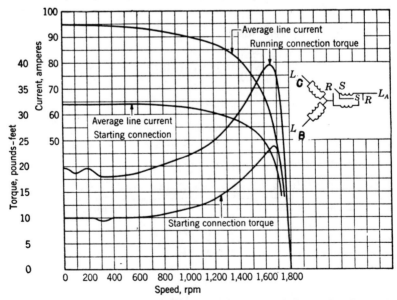

Fig. 8.22. Starting Performance of Three-Phase Motor with Split-Winding Connection.

SPLIT-WINDING STARTING

will be nearly the same as that of the positive-phase sequence currents alone. Furthermore, the less-than-normal flicker on the line supplying the lowest current to the motor may more than offset the higher voltage flicker on the other lines.

Assume a three-phase induction motor with symmetrical stator phase windings, A, B, and C, and let the winding of phase A be reconnected to have K times as many turns in series as the other two phases, Fig. 8.22. Also, let:

I_A, I_B, I_C = phase currents,

E_f, E_b, E_0 = symmetrical component voltages induced in the rotor, referred to the (unit) number of turns in the B or C phase (Sect. 3.15),

E_A, E_B, E_C = symmetrical three-phase impressed voltages for Δ-connected windings,

E_{AB}, E_{BC}, E_{CA} = symmetrical three-phase impressed voltages for Y-connected windings,

Z_f, Z_b, Z_0 = symmetrical impedances of stator, per phase, referred to unit turns.

Then, the sequence voltages produced by the phase currents are:

(8.104) $\qquad \sqrt{3} E_f = (KI_A + I_B\underline{|120} + I_C\underline{|240}) Z_f,$

(8.105) $\qquad \sqrt{3} E_b = (KI_A + I_B\underline{|240} + I_C\underline{|120}) Z_b,$

(8.106) $\qquad \sqrt{3} E_0 = (KI_A + I_B + I_C) Z_0.$

At speed, $Z_f = R_1 + \dfrac{R_2}{s} + j(X_1 + X_2)$, and $Z_b = R_1 + \dfrac{R_2}{2-s} + j(X_1 + X_2)$, neglecting the magnetizing current.

At standstill, $Z_f = Z_b = Z$, and, for Δ connection:

(8.107) $\quad E_A = E = K(E_f + E_b + E_0)/\sqrt{3} = [2K^2 I_A - K(I_B + I_C)]\dfrac{Z}{3}$

$$+ K(KI_A + I_B + I_C)\dfrac{Z_0}{3};$$

(8.108) $\quad E_B = E\underline{|240} = (E_f\underline{|240} + E_b\underline{|120} + E_0)/\sqrt{3}$

$$= (2I_B - KI_A - I_C)\dfrac{Z}{3} + (KI_A + I_B + I_C)\dfrac{Z_0}{3};$$

(8.109) $\quad E_C = E\underline{|120} = (E_f\underline{|120} + E_b\underline{|240} + E_0)/\sqrt{3}$

$$= (2I_C - KI_A - I_B)\dfrac{Z}{3} + (KI_A + I_B + I_C)\dfrac{Z_0}{3}.$$

Subtracting Eq. 8.108 from Eq. 8.109:

$$(8.110) \qquad I_C = I_B + \frac{jE\sqrt{3}}{Z}.$$

Multiplying Eqs. 8.108 and 8.109 by K, and adding their sum to Eq. 8.107:

$$(8.111) \qquad (1-K)E = (KI_A + I_B + I_C)KZ_0.$$

Combining Eqs. 8.110 and 8.111:

$$(8.112) \qquad Z(1-K)E = \left(KI_A + 2I_B + j\frac{\sqrt{3}E}{Z}\right)KZZ_0.$$

From Eqs. 8.107, 8.110, and 8.111:

$$(8.113) \qquad E(2+K+jK\sqrt{3}) = (2K^2 I_A + 2KI_B)Z.$$

From Eqs. 8.112 and 8.113:

$$(8.114) \qquad 3K^2 I_A ZZ_0 = E[Z(1-K) + Z_0(K+2)],$$

$$(8.115) \qquad 6KI_B ZZ_0 = -E[2(K-1)Z + (K+2)Z_0 + j3KZ_0\sqrt{3}],$$

$$(8.116) \qquad 6KI_C ZZ_0 = -E[2(K-1)Z + (K+2)Z_0 - j3KZ_0\sqrt{3}].$$

The line currents at standstill are:

$$(8.117) \qquad I_{AB} = I_B - I_A = \frac{-E}{6K^2 ZZ_0}[2(K-1)^2 Z + (K+2)^2 Z_0 + j3K^2 Z_0\sqrt{3}],$$

$$(8.118) \qquad I_{BC} = I_C - I_B = j\frac{E\sqrt{3}}{Z},$$

$$(8.119) \qquad I_{CA} = I_A - I_C = \frac{E}{6K^2 ZZ_0}[2(K-1)^2 Z + (K+2)^2 Z_0 - j3K^2 Z_0\sqrt{3}].$$

The rotor sequence voltages are:

$$(8.120) \qquad E_f = \frac{E}{\sqrt{3}K}(1+2K),$$

$$(8.121) \qquad E_b = -\frac{E}{\sqrt{3}K}(K-1),$$

$$(8.122) \qquad E_0 = -\frac{E}{\sqrt{3}K}(K-1).$$

The ratio of locked-rotor torque to normal locked-rotor torque (when $K=1$) is:

(8.123) $$T = \frac{E_f^2 - E_b^2}{3E^2} = \frac{2+K}{3K}.$$

The zero-phase sequence current (which produces a third harmonic air-gap field) is:

(8.124) $$I_0 = \frac{E_0}{Z_0} = \frac{E(1-K)}{\sqrt{3}KZ_0}.$$

This is not the same as the zero-phase sequence stator current, which is $(I_A + I_B + I_C)/\sqrt{3}$.

If the windings are connected in Y:

(8.125) $$E_{AB} = \sqrt{3}\,E\,\underline{/30}$$
$$= [K(E_f + E_b + E_0) - (E_f\underline{/240} + E_b\underline{/120} + E_0)]/\sqrt{3},$$

(8.126) $$E_{BC} = -jE\sqrt{3}$$
$$= (E_f\underline{/240} + E_b\underline{/120} + E_0 - E_f\underline{/120} - E_b\underline{/240} - E_0)/\sqrt{3},$$

(8.127) $$E_{CA} = \sqrt{3}\,E\,\underline{/150}$$
$$= [E_f\underline{/120} + E_b\underline{/240} + E_0 - K(E_f + E_b + E_0)]/\sqrt{3},$$

and

(8.128) $\quad I_A + I_B + I_C = 0.$

Combining Eqs. 8.104, 8.105, 8.106, 8.125, 8.126, and 8.127 gives, taking $Z_f = Z_b = Z$:

(8.129) $$\sqrt{3}\,E\,\underline{/30°} = [K(2K+1)I_A - (K+2)I_B - (K-1)I_C]\frac{Z}{3}$$
$$+ (K-1)(KI_A + I_B + I_C)\frac{Z_0}{3},$$

(8.130) $$-jE\sqrt{3} = (I_B - I_C)Z,$$

(8.131) $$-\sqrt{3}\,E\,\underline{/150} = (K-1)^2\frac{I_A Z_0}{3} - I_C Z + \frac{I_A Z}{3}(2K^2 + 2K - 1).$$

Solving these last equations for the line currents:

(8.132) $$I_A = \frac{9E}{M},$$

(8.133) $$I_B = -\frac{9E}{2M} - j\frac{E\sqrt{3}}{2Z},$$

(8.134) $$I_C = -\frac{9E}{2M} + j\frac{E\sqrt{3}}{2Z},$$

where

(8.135) $$M = (2K+1)^2 Z + 2(K-1)^2 Z_0.$$

Substituting in Eqs. 8.104, 8.105, and 8.106, the rotor-sequence voltages are:

(8.136) $$E_f = \frac{3\sqrt{3} E(2K+1)Z}{2M} + \frac{E\sqrt{3}}{2},$$

(8.137) $$E_b = \frac{3\sqrt{3} E(2K+1)Z}{2M} - \frac{E\sqrt{3}}{2},$$

(8.138) $$E_0 = \frac{3\sqrt{3} E(K-1)Z_0}{M}.$$

The locked-rotor torque ratio is:

(8.139) $$T = \frac{E_f^2 - E_b^2}{3E^2} = \frac{3(2K+1)Z}{M}.$$

The positive-phase sequence line current is:

(8.140) $$I_f = \frac{1}{\sqrt{3}}(I_A + I_B \underline{/120} + I_C \underline{/240})$$
$$= \frac{9\sqrt{3}E}{2M} + \frac{\sqrt{3}E}{2Z};$$

and the negative-phase sequence current is:

(8.141) $$I_b = \frac{9\sqrt{3}E}{2M} - \frac{\sqrt{3}E}{2Z}.$$

It will be found that I_f is always a little smaller than $\sqrt{3}$ times the arithmetic average of the three line currents.

The resulting values of locked-rotor torque and current for both Y and Δ connections are shown in Table 8.2 for various values of the turn ratio, K. The value of E for the Δ case is taken as $1/\sqrt{3}$ times E for the Y case, to give the same line current and torque base.

To sum up, this two-circuit in series Y (for one phase only) starting connection reduces the positive-phase sequence line currents from one to about two-thirds (actual line currents are 0.35, 0.88, and 0.88); and the torque from one to 0.57. Or, about half torque is obtained for two-thirds current, values that are distinctly preferable to the 0.33 ratios given by

Y-Δ starting. The cost of starting control is moderate, as only three half-current single-pole contactors are required in addition to the usual full voltage starting equipment. The current densities are lower than for full voltage starting, so the permissible stalled time is correspondingly increased.

TABLE 8.2

STARTING PERFORMANCE WITH SPLIT-WINDING CONNECTION OF FIG. 8.22.

K	$\dfrac{E_f}{\sqrt{3}}$	$\dfrac{E_b}{\sqrt{3}}$	$\dfrac{E_0}{\sqrt{3}}$	Line Currents			Torque
				Max	Min	Avg	
Δ Connection, $Z_0 = Z/2$							
1	1	0	0	1	1	1	1
1.5	0.889	0.111	0.111	1	0.755	0.837	0.778
2	0.833	0.167	0.167	1	0.694	0.796	0.667
∞	0.667	0.333	0.333	1	0.680	0.787	0.333
Δ Connection, $Z_0 = Z/4$							
1.5	0.889	0.111	0.111	1	0.787	0.858	0.778
2	0.833	0.167	0.167	1	0.764	0.843	0.667
∞	0.667	0.333	0.333	1	1.00	1.00	0.333
Y Connection, $Z_0 = Z/2$							
1	1	0	0	1	1	1	1
1.5	0.869	0.131	0.046	0.910	0.554	0.791	0.739
2	0.789	0.211	0.058	0.883	0.346	0.704	0.577
∞	0.500	0.500	0	0.866	0	0.577	0
Y Connection, $Z_0 = 0$							
1.5	0.875	0.125	0	0.911	0.562	0.795	0.750
2	0.800	0.200	0	0.884	0.360	0.709	0.600

Figure 8.22 shows the torque and current curves of a motor tested with this split-winding connection. The test ratios of the torque to normal averaged between 0.50 and 0.55 and of the current between 0.65 and 0.70, over the speed range, in close agreement with the above equations.

One difficulty which is inherent in this unbalanced starting method is the presence of a small amount of third harmonic field, due to the zero-phase sequence component of the mmf impressed on the rotor. The third harmonic field produces a dip in torque at one-third speed (Sect. 9.3),

302   SPEED-TORQUE-CURRENT RELATIONS

which may be serious if the winding pitch is far from $\frac{2}{3}$ and the Δ connection is used.

In the Y-connected motor, the zero-phase sequence voltage, from Table 8.2, is only about 0.05, so that no trouble will occur from this source. Another disadvantage of the Δ connection is that the current in one line remains at full value. This can be remedied by including external impedance in the B and C phases. This case is treated more fully in Sect. 13.3.

For these reasons, the Y connection is preferable. If the motor poles are a multiple of 4, it is possible to reconnect only half of one phase winding with circuits in series for starting, leaving the other half of this phase, and the other two phases, with circuits in parallel throughout. This corresponds to making $K = 1.50$, and gives for the Y case a per unit net torque of 0.74, a forward sequence current in the lines of 0.78, and a peak current of 0.91, with a zero-phase sequence voltage of 0.05 times the forward field (Table 8.2).

11 Pole Changing and Concatenation

Instead of varying the voltage or impedance to control the starting performance, the motor may be wound with two stator windings having different numbers of poles. When one winding is in use, the other is open-circuited, and is idle. The low-speed winding, with the larger number of poles, allows continuous operation at reduced speed, and also gives superior starting performance (greater starting torque per ampere). For example, three-phase windings of both 4 and 6 poles may be placed in a 72-slot stator, with a squirrel-cage rotor. For convenience in winding, it is usual to employ the same coil throw for both windings. A coil throw of 11 slots, for example, would give $\frac{11}{12}$ pitch for the six-pole and $\frac{11}{18}$ for the four-pole winding, both of which are satisfactory. The motor speed is changed by opening one winding and closing the other on the line, so that it is an adjustable-speed, and not a varying-speed, motor.

It is necessary in designing two winding motors to select the numbers of slots, circuits, and pole connections in such a way that no circulating currents are induced in the idle winding by the voltages developed by the active winding. For example, the voltages induced in the separate poles of the four-pole winding by the six-pole magnetic fluxes are 90° out of phase, so that large circulating currents would flow between circuits if all 4 poles were connected in parallel. By connecting the four-pole winding with 2 circuits, each having 2 like (alternate) poles in series, the voltage of each circuit will be zero under six-pole excitation, and the winding will be free of circulating currents.

Instead of using 2 separate windings, a single winding can be reconnected to give a 2/1 pole ratio. For example, a two-pole, 24-slot, 50 per cent pitch

POLE CHANGING AND CONCATENATION

stator winding may be connected with 2 circuits per phase, alternate poles in series. By reversing 1 circuit in each phase, as indicated in Fig. 8.23, the winding is converted into a four-pole, 100 per cent pitch winding, with 1 circuit per phase. On the two-pole connection, there is 1 north and 1 south pole (coil group) in each phase. On the four-pole connection,

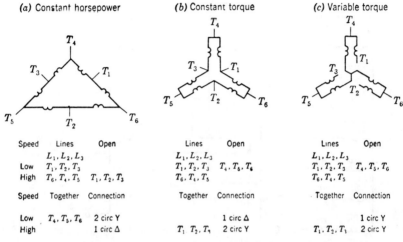

(a) Constant horsepower			(b) Constant torque			(c) Variable torque		
Speed	Lines	Open	Speed	Lines	Open	Speed	Lines	Open
	L_1, L_2, L_3			L_1, L_2, L_3			L_1, L_2, L_3	
Low	T_1, T_2, T_3		Low	T_1, T_2, T_3	T_4, T_5, T_6	Low	T_1, T_2, T_3	T_4, T_5, T_6
High	T_6, T_4, T_5	T_1, T_2, T_3	High	T_6, T_4, T_5		High	T_6, T_4, T_5	
Speed	Together	Connection	Speed	Together	Connection	Speed	Together	Connection
Low	T_4, T_5, T_6	2 circ Y	Low		1 circ Δ	Low		1 circ Y
High		1 circ Δ	High	T_1, T_2, T_3	2 circ Y	High	T_1, T_2, T_3	2 circ Y

FIG. 8.23. Three Phase, 2/1 Speed, Single-Voltage Windings.

there are 2 north (or south) poles in each phase, the opposite, or "consequent", poles being formed by flux returning in the spaces between coil groups.

On the two-pole connection, each phase belt of a normal three-phase winding spans 60 electrical degrees, while on the four-pole connection, the (same) phase belts are 120° wide. The phase sequences in successive slots are:

Two-Pole Connection

A A A A $\bar{B}$ $\bar{B}$ $\bar{B}$ $\bar{B}$ C C C C $\bar{A}$ $\bar{A}$ $\bar{A}$ $\bar{A}$ B B B B $\bar{C}$ $\bar{C}$ $\bar{C}$ $\bar{C}$

$\bar{B}$ $\bar{B}$ C C C C $\bar{A}$ $\bar{A}$ $\bar{A}$ $\bar{A}$ B B B B $\bar{C}$ $\bar{C}$ $\bar{C}$ $\bar{C}$ A A A A $\bar{B}$ $\bar{B}$

Four-Pole Connection

a a a a b b b b c c c c a a a a b b b b c c c c

$\bar{b}$ $\bar{b}$ $\bar{c}$ $\bar{c}$ $\bar{c}$ $\bar{c}$ $\bar{a}$ $\bar{a}$ $\bar{a}$ $\bar{a}$ $\bar{b}$ $\bar{b}$ $\bar{b}$ $\bar{b}$ $\bar{c}$ $\bar{c}$ $\bar{c}$ $\bar{c}$ $\bar{a}$ $\bar{a}$ $\bar{a}$ $\bar{a}$ $\bar{b}$ $\bar{b}$

Other speed ratios than 2/1 may be obtained, by connecting coils that are in the same phase relation on both speeds in separate circuits, and reconnecting these circuits appropriately. In general, with a three-phase winding, equal numbers of coils that are connected in the positive direction in phase *A*, *B*, or *C* on the high-speed connection, will be in each of the positive and the negative directions in phases *a*, *b*, and *c*, on the low-speed connection. Thus, the coils must be grouped in six distinct circuits for each phase, corresponding to Aa, $A\bar{a}$, Ab, $A\bar{b}$, Ac, $A\bar{c}$ combinations,

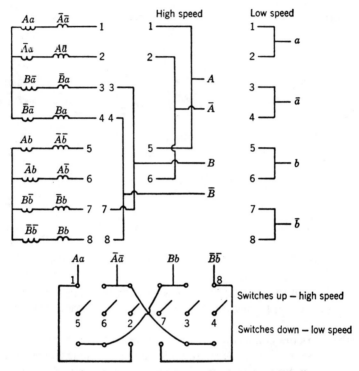

FIG. 8.24. Two Phase, 3/2 Speed, Single-Voltage Winding.

requiring in all 3 × 12, or 36 terminals in the general case. By using half as many circuits on the low-speed connection, by not bringing out the Y point, or by using 2 instead of 3 phases, the number of leads can be reduced. For example, a two-phase, 72-slot winding can be connected for either 4 or 6 poles by bringing out 8 leads and using 6 single-pole, double-throw switches. There are 8 full-width and 8 half-width groups, containing 6 and 3 coils, respectively, with the sequence:

Six poles	a	b	b	a	$\bar{b}$	a	a	b	$\bar{a}$	$\bar{b}$	$\bar{b}$	a	b	a	a	$\bar{b}$
Four poles	A	A	B	B	$\bar{A}$	$\bar{A}$	$\bar{B}$	$\bar{B}$	A	A	B	B	$\bar{A}$	$\bar{A}$	$\bar{B}$	$\bar{B}$
Slots/group	6	3	3	6	6	3	3	6	6	3	3	6	6	3	3	6

This gives a regular nine-slot-per-phase-belt, two-phase four-pole winding, or a regular six-slot-per-phase-belt, two-phase, six-pole winding. The connection diagram is shown in Fig. 8.24.

To avoid extra slip rings, such multispeed motors are usually of the squirrel-cage type, or have wound rotors that are internally short-circuited on the high-speed connection. For example, a twelve-pole rotor winding, with all poles in parallel, is short-circuited on 8 poles, as the voltages induced in the 12 parallel circuits are then 90° out of phase, and their vector sum is zero.

In all pole-changing windings, the coil pitch is a compromise between the most desirable values for the two numbers of poles. Generally, the width of phase belt on the low-speed winding is 120° instead of 60°. If the winding pitch of a 120° phase-belt, three-phase winding is not 100 per cent, even-numbered harmonics in the air-gap mmf wave will occur (Sect. 3.11), causing extra losses and crawling torques. These facts give multispeed windings somewhat inferior performance to that of a normal single-speed motor.

By bringing out more leads, and by allowing fractional numbers of slots per pole, many pole combinations can be obtained with a single winding, but the economic limitations of cost of contactors for changing connections, and the less-good motor performance make these schemes unattractive. Whenever the phases belts are dissimilar, additional harmonic fluxes are created, which may cause extra losses, or noise, or torque dips, requiring extended design studies, special constructions, and tests to assure satisfactory performance.

Another method for changing the motor speed is to employ two or more separate wound-rotor motors, all mechanically connected to the same shaft, or through gears. If the stator of one motor, with P_1 poles, is connected to the line, and its rotor slip rings are connected to the stator of a second motor, with P_2 poles, whose rotor winding is short-circuited; the common shaft will be driven at a speed corresponding to $P_1 + P_2$ poles.

For, if the shaft speed at synchronism (no load) is $1/n$ times two-pole synchronous speed:

The slip of the P_2-pole motor is 0;

The frequency of the stator voltage of the P_2-pole motor is $P_2 f/2n$;

The frequency of the secondary voltage of the P_1-pole motor is also $P_2 f/2n$;

The frequency of the primary voltage of the P_1-pole motor is

$$\frac{P_{2f}}{2n} + \frac{P_{1f}}{2n} = f.$$

Hence: $2n = P_1 + P_2$, and the shaft speed will be the same as for a single motor with $P_1 + P_2$ poles.

By interposing a gear ratio, or by putting three motors in series, and by other combinations, a large number of possible speeds can be obtained. The performance of these concatenated motors can be calculated from the equivalent circuit formed by putting the separate equivalent circuits of the several motors (Fig. 5.1) in series, and relating the different motor slip values in accordance with the shaft arrangement, or gear ratios.

Although formerly used, these "concatenated" motors are no longer employed. One reason for this is that the wound-rotor designs that are required (insulated to withstand continuous operation with a relatively high frequency and voltage in the rotor) cost much more than squirrel-cage or synchronous motors. Also, the magnetizing current required by the second motor must be carried through both the rotor and stator of the first motor, making the power factor of the set lower than the power factor of a single $P_1 + P_2$-pole motor. The synchronous motor, with its high power factor and relatively low cost for low-speed operation, is preferred when only a single speed is required. A synchronous motor can be concatenated with an induction motor to obtain two speeds. It is generally true, however, that d-c motors are preferable when several speeds are required, as their wide speed range, with rapid and accurate control, offers advantages over a two- or three-speed drive that will justify the increased cost.

If the slip rings of two or more similar motors are connected in parallel, while the stators are fed from a common power source, the slip frequencies, and, therefore, the speeds, of the several motors will be maintained in synchronism. This "power Selsyn" scheme is useful for operating the two ends of a lift bridge, or different elements of a sectionalized drive. Special problems arise with "hunting", or speed oscillations of the interconnected motors, and in determining the ability of the motors to maintain synchronism during transients.

By connecting the rotor and stator of a single wound-rotor motor in parallel, giving the "doubly fed motor", it is theoretically possible to obtain twice normal speed, or 7,200 rpm for a 2-pole, 60-cycle motor. However, the motor will not accelerate beyond half speed, and is extremely unstable at double speed, due to lack of damping torque.

12 Wound Rotor with Rheostatic Control

As outlined in Sect. 8.3, a high-resistance rotor enables high torque at low speed to be obtained, but has the disadvantage of high rotor heating, if low-speed operation is frequent or continuous.

By winding the rotor with an insulated winding brought out to slip rings, and connecting external rheostats to brushes running on the rings, the losses and heating are divorced from the motor. Further, it is then easy to vary the resistance by switching, or a controller, and so to vary the speed-torque curve of the motor over the full range indicated by Fig. 8.1. Slip-ring motors, therefore, are widely used for driving elevators, cranes, blowers, and mills of various sorts.

The extra cost of a wound-rotor over a squirrel-cage motor is due to the many extra operations required in manufacture, and to the small-quantity production, factors which decrease in relative importance as the motor size (and, therefore, the ratio of material to total cost) increases. Also, the need for limited starting current is greater the larger the motor size. The percentage of motors that are built of the wound-rotor design, therefore, varies from nearly zero for motors of less than 5-hp rating to nearly 100 per cent for motors of more than 1,000-hp rating.

From the design point of view, the wound rotor presents several interesting problems. To provide a balanced, regular winding, as necessary for low reactance and good performance, the rotor slots should be a multiple of the poles times phases. This gives rise to permeance locking torques at standstill, so that a high external rotor resistance is required to secure smooth starting, as well as to hold down the starting current (Sect. 9.4).

The admittance of the rotor winding to stator harmonic fields is much less than for a squirrel cage, due to the series-connected-out-of-phase voltages in successive coils of a phase belt. Also, the phase belts of the rotor winding create additional harmonic fields. These two conditions, plus the longer end windings and the deeper rotor slots, give the wound rotor a higher reactance than a squirrel-cage rotor in the same stator.

The need for insulation space in the rotor makes the rotor teeth narrower and longer than those of a squirrel cage. This, and the increased rotor flux pulsation due to the absence of induced rotor currents, lead to a considerably increased rotor core loss. On the other hand, the high-frequency rotor I^2R losses are greatly reduced, giving less stray-load loss. The high-frequency voltages appearing across the stator terminals are usually larger for a wound-rotor than for a squirrel-cage motor, with the same degree of slot skew (Sect. 10.8).

It is important to keep the resistances across the slip rings nearly equal under all conditions, as any unbalance in the rotor currents produces a backward-revolving component of the fundamental rotor field. This backward component, with a rotor frequency sf cycles per second, as the rotor is turning forward at a frequency $(1-s)f$, generates a stator frequency $(1-s)f-sf$, or $(1-2s)f$. Therefore, at half speed, when $1-2s=0$, this backward rotor field stands still with respect to the stator, marking a transition from motor to generator torque. Above half speed, the $1-2s$

frequency currents induced in the stator will cause a rapid dip in the torque curve (Fig. 8.25), and also flicker in the impressed stator voltage, which are very objectionable. At the extreme, if one rotor phase is open-circuited, the motor will operate stably at half speed, as if it had twice as many poles.

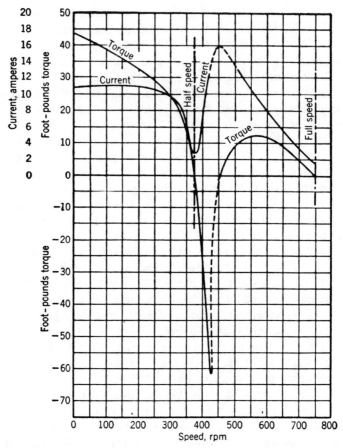

FIG. 8.25. Speed-Torque Curve for Three-Phase Wound Rotor Induction Motor with Single-Phase Secondary Four-Pole, 25-Cycle 5-hp Motor.

The effect may be visualized as the concatenation of two motors with equal numbers of poles, as explained in the preceding section. (See Fig. 9.6 and discussion.)

Another objection to an unbalanced or single-phase rotor is that very high voltages will appear across the rotor slip rings, if the stator winding is also single-phased, even momentarily; as when the contacts of the three line switches open or close at slightly different times. This is explained by the constant linkage theorem (Sect. 2.8). If both rotor and stator are single-phase, the mutual reactance of the two windings pulsates each quarter cycle between 0 and X_M, forcing very rapid flux changes, with consequent high pulsating torques as well as high voltages.

A fixed, high value of the resistances across the slip rings gives a nearly straight line speed-torque curve (Fig. 8.1), with consequently a large speed change for a small change in load.

The large power losses in the secondary resistances at high values of slip preclude the use of this simple form of speed control for continuous low-speed operation. By automatic control of the secondary resistance, or by including a saturable core reactor with feedback control in the rotor circuit also, the speed regulation can be improved, and a wide variety of speed-torque curves can be obtained. Unbalancing the primary voltages, giving a reverse torque component, widens the range of possibilities even further.

By supplying direct current to the slip rings, the machine will operate as a synchronous motor with high or leading power factor. This type has met with some favor abroad, but it has a severe cost disadvantage as compared to the squirrel-cage motor, and has more losses and less field capacity than a salient pole synchronous motor. It is very seldom used in the United States (Sect. 13.8).

13 Wound Rotor With Saturistor

By connecting a Saturistor (Sect. 1.10) across the secondary of a wound-rotor motor, a nearly constant torque and current can be obtained, all the way from full speed backward to full speed forward, as shown in Fig. 8.26 which gives test results on a 5-hp, 4-pole motor. Suppose it is desired to obtain a uniform accelerating torque of 0.50 per unit, and the motor equivalent circuit constants are $R_1 = 0.02$, $X_1 + X_2 = 0.15$, $R_2 = 0.02 + 0.05$, and $X_m = 2.50$ in per unit terms. The extra 0.05 in R_2 is a liberal allowance for the resistance of the Saturistor winding. Then, the ratio of volt amperes input at rated load to rated output will be about 1.20. By designing the Saturistor to operate near the peak of its impedance, Fig. 1.15, we can assume $R_h = X_h$, so that the motor impedance will be, neglecting magnetizing current:

(8.142) $\qquad Z = 0.02 + j0.15 + R_h(1+j) + 0.07/s.$

If R_h is chosen to be 1.0 per unit, and an external primary reactance of 0.05 is included, this gives for the impedance at locked rotor and at 0.10 slip, when the magnetizing current is included:

(8.143) $\quad Z = 0.50 + j1.02, \; s = 1; \text{ and } Z = 0.70 + j1.20, \; s = 0.10.$

310 SPEED-TORQUE-CURRENT RELATIONS

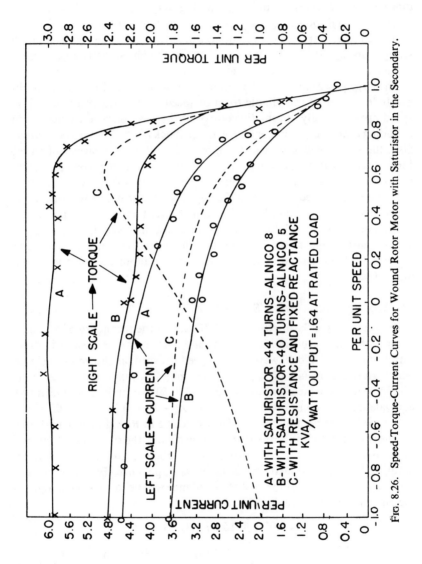

Fig. 8.26. Speed-Torque-Current Curves for Wound Rotor Motor with Saturistor in the Secondary.

The corresponding currents are 0.88 and 0.72, and the torques are 0.44 and 0.42 per unit. The actual watts dissipated in the Alnico at locked rotor are 0.44 (1.00/1.07) (746), or 307 watts per hp of motor rating. From Fig. 1.15, this will require about 307/250, or 1.25 cubic inches of Alnico 5 per hp, or the equivalent. At 0.10 slip, the watts are 0.42 (1.00/1.70) (746) (0.10) = 18 watts per hp only.

Fig. 8.27 shows calculated speed-torque curves for various sizes of Saturistor connected across the secondary of a wound rotor motor with the circuit constants given above, and Fig. 8.28 shows corresponding current curves. A uniform accelerating torque of any desired amount up to perhaps 150 per cent of rated can be obtained, with a current of less than twice the per unit torque, by choosing the right size of Saturistor.

The rate of temperature rise of Alnico with 300 watts per cu in. being absorbed without heat loss is about 6 C per second. By suitable aluminum conducting strips to carry the heat into the laminated core of the Saturistor, the effective heat capacity of the Alnico can be multiplied about 6 times, making the net expected rise about 1 C per second. This will enable a one-minute stalled time to be held without harm.

It is perfectly feasible to construct the Saturistor in circular form, with the Alnico blocks forming radial "poles", and to mount this on the rotor, thus eliminating the need for slip rings and brushes, although this will increase the rotor inertia. And, by providing a d-c control winding on the Saturistor, its impedance can be varied over a wide range, with feedback control.

By these means, the Saturistor enables a wound rotor motor to be given nearly ideal characteristics for starting and operating conveyors, pumps, motor-generator sets, and many other drives. Usually, it will be desirable to short circuit the Saturistor when the motor attains full speed, to limit heating of the Saturistor winding, and to provide greater breakdown torque. If so, however, the control should be arranged to immediately remove the short circuit whenever the voltage goes off, so as to limit the current inrush on reclosing.

For high torque, for reversing drives, and when repeated starting is required, it is desirable to connect a resistor across the slip rings also, in parallel with the Saturistor. Then, the Saturistor provides a "base" torque, of perhaps $\frac{3}{4}$ of rated value over the entire speed range, and the resistor provides as much more torque as desired. This is very important in reversed speed operation, as the watts in the Alnico increase proportionally with the frequency, and would reach prohibitive values if the Saturistor alone were relied upon when the slip is greater than one. Since the impedance of the resistor, when referred to the primary, varies in proportion to $1/s$, it draws more and more current as s increases, thereby increasing

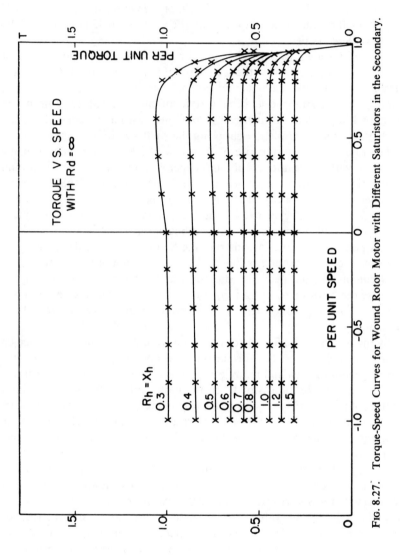

FIG. 8.27. Torque-Speed Curves for Wound Rotor Motor with Different Saturistors in the Secondary.

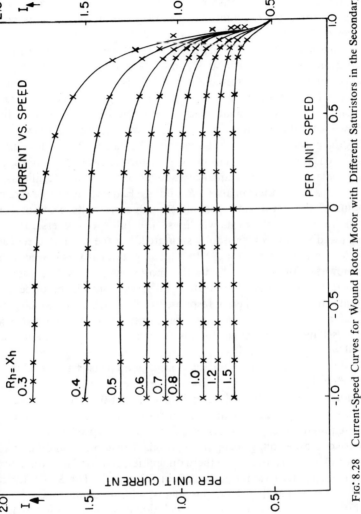

FIG. 8.28 Current-Speed Curves for Wound Rotor Motor with Different Saturistors in the Secondary.

the voltage drop in the primary, and decreasing the voltage on the Saturistor. Figures 8.29, 8.30, and 8.31 show the torque, current, and Saturistor watts for a wound rotor motor with the equivalent circuit constants given above, with a fixed Saturistor having $R_h = X_h = 0.4$ per unit, and various sizes of resistor connected in parallel across the secondary. If the resistor is 0.4 per unit, for example, the motor develops per unit torques of 2.15, 1.90, and 0.86 at slips of 2, 1, and 0.10, respectively, with corresponding currents of 3.45, 2.60, and 1.08 per unit. The watts dissipated in the Alnico are 0.76, 0.52, and 0.02 at these three speeds.

This performance is suitable for a-c hoist drive. One of the very desirable features of an a-c hoist drive is stepless speed control and reversing. The Foote scheme, using a saturable 1 to 1.1 ratio transformer and two saturable reactors in the primary circuit, accomplishes this very well (Fig. 8.32), To go forward, reactors $SX1$ and $SX2$ are fully saturated, and there is no control current in the transformer control windings, giving nearly balanced forward sequence voltages on the motor terminals, with lines 1, 2, and 3 on terminals C, B, and A. Extra turns on the secondary winding of the transformer provide enough extra voltage to compensate for the reactance drop in its windings. For full speed reverse, the transformer and reactor $SX2$ are fully saturated, but there is no current in the control windings of reactor $SX1$. This applies nearly balanced voltages with reversed phase rotation to the motor, with lines 1, 2, and 3 on terminals C, A, and B. For torque and speed control, the three control currents are varied together, in response to a feedback circuit which matches the actual motor speed given by a tachometer against a control voltage that is set by the operator. In this manner, the motor voltages can be varied smoothly from full value forward down to nearly zero and up to full value backward, preserving reasonably good phase balance all the way.

As first applied, a fixed (closed core) reactor in series with a low resistance, all in parallel with a resistor of about 0.25 per unit, were used in the secondary circuit of the wound rotor motor with the Foote drive. This required either an excessive current at full speed reversed, with a consequent large voltage drop in the primary reactor, or undesirably high slip at full load. To overcome these difficulties, a secondary contactor has been used, to decrease the resistance at full speed. The Saturistor offers another way of meeting the requirements, without contactors.

If only a small amount of speed regulation is required, as for some large fans, good performance can be obtained by operating a wound rotor motor at full voltage, with a d-c control winding used to vary the impedance of a Saturistor (such as Fig. 1.16) in the secondary circuit. Figures 8.33 and 8.34 show the torque and current versus speed curves for a 5-hp motor with this type of speed control. As noted in 1-10, the power factor

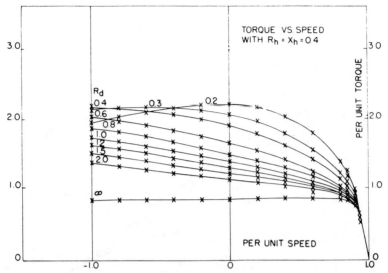

FIG. 8.29. Torque-Speed Curves for Wound Rotor Motor with a Saturistor and Different Resistors in the Secondary.

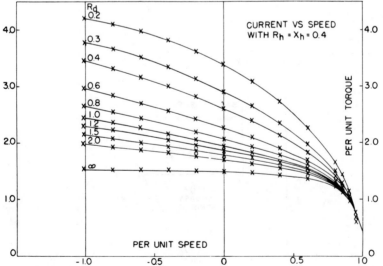

FIG. 8.30. Current-Speed Curves for Wound Rotor Motor with a Saturistor and Different Resistors in the Secondary.

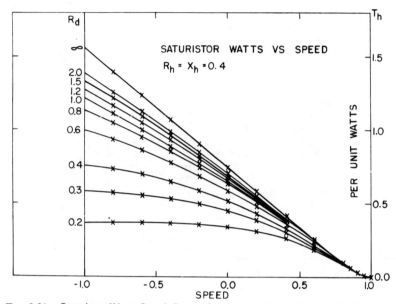

Fig. 8.31. Saturistor Watts-Speed Curves for Wound Rotor Motor with Saturistor and Different Resistors in the Secondary.

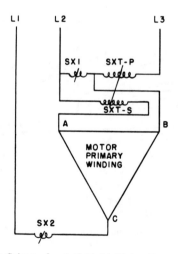

Fig. 8.32. The Foote Scheme for A-C Hoist Motor Reversal without Contactors.

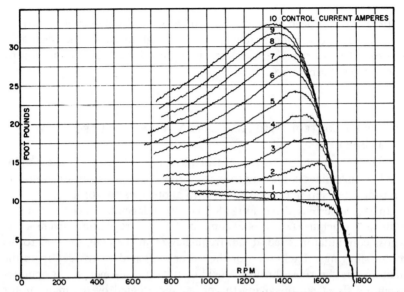

FIG 8.33. Full-Voltage Torque versus Speed Curves for 5-HP Wound Rotor Motor with Various Saturistor Control Currents.

Full-load torque=16 ft-lb (at 1650 RPM).

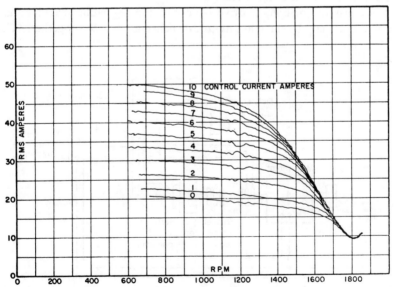

FIG. 8.34. Full-Voltage Current versus Speed Curves for 5-HP Wound Rotor Motor with Various Saturistor Control Currents.

Full-load current=18 amperes.

of the Saturistor becomes very low at high slips with high control current, so continuous operation at low speeds and high torques is not desirable, but excellent starting performance is obtained by keeping the control current low during acceleration.

14 Speed Control with Auxiliary Machines or Frequency Converters

Several methods of speed control have been developed that employ auxiliary machines connected to the slip rings of a wound rotor motor, to transform the slip-frequency power into another useful form.

In the Scherbius drive (Fig. 8.35), an independent synchronous or induction machine D drives a commutator-type regulating machine, B, whose commutator is fed at slip frequency from the main (wound rotor) motor slip rings. Excitation for the stator of B is also provided from the main motor slip rings through a field control transformer, C. B operates as an a-c shunt motor, giving the power represented by the main motor slip back to the line through D acting as a generator. This gives a constant-torque drive, assuming that the main motor, A, has adequate ventilation for constant current at all speeds.

If a constant-horsepower drive is desired, the regulating machine may be driven directly by the main motor, dispensing with D altogether. The

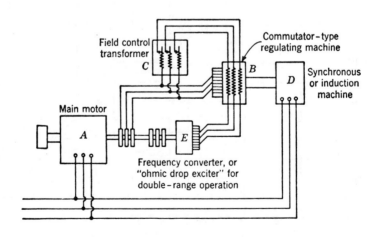

Fig. 8.35. Scherbius Adjustable-Speed Drive.

slip-frequency power is then converted into mechanical power on the main motor shaft. The size of the regulating machine may be nearly cut in two by operating the main motor both above and below synchronism, instead of only below. To accomplish this, a small "ohmic drop exciter" or frequency converter, E, is added, driven by the main motor, to provide additional excitation for the regulating machine. Without E, the excitation and torque of B would fall to zero at the synchronous speed of the main motor. At synchronous speed, E has to supply only enough voltage to overcome the ohmic drop; hence its name.

All Scherbius drives are limited in speed range by the low frequency, not more than 10 or 15 cycles, that the regulating machine is able to commutate. They are most useful when only a small change of speed is desired, as for frequency changers connecting two power systems that have a varying frequency ratio.

The Krämer system substitutes a rotary converter for the regulating machine and its driving motor, and provides a d-c motor on the main motor shaft to reconvert the slip-frequency power into mechanical energy. This drive is, therefore, best suited to constant-horsepower drive. It is adapted to a wider speed range than the Scherbius system, because a rotary converter can commutate materially higher frequencies than an a-c shunt motor or generator.

Both these drives are relatively costly for the widely varying speeds desired in modern rolling mills, so that d-c drives have supplanted them.

A third system (Fig. 8.36), developed in recent years for speed control of large fans for wind tunnels, employs a synchronous machine to absorb the slip-frequency power. The synchronous machine drives a d-c generator, which in turn drives a d-c synchronous motor-generator set that returns power to the line. This system is single range, operating only below synchronous speed of the main motor. It is free of a-c commutation difficulties, but it requires careful design and a sensitive, speedy control system to avoid difficulties from "hunting", or speed oscillations of the synchronous motor.

Recently, the advent of the silicon controlled rectifier, or Thyristor, has opened the way to a wide variety of stepless methods of motor speed and torque control. The simplest scheme uses two back to back SCRs in each line. Figure 12.8, with control circuits that enable the angle of firing of each SCR to be retarded, giving the voltage waveform shown in Fig. 12.9. By varying the firing angle the effective motor voltage, and therefore the torque and speed, can be adjusted over a wide range. The harmonics in the voltage waveform create harmonic currents in the motor, making the performance not quite as good as with sine wave voltage. But its speed of response and ease of control promise to make this scheme

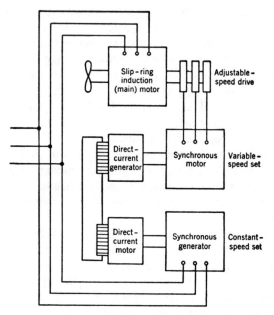

Fig. 8.36. Modified Krämer (Clymer) Drive.

widely useful. Another scheme is to employ SCR frequency converters to vary the frequency, enabling squirrel-cage motors to be used for vehicle drive, the frequency being controlled to maintain a desired small value of slip over the entire speed range. This promises to become much more important in future years.

BIBLIOGRAPHY

Reference to Section

1. "New Method for Part-Winding Starting of Polyphase Motors," L. M. Agacinsky and P. L. Alger, *A.I.E.E. Trans.*, Vol. 74, Part III, 1955, pp. 1455–62. 9
2. "The GM High Performance Induction Motor Drive System," P. D. Agarwal, *I.E.E.E. Trans.* 1968, Paper No. 68 TP 107 PWR. 6
3. "The Development of Low-Starting Current Induction Motors," P. L. Alger, *G. E. Rev.*, Vol. 28, July 1925, pp. 499–508. 6
4. "Performance Calculations for Part-Winding Starting of Three-Phase Motors," P. L. Alger, *A.I.E.E. Trans.*, Vol. 75, Part III, 1956, pp. 1535–43. 9
5. "Saturistors and Low Starting Current Induction Motors," P. L. Alger, G. Angst, and W. M. Schweder, *I.E.E.E. Trans.*, Power Apparatus and Systems, Vol. 82, June 1964, pp. 291–8. 13
6. "Stepless Starting of Wound Rotor Induction Motors," P. L. Alger and Jalaluddin, *A.I.E.E. Trans.*, Vol. 81, Part II, 1962, pp. 262–70. 12
7. "Speed-Torque Calculations for Induction Motors with Part Windings," P. L. Alger, Y. H. Ku, and C. H. T. Pan, *A.I.E.E. Trans.*, Vol. 73, Part III, 1954, pp. 151–9.
8. "Split-Winding Starting of Three-Phase Motors," P. L. Alger, H. C. Ward, Jr., and F. H. Wright, *A.I.E.E. Trans.*, Vol. 80, Part III, 1961, pp. 902–10.

BIBLIOGRAPHY

Reference to Section

9. "Speed Control of Induction Motors Using Saturable Reactors," P. L. Alger and Y. H. Ku, *A.I.E.E. Trans.*, Vol. 75, Part III, 1956, pp. 1335–41. — 12
10. "Speed Control of Wound Rotor Motors with SCRs and Saturistors," Philip L. Alger, William A. Coelho, and Mukund R. Patel, *I.E.E.E. TRANS. ON INDUSTRY AND GENERAL APPLICATIONS*, Vol. IGA-4, No. 5, September/October 1968, pp. 477–485. — 9
11. "A New Wide-Range Speed Control System for Induction Motors," P. L. Alger, E. A. De Meo, and H. C. Ward, Jr., *I.E.E.E.*, 1963.
12. "Split-Winding Starting of Three-Phase Motors," P. L. Alger, H. C. Ward, Jr., and F. H. Wright, *A.I.E.E. Trans.*, Vol. 70, Part I, 1951, pp. 867–72. — 10
13. "Polyphase Induction Motor with Solid Rotor," G. Angst, *A.I.E.E. Trans.*, Vol. 80, Part III, 1961, pp. 902–10. — 3
14. "Circuit Analysis Method of Determination of A-C Impedances of Machine Conductors," D. S. Babb and J. E. Williams, *A.I.E.E. Trans.*, Vol. 70, Part I, 1951, pp. 661–5. — 6
15. "A New Equivalent Circuit for Double-Cage Induction Motors," A. K. Bandyopadhyay, I.E.E.E. PWR Paper 70–TP 182, Jan. 1970.
16. "Large Adjustable-Speed Wind-Tunnel Drive," C. C. Clymer, *A.I.E.E. Trans.*, Vol. 61, 1942, pp. 156–8. — 14
17. "The Doubly Fed Machine," C. Concordia, S. B. Crary, and G. Kron, *A.I.E.E. Trans.*, Vol. 61, 1942, pp. 286–9. — 14
18. "Damping and Synchronizing Torques of Power Selsyns," C. Concordia and G. Kron, *A.I.E.E. Trans.*, Vol. 64, 1945, pp. 366–71. — 14
19. 'An Induction Motor with Paralleled Rotor and Stator," A. G. Conrad and R. G. Warner, *A.I.E.E. Trans.*, Vol. 51, June 1932, pp. 418–22. — 14
21. "Adjustable Frequency Inverters and Their Application to Various Speed Drives," Bradley, Clarke, Davis, and Jones, *PROCEEDINGS I.E.E.E.*, Vol. 111, No. 11, Nov. 1964, pp. 1833-1846. — 9
20. "Ten Part-Winding Arrangements in Sample 4-Pole Induction Motor," J. J. Courtin, *A.I.E.E. Trans.*, Vol. 74, Part III, 1955, pp. 1248–53.
22. "The Double-Delta Reduced Voltage Starting Method," J. J. Courtin, *A.I.E.E. Conference Paper 62–1413*. — 9
23. "Some Developments in Multi-Speed Cascade Induction Motors," F. Creedy, *The Jour. of the I.E.E.*, Vol. 59, 1921, pp. 511–37. — 11
24. U.S. Patent 427,978, Alternating-Current Motor, M. von Dolivo-Dobrowolsky, May 13, 1890. — 6
25. "Nonuniform Torque in Induction Motors Caused by Unbalanced Rotor Impedance," O. I. Elgerd, *A.I.E.E. Trans.*, Vol. 73, Part IIIB, 1954, pp. 1481–4. — 12
26. "Inverter Drive of an Induction Motor," M. S. Erlicki, *I.E.E.E. Trans.*, Vol. PAS-84, No. 11, November 1965, pp. 1011-1016.
27. "Eddy Currents in Large Slot-Wound Conductors," A. B. Field, *A.I.E.E. Trans.*, Vol. 24, 1905, pp. 761–88. — 4
28. "Eddy-Current Losses in Armature Conductors," R. E. Gilman, *A.I.E.E. Trans.*, Vol. 39, Part I, 1920, pp. 997–1048. — 4
29. "Improved Starting Performance of Wound Rotor Motors Using Saturistors," C. E. Gunn, *I.E.E.E. Trans.*, Power Apparatus and Systems, Vol. 82, June 1964, pp. 298–302. — 13
30. "Theory of Speed and Power Factor Control of Large Induction Motors by Neutralized Polyphase Alternating-Current Commutator Machines," J. I. Hull, *A.I.E.E. Trans.*, Vol. 39, Part II, 1920, pp. 1135–69. — 14

31. "Synthesis of Double-Cage Induction Motor Design," H. E. Jordan, *A.I.E.E. Trans.*, Vol. 78, Part II, 1959, pp. 691–5. 6
32. "Line-Start Induction Motors," C. J. Koch, *A.I.E.E. Trans.*, Vol. 48, April 1929, pp. 633–40. 6
33. "Equivalent Circuit for the Concatenation of Induction Motors," Y. H. Ku, *A.I.E.E. Trans.*, Vol. 74, Part III, 1955, pp. 1214–18. 11
34. "Heat Losses in the Conductors of Alternating-Current Machines," W. V. Lyon, *A.I.E.E. Trans.*, Vol. 40, 1921, pp. 1361–95. 4
35. "Overvoltages in Polyphase Induction Motors during Single-Phase Operation," C. Macmillan and G. K. Carter, *A.I.E.E. Trans.*, Vol. 60, 1941, pp. 819–23. 12
36. "The Liquid Rheostat for Speed Control of Wound Rotor Induction Motors," G. L. McFarland and W. Alvarez, *A.I.E.E. Trans.*, Vol. 67, Part I, 1948, pp. 603–9. 12
37. "Induction Motors as Selsyn Drives," L. M. Nowacki, *A.I.E.E. Trans.*, Vol. 53, 1934, pp. 1721–6. 14
38. "Concatenated Induction Motors for Rolling Mill Drive," W. O. Oschmann, *A.I.E.E. Trans.*, Vol. 33, Part I, 1914, pp. 899–920. 11
39. "Speed-Changing Induction Motors," G. H. Rawcliffe and W. Fong, *I.E.E. Proc.*, Vol. 108, Part A, Oct. 1961, pp. 357–68. 11
40. "Multispeed Induction Motors," H. G. Reist and H. Maxwell, *A.I.E.E. Trans.*, Vol. 28, Part I, 1909, pp. 601–9. 11
41. "Rotor Impedance Control of the Wound-Rotor Induction Motor," W. Shepherd and G. R. Slemon, *A.I.E.E. Trans.*, Vol. 78, Part IIIA, 1959, pp. 807–14. 3
42. "Wide-Range Reversible Voltage Controllers for Polyphase Induction Motors," W. Shepherd and N. Zagalsky, *A.I.E.E. Trans.* 1962, Paper No. 62-22.
43. "Eddy Currents in Stator Windings," H. W. Taylor, *The Jour. of the I.E.E.*, Vol. 58, April 1920, pp. 279–98. 4
44. "General Theory of the Electric Shaft," Y. Wallach and M. S. Erlicki, *I.E.E.E.* Conference Paper, 1969, Paper No. 69 CP 145-PWR.

CONTENTS—CHAPTER 9

CRAWLING, LOCKING, AND STRAY LOSSES

		PAGE
1	Effects Produced by Harmonic Fields	325
2	Permeance Waves	328
3	Asynchronous Crawling	331
4	Stray Losses	339
5.	Standstill Locking	346
6	Synchronous Crawling	355
7	Unbalanced Magnetic Pull	356

9
CRAWLING, LOCKING, AND STRAY LOSSES

1 Effects Produced by Harmonic Fields

Superposed upon the currents and forces due to the fundamental sine-wave field of an induction machine, there are many smaller currents and forces produced by the myriad of harmonic fields that are also present—just as every ocean wave is surmounted by many ripples. The harmonic field effects are of great importance to the educator, who finds them a key to understanding and a fruitful subject for experiments; and to the designer, who must keep them under control as a *sine qua non* of good motor performance. These effects are of four kinds.

Asynchronous Crawling (Sect. 9.3). First, the space harmonics of the air gap permeance and the winding mmf, Fig. 3.11, create rotating fields which induce secondary currents, and produce torques similar to those of the fundamental, but which have more poles and, therefore, lower synchronous speeds. As the motor accelerates through the synchronous speed of one of these harmonics, its torque reverses, causing a dip in the resultant motor torque-speed curve (Fig. 9.1). Unless minimized by good design, the consequent "asynchronous crawling" may seriously impair the motor's starting ability.

The harmonic fields may be thought of as separate low-power motors that are direct-coupled to the same shaft as the fundamental, and are electrically connected in series with it (Fig. 3.15). At speeds above their respective synchronous values, the forward harmonics produce braking torques, as the backward harmonics do at all forward speeds. They create stray-load losses and increase the motor heating.

Locking and Synchronous Crawling (Sects. 9.5 and 9.6). Second, if any two of the separate harmonic fields have the same number of poles, pulsating torques will be produced as they slip past each other. When their speeds coincide, the two like fields will synchronize, and a corresponding locking or "synchronous crawling" torque will be observed (Fig. 9.2).

Magnetic Noise and Vibration (Chapter 10). Third, if two harmonic fields, with numbers of poles differing by 2, coexist in the air gap, they will produce unbalanced radial magnetic forces, and consequent radial vibration of the rotor as a whole (Fig. 9.3). Also, symmetrical radial forces of

326 CRAWLING, LOCKING, AND STRAY LOSSES

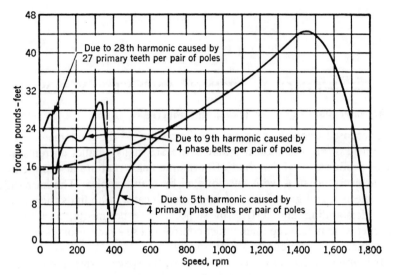

Fig. 9.1. Asynchronous Crawling of Two-Phase Induction Motor.

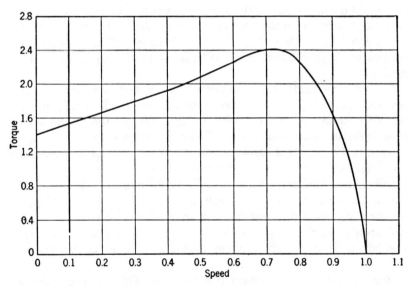

Fig. 9.2. Speed-Torque Curve of Three-Phase, 60-Cycle Motor with 54 Stator and 60 Rotor Slots, Showing Synchronous Crawling Torque, at 0.1 Speed.

high frequency are produced by the superposition of rotating magnetic fields of different pole numbers. These phenomena create stator vibration and magnetic noise.

Voltage Ripples (Chapter 10). Fourth, the harmonic fields produced by the stator current induce currents in the rotor which reflect back into the stator additional harmonic fields, giving rise to terminal voltage ripples

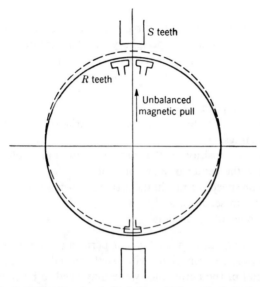

FIG. 9.3. Radial Vibration due to Unbalanced Magnetic Pull when $R - S = 1$.

and to extra core losses. These voltage ripples, in turn, produce high-frequency currents in the supply lines, which may create inductive interference with communication circuits.

These phenomena are caused in part by the permeance variations in the air gap due to slot openings, and in part by the mmf harmonics due to the winding distribution. To calculate them, it is convenient first to express the air-gap permeance as the sum of a constant (average) value and a series of sine-wave variations, then to express the mmf as a series of harmonic fields (Eq. 3.34), and, finally, to multiply these series, term by term, to obtain the resultant air-gap magnetic field.

Any one who is seriously interested in these harmonic field effects should read the classic (1919) paper by Wilhelm Stiel, presenting the results of his extensive tests on squirrel-cage rotors having different numbers of slots. He used a 4-pole, 50-cycle, 24-slot stator, and tested with it rotors having 18, 19, 20, 22, 25, 27, 28, 29, 38, 41, and 42 slots. His paper gives accurate

328   CRAWLING, LOCKING, AND STRAY LOSSES

torque curves taken on each of these rotors, with also some locking torque position curves and a table of the relative noise produced by each. In 1930 Möller published 58 carefully measured speed-torque curves taken with three different 4-pole stators, having 24, 36, and 48 slots, with 19 different rotors. It is believed that all these published results are consistent with the theories presented in this and the following chapter.

2   Permeance Waves

If a rotor has R open slots, the air-gap permeance will have R complete cycles of variation around the periphery. If the variation is taken to be sinusoidal, the permeance as viewed from the center of a stator tooth is:

$$(9.1) \qquad P_g = P_0 + P_R \cos R(x - Nt),$$

where P_0 = average permeance,
P_R = half amplitude of permeance variation due to rotor slot openings,
x = angular position of the rotor, measured in mechanical radians from the center of a stator tooth,
N = rotor speed, in mechanical radians per second,
t = time, in seconds, and
R = number of rotor slots.

Only the fundamental term of the slot permeance variation is included in Fig. 9.1. In practice, especially with partly closed slots, the second and higher harmonics of the permeance wave may need to be considered also. More generally, therefore, we should add a series of $\cos mR(x - Nt)$ terms to Eq. 9.1.

The stator slot openings introduce another, stationary, permeance wave. If there are S stator slots, the half amplitude of stator permeance variation is P_S, and the center of a stator tooth is taken as the reference point, where $x = 0$; the combined air-gap permeance will be the product of Eq. 9.1 by a similar expression for the stator:

$$(9.2) \qquad P_g = P_0 \left[1 + \frac{P_R}{P_0} \cos R(x - Nt)\right]\left(1 + \frac{P_S}{P_0} \cos Sx\right)$$

$$= P_0 \left[1 + \frac{P_S}{P_0} \cos Sx + \frac{P_R}{P_0} \cos R(x - Nt) \right.$$

$$\left. + \frac{P_R P_S}{P_0^2} \cos Sx \cos R(x - Nt)\right].$$

When $t = 0$, the center of a rotor tooth coincides with the center of the stator tooth at which $x = 0$.

The stator mmf is given by Eq. 3.34 and the rotor mmf by a similar expression. The air-gap flux distribution is the resultant obtained by multiplying the sum of the stator and rotor mmfs by Eq. 9.2.

The results we are interested in will appear most clearly if we focus our attention on the several terms of Eq. 9.2 separately, and neglect higher-order effects for the time being.

Expanding the fourth term of Eq. 9.2, using the trigonometric identity $2 \cos a \cos b = \cos (a-b) + \cos (a+b)$:

$$(9.3) \quad P_g = P_0 \left(1 + \frac{P_S}{P_0} \cos Sx + \frac{P_R}{P_0} \cos R(x-Nt) \right.$$

$$\left. + \frac{P_R P_S}{2P_0^2} \{ \cos [(R-S)x - RNt] + \cos [(R+S)x - RNt] \} \right).$$

The $\cos [(R-S)x - RNt]$ term in Eq. 9.3 is the important one, as this corresponds to a field of a small number of poles, $2(R-S)$. Its importance may be visualized by referring to Fig. 9.3, but considering a motor with $R-S=2$ instead of 1. At the two opposite ends of the diameter at which a stator tooth is in line with a rotor tooth, the main flux permeance (averaged over one tooth pitch) is a maximum, and the zigzag-flux permeance is a minimum. At the two points 90° away, a stator-tooth center is opposite a rotor slot, giving minimum fundamental and maximum zigzag permeance. The intermediate air-gap permeance, summed over one slot pitch, varies slowly between these extremes. Thus, there is a long pitch permeance wave, having $(R-S)$ cycles around the periphery. The permeance wave revolves much faster than the rotor, in the same direction if $R>S$, oppositely if $R<S$. It moves through one rotor slot pitch while the rotor itself moves through the difference between one rotor and one stator slot pitch.

 • The fifth term of Eq. 9.3, in $\cos [(R+S)x - RNt]$, is unimportant, as its number of poles is too great to produce any large torque or force variations. The second and third terms of Eq. 9.3 are the short pitch permeance ripples due to the stator and rotor slot openings. The sine-wave fundamental mmf acting on these permeance ripples produces air-gap fields of exactly the same kind as are produced by the corresponding slot ripples in the stator and rotor mmf waves acting on the uniform gap permeance. They are major sources of stray losses.

Instead of taking the permeance to be the product of the rotor and stator factors, it may be taken as the sum. As just indicated, the product equation is helpful in visualizing the causes of noise and vibration, but the sum equation enables the flow of energy to be understood better. In practice, the rotor slot openings usually are so small that P_R can be taken as zero.

It is true that the presence of currents in the rotor slots causes magnetic saturation of the rotor tooth tips, but the permeance effects of this can be considered as modifying the mmf waves associated with the currents. Taking P_R to be zero, therefore, the effective air-gap permeance is:

$$(9.4) \qquad P_g = P_0 + P_1 \cos Sx,$$

where $P_1 = P_S/P_0$.

From the analysis in Sects. 3.7 to 3.12, the principal harmonics in the mmf of a symmetrical stator winding with q phase belts per pole, P pairs of poles, and S slots are the two phase-belt harmonics of orders $2q-1$ and $2q+1$, and the two slot harmonics of orders $(S/P)-1$ and $(S/P)+1$. The rotor winding will produce additional mmf waves of orders $(R/P)-1$ and $(R/P)+1$. Neglecting all harmonics except these, the total air-gap mmf wave is, from Eq. 3.34:

$$(9.5) \qquad \begin{aligned} M = &\ A \cos(Px - \omega t) + B \cos[(2q-1)Px + \omega t] \\ &+ C \cos[(2q+1)Px - \omega t] \\ &+ D \cos[(S-P)x + \omega t] \\ &+ E \cos[(S+P)x - \omega t] \\ &+ F \cos[(R-P)(x - Nt) + \omega st] \\ &+ G \cos[(R+P)(x - Nt) - \omega st]. \end{aligned}$$

A, B, C, etc., are numerical coefficients that depend on the currents, winding pitch, slots per pole, slot openings, etc. A is proportional to the phasor sum of stator and rotor fundamental currents; B, C, D, and E are proportional to the stator current only (neglecting the induced opposing harmonic currents); and F and G to the rotor current only. $N = (1-s)(\omega/P) = $ rotor speed, in mechanical radians per second.

Evidently, the ratios of the harmonic fields, B, C, ..., to the fundamental A, are least at full speed, no load. As the stator and the opposing rotor currents increase with increase in load, the harmonic fields tend to increase proportionally, while the net fundamental field, A, decreases slightly. For this reason, the harmonics have their maximum effects during the starting period, when the current is high. Therefore, the problem of designing a squirrel-cage motor with good starting performance reduces to that of controlling the harmonic fields.

Multiplying Eq. 9.4 by Eq. 9.5, term by term, gives the total air-gap flux wave. This comprises seven groups of harmonic fields:

1. The AP_0 term is the fundamental field:

$$AP_0 \cos(Px - \omega t).$$

2. The AP_1 term gives the stator slot permeance harmonic fields:

$$AP_1 \cos(Px - \omega t)\cos Sx$$
$$= (AP_1/2)\{\cos[(S-P)x + \omega t] + \cos[(S+P)x - \omega t]\}.$$

3. The BP_0 and CP_0 terms are the phase-belt harmonic fields:

$$BP_0 \cos[(2q-1)Px + \omega t] \quad \text{and} \quad CP_0 \cos[(2q+1)Px - \omega t].$$

4. The DP_0 and EP_0 terms are the stator slot mmf harmonic fields:

$$DP_0 \cos[(S-P)x + \omega t], \quad EP_0 \cos[(S+P)x - \omega t].$$

5. The FP_0 and GP_0 terms are the rotor slot mmf harmonic fields:

$$FP_0 \cos[(R-P)x + (\omega - RN)t],$$
$$GP_0 \cos[(R+P)x - (\omega + RN)t].$$

6, 7. The D to G times P_1 terms are higher-order fields that may be neglected for present purposes.

If there is any backward-rotating fundamental mmf, as in a single-phase motor, this produces a similar set of six harmonic fields with reversed rotation. If the rotor has large slot openings or salient poles, these will produce still other sets of harmonic fields. Here, we shall consider only the forward mmf, with negligible rotor slot openings.

The permeance harmonics (2) are due to the presence of slot openings, while the mmf harmonics (4) are due to the concentration of the mmf in slots. These different fields have similar effects, so far as air-gap forces and torques are concerned, so that either set of fields may be considered merely an augmentation of the other. Nevertheless it is necessary to consider them separately in the analysis.

It is generally assumed that no unidirectional, or homopolar, flux can flow from stator to rotor of an induction machine; i.e., the total flux crossing the air gap in each direction is the same. The only path for homopolar flux to return from the rotor to the stator is through the shaft, bearings, end shields, and stator frame. This path has a very high reluctance, especially for an alternating flux. Therefore, no harmonic field with zero poles can exist. Whenever such terms are found in multiplying the mmf by the permeance waves, they are to be neglected (just as the triple harmonic currents are taken to be zero in a Y-connected motor).

3 Asynchronous Crawling

The forward-revolving slot and phase-belt harmonics due to the stator winding (the CP_0, EP_0, and part of the AP_1 terms) produce magnetic fields of $(S+P)$ and $(2q+1)P$ pairs of poles, whose mmfs are in direct proportion to the primary current. They induce voltages in the rotor winding, which in turn produce rotor currents and torques, the phenomena being in all respects similar to those of the fundamental field. When the rotor, during acceleration from rest, reaches the synchronous speed of one of these harmonics, that is $P/(S+P)$ or $1/(2q+1)$ times fundamental synchronous speed, the corresponding harmonic torque falls to zero and at higher speeds it reverses, becoming thereby a brake on the rotor. The resultant speed-torque curve of a squirrel-cage motor may have the shape of Fig. 9.1, and "asynchronous crawling" will occur just above the harmonic synchronous speed, if the load torque is greater than the minimum accelerating torque.

The backward-revolving harmonics, with $S-P$ and $(2q-1)P$ pairs of poles (the BP_0, DP_0, and part of the AP_1 terms) are similar in all respects, except that their synchronous speeds are reached when the motor is driven backward at speeds $P/(S-P)$ or $1/(2q-1)$ times fundamental synchronous speed. For all forward rotation, these backward harmonics produce braking torque. At standstill, the net torque due to all the harmonics is normally negative, as the backward harmonics, with fewer numbers of poles, generally produce more braking torque than the accelerating torque produced by the forward harmonics.

The effects of these stator harmonic fields can be taken into account by considering each harmonic field as if it were a separate motor, connected in series with the fundamental, but creating its own independent rotor currents and torque. On this basis, the equivalent circuit of Fig. 5.1 is expanded to that of Fig. 3.15. For each separate harmonic, the magnetizing reactance is calculated by Eq. 7.1, putting K_{dm} and K_{pm} from Eqs. 3.16 and 3.17 in place of K_{d1} and K_{p1}, and putting mP in place of P. For the seventh harmonic of a $\frac{7}{9}$-pitch, three-phase, 60° phase-belt stator winding, with 9 slots per pole, for example:

$$K_{d7} = \frac{\sin\frac{7\pi}{6}}{3\sin\frac{7\pi}{18}} = 0.177, \qquad K_{d1} = \frac{\sin\frac{\pi}{6}}{3\sin\frac{\pi}{18}} = 0.958,$$

$$K_{p7} = \sin\frac{49\pi}{18} = 0.766, \qquad K_{p1} = \sin\frac{7\pi}{18} = 0.940,$$

so that: $\quad X_{M7} = \dfrac{X_{M1}}{49}\left(\dfrac{0.177 \times 0.766}{0.958 \times 0.940}\right)^2 = 0.00046 X_{M1}.$

That is, the magnetizing reactance of the seventh harmonic in this case is 0.046 per cent of that of the fundamental. Similarly, for the nineteenth harmonic:

$$K_{d19} = \frac{\sin\frac{19\pi}{6}}{3\sin\frac{19\pi}{18}} = 0.958, \quad K_{p19} = \sin\frac{133\pi}{18} = 0.940,$$

so that $X_{M19} = \frac{X_{M1}}{361}(1) = 0.0028\, X_{M1}$.

The magnitude of any particular phase-belt harmonic can be controlled by selecting the pitch. When this is not convenient, C. Macmillan showed that it is also possible to reduce the harmonic by interchanging the end coils of adjacent phase belts, or "interspersing". This changes the phase

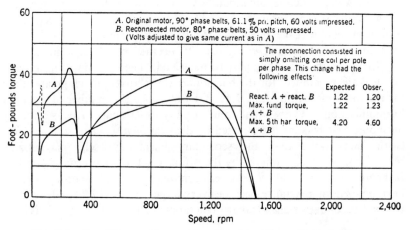

FIG. 9.4. Speed-Torque Curves, Two-Pole, Two-Phase, 25-Cycle Motor.

angle of the outer coils of each phase belt by one slot pitch, thus reducing the $(2q-1)$th harmonic and increasing the $(2q+1)$th harmonic. Or one outer coil of each phase belt can simply be omitted with a similar effect. Fig. 9.4 shows the effect on the torque curve of such a change in the case of a two-pole, two-phase motor. Here, the forward-revolving fifth harmonic was reduced by omitting one coil per pole per phase, at the outer end of each phase belt.

The values of K_p and K_d are identically the same for a slot harmonic as for the fundamental, since the electrical angle between slots is

$$\frac{2\pi P}{S}\left(\frac{S}{P} \pm 1\right) = 2\pi \pm \frac{2\pi P}{S}$$

for the former and $\frac{2\pi P}{S}$ for the latter.

334   CRAWLING, LOCKING, AND STRAY LOSSES

The sum of all these harmonic mmf magnetizing reactances constitutes the differential leakage reactance (Sect. 7.6). As shown in Fig. 3.15, each of them is shunted by the secondary winding impedance corresponding to the particular harmonic. The secondary resistance is given by Eq. 6.11 or Eq. 6.14, substituting K_{pm} and K_{dm} for K_{p1} and K_{d1}, and mP for P in the end-ring resistance term. Just as in a transformer, the secondary resistance of the rotor referred to the primary varies as the square of the effective number of primary turns:

(9.6)
$$\frac{R_2 \text{ for } m\text{th harmonic}}{R_2 \text{ for fundamental}} = \frac{K_{pm}^2 K_{dm}^2}{K_{p1}^2 K_{d1}^2}.$$

By the same principle, the secondary slot reactance for any harmonic is given by the second term of Eq. 7.11, putting $K_{pm} K_{dm}$ for $K_{p1} K_{d1}$. The secondary end leakage reactance for the mth harmonic is still further reduced by the factor $1/m^2$ (since P^2 occurs in the denominator of Eq. 7.102), so this normally may be neglected. The secondary zigzag- and belt-leakage reactances for the harmonics are relatively much greater than for the fundamental, by Eqs. 7.47 and 7.66, since the number of slots per harmonic pole is very small and usually fractional. This secondary harmonic belt leakage is expressed by the sum of a series of subharmonic (with fewer poles than the secondary harmonic itself) magnetizing reactances, which are shunted by the primary acting as a tertiary winding. If the number of secondary slots per harmonic pole is of the order of $\frac{1}{2}$, the harmonic zigzag reactance becomes very large. By Eqs. 7.29 and 7.32, the per unit secondary zigzag-leakage reactance for the mth harmonic is:

(9.7)
$$\frac{L_t - L_s}{L_s} = \frac{m^2 \pi^2}{4 s_2^2} \csc^2 \frac{m\pi}{2 s_2} - 1 = X_{2Zm},$$

where s_2 = squirrel-cage slots per fundamental pole. Figure 9.5 shows this relation for small values of s_2/m. For large values (>3), Eq. 9.7 reduces to Eq. 7.33 with an additional m^2 factor:

(9.8)
$$\frac{L_t - L_s}{L_s} = \frac{m^2 \pi^2}{12 s_2^2}.$$

It is evident from Fig. 9.5 that the secondary leakage reactance to any harmonic can be made indefinitely large by choosing the number of secondary slots to be nearly one per pair of harmonic poles. If s_2/m is exactly $\frac{1}{2}$, there are 360 electrical degrees between adjacent rotor slots, so that all the rotor voltages induced by the harmonic are in time phase. No current can flow under these conditions (unless there is a "Gramme ring" effect due to stray currents flowing from the ends of uninsulated bars radially down the end laminations of the rotor, and back along the rotor spider or shaft). In this case, or when an insulated phase winding is used that is open-circuited to the harmonic, no harmonic torque is produced. In practice, the number of secondary slots is usually chosen to be between 0.6 and 0.75 per slot harmonic pole, to avoid the difficulties described in the following sections, while staying up as high on the reactance curve of Fig. 9.5 as possible. This means that normally the stator slot harmonics

$$\frac{L_t - L_s}{L_s} = \frac{m^2 \pi^2}{12 s_2^2}.$$

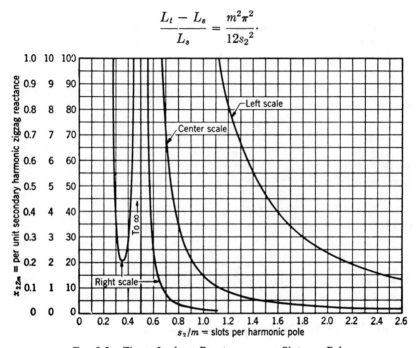

FIG. 9.5. Zigzag-Leakage Reactance versus Slots per Pole.

are nearly open-circuited by the rotor; and the stator phase-belt harmonics are substantially short-circuited.

The subharmonic fields, produced by the induced harmonic currents in the rotor, create high-frequency voltages in the stator winding (Sect. 10.8), which are short-circuited through the primary source of power supply (Fig. 9.6). Even though the stator-winding factors for these subharmonic fields are usually small, they may be large enough to allow some currents to flow, thus reducing the secondary harmonic reactance below the curve of Fig. 9.5.

If the rotor is skewed (spiraled) through nearly one stator slot pitch, the slot harmonic voltages induced in one rotor bar vary through nearly 360°

over the slot length, thus greatly increasing the effective rotor reactance (Fig. 7.8). If the angle of skew is σ times one stator slot pitch, the winding factor for the $\left(\dfrac{S}{P}\pm 1\right)$th stator slot harmonic, for example, is reduced by a factor:

(9.9) $$K_\sigma = \dfrac{S \sin\left(1-\sigma\mp\dfrac{\sigma P}{S}\right)\pi}{\pi\sigma(S\pm P)},$$

and the effective impedance referred to the stator is increased by the factor $1/K_\sigma^2$.

The secondary resistance for each harmonic, referred to the stator, must be divided by the slip referred to the harmonic synchronous speed, as explained in Sect. 4.3. For the mth harmonic revolving forward, per unit synchronous speed is:

$$N_{sm} = \dfrac{1}{m},$$

so that, at any value of fundamental slip s, or per unit speed $1-s$, the harmonic slip is:

(9.10) $$s_m = \dfrac{\dfrac{1}{m}-(1-s)}{\dfrac{1}{m}} = 1-m(1-s) = 1-m+ms.$$

This is zero when $s = (m-1)/m$, and is negative for all smaller values of fundamental slip. The harmonic torque adds to the fundamental torque for rotor speeds between zero and $1/m$, and subtracts from it for all higher speeds. For the mth backward-revolving harmonic:

$$N_{sm} = -\dfrac{1}{m},$$

and

(9.11) $$s_m = \dfrac{-\dfrac{1}{m}-(1-s)}{-\dfrac{1}{m}} = 1+m(1-s) = 1+m-ms.$$

To sum up, in the mth stator mmf harmonic circuit of Fig. 3.15, the secondary impedances are:

(9.12) $X_{Mm} = \dfrac{K_{pm}^2 K_{dm}^2 X_M}{m^2 K_{p1}^2 K_{d1}^2}$ for phase belt harmonics, from Eq. 7.1,

(9.13) $X_{Mm} = \dfrac{X_M}{m^2}$ for slot mmf harmonics,

and m is to be taken as negative for backward-revolving fields.

(9.14) $R_{2m} = \dfrac{K_{pm}^2 K_{dm}^2}{K_\sigma^2 K_{p1}^2 K_{d1}^2} R_2$ from Eqs. 9.6 and 9.9,

(9.15) $X_{2m} = \dfrac{K_m^2}{K_\sigma^2} X_{Mm}$ approximately, from Eqs. 9.7 and 9.9,

and K_m^2 is given by Fig. 9.5.

The end-ring resistance may be neglected in calculating R_{2m}, but the extra loss factor due to skin effect in the bars must be allowed for, based on the rotor frequency for each harmonic. From Eq. 8.23, this loss factor, at synchronous speed for the fundamental, is approximately:

$$\frac{R_{ac}}{R_{dc}} = \alpha d = 2.70d \sqrt{\left[\frac{(m \pm 1)f}{60}\right]},$$

where $m \pm 1 = 6$ for the fifth and seventh harmonics, or 12 for the eleventh and thirteenth, etc., and d = depth in inches of the rotor bars (copper). If the bars are aluminum, of 50 per cent conductivity, the 2.70 becomes 1.84.

Calculations made in this way give reasonably accurate values of torque and losses for the phase-belt harmonics. For the slot harmonics they are less accurate, because of the difficulty of evaluating X_{2Z} and the effects of the slot permeance harmonics considered in the next section.

Figure 9.6 shows the equivalent circuit of Fig. 3.15 expanded to include the rotor winding harmonic fields, and their induced current paths in the stator and supply circuits. Only one forward and one backward harmonic are shown, produced by the fundamental, slip-frequency, rotor current; and similar pairs of harmonics for the rotor currents induced by one forward- and one backward-revolving stator harmonic field. Actually, there is an infinite series of such harmonics produced by each rotor current, of orders $2q \pm 1$ and $(R/P) \pm 1$. The Z_i branches allow for eddy currents in the rotor laminations.

The circuit is derived by separating from the secondary reactance, X_{2m}, for each rotor circuit, its differential leakage components, X_{2Zm}, due to the rotor winding harmonics, and shunting these harmonic magnetizing reactances by their appropriate stator impedances. The magnetizing reactance for the kth slot harmonic is equal to the fundamental magnetizing reactance divided by k^2, since $K_p K_d$ for any slot harmonic is the same as for the fundamental. The stator impedance values, R_{1mk} and X_{1mk}, are found by the same procedure used in calculating the rotor impedances for the stator harmonic fields, simply interchanging stator and rotor winding coefficients. Each stator circuit resistance so found must be divided by the

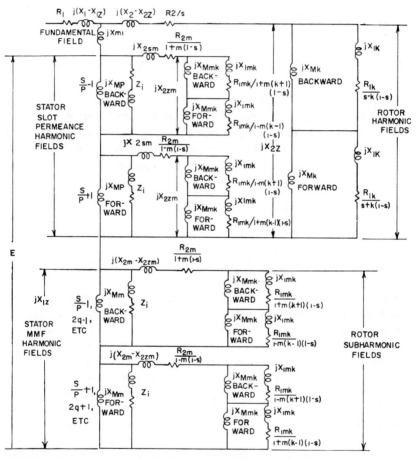

FIG. 9.6. Expanded Equivalent Circuit of Polyphase Induction Motor, Showing Circuits of Harmonic Fields of Both Rotor and Stator.

per unit frequency of the actual currents for the particular harmonic. For example, considering the kth forward rotor harmonic due to the rotor current induced by the mth backward stator harmonic field, the speed with respect to the stator of this mth harmonic stator field is:

$$-\frac{\omega}{mP},$$

and its speed with respect to the rotor is:

$$-\frac{\omega}{mP}-(1-s)\frac{\omega}{P} = \frac{\omega}{mP}[ms-(m+1)].$$

The speed of the kth forward rotor harmonic of this field, with respect to the rotor, is:

$$\frac{\omega}{mkP}[ms-(m+1)],$$

and the speed of this mkth harmonic with respect to the stator is:

$$\frac{\omega}{mkP}[ms-(m+1)]+\frac{(1-s)\omega}{P} = \frac{\omega}{mkP}[m(k-1)(1-s)-1].$$

The per unit speed of this field is found by dividing this last expression by its speed when $s=1$, or $-\omega/mkP$, giving:

$$1-m(k-1)(1-s).$$

Accordingly, this per unit speed appears as the denominator of the stator resistance term in the kth forward field circuit of the mth backward stator harmonic field in Fig. 9.6. The resistance and reactance of the supply circuits, through which the induced harmonic currents flow, must be included in the values of R_{1mk} and X_{1mk}.

By putting $k=1$ and neglecting all the mth harmonics in Fig. 9.6, the circuit represents the interesting case of a polyphase stator with single-phase rotor, which was first analyzed by Dreyfus (Fig. 8.16). The forward rotor harmonic field branch of the circuit drops out in this case, as for $k=1$ it is the fundamental field itself, already included in the circuit by R_1+jX_1 in parallel with X_M. That is, the kth harmonics included in the secondary circuits are only those fluxes occurring in addition to the forward fundamental field. The backward rotor harmonic field impedance becomes $R_1/(2s-1)+jX_1$, as mentioned in Sect. 8.12.

4 Stray Losses

When the motor is at full speed, the frequency of the mth stator harmonic field in the rotor approaches $m\pm 1$ times line frequency, so that the resistance of the harmonic circuit becomes very small compared with its reactance. Therefore, in Fig. 9.6, the secondary current of each harmonic is nearly equal to:

(9.16) $\quad I_{2m} = I_1 X_{mm}/(X_{2m}+X_{mm}),\quad$ slip < 0.2, approximately.

For the phase belt harmonics, $m=2q\pm 1$, the secondary reactance X_{2m} is small in comparison with the magnetizing reactance, X_{mm}, so that for these fields we may assume $I_{2mq} = I_1$. Hence, the stray-load losses due to the phase belt harmonics are approximately equal to:

$$(9.17) \qquad W_B = qk_{mq} R_{2b} (K_{2q-1}^2 + K_{2q+1}^2)/K_1^2 \text{ watts,}$$

where k_{mq} = skin effect ratio for the rotor bars at the phase-belt frequency ($2qf$ when $s = 0$).

$K_{2q\pm1}$ = pitch factor times distribution factor for the stator winding for the $2q\pm1$ harmonics (Eqs. 3.16, 3.17), q = number of phases, I_1 = stator amperes per phase, and R_{2b} = rotor bar resistance in ohms per phase referred to the stator.

For the stator slot mmf harmonic fields, $m = (S\pm P)/P$, the zigzag reactance, X_{2Zm}, is much the largest part of the secondary reactance, X_{2m}. The secondary current for these harmonics, therefore, is approximately equal to:

$$(9.18) \qquad I_{2m(S\pm P)} = \frac{4s_2^2 P^2 I_1}{(S\pm P)^2 \pi^2} \sin^2 \frac{(S\pm P)\pi}{2Ps_2}$$

from Eqs. 9.7 and 9.16.

If the rotor slots per harmonic pole, s_2/m, are large, X_{2Zm} becomes very small, as indicated by Fig. 9.5, and $I_{2m(S\pm P)}$ is nearly equal to I_1. Since the rotor frequency for these slot harmonics is high, Sf/P at synchronous speed, the skin effect factor for the rotor bar resistance is high. The bar resistance for the stator slot harmonics is the same as for the fundamental, $R_{2ms} = R_2$, since the pitch and distribution factors for the slot harmonics are the same as for the fundamental. Hence, if s_2/m is large, the stray-load loss due to the stator slot mmf harmonics may be several times as large as the $I^2 R_2$ loss of the fundamental field.

To avoid such high losses, it is essential to make s_2/m not far different from unity. Usually the number of rotor slots is chosen to be either 0.75 to 0.85, or 1.2 to 1.35, times the number of stator slots. In this case, Eq. 9.18 holds, and the stray-load loss for the stator slot mmf harmonic fields becomes:

$$(9.19) \qquad W_Z = qC_m I_1^2 k_{ms} R_{2b} \text{ watts,}$$

where k_{ms} is the skin effect factor for the rotor bar resistance at the stator slot frequency (Sf/P at synchronous speed), and C_m is the loss factor:

$$(9.20) \qquad C_m = (I_{2m(S-P)}^2 + I_{2m(S+P)}^2)/I_1^2.$$

This is calculated from Eq. 9.18 for the $(S-P)/P$ and $(S+P)/P$ harmonics, and is shown by the dotted lines on Figs. 9.7, 9.8, and 9.9 for the cases of 9, 12, and 27 stator slots per pole.

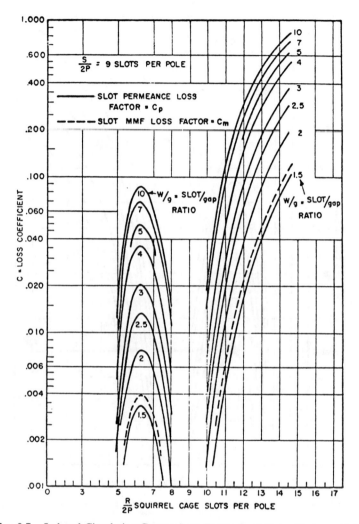

FIG. 9.7. Induced Circulating Current Loss Factor for 9 Stator Slots per Pole.

W_Z is likely to be very large for rotors with double squirrel cages, or deep bars, which have a high skin effect ratio, and when the rotor to stator slot ratio exceeds about 1.3.

The stator slot permeance harmonics are a third major cause of stray losses, which may be the largest of all if the stator slot opening to air gap ratio is large. These permeance harmonics are proportional to the current $I_0 = I_{1M}$ in the magnetizing branch of the circuit. Their rotor circuits are identical with those of the slot mmf harmonics, but their magnetizing

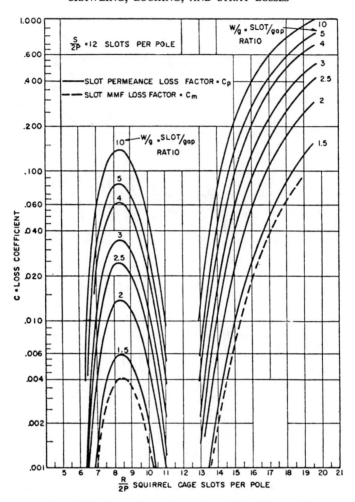

FIG. 9.8. Induced Circulating Current Loss Factor for 12 Stator Slots per Pole.

reactances, X_{MP}, are much larger than the X_{Ms} values. Hence, the ratio I_{2mP}/I_{1M} is much larger than I_{2ms}/I_1, and this may more than make up for the fact that I_{1M} is smaller than I_1.

To determine the value of X_{MP}, we note that the ratio of the single amplitude of the flux pulsation due to the slot openings to the amplitude of the fundamental field is:

$$(9.21) \qquad K = \frac{\text{permeance ripple}}{\text{fundamental field}} = \frac{\beta/2}{1 - \beta/2} = \frac{\beta}{2 - \beta},$$

STRAY LOSSES

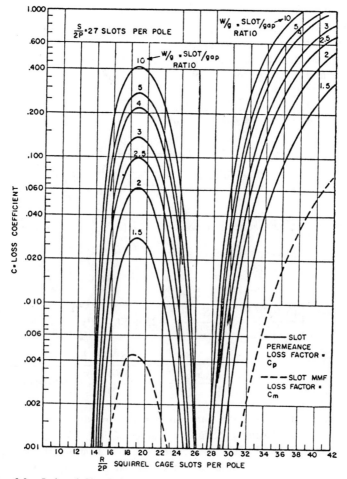

FIG. 9.9. Induced Circulating Current Loss Factor for 27 Stator Slots per Pole.

where β is the ordinate of Fig. 6.11. Half of this dip is due to the $(S+P)$ and half to the $(S-P)$ harmonic. The corresponding ratio for the stator slot mmf ripple is:

$$(9.22) \qquad \frac{\text{slot mmf ripple}}{\text{fundamental field}} = \frac{P}{(S \pm P)}.$$

Hence, the ratio of the magnetizing reactance for each of the permeance harmonics to that for the slot mmf harmonics is:

$$(9.23) \qquad X_{MP}/X_{Ms} = \frac{\beta(S \pm P)}{2P(2-\beta)} = \frac{K(S \pm P)}{2P}.$$

Using this value of X_{MP} in place of X_{Mm} in Eq. 9.16, and putting X_{2Zm} from Eq. 9.7 in place of X_{2m}, the values of the secondary permeance harmonic currents can be calculated.

$I_{2P(S \pm P)}$ is equal to I_{1M} times the ratio of the combined impedance of $X_{MP(S \pm P)}$ and $X_{2m(S \pm P)}$ in parallel to $X_{2m(S \pm P)}$, or:

$$(9.24) \quad X_{2P(S \pm P)} = \frac{I_{1M} X_{MP(S \pm P)}}{X_{2m(S \pm P)} + X_{MP(S \pm P)}}$$

$$= \frac{I_{1M} \left[\dfrac{K(S \pm P)}{2P} \right]}{\dfrac{(S \pm P)^2 \pi^2}{4P^2 s_2^2} \csc^2 \dfrac{(S \pm P)\pi}{2Ps_2} - 1 + \dfrac{K(S \pm P)}{2P}}.$$

Therefore, the permeance harmonic loss factor, C_0, is the ratio:

$$(9.25) \quad C_0 = \frac{I_{2P(S-P)}^2 + I_{2P(S+P)}^2}{I_{1M}^2}$$

$$= \frac{K^2(S-P)^2}{\left[\dfrac{(S-P)^2 \pi^2}{2Ps_2^2} \csc^2 \dfrac{(S-P)\pi}{2Ps_2} - 2P + K(S-P) \right]^2}$$

$$+ \frac{K^2(S+P)^2}{\left[\dfrac{(S+P)^2 \pi^2}{2Ps_2^2} \csc^2 \dfrac{(S+P)\pi}{2Ps_2} - 2P + K(S+P) \right]^2}.$$

$2Ps_2$ in these equations is the total number of rotor slots.

Values of C_0 are given by the full-line curves of Figs. 9.7, 9.8, and 9.9 for various values of β and of slot ratio; for 9, 12, and 27 stator slots per pole. The corresponding stray loss, which occurs at no load, and varies in proportion to the square of the magnetizing current, is:

$$(9.26) \quad W_P = q C_0 I_{1M}^2 k_{ms} R_{2b} \text{ watts,}$$

which is the same as Eq. 9.19, except for the values of C and I.

There are two ways in which these stray losses can be strongly modified. In the first place, if the rotor slots are closed, with a bridge sufficiently deep to carry one-half the flux per pole of the permeance ripple without saturation, the effective reactance X_{2m} will be very large, and the induced rotor currents will be very small. In this way, the no-load stray loss may be held to a low value, even though the ratio of rotor to stator slots is high. However, when load comes on, the rotor slot bridges will become magnetically saturated, and the effective reactance for the harmonics will be reduced, so that the full amount of the permeance wave loss, W_P, will appear. Thus, a motor with closed rotor slots usually will have a smaller no-load loss and a greater stray-load loss than if the rotor slots are open.

This saturation effect depends on the relative space phase angles of the peak rotor (load) current (where the rotor slot openings are most saturated), of the peak stator current (where the stator slot mmfs are greatest), and of the peak fundamental flux (where the stator permeance harmonics are a maximum).

Since the slot permeance and slot mmf harmonics have the same frequency and the same number of poles, they combine into a single resultant field in the rotor. If they were in space and time phase, their combined loss would be proportional to the square of their sum. If, on the other hand, they were displaced 90°, the loss would be the sum of their separate losses. The two $S-P$ fields add, and the $S+P$ fields subtract, at no load, and they move out of phase under load. Hence, it is customary to add the calculated losses without regard to the phase displacements of the two kinds of field.

In the second place, if the rotor slots are skewed through one stator slot pitch, the induced harmonic currents can be greatly reduced, as shown in Sect. 9.3. However, with a 360 degree skew, the harmonic voltage between midpoints of adjacent bars is still half the bar voltage, even though the voltage over the whole bar is zero. In consequence, a current is created from bar to bar through the laminations, in spite of the high contact resistance. Depending on the resistance, the resultant loss may be more or less than the loss without skew. If the slots are insulated, the bar to bar currents can be held down, but practical insulating methods are not very effective. Hence, the actual losses in skewed rotors usually are not very different from those without skew. One important benefit from skew is that it markedly reduces the torque dip in asynchronous crawling, although it spreads the low torque over a much wider speed range.

The skew also creates additional stray losses. For, there is a phase displacement of the fundamental mmf waves at the two ends of the core, equal to the angle of skew. Assuming the load component of the stator mmf exactly offsets the rotor mmf at the center of the core, the air-gap flux at this point will be the same as at no load. As the distance from the center increases, the angle of phase difference increases linearly, and there is a corresponding increase in radial flux density. Since the load component of stator current is displaced nearly 90° from the magnetizing component, the increase in flux due to skew will be greatest nearly along the axial line of zero fundamental flux. Hence, the losses due to skew leakage flux may be considered independent of the no-load flux, to the first order of accuracy.

With S/P slots per pole pair in the stator, and one slot pitch skew, the ratio of the extra mmf at each end of the core to no-load mmf will be:

$$(9.27) \quad \frac{\text{skew mmf}}{\text{fundamental mmf}} = (2\sin \pi P/2S)(I'/I_0), \text{ approximately,}$$

where I' is the phasor difference between the load current, I_1, and the magnetizing component of this current, I_{1M}.

The loss at any distance x from the core ends varies as x^2, so the average loss over the whole core length will be one-third of that at the ends. Therefore, the total stray loss due to skew is:

$$(9.28) \quad W_K = (\sigma \pi I'P)^2/3S^2 I_0^2 \text{ (stator iron loss plus rotor surface loss at no load) watts,}$$

where σ = ratio of skew to one stator slot pitch.

This assumes that the loss per inch of core length is the same for a given air-gap flux density, and that there is no magnetic saturation. Actually, there is severe saturation at high loads and during starting. This, and the inaccuracies of the other assumptions make Eq. 9.28 only a rough estimate of the loss.

Still another factor affecting the stray losses is the presence of eddy currents in the rotor laminations, as indicated by the Z_i branches in the circuit of Fig. 9.6. Oddly enough, the use of thick laminations may increase the losses more because they reduce the high-frequency reactance, and so allow more current to flow in the bars, than because of the eddy-current losses in the iron itself.

5 STANDSTILL LOCKING

If the rotor is turned slowly, with voltage impressed on the stator, the zigzag leakage flux from each tooth tip will vary up and down as the teeth pass by each other, causing changes in the magnetic energy stored in the air gap. With many slot combinations, this will cause "cogging", or variations in the motor torque as it turns.

The nature of this standstill locking can be understood by considering all seven types of air-gap fields listed in Sect. 9.2. These are:

Type	Forward Fields
1	$AP_0 \cos(Px - \omega t)$
2	$(AP_1/2) \cos[(S + P)x - \omega_i]$
3	$CP_0 \cos[(2q + 1)x - \omega t]$
4	$EP_0 \cos[(S + P)x - \omega t]$
5	$GP_0 \cos[(R + P)x - (\omega + RN)t]$
6	$EP_R \cos[(S + P)x - \omega t] \cos R(x - Nt)$
7	$GP_S \cos[(R + P)x - (\omega + RN)t] \cos Sx$

STANDSTILL LOCKING 347

Type	Backward Fields
2	$(AP_1/2)\cos[(S-P)x + \omega t]$
3	$BP_0 \cos[(2q-1)x + \omega t]$
4	$DP_0 \cos[(S-P)x + \omega t]$
5	$FP_0 \cos[(R-P)x + (\omega - RN)t]$
6	$DP_R \cos[(S-P)x + \omega t] \cos R(x - Nt)$
7	$FP_S \cos[(R-P)x + (\omega - RN)t] \cos Sx$

The type 6 and 7 terms can be replaced by cosine terms in the sums and differences of the angles. Only the terms with $(R-S)x$ in their angles need be considered, as those with $(R+S)x$ have too many poles to have much effect. The type 6 and 7 terms then become:

Type	Forward Fields
6	$(EP_R/2)\cos[(S - R + P)x - (\omega - RN)t]$
7	$(GP_S/2)\cos[(R - S + P)x - (\omega + RN)t]$

Type	Backward Fields
6	$(DP_R/2)\cos[(S - R - P)x + (\omega + RN)t]$
7	$(FP_S/2)\cos[(R - S - P)x + (\omega - RN)t]$

At standstill, RNt may be replaced by the position angle, $R\theta$. Since P_R usually is very small, the type 6 terms may often be neglected, but they are increased by the effects of magnetic saturation. It should be remembered that whenever the coefficient of x in the angles becomes negative, the direction of rotation of the field is reversed.

If any of the fields of types 1 to 5 have the same numbers of poles and the same direction and speed of rotation as any of the fields of types 6 or 7, there will be two independently produced fields rotating in synchronism, with a phase displacement $R\theta$ that varies with the relative positions of stator and rotor teeth. The motor torque will then vary widely as the two fields pass by each other (Eq. 3.5).

The conditions that cause this locking at standstill are found by equating the coefficients of x in the types 6 and 7 fields to those of the other fields with the same directions of rotation. Locking with the fundamental field will occur if:

(9.29) $\quad\quad\quad$ Type 6 $\quad\quad$ Type 7
$\quad\quad\quad\quad\quad S-R+P \ $ or $\ R-S+P = P; \ $ or $\ R-S = 0.$

Locking will occur with the $2q-1$ harmonic fields if:

(9.30) $\quad S-R-P \quad$ or $\quad R-S-P = (2q-1)P; \quad$ or $\quad R-S = \pm 2qP.$

Locking will occur with the $2q+1$ harmonic field if:

(9.31) $\quad S-R+P \quad$ or $\quad R-S+P = (2q+1)P; \quad$ or $\quad R-S = \pm 2qP.$

Thus, if $R = S$ there will be very severe locking at standstill. And, in any three-phase motor ($q = 3$) standstill locking will occur if the difference between stator and rotor slots is 3 per pole. Since P_R usually is small, the locking should be less severe when the rotor slots are less than the stator slots than when the rotor slots are more.

More generally, taking into account the mth harmonic of rotor permeance, the nth harmonic of stator permeance variation, and the kth harmonic of the phase-belt variation, any slot combination having:

$$mR - nS = \pm 2kqP$$

will have a locking tendency at standstill.

This means that some locking tendency will be present in every motor, the torque variation being greater the larger the common factors of R and S are, and the smaller R, S, and q are. The $(2q \pm 1)$ harmonic locking can be minimized by using a fractional-pitch winding, preferably $\frac{5}{6}$ for a 60° phase-belt winding, as this has almost the same effect as making the phase belts half as wide, or making q twice as large. A similar advantage can be gained with a single-layer winding by using different numbers of turns in each coil of a phase belt.

The locking torque due to the varying reluctance of the air gap, or saliency of the rotor and stator teeth, is of the same character as the torque that turns a salient pole rotor into line with an exciting mmf. Even though the rotor and stator slots are closed, giving a perfectly uniform air gap, the locking torque will still appear, because the saturation that occurs in the slot bridges will create a varying permeance similar to that with open slots. Any eccentricity of the air gap will introduce additional harmonic fields and more locking.

At any rotor speed except zero, the pairs of like fields rotate past each other, creating pulsating torques which produce small oscillations of the rotor speed. If these torques are large, they may produce objectionable chatter of gears or couplings. Or the torque reactions transmitted through the stator feet may cause vibration of surrounding structures, and noise.

It is impractical to calculate the magnitude of locking torque by the permeance wave method alone, because of the slow convergence of the infinite series of harmonic fields that contribute to the locking. Instead, a summation process may be used, proposed by H. R. West, based on the tooth overlap method of zigzag-reactance calculation.

Equations 7.50 and 7.52 give the total magnetic energy, W, stored in the air-gap space, due to the zigzag-leakage flux of one rotor-tooth pitch, as a function of rotor position. The differential of this energy with respect to the rotor displacement is a measure of the torque due to zigzag leakage, or the locking torque (Eq. 3.8). If the stator and rotor slot numbers were equal, and there were no skew, the locking torque would follow exactly this pattern of the energy differential for a single slot pitch. In practice, however, R and S are not equal, so that the curves for successive slot pitches must be superposed out of phase.

If the ratio of R to S, reduced to its lowest terms, is r/s, the periphery can be divided into R/r identical sectors, each containing s stator slot pitches, and the actual locking-torque curve of the motor will be obtained by superposing r of the energy differential curves of one stator slot pitch, each displaced $360°/r$ (or s of the curves for one rotor slot pitch, each displaced $360°/s$). $360°$ is taken as the span of the torque cycle of a single tooth, and is equal to the rotor slot pitch for a stator tooth, or the stator slot pitch for a rotor tooth.

The locking torque, therefore, goes through one cycle each time the rotor is advanced $1/s$th of one rotor-tooth pitch, and the total number of cycles of locking torque as the rotor is turned through one complete revolution is Rs (or rS). If the torque due to each tooth varied sinusoidally, the variation would be completely smoothed out by making s equal to 2 or any larger number. Actually, the torque changes very sharply as the edge of a rotor tooth approaches that of a stator tooth; i.e., the higher harmonics may be much larger than the fundamental of the slot pitch variation. Therefore, the locking torque varies inversely with s, but may be still quite large, even when s is 10 or more.

If the air gap is increased, the energy differential, or torque, decreases rapidly, because of the more gradual variation of flux density at the tooth edges, and the harmonics are also markedly decreased with respect to the fundamental. Therefore, the larger the air gap, the more rapidly the locking torque decreases with increase in s.

It appears at first glance that skewing the rotor one stator slot pitch should completely eliminate locking. This is not the case, however. If the angle of skew is α electrical degrees, the stator current at one end of the rotor is $\theta - \alpha/2$ ahead of the opposite rotor bar current, and that at the other end of the rotor is $\theta + \alpha/2$ behind, assuming that the stator current lags θ behind phase opposition at the center of the core (Fig. 9.10), θ being the phase angle due to the magnetizing current excess of stator over rotor current (Fig. 4.4). At standstill, θ is small, of the order of 2 or 3°, since the locked-rotor current is 6 or more times normal, and the magnetizing current under locked-rotor conditions is only about half of its no-load value, or of the order of one-fourth of rated current. For 12 stator slots per pole, and a full slot pitch spiral, $\alpha = 15°$. Thus, the phasor difference between the stator and the opposing rotor current, which is available to produce air-gap flux, may be several times as much at one end of the core as in the center, and at the other end it may be reversed in phase. This extra flux, varying over the core length, markedly changes the air-gap energy distribution,

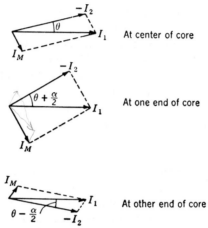

FIG. 9.10. Axial Variation of Effective Magnetizing Current due to Skew.

and also introduces a varying degree of magnetic saturation, so that some degree of locking torque will remain whatever the angle of skew.

It is possible to calculate the locking torque, following the general procedure of Sect. 7.7, first deriving an expression for the magnetic energy of each rotor tooth as a function of its position, and differentiating this with respect to the rotor position angle to derive the torque. However, a complete analysis of this problem requires taking account independently of the values of R, S, and P, of the position of the mmf wave with respect to a stator tooth, the relative position of the rotor and stator teeth, the exact slot and tooth dimensions, and the variations in air-gap flux fringing as the teeth move by each other. This is far too extensive a program to give here. Therefore, the present analysis will be confined to the idealized case considered in Sect. 7.7, with indefinitely large values of R/P and S/P, and with "lumped" fringing as in Fig. 7.5 (c).

Instead of following exactly the former procedure, we shall consider a motor with open stator and semi-closed rotor slots, and $R > S$, as shown in Fig. 9.11. Taking into account the magnetic energy stored in the air gap

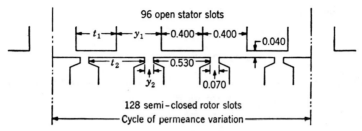

FIG. 9.11. Slot Configuration.

opposite one rotor-tooth face, there are three regions to be considered. Starting with tooth opposite tooth, the first region extends to the point where the rotor tooth has moved until its left edge lines up with the left edge of the stator tooth. By analogy with Eq. 7.50, the energy per stator tooth in this (*a*) region is:

$$W_a = K t_1 x^2, \quad 0 < x < \frac{t_2 - t_1}{2}, \tag{9.32}$$

where K is a constant proportional to $(n_2 i_2)^2 / g w_2^2$, which will be evaluated later.

The torque in this region is:

$$T_a = \frac{dW_a}{dx} = 2K t_1 x, \quad 0 < x < \frac{t_2 - t_1}{2}. \tag{9.33}$$

The second region extends to the point where the right-hand edge of the rotor tooth reaches the left-hand edge of the next stator tooth. By analogy with Eq. 7.52, the energy is:

$$W_b = K x^2 \left(\frac{t_1 + t_2}{2} - x \right), \quad \frac{t_2 - t_1}{2} < x < y_1 - \frac{t_2 - t_1}{2}, \tag{9.34}$$

and the torque is:

$$T_b = \frac{dW_b}{dx} = Kx(t_1 + t_2 - 3x). \tag{9.35}$$

The third region extends to the point where the center of the rotor tooth reaches the midpoint of the stator slot, when the travel is one-half the stator tooth pitch. In this (*c*) region, the energy is made up of two parts, one due to flux from the left stator tooth, and the other due to flux from the next stator tooth on the right. The total energy is:

$$W_c = K x^2 \left(\frac{t_1 + t_2}{2} - x \right) + K(w_1 - x)^2 \left(x - y_1 + \frac{t_2 - t_1}{2} \right), \tag{9.36}$$

$$y_1 - \frac{t_2 - t_1}{2} < x < \frac{w_1}{2},$$

whence the torque is:

$$T_c = Kx(2t_2 - 4t_1 - 6y_1) + Kw_1(2w_1 + y_1 - t_2)$$
$$= K(w_1 - 2x)(2w_1 + y_1 - t_2).$$

Taking the values $y_1 = 0.400$, $t_1 = 0.400$, $y_2 = 0.070$, $t_2 = 0.530$, $g = 0.040$, for a particular motor with $R/S = 4/3$, and neglecting fringing altogether, these torque equations become:

(9.38) $\qquad T_a = 0.8Kx, \quad 0 < x < 0.065;$

(9.39) $\qquad T_b = Kx(0.930 - 3x), \quad 0.065 < x < 0.335;$

(9.40) $\qquad T_c = 2.94K(0.400 - x), \quad 0.335 < x < 0.400.$

These values are plotted in Fig. 9.12 in dotted lines. The abrupt jump in torque when the rotor tooth first bridges the stator slot is conspicuous. When allowance is made for fringing, the torque jump is much less abrupt, since the flux builds up gradually as the tooth edges approach each other. The full-line curve in the figure gives a rough idea of the true torque curve.

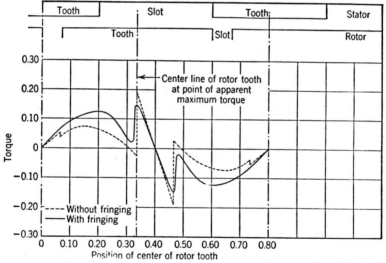

FIG. 9.12. Locking Torque at Standstill due to One Rotor Tooth.

Figure 9.12 gives the torque-position curve of a single rotor tooth. The entire periphery is made up of similar groups of three stator and four rotor teeth. The resultant torque curve of the motor, Fig. 9.13, is found, therefore, by adding four of the curves of Fig. 9.12, each displaced one-quarter of a stator slot pitch, and multiplying the result by $R/4$. Actually, the individual torque curves of the four teeth, and the resultant torques of each group of four teeth are not the same, because of their dependence on the number of slots per pole, unless R and S are very large, as here assumed. The full-line curve in Fig. 9.13 indicates the approximate resultant torque-position when fringing is allowed for.

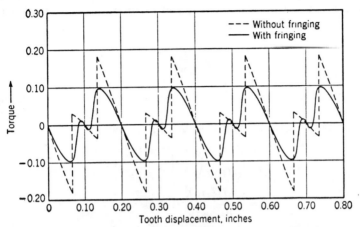

FIG. 9.13. Resultant Locking Torque at Standstill.

If the curve of Fig. 9.12 were sinusoidal, that of Fig. 9.13 would be zero in all positions. The net torque is due to only those harmonics of the single tooth torque whose orders are multiples of M times the rotor slot pitch, when M is the numerator of the fraction $R/(R-S)$ reduced to its lowest terms (in this case $M = 4$).

It is evident from Fig. 9.13 that the resultant torque is critically dependent on the fringing.

To evaluate K, we know from Eq. 1.25 that at 6.45×10^{-4} webers per sq in. density there are 4.81 ft-lb of energy stored in each cubic inch of air gap. The rms air-gap density in region (a) is:

$$B = \frac{3.19 I_s}{g 10^8}\left(\frac{x}{w_1}\right) \text{ webers per sq in.,}$$

where I_s is the rms current in one stator slot, g is the air-gap length, and w_1 is the stator slot pitch. The energy per rotor tooth in this region is, therefore:

(9.41) $\quad$ Energy per tooth $= \left(\frac{3.19}{6.45}\right)^2 (4.81)\left(\frac{I_s}{g}\right)^2 \left(\frac{x}{w_1}\right)^2 (g t_1 L) 10^{-8}$

$$= \frac{1.18}{10^8} \frac{I_s^2 L}{g w_1^2} (x^2 t_1) \text{ ft-lb per tooth,}$$

where L is the core length in inches.

Multiplying by R, the energy for the complete motor in region (a) is:

$$\frac{1.18}{10^8} \frac{I_s^2 LR}{gw_1^2}(x^2 t_1) \text{ ft-lb.}$$

Differentiating with respect to x, and multiplying by 12, since x is in inches, the tangential force is:

$$\frac{2.93 I_s^2 LRxt_1}{10^7 gw_1^2} \text{ lb.}$$

The torque in foot pounds is $D/24$ times this, where D is the gap diameter in inches.

Equating this to Eq. 9.33, the value of K is found to be:

(9.42) $$K = \left(\frac{1}{2t_1 x}\right)\left(\frac{D}{24}\right)\left(\frac{2.93 I_s^2 LRxt_1}{10^7 gw_1^2}\right)$$

$$= \frac{0.61 I_s^2 DLR}{10^8 gw_1^2} \text{ ft-lb per sq in.}$$

To get an approximate formula for the net locking torque, we may first neglect fringing, and then apply an empirical correction factor to allow for it. The maximum value of torque due to a single rotor tooth, Fig. 9.12, occurs at the start of region (c), when $x = y_1 - \frac{t_2 - t_1}{2}$. The magnitude of the jump in torque at this point is the difference between Eqs. 9.37 and 9.35 at this value of x, which is:

(9.43) $$T_0 - T_b \bigg]_{x = y_1 - \frac{t_2 - t_1}{2}} = \frac{K(t_1 + t_2)^2}{4}.$$

When M of the individual torque curves of Fig. 9.12 are added out of phase to give the resultant torque, Fig. 9.13, this jump in torque remains the same, but occurs at M points per stator slot pitch. Equation 9.43 gives the total change in torque. The minimum motor torque will be given by subtracting only half of this value from the normal $I^2 R$ torque. The final expression for the loss in torque due to standstill locking (or cogging) is, therefore:

(9.44) Loss in torque due to locking $= \dfrac{K(t_1 + t_2)^2}{8M}$

$$= \frac{0.75 C I_s^2 DLR(t_1 + t_2)^2}{gw_1^2 M 10^9} \text{ ft-lb,}$$

where C is the factor to allow for fringing. A usual value of C is of the order of 0.5, as indicated by comparing the full- with the dotted-line curve of Fig. 9.13, but it may vary widely, depending on details of the slot and air-gap configuration.

Throughout the derivation of Eq. 9.44, we have assumed stator and rotor fundamental mmfs to be exactly equal and opposite; i.e., there is no fundamental field. Therefore, Eq. 9.44 does not include the locking between the fundamental and harmonic fields that was considered in deriving Eq. 9.31. A complete theory of locking requires the consideration of a fundamental field superposed on the zigzag leakage fields.

An additional factor, dependent on P and $R-S$, must be included in Eq. 9.44 to take account of the fact that the numbers of slots per pole are finite, instead of being indefinitely large, as assumed in this analysis. If this factor is correctly evaluated, it will make the locking torque depend on the value of $mR-nS$, as indicated by Eq. 9.31. Still another reduction factor that must be allowed for to check test values is the effect of magnetic saturation, which may be very large under full-voltage standstill conditions.

And, finally, just as the factor F_{sc} occurs in Eq. 7.47 for the zigzag reactance, the effects of rotor currents induced by the harmonic fields should be taken into account.

6 Synchronous Crawling

As the rotor speed increases, the speeds of the harmonic fields due to the rotor change also. At the speed when the initially backward rotor harmonic is rotating forward at the same speed as the forward stator field of the same number of poles, the two fields will lock into step, and synchronous crawling will occur, as shown in Fig. 9.2.

The fields of both types 6 and 7 will lock with the fundamental field when:

$$S-R-P = P; \text{ or } S-R = 2P \text{ if } \omega + RN = -\omega; \text{ or } N = -2\omega/R$$

and when

$$R-S-P = P; \text{ or } R-S = 2P \text{ if } \omega - RN = -\omega; \text{ or } N = 2\omega/R.$$

They will both lock with the $2q-1$ harmonic field when:

$$S-R+P = (2q-1)P; \text{ or } S-R = 2(q-1)P \text{ if } -(\omega - RN) = \omega,$$
$$\text{or } N = 2\omega/R$$

and when

$$R-S+P = (2q-1)P; \text{ or } R-S = 2(q-1)P \text{ if } -(\omega + RN) = \omega,$$
$$\text{or } N = -2\omega/R.$$

They will both lock with the $2q+1$ harmonic field when:

$$S-R-P = (2q+1)P; \text{ or } S-R = 2(q+1)P \text{ if } \omega+RN = -\omega;$$
$$\text{or } N = -2\omega/R$$

and when

$$R-S-P = (2q-1)P; \text{ or } R-S = 2(q+1)P \text{ if } \omega-RN = -\omega;$$
$$\text{or } N = 2\omega/R.$$

Thus, synchronous crawling torques will be present at a speed $\pm$ synchronous speed divided by rotor slots per pole in a three-phase motor ($q = 3$), if the difference between stator and rotor slots is 1, 2, or 4 per pole; while locking at standstill will occur if the difference is 3 per pole. If the difference in slots is one per pole, the crawling will be very severe, with a peak crawling torque that may be much larger than the maximum torque of the fundamental. This behavior is utilized sometimes to make small "subsynchronous" motors for timers, with outputs of a few watts.

If the difference in slots is 2 per pole, for a three-phase motor, the crawling is due only to the 5th and 7th harmonics, which can be made small by choosing a winding pitch nearly $\frac{5}{6}$, so that this slot combination is quite satisfactory from the torque viewpoint. Skewing the rotor slots greatly reduces the synchronous crawling torques due to the harmonic fields, but it is not effective in eliminating that due to the fundamental, when the slot difference is only one per pole.

Even though the crawling tendency is present, such a motor may repeatedly bring a low-inertia load to full speed, and only occasionally stick at low speed, as the motor may accelerate through the sticking speed at a moment when the two fields are out of phase. However, if the load inertia is high, the motor will always stick at the crawling speed. If the motor is belted to its load, and the belt is struck sharply while it is crawling, the motor may break out of synchronism and accelerate to full speed.

7 Unbalanced Magnetic Pull

In the extreme case when the stator and rotor slot numbers differ by only $1 (R-S = \pm 1)$, the radial magnetic pull during the starting period is a maximum at the point of tooth opposite slot, and a minimum at the opposite end of the same diameter, where tooth is opposite tooth. This is so because the zigzag-leakage flux is largest when tooth is opposite slot, and this leakage flux has a much higher air-gap density than the fundamental flux during starting, when the stator and rotor currents are each many times as large as the no-load magnetizing current. Around the periphery between the two extremes, the magnetic pull varies gradually, forming a two-node force wave (Fig. 9.3). The force wave rotates at a speed R times the speed of the rotor itself, since it moves forward (or backward if $R<S$) a whole revolution for each advance of the rotor through $1/R$ revolution.

Putting $R-S=1$ in the fields of Sect. 9.5, and adding the fundamental field term, the three principal components of the air-gap field of this motor (types 1, 6, and 7) are:

$$(9.45) \quad B_{gap} = AP_0 \cos(Px-\omega t) + C_1 \cos[(P-1)x-\omega t + RNt]$$
$$+ C_2 \cos[(P+1)x-\omega t - RNt],$$

where $\quad C_1 = \dfrac{EP_R + FP_s}{2} \quad$ and $\quad C_2 = \dfrac{DP_R + GP_s}{2}.$

The radial magnetic force along a single stator radius is proportional to the square of the flux density. Squaring Eq. 9.45, and neglecting the terms in P_R^2, which are relatively small:

$$(9.46) \quad B_{gap}^2 = \dfrac{A^2 P_0^2}{2}[1+\cos 2(Px-\omega t)]$$
$$+ A^2 P_0 C_1 \{\cos[(2P-1)x - 2\omega t + RNt] + \cos(x - RNt)\}$$
$$+ A^2 P_0 C_2 \{\cos[(2P+1)x - 2\omega t + RNt] + \cos(x - RNt)\}.$$

The significance of the several terms in this force wave is considered in the next chapter. Here, we are interested only in the resultant total force on the stator (or rotor), which is obtained by integrating the projection of the radial force component along a fixed axis. If $P>1$:

$$(9.47) \quad F = \int_0^\pi B_{gap}^2 \cos(x+\theta)\,dx$$
$$= A^2 P_0 (C_1 + C_2) \int_0^\pi \cos(x+\theta)\cos(x-RNt)\,dx$$
$$= \dfrac{\pi A^2 P_0 (C_1 + C_2)}{2} \cos(RNt + \theta).$$

Returning to Eq. 9.45, in the case of a two-pole motor, $P=1$, and the first permeance harmonic is homopolar, since $P-1=0$. With usual motor construction, the homopolar flux must be very small, because of the high reluctance of the return path through the shaft and bearings (Sect. 9.2). Taking this field as zero, the unbalanced pull for a two-pole motor becomes equal to only half of the force value given by Eq. 9.47.

Equation 9.47 represents an alternating radial force at rotor-tooth frequency, which, therefore, tends to bend the rotor shaft and cause oscillations of the shaft in the bearing clearance. If the shaft is sufficiently stiff, and the bearing clearance is small, the force will cause some vibration and noise during the starting period, but will not materially affect full-speed operation (Sect. 10.6 and peak I on Fig. 10.4).

Normally, however, the natural frequency of the shaft in radial bending between the bearing centers is far below the full-speed rotor-tooth frequency. Therefore, the rotor must go through a critical speed during acceleration. This will be accompanied by a dip in torque (Fig. 9.14) and severe noise and vibration.

The permeance harmonic fields with $(P-R+S)$ pairs of poles, which cause most of the unbalanced pull, go through synchronism with respect to the stator at a speed P/R times synchronous, and with respect to the

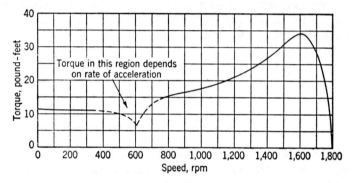

FIG. 9.14. Speed-Torque Curve of Induction Motor with Offset Air Gap.

rotor at a speed $P/(P+S)$ times synchronous. In this region, therefore, the flux of this field is unopposed by induced currents and is high, so the unbalanced pull is correspondingly large. If the shaft critical occurs in this same speed region, severe difficulties may be looked for in acceleration. If, however, the rotor-tooth frequency does not become equal to the shaft critical frequency until the speed is well above this region, the harmonic flux causing the unbalanced pull will be reduced, and the motor should accelerate without difficulty.

In any case, the magnitude of the torque dip due to unbalanced pull will be greatly dependent on the rate of acceleration. If the motor accelerates slowly, time will be given for the amplitude of shaft vibration to build up, producing severe cramping forces in the bearings, and in extreme cases causing the rotor to rub on the stator, with a corresponding decrease in effective torque. If, on the other hand, the accelerating torque is high, the motor will pass through resonance so quickly that the vibration amplitude will be small.

So far, we have considered only the case of $R-S=1$. The same pheno-

UNBALANCED MAGNETIC PULL 359

mena of unbalanced pull and consequent torque dip occur at a speed corresponding to the critical speed of the shaft, whenever a harmonic field with two more or less poles than the fundamental field exists. This will happen, as shown above, when $R-S=1$; and also when $R-S=2P-1$, $-2P+1$, -1, $2P-1$, or $2P+1$. Putting these values successively in the fields of Sect. 9.5, and by analogy with Eqs. 9.45 to 9.47, the corresponding magnitudes and rotational speeds of the unbalanced pull can be found. The results are given in Table 9.1:

TABLE 9.1

SPEEDS OF RESONANT SHAFT VIBRATION

$R-S$	Unbalanced Force Factor (Eq. 9.47) $\dfrac{4F}{\pi A^2 P_0 P_1}$	Speed of Rotation of Unbalanced Pull, rev. per sec.	Fraction of Synchronous Speed at Which Torque Dip Occurs $1-s$
$-2P-1$	$\cos(x + 2\omega t + RNt)$	$-\dfrac{R(1-s)f}{P} - 2f$	$-\dfrac{2P}{R} \pm \dfrac{Pf_c}{Rf}$
$-2P+1$	$\cos(x - 2\omega t - RNt)$	$\dfrac{R(1-s)f}{P} + 2f$	$-\dfrac{2P}{R} \pm \dfrac{Pf_c}{Rf}$
-1	$2\cos(x + RNt)$	$-\dfrac{R(1-s)f}{P}$	$\pm \dfrac{Pf_c}{Rf}$
1	$2\cos(x - RNt)$	$\dfrac{R(1-s)f}{P}$	$\pm \dfrac{Pf_c}{Rf}$
$2P-1$	$\cos(x - 2\omega t + RNt)$	$2f - \dfrac{R(1-s)f}{P}$	$\dfrac{2P}{R} \pm \dfrac{Pf_c}{Rf}$
$2P+1$	$\cos(x + 2\omega t - RNt)$	$\dfrac{R(1-s)f}{P} - 2f$	$\dfrac{2P}{R} \pm \dfrac{Pf_c}{Rf}$

In the table, f_c is the frequency of shaft resonance, or frequency corresponding to the critical speed of the shaft in mechanical rotation. f_c depends on the weight of the rotor and the direct-connected load, the shaft diameter, the distance between bearings, and the flexibility of the bearing supports. If, as usually is the case, the horizontal and vertical bearing supports are unequally stiff, there will be two values of f_c for each mode of shaft deflection. Generally, Pf_c will be of the order of $10f$ or more.

Since the rotating wave of unbalanced force has two poles, its frequency in the stator is equal to its speed in revolutions per second, given in the third column of the table. The speed at which a torque dip will occur, given in the fourth column of the table, is found by equating the values in the third column to $\pm f_c$.

For example, a 19-slot rotor in a 4-pole, 24-slot stator, and having a critical shaft frequency 5.3 times the line frequency, will have a torque dip with severe noise and vibration at a speed:

$$1-s = -\frac{4 \pm 2(5.4)}{19} = 0.36$$

or -0.78 times synchronous speed.

The corresponding speeds with a 25-slot rotor will be $\pm\frac{2(5.4)}{25} = 0.43$ and -0.43 times synchronous speed, and for a 27-slot rotor will be $\frac{4 \pm 2(5.4)}{27} = 0.55$ and -0.25 times synchronous speed. These values are in good accord with those reported by Stiel and Möller.

Smaller torque dips and accompanying noise will occur at other speeds with any odd value of $R-S$, since there will then be harmonic fields with poles differing by 2.

When the rotor is offset, giving a non-concentric air gap, there will be a two-pole permeance term added to Eq. 9.1, and this will give rise to additional harmonic fields, such as would occur if $R-S$ were increased or decreased by 1. Figure 9.14 shows the torque curve of a motor with an even value of $R-S$, but a purposely offset air gap. As there is always some eccentricity, all motors will have in some degree the locking and synchronous crawling tendencies associated with even values of $R-S$, and also in some degree the noise and vibration tendencies associated with odd values of $R-S$. For this reason, it is always desirable to make the shaft as stiff, the bearing centers as close, and the air gap as large, as permitted by other considerations.

Richter and others have given detailed recommendations for choosing the number of rotor slots, for given values of P and S. At best, however, the choice is a compromise, and even the worst-appearing slot combinations may be highly successful if the designer exerts sufficient skill in providing shaft stiffness or stator yoke depth, or other features that will reduce the undesirable characteristics to a low enough level.

Since the unbalanced magnetic pull is due to inequalities between the flux densities on opposite sides of the air gap, anything that reduces the inequality is helpful. The simplest, and generally used, method of maintaining this equality is to connect the stator winding in two or more parallel circuits. With the same voltage impressed on each circuit, any difference that appears in their flux-induced voltages is immediately opposed by changes in their magnetizing currents, which restore their flux linkages to equality.

If one circuit includes only north poles and the other only south poles (alternate poles in series), the even harmonics of the air-gap flux which are the principal cause of unbalanced force, are greatly reduced. Best results are obtained by connecting all the poles in parallel, provided this does not require an excessive number of winding turns. By connecting "equalizers" between corresponding coils in the parallel circuits, the dissymmetries of air-gap flux can be still further reduced, but circulating current losses due to rotor harmonic fluxes may then be incurred.

BIBLIOGRAPHY *Reference to Section*

1. "Equivalent Circuits and Performance Calculations of Canned Motors," P. D. Agarwal, *A.I.E.E. Trans.*, Vol. 79, Part III, 1960, pp. 635–42. — 1
2. "Some Effects of Eccentric Air Gaps in Induction Machines," J. R. M. Alger, *M.I.T. Thesis* for M.S. in E.E., June 1950. — 7
3. "Induced High-Frequency Currents in Squirrel-Cage Windings," P. L. Alger, *A.I.E.E. Trans.*, Vol. 76, Part III, 1957, pp. 724–9. — 1
4. "Uber die Phasengleiche Addition der Nutharmonischen Wicklungscherfelder und der Nutungsoberfelder bei Phasenreinen Mehrphasen Wicklungen," H. W. Bolter and H. Jordan, *ETZ-A*. 1963, p. 235.
5. "Origins of Stray Load Losses in Induction Motors with Cast Aluminum Rotors, N. Christofides, *Proceedings I.E.E.*, Vol. 112, No. 12, Dec. 1965, pp. 2317-2332.
6. "Synchronous Motor Effects in Induction Machines," E. Dreese, *A.I.E.E. Trans.*, Vol. 49, 1930, pp. 1033–40. — 6
7. "Rotor-Bar Currents in Squirrel-Cage Induction Motors," J. S. Gault, *A.I.E.E. Trans.*, Vol. 60, 1941, pp. 784–91. — 1
8. "The Calculation of Harmonics due to Slotting," E. M. Freeman, *Proc. I.E.E.*, Vol. 109, Part C, 1962, p. 580.
9. "Magnetomotive Force Wave of Polyphase Windings with Special Reference to Subsynchronous Machines." Q. Graham, *A.I.E.E. Trans.*, Vol. 46, 1927, pp. 19–28. — 1
10. "Dead Points in Squirrel-Cage Motors," Q. Graham, *A.I.E.E. Trans.*, Vol. 59, 1940, pp. 637–42. — 5
11. *Magnetic Control of Industrial Motors*, Second Edition, G. W. Heumann, John Wiley, 1947.
12. "Induction Motor Slot Combinations," G. Kron, *A.I.E.E. Trans.*, Vol. 50, June 1931, pp. 757–67 — 1
13. "Starting Torque in Three-Phase Squirrel-Cage Motors," H. Möller, *Arch. Für Elek.*, Vol. 24, 1930, pp. 401–24. — 1
14. "The Calculation of Unbalanced Magnetic Pull in Synchronous and Induction Motors," R. C. Robinson, *A.I.E.E. Trans.*, Vol. 62, 1943, pp. 620–4. — 7
15. "Survey of Basic Stray Losses in Squirrel Cage Induction Motors," K. K. Schwarz, *Proceedings I.E.E.*, Vol. 111, No. 9, September 1964, pp. 1565-1574.
16. "Zusatzverluste von Asynchronmaschinen," Frank Taegen, *ACTA TECHNICA CSAV* (Czechoslovakia), 1968, No. 1.
17. "Field Harmonics in Induction Motors," P. H. Trickey, *Elec. Engineering*, Vol. 50, Dec. 1931, pp. 939–41. — 1
18. "Spatial Harmonic Magnetomotive Forces in Irregular Windings and Special Connections of Polyphase Windings," C. G. Veinott, *I.E.E.E. Trans.*, Vol. 83, No. 12, December 1964, pp. 1246-1252.
19. "Harmonics due to Slot Openings," C. A. M. Weber and F. W. Lee, *A.I.E.E. Trans.*, Vol. 43, 1924, pp. 687–93. — 1

CONTENTS—CHAPTER 10

MAGNETIC NOISE AND VOLTAGE RIPPLES

		PAGE
1	The Nature of Magnetic Noise	365
2	Stator Frame Vibration	367
3	Sound Radiation	370
4	Noise Calculations	372
5	Resonant Frequencies	376
6	Slot Frequency Force Waves	377
7	Effects of Dissymmetry	382
8	Induced Voltage Ripples	383
9	Effects of Harmonics in the Impressed Voltage	388

10

MAGNETIC NOISE AND VOLTAGE RIPPLES

1 The Nature of Magnetic Noise

An extremely small amplitude of vibration of a motor frame, of the order of half a micron at usual slot frequencies around 1,000 cycles per second, will produce objectionable noise. To avoid this, it is necessary to "balance out" the magnetic forces accurately, to make the laminations, shaft, and frame adequately stiff, and to prevent any natural (resonant) frequencies of the mechanical parts from coinciding with frequencies of the impressed magnetic forces.

As the scale of audibility for the human ear is logarithmic, the intensity of a sound wave must be reduced to about one-tenth for its apparent loudness to be reduced to one-half. For this reason, sound intensities are measured in decibels, defined by the relation:

(10.1) $\qquad$ Sound intensity in decibels $= 10 \log_{10}\left(\dfrac{I}{I_0}\right),$

where $I =$ actual sound intensity in watts per square centimeter in the direction of propagation, and $I_0 =$ reference sound intensity $= 10^{-16}$ watts per cm^2.

The sound intensity, I, is proportional to the square of the sound pressure, as given by the equation:

$$I = \frac{P^2 10^{-7}}{\rho} \text{ watts per sq cm,}$$

where $P =$ rms sound pressure in dynes per square centimeter, and

$\rho =$ density of the sound medium in grams per cubic centimeter.

The reference sound intensity corresponds to a sinusoidal double amplitude displacement of the air, at normal atmospheric pressure and temperature, equal to $(2.22 \times 10^{-6})/f$ cm, where f is the frequency of the sound wave; and to a root-mean-square pressure

$$P = 0.000207 \sqrt{\left[\frac{H}{76}\sqrt{\left(\frac{273}{T}\right)}\right]} \text{ dynes per sq cm,}$$

where $H =$ barometric height in centimeters, and

$T = 273 +$ temperature in degrees C.

From these relations, it follows that a sound intensity of one microwatt per cm² produces 100 db, and a sound pressure of 1 dyne per cm² (1 microbar) corresponds to approximately 74 db. Also, the sound intensity of a plane wave can be estimated by the formula:

$$\text{Sound intensity} = 1.3 \times 10^{-4}(2d)^2 f^2 \text{ watts per cm}^2,$$

or

(10.2) $\text{Sound intensity in decibels} = 121 + 20 \log_{10}(2df),$

where $2d$ = double amplitude of vibration in inches, and f = frequency.

A sinusoidal double vibration amplitude of 0.28×10^{-6} in. at 1,000 cycles, or one-quarter of a micron, will, therefore, create a 50-db sound intensity in the immediate vicinity of the vibrating surface. As explained later on, the actual sound observable some distance away will depend very markedly

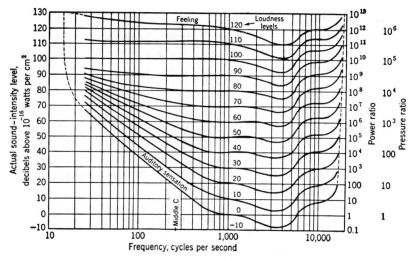

FIG. 10.1. Loudness Contours.

on the area and vibration pattern of the surface. To calculate the sound level at any point, therefore, it is necessary first to determine the amplitude and mode of vibration of the surface that gives rise to the noise; second, to calculate the resultant sound intensity at the surface, and then to calculate the sound intensity at the (distant) point of measurement as determined by the radiation pattern for the particular conditions; and, finally, to find the sound level, by making allowance for the audibility versus frequency curve of hearing.

The contribution of a particular motor or device to the overall sound level of a room is determined by the total sound-power given out. Sound-power is calculated from measurements of the sound level at a number of points arranged on a hemispherical surface surrounding the noise source, as prescribed in an A.I.E.E. Test Code. The unit of sound-power is generally taken to be 10^{-12} watts.

The human ear has its peak sensitivity in the range of 2,000 to 5,000 cycles, and is quite insensitive to frequencies of less than 50 or more than 50,000 cycles (Fig. 10.1). Sound-level meters, used to measure noise, take this into account by frequency networks, which weight the different frequency components in proportion to their audibility. The decibels sound *level* given by the meter corresponds to the *weighted* rms combination of the sound *intensities* of different frequencies actually existing. Thus, the decibels calculated by Eq. 10.2 agree fairly well with sound-level meter readings at 1,000 cycles, but will vary considerably from these readings at other frequencies.

Typical sound levels are:

A quiet home	30 db
A quiet office	40
An average office	50
Conversation	60
A typical factory	70
Heavy street traffic	80
An automobile horn at 20 ft	90–100

2 Stator Frame Vibration

The radial forces due to the air-gap field are by far the largest sources of magnetic vibration and noise in electrical machines. The peak air-gap flux density under no-load full-speed conditions is usually of the order of 4×10^{-4} webers per sq in., giving a radial magnetic pull of about 22 psi (Eq. 1.32). If the flux were sinusoidally distributed, and there were no flux pulsations, the magnetic force wave around the periphery would be a $\sin^2 Px$ curve—that is, a fully displaced sinusoid with twice as many poles of force as there are magnetic poles. Actually, however, as previously shown, the phase-belt and slot harmonic mmfs, and the permeance waves, create a numerous train of harmonic fields, superposed on the fundamental flux wave, giving rise to high-frequency pulsations in the radial magnetic forces. These may be resolved into a series of sinusoidal force waves with different numbers of poles, revolving at different speeds, each force wave having twice as many poles as the magnetic field that produces it.

Under these radial forces, the stator core and frame are set into vibration, in the same manner that a hoop of steel, or a cylindrical bell, responds when struck, Fig. 6.8. If the force-producing magnetic field has two poles, there will be two opposite centers of maximum pull at the poles, and two intermediate points of zero force. The stator will, therefore, be pulled into an elliptical shape, the short axis of the ellipse coinciding with the pole axis, and revolving synchronously with it. This will give a four-node vibration. Similarly, a 2P-pole magnetic field produces a vibration with 4P nodes.

The uniform pull due to the average displacement (or unidirectional component) of the magnetic pull curve may be neglected, as the peripheral compression it causes is too small to have any bearing on noise production. Therefore, the noise-producing force curve for a 2P-pole field is a symmetrical 4P-pole force wave, with a peak value in each direction equal to one-half the actual maximum pull at the center of a magnetic pole. The resulting displacement curve will have a point of inflection at each node, so that there will be zero bending movement at these points. The deflection of the stator can, therefore, be approximated by formulas for the deflection of a beam freely supported at each end, and carrying a sinusoidally distributed load.

This approximation will be more nearly correct the larger the stator diameter, the more the poles, and the shallower the radial depth of yoke. For simplicity, the stator "beam" strength will be assumed the same as that of a solid steel beam with the same cross section as the stator yoke, the stiffening due to the stator teeth and frame being neglected.

The formula for the deflection of a uniform beam freely supported at both ends and weighted symmetrically with a sinusoidally distributed load is:

$$(10.3) \qquad d = \frac{WL^3}{2\pi^3 EI} = \frac{0.54 W L^3 10^{-9}}{I} \text{ in.,}$$

where L is the distance between beam supports,
W is the total load in pounds,
E is the modulus of elasticity $= 3 \times 10^7$ for steel, and
I is the moment of inertia of the beam section about its center line, all in inch units. The coefficient $1/2\pi^3$ becomes $\frac{1}{48}$ for a concentrated load, and $\frac{5}{384}$ for a uniformly distributed load.

To adapt this formula for our purposes, we shall substitute the mean peripheral distance between nodes of the stator core for L, and express I in terms of the stator core depth. Let:

h = radial depth of stator core behind the slots, in inches,
D_s = mean diameter of the stator core in inches
 = D at gap + 2(slot depth) + h,
m = one-half the number of nodes of core flexure
 = $2P$ = the number of poles of the force producing magnetic field.

Then:

$$L = \frac{\pi D_s}{2m} = \frac{\pi D_s}{4P} \text{ in.,}$$

$$I = \frac{h^3}{12} \text{ per inch of core stacking.}$$

Substituting these values in Eq. 10.3, the radial deflection is:

(10.4) $$d = \frac{3WD_s^3}{4Em^3h^3} = \frac{0.25WD_s^310^{-7}}{m^3h^3} \text{ in.}$$

This approximate formula, given by familiar beam theory, has been derived in order to give a clear understanding of the problem. The true formulas for the deflection of a curved ring under a sinusoidal radially applied force are:

For 4 nodes, $m = 2$:

(10.5) $$d = \frac{WD_s^3}{6Eh^3}.$$

For 6 nodes, $m = 3$:

(10.6) $$d = \frac{9WD_s^3}{256Eh^3}.$$

For 8 nodes, $m = 4$:

(10.7) $$d = \frac{WD_s^3}{75Eh^3}.$$

It will be observed that, as m increases, the approximate formula, Eq. 10.4, approaches the correct value, the ratios for 4, 6, and 8 nodes being 0.56, 0.79, and 0.88. Thus, Eq. 10.4 is true only when m is large. All the values given are for static forces; the amplitudes increase as the resonant frequency is approached.

Since we are interested only in magnetic forces, we can obtain a more directly useful formula by substituting for W its value in terms of the peak air-gap flux density, P, from Eq. 1.32. The value of W is half the peak magnetic pull intensity, in pounds per square inch, multiplied by $2/\pi$ and by the pole area. This gives:

$$W = \tfrac{1}{2}(1.39B^2 10^8)\left(\frac{2}{\pi}\right)\left(\frac{\pi D}{4P}\right)$$

(10.8) $$= 0.347B^2 10^8 \left(\frac{D}{P}\right) \text{ lb per in. of core stacking,}$$

where B is in webers per square inch.

Equation 10.4 then becomes:

$$d = \frac{0.25(0.347)B^2 10^8 DD_s^3 10^{-7}}{Pm^3 h^3},$$

or, since $m = 2P$:

(10.9) $$d = \frac{0.108 B^2 DD_s^3}{P^4 h^3} \text{ in.}$$

In accordance with Eqs. 10.5, 10.6, and 10.7, the coefficient 0.108 should be replaced by 0.193 for $P = 1$, by 0.137 for $P = 2$, and by 0.123 for $P = 3$.

Equation 10.9 gives the single amplitude of (non-resonant) radial vibration of the stator surface due to the force wave of a $2P$-pole magnetic field. With given values of B, h, and D, the vibration amplitude varies inversely as the fourth power of the number of poles. Hence, in a given motor, only the fields with fewest poles are factors in noise production, in the absence of resonance.

3 Sound Radiation

The theory of radiation of sound from circular cylinders has been developed by Morse and others. The solution of the wave equation, which represents the sound wave pattern created by a cylindrical surface (of indefinite axial length), vibrating peripherally in $2m$ nodes, Fig. 6.8, gives a relation between the sound pressure, A_0, and the radial displacement, d, at the radius, r_0, of the surface of the cylinder, of the form:

$$\text{Radial displacement} = d \sin m\theta$$

$$= \frac{A_0 \sin m\theta}{\gamma p k} \left[\frac{J'_m(kr_0) + jN'_m(kr_0)}{J_m(kr_0) + jN_m(kr_0)} \right];$$

or

(10.10) $$|d| = \frac{A_0 C'_m(kr_0)}{\gamma p k C_m(kr_0)} \text{ ft,}$$

where $\sin m\theta$ is a factor representing the $2m$-node distribution of the radial motion around the cylinder;

J_m is a Bessel function of the first kind and the mth order;

N_m is a Neumann function (or a Bessel function of the second kind) of the mth order;

C_m is the absolute value of $J_m + jN_m$, called the Hankel function;

J'_m, N'_m, and C'_m are the first derivatives of J_m, N_m, and C_m with respect to r_0;

$j = \sqrt{-1}$, indicating a 90° phase displacement in time of the J and N terms at any point in space;

$k = 2\pi f/c = 0.00559f$ ft^{-1}, where c = velocity of sound in air = 1,125 fps at 20 C and 760 mm mercury pressure;
$\gamma = 1.4$ = the ratio of the specific heats of air at constant pressure and constant volume;
p is the atmospheric pressure in the same units as A_0; and r_0 is the radius of the cylinder in feet.

Also, the relation between the sound pressure, A, created at any external radius, r, by the pressure A_0 at the surface of the cylinder, is given by the expression:

(10.11) $$\frac{A}{A_0} = \frac{J_m(kr)+jN_m(kr)}{J_m(kr_0)+jN_m(kr_0)} = \frac{C_m(kr)}{C_m(kr_0)}.$$

The value of kr in the above expressions is a numeric, equal to the ratio of the cylinder periphery to the wave length of sound in air, at the frequency f. When this ratio is small, neighboring points on the cylinder are so nearly in phase from the viewpoint of the normal wave length in air that their opposing motions nearly cancel, and the sound will attenuate rapidly with distance.

When kr is small, C_m approaches as the limit:

(10.12) $$C_m(kr) = \frac{(m-1)!\,2^m}{\pi(kr)^m}, \quad kr < \sqrt{(2m+1)},$$

and

(10.13) $$C'_m(kr) = \frac{m!\,2^m}{\pi(kr)^{m+1}}, \quad kr < \sqrt{(2m+1)}.$$

When kr is large:

(10.14) $$C_m(kr) = \sqrt{\left(\frac{2}{\pi kr}\right)}, \quad kr > m^2$$

and

(10.15) $$C'_m = C_m, \quad kr > m^2.$$

Combining Eqs. 10.10 and 10.11, the equivalent displacement, d_e, of a plane sound wave at the radius r to give the same sound intensity as the actual pressure, is given by the expression:

(10.16) $$d_e = \frac{A}{\gamma pk} = \frac{dC_m(kr)}{C'_m(kr_0)}$$

$$= d\left[\frac{C_m(kr)}{C_m(kr_0)}\right]\left[\frac{C_m(kr_0)}{C'_m(kr_0)}\right],$$

where d = the single amplitude of radial vibration of the cylinder, Eq. 10.9.

From Eq. 10.2, therefore, putting d_e for d, the sound intensity in decibels, at a radius r, is equal to:

$$\text{(10.17)} \quad SI = 121 + 20 \log_{10}\left[\frac{C_m(kr)}{C_m(kr_0)}\right]$$

$$+ 20 \log_{10}\left[\frac{C_m(kr_0)}{C'_m(kr_0)}\right] + 20 \log_{10}(2df).$$

The second and third terms of this equation are charted on Figs. 10.2 and 10.3, respectively.

Inspection of Figs. 10.2 and 10.3 indicates that when kr_0 is small, or when the number of nodes, $2m$, is large, the sound intensity decreases with great rapidity as r increases. This again indicates that only magnetic fields with few poles, which produce vibrations with small numbers of nodes, can be of any importance in noise production. One way of minimizing motor noise, therefore, is so to choose the slot and pole numbers that there are no force waves with few nodes.

4 Noise Calculations

A normal induction-motor design will have a mean core diameter about $1.4D$, a stator-tooth cross section about one-third of the total air-gap area, and a core cross section about 0.35 of the tooth section per pole, giving a peak density in the core about 90 per cent of that in the teeth. The normal value of h, therefore, is equal to:

$$0.33 \times 0.35 \times \frac{\pi D}{2P} = \frac{D}{5P}, \quad \text{very roughly.}$$

Putting these values in Eq. 10.9:

$$\text{(10.18)} \quad d = 0.108\left(\frac{5P}{D}\right)^3 (B)^2 \left(\frac{D}{P}\right)^4 (1.4)^3$$

$$= 37B^2\left(\frac{D}{P}\right), \text{ approximately, times 1.78, 1.26, or 1.14,}$$

$$\text{for } P = 1, 2, \text{ or } 3.$$

For example, taking $B = 3 \times 10^{-4}$ webers per sq in., $D = 8$ in., and $P = 2$, representative of a normal 4-pole, 1,800-rpm, 60-cycle motor of about 15-hp rating:

$$d = (1.26)(37)(9)(8/2)(10^{-8}) = 16.8 \times 10^{-6} \text{ in.}$$

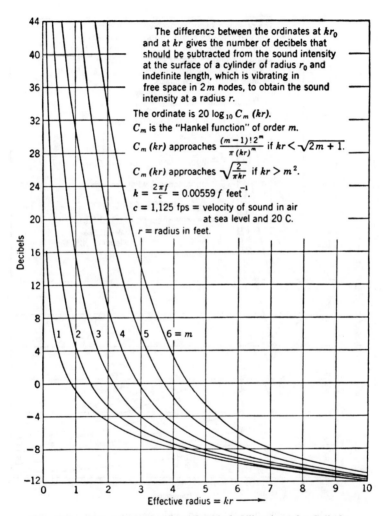

FIG. 10.2. Sound Radiation from $2m$-Node Vibration of a Cylinder.

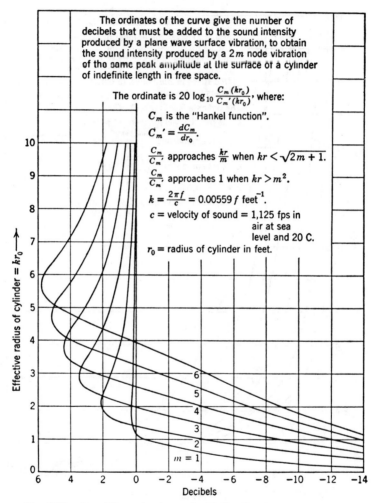

FIG. 10.3. Sound Intensity due to $2m$-Node Vibration of a Cylinder.

NOISE CALCULATIONS 375

This radial displacement of the stator bore, due to the fundamental four-pole flux wave, occurs at a frequency of 120 cycles (8 poles of force revolving at 1,800 rpm). At the stator surface, taking the outer core diameter as 12 in., $kr_0 = 0.00559\,(120)\,(12/24) = 0.33$, and at an 18-in. radius, $kr = 1$. From Eqs. 10.12 and 10.13, since kr is small and $m = 4$:

$$\frac{C_m(kr_0)}{C'_m(kr_0)} = \frac{kr_0}{m} = \frac{0.33}{4} = 0.083,$$

and

$$\frac{C_m(kr)}{C_m(kr_0)} = \left(\frac{r_0}{r}\right) = \left(\frac{1}{3}\right) = 0.0123.$$

The sound intensity at an 18-in. radius, therefore, from Eq. 10.17, is:

$$SI = 121 + 20\log 0.0123 + 20\log 0.083$$
$$+ 20\log(2 \times 120 \times 16.8 \times 10^{-6})$$
$$= 121 - 38.2 - 21.6 - 47.9 = 13 \text{ db, roughly.}$$

This is far below the limit of hearing at 120 cycles. We conclude, therefore, that the 120-cycle noise of a small polyphase motor is inaudible, unless mechanical resonance occurs or there is some dissymmetry.

For a two-pole turbine generator of 50,000-kw rating, however, $P = 1$, D may be 40, and B may be 5×10^{-4}, giving from Eq. 10.14:

$$d = (1.78)(37)(25)(40)(10^{-8}) = 6.60 \times 10^{-4} \text{ in.}$$

To find the 120-cycle sound intensity at a radius of 10 ft, taking the outer stator diameter as 5 ft, we have $kr_0 = 0.00559\,(120)\,(2.5) = 1.67$, and $kr = 6.7$, with $m = 2$. From Eq. 10.17 and Figs. 10.2 and 10.3, the sound intensity is:

$$SI = 121 + (1.2 - 10.0) + 1.7$$
$$+ 20\log(2 \times 6.60 \times 120 \times 10^{-4})$$
$$= 121 - 8.8 + 1.7 - 16.0 = 97.1 \text{ db.}$$

From Fig. 10.1, the corresponding sound level is 98 db. Such a loud 120-cycle noise can be a cause of serious annoyance in large two-pole machines. To overcome this, it is customary to mount the cores of such machines elastically, so that the core vibration will not be transmitted to the frame or foundations. Also, the very large machines are totally enclosed and hydrogen cooled, so that there is no air-borne noise.

Similar calculations indicate that typical three-phase, 60-cycle induction machines will have radial core deflections due to the fundamental field equal to roughly 5, 1.2, 0.7, and 0.5 millionths of an inch per inch of outside diameter. To give a 120-cycle sound level of 60 db at a radius twice that of the core, the outside diameter must be about 24 inches for 2 poles and 68 inches for 4 poles.

5 RESONANT FREQUENCIES

It is useful to check the possibilities of mechanical resonance, which, when present, greatly amplifies the noise. It is well known that the natural frequency of vibration of any structure is given by the formula:

(10.19) $$f = \frac{1}{2\pi}\sqrt{\left(\frac{g}{d_0}\right)} = \frac{3.13}{\sqrt{d_0}} \text{ cycles per second,}$$

where g = acceleration of gravity = 386 ips squared, and d_0 = deflection in inches due to a force equal to the weight of the structure, acting in the direction of the vibratory motion.

The weight of the stator laminations is uniformly distributed, and is equal to ρKhL pounds per inch of core length, for the distance between nodes (one-half the magnetic pole pitch), where ρ is the weight in pounds per cubic inch (0.29 for steel), K is the ratio of gross weight of core and teeth to core alone, and h and L are as previously defined. Hence, substituting the coefficient for a uniformly distributed load, $\frac{5}{384}$, in Eq. 10.3:

(10.20) $$d_0 = \frac{5\rho K h L^4}{384 EI} \text{ in.}$$

Substituting this in Eq. 10.19, the expected natural frequency of radial vibration in $2m$ nodes is:

(10.21) $$f = \frac{1}{2\pi}\sqrt{\left(\frac{384gEI}{5\rho KhL^4}\right)}$$

$$= \frac{1}{2\pi}\sqrt{\left[\frac{384gE(h^2)(2m)^4}{5\rho K(12\pi^4 D_s^4)}\right]}$$

$$= \frac{8\sqrt{2}\,m^2 h}{\pi^3 \sqrt{5}\,D_s^2}\sqrt{\left(\frac{gE}{\rho K}\right)} \text{ approximately, cycles per second.}$$

The exact formula, derived by elasticity theory, for flexural vibration of a ring, with fixed peripheral length, and neglecting shear, is:

(10.22) $$f = \frac{m(m^2-1)h}{\sqrt{3}\,\pi D_s^2 \sqrt{(m^2+1)}}\sqrt{\left(\frac{gE}{\rho K}\right)} \text{ cycles per second.}$$

The approximate formula, Eq. 10.21, gives values 1.32, 1.05, 0.98, and 0.89 times the exact formula, for $m = 2, 3, 4,$ and ∞, respectively.

Substituting $E = 3 \times 10^7$, $g = 386$, $\rho = 0.29$, in Eq. 10.22, we obtain for a steel ring:

(10.23) $$f = \frac{36{,}700 m(m^2-1)h}{D_s^2 \sqrt{[K(m^2+1)]}} \text{ cycles per second.}$$

For the four-pole, 8-in.-gap-diameter motor previously considered, and taking $K = 1.25$, we have $P = 2$, $m = 4$, $h = 0.80$, $D_s = 11.2$, giving:

$$f = \frac{36{,}700(4)(15)(0.80)}{(11.2)^2 \sqrt{[1.25(17)]}} = 3{,}050,$$

and for the two-pole, 40-in.-gap-diameter generator, $P = 1$, $m = 2$, $h = 8$, $D_s = 56$, giving $f = 224$.

It is evident that the frequency of resonance in a given number of nodes for stator cores of given proportions decreases in inverse proportion to the diameter, i.e., fD tends to be a constant (Sect. 2.13).

On the other hand, the number of slots per pole, and hence the impressed slot frequencies, increase with the size of machine. In the larger motors, therefore, resonant frequencies in the smaller numbers of nodes occur well below the slot frequencies at full speed. During acceleration, the stator laminations must pass through one, two, or even more resonances. This accounts for the several cycles of rising and falling sound that are generally heard while a large motor is accelerating to full speed. However, the numbers of nodes of the impressed magnetic force waves also increase with the numbers of poles and slots, so that the stimulus to vibration when going through the few nodes resonances is slight with the usual slot ratios.

The amplitude of vibration in all cases is increased by a factor:

$$\frac{1}{1-(f/f_0)^2}$$

as resonance is approached, where f is the impressed and f_0 the resonant frequency, and damping is neglected. At $f = f_0$, the amplitude is limited by damping forces only, and above f_0 the amplitude decreases rapidly.

6 Slot Frequency Force Waves

The air-gap flux density in a polyphase induction motor consists of 5 principal rotating fields:

(10.24) $B = B_0 \cos(Px - \omega t)$

$\qquad + B_1 \cos[(S-P)x + \omega t] + B_2 \cos[(S-P)x - \omega t]$

$\qquad + B_3 \cos\{(R-P)x + \omega t[1 - R(1-s)/P]\}$

$\qquad + B_4 \cos\{(R+P)x - \omega t[1 + R(1-s)/P]\}.$

The radial force across the air gap, from Eq. 1.32, is proportional to B^2. Squaring Eq. 10.24 and keeping only the terms which have relatively few poles, we have:

(10.25) $$B^2 = \frac{B_0^2}{2} \cos 2(Px - \omega t)$$
$$+ B_2 B_3 \cos\{(R - S - 2P)x + \omega t[2 - R(1 - s)/P]\}$$
$$+ (B_1 B_3 + B_2 B_4) \cos[(R - S)x - \omega t R(1 - s)/P]$$
$$+ B_1 B_4 \cos\{(R - S + 2P)x - \omega t[2 + R(1 - s)/P]\} + \ldots.$$

The B_0^2 term is the fundamental field, with $4P$ poles of force. The other three terms have $2(R-S-2P)$, $2(R-S)$, and $2(R-S+2P)$ force poles, respectively, with frequencies in the stator for 60-cycle machines of $60R(1-s)/P - 120$; $60R(1-s)/P$; and $60R(1-s)/P + 120$ cycles, respectively. If the rotor slots are more than the stator slots, $R > S$, the $(R-S-2P)$ term is the important one; if $R < S$, the $(R-S+2P)$ term is the important one, since these have the fewest poles.

As shown in Sect. 9.2, the second and third terms in Eq. 10.24 actually are the phasor sums of the stator slot permeance and slot mmf harmonics. Since the rotor slot openings normally are very small, the third and fourth terms are the rotor slot mmf harmonics only. As shown in Sect. 9.4, the magnitudes of B_1, B_2, B_3, and B_4 are:

10.26) $$B_1 = B_0 [P(L+k)/(S-P)]$$
$$B_2 = B_0 [P(L+k)/(S+P)]$$
$$B_3 = B_0 PL/(R-P)$$
$$B_4 = B_0 PL/(R+P),$$

where L = ratio of load component of current to no-load (magnetizing) current, and

k = ratio of stator slot permeance harmonic to stator slot mmf harmonic (Eq. 9.23).

L and k must be combined by phasor addition. Usually they are assumed to be 90° apart.

For best overall performance, with especial regard to noise, the number of rotor slots for squirrel-cage motors is most often chosen to be $R = S \pm 4P$, except that for 2-pole machines, normally, $R = S - 8P = S - 8$. Taking $R = S - 4P$ as typical, the most important noise producing force wave has $2(R - S \pm 2P) = 4P$ poles, the same as for the fundamental field, and its

SLOT FREQUENCY FORCE WAVES

frequency at synchronous speed is $\dfrac{60R}{P}+120$ cycles per second. From Eqs. 10.25 and 10.26, the ratio of the force due to this wave to that of the fundamental field is:

(10.27) $\quad W_T/W_0 = \dfrac{2B_1 B_4}{B_0^2}$

$= \dfrac{2P^2 L(L+k)}{(S-P)(R+P)} = \dfrac{2P^2 L(L+k)}{RS}$, approximately.

The value of L usually is about 3 for motors with 2 to 6 poles, and k may be about 5. Putting these numbers in Eq. 10.27, and assuming $S/P = 30$ slots per pair of poles, gives a force ratio of $\dfrac{2(3)(3+5j)}{(29)(27)} = 0.045$. As the frequency for this slot harmonic is 14 times that of the fundamental, the product df, on which the sound intensity depends (Eq. 10.2), is 0.63 as great for the slot harmonic as for the fundamental field. Fig. 10.1 shows that the sound level at 1,800 cycles is 60 db for the same vibration amplitude that gives only 40 db at 120 cycles. Therefore, the slot frequency noise is far more important than that at 120 cycles for usual industrial motors.

Figure 10.4 shows the observed noise spectrum of a four-pole, 60-cycle motor with 60 stator and 62 rotor (squirrel-cage) slots, without skew, when operating at normal voltage, 60 cycles, and full load. The spectrum is drawn, for convenience, in three separate curves to different scales, for the low-, middle-, and high-frequency ranges. The test data were obtained on a non-linear scale, and, therefore, the frequencies shown on the curve in the higher ranges are only approximate. The curve shows the existence of frequencies and magnitudes corresponding to the observed peaks. The connecting parts of the curve between peaks have no significance. If a perfect sound filter were employed, having zero band width, the curve would have zero height at all points between peaks.

In this extreme case, having $R-S=2$ and $P=2$, there are a number of four- and eight-node force waves, and, therefore, there is a very marked frame vibration at each of the corresponding frequencies. The expected force waves, from Eq. 10.25, are:

(a) A $2R-2S-4P =$ four-node wave at a frequency

$$\left(\dfrac{R}{P}-2\right)f = 30 \times 62 - 120 = 1{,}740 \text{ cycles.}$$

This checks the first peak in curve II of the figure.

(b) A $2(R-S) =$ four-node wave at a frequency $Rf/P = 1{,}860$ cycles, which checks the second peak in curve II.

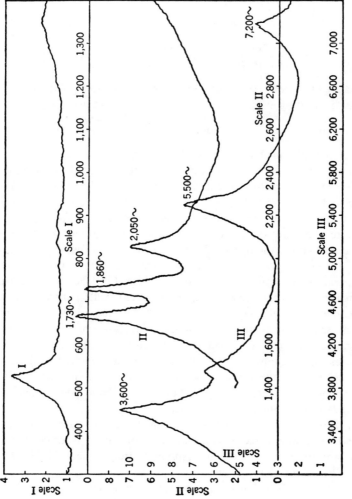

Fig. 10.4. Noise Spectrum of Three-Phase, Four-Pole Motor with 60 Stator and 62 Rotor Slots Operating at Full Load, Normal Voltage, 60 Cycles.

(c) A $2R-2S+4P =$ twelve-node wave at a frequency

$$\left(\frac{R}{P}+2\right)f = 1{,}980 \text{ cycles.}$$

This twelve-node wave should produce too small a deflection to be observed, and this is confirmed by the absence of any 1,980-cycle peak in curve II.

Besides these major waves:

(d) The second harmonic of the permeance wave, interacting with the $2P$-pole mmf, produces a force wave analogous to (b) above, having $4(R-S) = 8$ nodes at a frequency $2RF/P = 3{,}720$ cycles. This roughly checks the first peak in curve III.

(e) The third harmonic of the permeance wave, interacting with the $2P$-pole mmf, produces a force wave analogous to (a) above, having $6(R-S)-4P = 4$ nodes at a frequency $\left(\frac{3R}{P}-2\right)f = 5{,}460$ cycles. This checks the second peak in curve III.

(f) The fourth harmonic of the permeance wave, interacting with the $2P$-pole mmf, produces a force wave analogous to (a) above, having $8(R-S)-4P = 8$ nodes at a frequency $\left(\frac{4R}{P}-2\right)f = 7{,}320$ cycles, which roughly checks the third peak in curve III.

This slot combination produces no two- or six-node force waves, unless there is an offset air gap or other magnetic dissymmetry. The force waves listed are the only four- and eight-node waves produced by the fundamental mmf. The 520-cycle peak in curve I represents a slight two-node vibration of the whole motor, due to air-gap dissymmetry, occurring at the frequency corresponding to the critical speed of the shaft (Sect. 9.7). The third peak in curve II at about 2,050 cycles is probably due to a force wave with $2P-R+S = 2$ pairs of nodes, and a frequency $\frac{Rf}{P}+4f = 2{,}100$, produced by the interaction of the fundamental four-pole field with a backward-revolving triple harmonic field caused by magnetic saturation. Normally, all the saturation harmonics of the air-gap field of a polyphase motor revolve forward in step with the fundamental; but in a Δ-connected motor, as this was, triple-frequency (single-phase) circulating currents are induced in the closed delta, and these create backward as well as forward triple harmonics.

The value of this method of analysis is in the perspective it affords of the various design factors that influence the noise. There are too many approximations in the method for the calculated values to be at all accurate. For example, the effects of all secondary induced currents, except for the fundamental slip-frequency current, have been neglected.

The secondary currents induced by, and opposing, the stator slot harmonic fields themselves produce additional rotor slot harmonics of exactly the same character as those given by Eq. 9.3 and included in Eq. 10.25. The simplest way to allow for them is to increase the value of P_R in Eq. 10.25. This effect explains why the simple use of closed rotor slots, which apparently reduces P_R to a very small value, is of little help as a means of reducing noise under load.

Two obvious ways of reducing the magnetic noise are to skew the rotor, and to make the stator core radially deep. If these are not sufficient, the air gap may be increased, or a different slot combination used. Evidently, the more nearly R and S are equal, the smaller the number of poles in the force wave, the greater the deflection of the yoke, and the louder the noise. This is the reason why it is usual practice to make the rotor and stator slots different by at least 15 per cent. Of course, care is also taken to avoid mechanical resonance of all parts of the stator frame and end shields in the frequency band $\left(\dfrac{R}{P} \pm 2\right)$ times line frequency, as otherwise objectionable noise may be produced.

In an extreme case, such as 4 poles, 24 stator and 25 rotor slots, or 8 poles, 72 stator and 69 rotor slots, the unbalanced radial force (Sect. 9.7) during acceleration may be so great that the shaft will be set in powerful vibration. When, during acceleration, the rotor reaches the speed at which the force frequency coincides with shaft resonance, the amplitude of shaft bending will increase so much that the rotor may even strike the stator, and the magnetic noise may be very loud. An oversized shaft is, therefore, helpful in ensuring quiet operation.

The pulsating torques due to two separate fields of the same number of poles are also a source of noise, as previously mentioned (Sect. 9.6). This type of noise is especially prominent in motors having unbalanced currents, and in single-phase motors. One remedy is an elastic mounting, such as a rubber ring around the bearing housing, providing torsional elasticity with a maximum of radial stiffness.

The purpose of studying motor vibration is to be able to reduce it to an insignificant level, so it is not necessary to calculate the amplitude with precision. And, in view of the many possibilities of resonance, and the difficulties in predicting resonant frequencies accurately, it is best for the practical designer to rely on experimental methods of determining resonant frequencies, after he has gained an understanding of their nature. Then, he has the choice of modifying the structure to bring the resonance away from the frequencies generated by the poles and slots of the motor, or of changing the slot numbers to avoid the resonant frequency.

7 Effects of Dissymmetry

If the air gap is eccentric, the air-gap permeance, Eq. 9.1, will have an extra term of the form $\varepsilon P_0 \sin x$ or $\varepsilon P_0 \sin(x - Nt)$, depending on whether the rotor shaft is off center, or the rotor is eccentric on the shaft. In either case, this eccentricity term, when multiplied by the mmf, gives rise to a series of air-gap fields with $P \pm 1$, $3P \pm 1$, $5P \pm 1$, etc., pairs of poles, and with amplitudes $\varepsilon/2$ times the amplitudes of the fields with symmetrical air gap, where $1 + \varepsilon$ is the ratio of the maximum to the average air gap.

When these extra terms are added to the air-gap flux density, Eq. 10.24, and this is squared, new terms with only 2 poles appear in the air-gap force, of the form $\varepsilon F \cos(mt + \theta)$ and $\varepsilon F \cos(mNt + \theta)$. There are many of these unbalanced force terms, with frequencies spaced only 30 or 60 cycles apart in the case of a single-phase, 60-cycle motor.

Ordinarily, the forces are small, but they may produce a great deal of noise whenever the natural frequency of the shaft coincides with that of one of the unbalanced force terms. Motors in industrial sizes, of 1-hp and larger, have shaft critical speeds well below 10,000 rpm, or 200 cycles, so that the audible noise due to small dissymmetries in manufacture usually is negligible, even though there is appreciable vibration, as indicated by the first peak in Fig. 10.4 (unless the difference in slots, $R - S$, is 1, 3, or 5, as discussed in Sect. 9.6).

However, fractional-horsepower motors in the $\frac{1}{8}$- to $\frac{1}{3}$-hp range usually have a shaft critical speed in the range of 20,000 rpm or more, or 400 to 800 cycles, where only a small vibration amplitude is required to produce a good deal of noise (Fig. 10.1). Also, motors in these sizes are normally single phase, which usually have appreciable third and fifth harmonic frequency components in their line currents, due to magnetic saturation, that are absent in polyphase motors. These additional mmfs combined with those due to dissymmetry often cause appreciable radial vibration at the critical frequency of the shaft. These effects, as well as the double-frequency torque (Sect. 12.7), make small single-phase motors inherently noisy. Skewed slots, stiff shafts, and great care in manufacture to minimize dissymmetry are required to obtain the low sound levels required for modern household appliances.

8 Induced Voltage Ripples

When an alternating voltage is applied to a transformer with open-circuited secondary, the magnetizing current creates an alternating flux whose waveform corresponds exactly to the impressed voltage, except for the small difference required to overcome the resistance and leakage-reactance drop of the magnetizing current:

(10.28) $$E = -N\frac{d\phi}{dt} = V - I_M(R_1 + jX_1).$$

If the magnetic circuit is highly saturated, the magnetizing current will have a peaked waveform, and its impedance drop will accordingly include 3rd, 5th, 7th, etc., harmonics, besides the fundamental voltage component. By connecting three single-phase transformers in Y or Δ, the triple frequency currents can be eliminated from the power lines, but with a Y connection, there will be a third harmonic voltage between the terminals and ground. This voltage is also eliminated by the Δ connection, allowing the triple frequency currents, which are required to maintain a sinusoidal flux waveform, to flow in the delta.

These rules apply also to an induction motor. However, the presence of the air gap makes the motor current-flux relation much more nearly linear than in a closed magnetic circuit transformer, so that the saturation-produced harmonic currents are less important than in either transformers or in salient pole synchronous machines.

However, the third harmonic voltages can cause two kinds of trouble. If the stator winding is Δ-connected, and the motor is operated at over-voltage, with high magnetic densities, the third harmonic current flowing in the Δ causes extra no-load losses and heating. If only a part of the winding is Δ, and the rest Y, the heating of the Δ winding may burn it out, even at no load. If the winding is Y-connected, and if the Y point is connected to ground, triple frequency currents will flow in the motor lines and a ground return path, and these may create objectionable noise in telephone circuits that are inductively linked with the power circuits.

If the induction motor is so designed that no harmonic frequency currents can flow in the rotor, the rotor slot openings are negligibly small, and the rotor mmf wave is sinusoidal, there is no difference with respect to voltage waveform between the motor and a transformer. There may be large stator slot harmonic fluxes in the air gap, but these produce only line-frequency voltages in the stator.

At the other extreme is a motor with open stator slots, and a squirrel-cage rotor with a relatively large number of slots, without skew. In this case, the stator harmonic fluxes induce large high-frequency secondary currents, whose "armature reactions" oppose the stator mmf, reducing the stator harmonic fluxes to a small value. Since the rotor slots per stator harmonic pole are few in number, the space waveform of the harmonic armature reaction will be very ragged. Thus, large low-order harmonics of the slot harmonic fields will be created (the zigzag-leakage flux of the slot harmonic fields, Fig. 9.5), and these in turn will induce high-frequency voltages in the stator winding, thus giving rise to voltage ripples at the primary terminals. And these voltages will produce currents in the supply lines, which will induce high-frequency voltages in any adjacent circuits, giving "telephone interference".

For example, consider a 4-pole, 60-cycle, 3-phase motor with 36 stator and 51 squirrel-cage slots, running at full speed, no load. The largest stator slot harmonic will be the 17th, $\left(\dfrac{S}{P}-1\right)$, having 68 poles, revolving backward at $\frac{1}{17}$ of synchronous speed, and cutting the rotor at $\frac{18}{17}$ of full speed (Eq. 9.5, D term). The no-load induced rotor frequency will, therefore, be $18 \times 60 = 1{,}080$ cycles. The forward-revolving 19th harmonic will cut the rotor at $\frac{18}{19}$ of full speed, and will also induce rotor voltages of 1,080-cycle frequency. The 5th and 7th phase belt harmonics will similarly induce 360-cycle rotor voltages.

By analogy with Eq. 9.5, the three principal fields produced by the induced 68-pole currents flowing in the 51 rotor bars are given by making $P = 34$, $R = 51$ in the D, F, and G terms. The impressed 68-pole, or 17th-harmonic, field of the stator, and the opposing armature reaction field produced by the induced rotor currents, at a slip s, are both expressed by:

$D\cos(Px+\omega t) = D\cos(34x+\omega t)$ referred to the stator or

$$D\cos\{34x+[1+17(1-s)]\,\omega t\}$$

$$= D\cos[34x+(18-17s)\omega t]\text{ referred to the rotor.}$$

The induced currents of this $(18-17s)$ 60-cycle, 68-pole field, revolving backward on the rotor, will produce a reversed rotation $(R-P)$ rotor slot harmonic, which will, therefore, rotate forward on the rotor, but will have the same rotor frequency. It will be:

$$F\cos[(R-P)x-(18-17s)\omega t] = F\cos[17x-(18-17s)\omega t]$$

referred to the rotor.

This is a 34-pole field revolving forward on the rotor at a speed $\dfrac{2(18-17s)}{17}$ times synchronous rotor speed.

To refer this to the stator, we must add to $(18-17s)\omega t$ the value corresponding to the additional frequency induced by the speed of the rotor, $\dfrac{(1-s)\omega t}{2}$, cutting this 17-pairs-of-poles field, or $\dfrac{17(1-s)\omega t}{2}$, giving:

(10.29) $$F\cos\left[17x-\left(\dfrac{53-51s}{2}\right)\omega t\right].$$

This field, therefore, induces a stator frequency of $\left(\dfrac{53-51s}{2}\right)$ 60 cycles, varying from 60 cycles with the motor at rest to 1,590 cycles at full speed, no load.

Similarly, the forward-revolving 19th stator slot harmonic is:

$E\cos(38x-\omega t)$ referred to the stator,

or $\qquad E\cos[38x+(18-19s)\omega t]$ referred to the rotor.

The rotor currents induced by the harmonic produce a principal slot harmonic of order $R-P$, and reversed rotation. Since, here, $R = 51$ and $P = 38$, the harmonic is:

(10.30) $\qquad G \cos[13x - (18 - 19s)\omega t]$ referred to the rotor,

or $\qquad G \cos\left[13x - \left(\dfrac{49 - 51s}{2}\right)\omega t\right]$ referred to the stator.

The corresponding stator frequency at no load is 1,470 cycles.

Equations 10.28 and 10.29 agree with Eq. 9.5, showing that the harmonic fields produced by the induced rotor currents due to stator fields are of exactly the same character as those produced by the stator mmf acting on the rotor permeance variations. All the rotor induced currents due to the fundamental and the phase belts, as well as the slot harmonics, give rise to the same stator frequencies as given by Eqs. 10.29 and 10.30.

Thus, rotor slot permeance variations, and rotor slot current concentrations combine to produce ripple voltages in the stator, whose frequencies are:

(10.31) $\qquad f_r = \left(\dfrac{R}{P} \pm 1\right)f.$

The magnitudes of the various components of the ripple voltages appearing across the stator terminals, are proportional to the stator winding pitch and distribution factors for fields of their respective numbers of poles. The induced rotor currents to which they are chiefly due can be calculated by the equivalent circuit of Fig. 3.15, as outlined in Sect. 9.3. They can be reduced by skewing the slots, as well as by proper selection of the stator winding arrangements and the slot ratio. Generally,' with a squirrel-cage rotor, the stator winding admittance to the rotor slot harmonics is small. If the power supply circuit has sufficiently low impedance, the ripple voltages will create high-frequency line currents that will in turn damp out the voltage ripples, but will themselves cause induction interference voltages in nearby communication circuits.

A much more important source of voltage ripples, which occurs in wound-rotor (or salient pole) machines only, is the presence of high-frequency induced rotor currents of the fundamental number of poles, which reflect the stator slot frequency back into the stator. In a wound-rotor motor, the rotor slots per pole are usually integral, and differ from the stator slots per pole by a multiple of the number of phases. Also, the rotor winding is usually full pitch. Therefore, the rotor winding factor, $K_p K_d$, is quite high for the stator slot harmonics, and fairly large currents induced by them flow in the rotor. These currents necessarily produce a field of the fundamental number of poles, as well as opposing harmonic fields. And these high-frequency rotor fields with the fundamental number of poles generate voltages in the stator winding of stator slot frequency. These effects correspond to fractional values of k in the rotor subharmonic field circuits of Fig. 9.6.

For example, a three-phase, six-pole, wound-rotor motor, with 54 stator and 72 rotor slots, has a backward revolving $S-P$ field, Eq. 9.5:

(10.22) $\qquad D\cos(51x+\omega t),$

which at synchronous speed of the rotor, cuts the rotor at $\frac{18}{17}$ times full speed and induces rotor currents:

(10.33) $\qquad D\cos[51x+(18-17s)\omega t].$

But these currents, flowing through the complete rotor winding, also produce a reverse field with the fundamental number of poles, but the same rotor frequency.

(10.34) $\qquad D'\cos[3x-(18-17s)\omega t].$

To refer this forward-revolving rotor field to the stator, we must add $(1-s)$ to the coefficient of ωt, representing the frequency due to the rotor speed and the fundamental poles, obtaining:

(10.35) $\qquad D'\cos[3x-(19-18s)\omega t],$

which is analogous to Eq. 10.24. This (Eq. 10.34) field produces a stator voltage of frequency $(19-18s)f$. Similarly, the forward-revolving 19th stator slot harmonic gives rise to a fundamental field of rotor frequency $(18-19s)f$, and stator frequency $(17-18s)f$. Thus the $\frac{S}{P}-1$ stator slot harmonic produces a reflected stator voltage ripple of frequency $\left(\frac{S}{P}+1\right)f$, and the $\frac{S}{P}+1$ harmonic gives rise to a $\left(\frac{S}{P}-1\right)f$ frequency ripple. These ripples can be eliminated by skewing, or by use of a fractional number of slots per pole on either stator or rotor. Or they may be reduced to any desired value by connecting a tuned capacitance filter across the terminals, providing a short-circuit path for the harmonic currents.

By measuring various quantities such as the voltage pulsation in one stator tooth, and that across the winding terminals, the noise spectra at different speeds and loads, the running-light slip, the core loss, the load loss, and the harmonic components of torque, it is possible to piece together a fairly complete picture of what is going on in an induction machine. With experience, measurement of just a few of these quantities gives a pretty good idea of what the others are.

9 Effects of Harmonics in the Impressed Voltage

The voltage impressed on a motor often includes harmonics with frequencies, 2, 3, 4, ... times the fundamental frequency, as when the voltage source is a static frequency changer, or inverter. Aside from heating, magnetic saturation, and third-order effects, the nth harmonic voltage produces only an nth harmonic current in the motor, so that each harmonic can be considered independently in terms of its own equivalent circuit.

Considering a three-phase motor, for example, with symmetrical applied voltages that include an nth harmonic component, the equivalent circuit for the harmonic will have the same resistances (neglecting skin effect) and n times the reactances of the fundamental circuit, Fig. 5.1. With the motor running at a speed $(1-s)N_s$, corresponding to a fundamental slip s, the slip, s_n, for the nth harmonic will be $(n-1+s)/n$ if the harmonic rotates forward, or $(n+1-s)/n$ if it rotates backward. Since n is 2 or more, and the circuit impedance is nearly the same for all values of s_n greater than 0.5, it usually is sufficient to assume $s_n = 1$ for all the impressed harmonics, and for all motor speeds. With $s_n = 1$, the current in the magnetizing branch of the harmonic circuit can be neglected, so that the circuit impedance for t e nth harmonic becomes simply:

(10.36) $\qquad Z_n = R_1 + R_2 + jn(X_1 + X_2)$ ohms per phase.

Since R is small in comparison with nX, the nth harmonic current due to the nth harmonic voltage, E_n, will be:

10.37) $\qquad I_n = E_n/n(X_1 + X_2)$ amperes.

The torque of the nth harmonic, expressed in synchronous watts at the fundamental synchronous speed, will be:

(10.38) $\qquad T_n = mI_n^2 K_n R_2/n$ watts,

where m is the number of phases and K_n is the skin effect factor for the secondary resistance. If values are given in per unit terms, the factor m should be omitted.

The phase displacement between the fundamental voltage of phases A and B is 120°, so that between the nth harmonic voltages is $120n°$. If n is 2, 5, 8, 11, 14, ..., the phase rotation of the harmonic is opposite to the fundamental, and the harmonic torques must be subtracted from the fundamental torque. If n is 4, 7, 10, 13, ..., the harmonic torque adds to the fundamental. If n is a multiple of 3, and the neutral is open, there are no harmonic currents; while if the neutral is closed, the harmonic currents will be single phase, producing both forward and backward torques.

The watts loss in the motor due to the nth harmonic will be:

(10.39) $$W_n = mI_n^2(R_1 + K_n R_2) \text{ watts.}$$

In the case of a normal industrial motor, the locked rotor current is about 6 per unit, $R_1 = 0.03$, approximately, and $K_n R_2$ may be taken as 0.08. If the fifth harmonic voltage is 0.20 per unit, the harmonic current will be about 0.24, the torque will be about 0.001, and the watts loss will be about 0.007, all in per unit terms. Thus, if none of the harmonics in the impressed voltage exceed 20 per cent of the fundamental, their effects on total current and torque are hardly appreciable. The heating due to the extra I^2R losses may be noticeable for a 20 per cent harmonic, and will become important if the voltage harmonics are much larger.

When silicon controlled rectifiers with retarded firing angles are used to vary the effective motor voltage (Sect. 8.14), the harmonics in the voltage wave become large when the firing angle is delayed more than 90°. At 120° delay, for example, with an extinction angle of 220°, the third harmonic is 1.30 times the fundamental component of the voltage.

None of the triple harmonics in the voltage can produce any currents in a balanced three-phase motor without neutral connection, so that it is highly desirable not to connect the neutral, whenever the harmonic content of the voltage is large. To analyze a specific case, the currents due to assumed harmonic voltages are calculated, the voltages are then adjusted to allow for the reactance drops of the currents, and the currents are recalculated. The total rms current is the square root of the sum of the squares of the I_n values and I_1^2.

BIBLIOGRAPHY

		Reference to Section
1.	*American Standard Test Code for Apparatus Noise Measurement*, A.S.A. Z24.7, 1950.	3, 4
2.	"Test Procedures for Noise Measurements on Rotating Electric Machinery," *A.I.E.E. Trans.*, Vol. 77, Part III, 1958, pp. 1615–28.	3, 4
3.	"Proposed Test Procedure for Noise Measurements on Rotating Electric Machinery," *A.I.E.E. No. 85*, Nov. 1960.	3, 7
4.	"The Magnetic Noise of Polyphase Induction Motors," P. L. Alger, *A.I.E.E. Trans.*, Vol. 73, Part IIIB, 1954, pp. 118–24.	2
5.	"Calculation of the Magnetic Noise of Polyphase Induction Motors," P. L. Alger and Edward Erdelyi, *Acoustical Society of Am. Jour.*, Vol. 28, Nov. 1956, pp. 1063–7.	4
6.	"The Cause and Elimination of Noise in Small Motors," W. R. Appleman, *A.I.E.E. Trans.*, Vol. 56, 1937, pp. 1359–67.	1
7.	"Magnetic Vibration in A-C Generator Stators," R. A. Baudry, P. R. Heller, and L. P. Curtis, *A.I.E.E. Trans.*, Vol. 73, Part IIIA, 1954, pp. 508–515.	2

	Reference to Section
8. "The Analysis and Measurement of the Noise Emitted by Machinery," B. A. G. Churcher and A. J. King, *The Jour. of the I.E.E.*, Vol. 68, 1930, pp. 97–125.	3, 4
9. "Predetermination of Sound Pressure Levels of Magnetic Noise of Polyphase Induction Motors," E. Erdelyi, *A.I.E.E. Trans.*, Vol. 74, Part III, 1955, pp. 1269–80.	2
10. "Einseitige Magnetische Zugkräfte in Drehstrommaschnen," W. Frese and H. Jordan, *ETZ-A*, 1962, p. 299.	1
11. "Noise Created by Electrical Machinery," H. Fritze, *Arch. Für. Elek.*, Vol. 10, 1921, pp. 73–95.	1
12. "Magnetic Noise in Synchronous Machines," Q. Graham, S. Beckwith, and F. H. Milliken, *A.I.E.E. Trans.*, Vol. 50, Sept. 1931, pp. 1056–62.	1
13. "Quiet Induction Motors," L. E. Hildebrand, *A.I.E.E. Trans.*, Vol. 49, July 1930, pp. 848–52.	1
14. "Entstehung und Bekämpfung der Gerausche Electrischer Maschinen," G. Hubner, *ETZ-A*, 1961, p. 771.	
15. "The Effect of Voltage Wave Shape on the Performance of a Three Phase Induction Motor," G. C. Jain, *I.E.E.E. Trans.*, Vol. 83, No. 6, June 1964, pp. 561-566.	
16. "Construction of Low-Noise Electric Motors." H. Jordan, *Elektro-Anzeiger*, No. 10/11, March 18, 1950, pp. 11-13.	1
17. "Akustische Wirkung des Schrägung bei Dreistromasynchronmaschinen mit Käfiganker," H. Jordan, and H. Müller-Tornfelde, *ETZ-A*, 1961, p. 788.	
18. *Vibration Prevention in Engineering*, A. L. Kimball, John Wiley, 1932.	1
19. *Scattering and Radiation from Circular Cylinders and Spheres, Tables of Amplitudes and Phase Angles*, A. N. Lowan, P. M. Morse, H. Feshbach, and M. Lax, Published by U.S. Navy Dept., Office of Research and Inventions, July 1946.	3, 4
20. "Harmonic Theory of Noise in Induction Motors," W. J. Morrill, *A.I.E.E. Trans.*, Vol. 59, 1940, pp. 474-80.	1
21. *Vibration and Sound*, Second Edition, P. M. Morse, McGraw-Hill, 1948.	3, 4
22. "Single-Phase Induction Motor Noise due to Dissymmetry Harmonics," D. F. Muster and G. L. Wolfert, *A.I.E.E. Trans.*, Vol. 74, Part III, 1955, pp. 1365-72.	7
23. "Storungen in Netzkommandanlagen durch Oberstrome von Asynchronmotoren, K. Oberretl, *Bulletin des Schweizerischen Vereins Bd.* 56 (1965) Nr. 12, S.453-463.	
24. "Central Forces in Asynchronous Motors," J. Pestarini, *Revue Générale de l'Eléctricité*, Vol. 24, 1928, pp. 517-23.	1
25. "Line-Frequency Magnetic Vibration of A-C Machines," R. C. Robinson, *I.E.E.E. Trans.*, Power Apparatus, 1963, p. 675.	
26. "Vibration in 2-Pole Induction Motors Related to Slip Frequency," E. W. Summers, *A.I.E.E. Trans.*, Vol. 74, Part III, 1954, pp. 69-72.	7
27. "Noise Rating of Large Electric Rotating Machines," M. E. Talaat, *A.I.E.E. Trans.*, Vol. 77, Part IIIB, 1954, pp. 508-15.	1
28. "Calculation of Windage Noise Power Level in Large Induction Machines," M. E. Talaat, *A.I.E.E. Trans.*, Vol. 76, Part III, 1957, pp. 44-53.	4
29. "Sonance Design of Large Induction Motors," R. L. Wall, *A.I.E.E. Trans.*, Vol. 74, Part III, 1955, pp. 1189-92.	1

CONTENTS—CHAPTER 11

Kron's Generalized Machine

		PAGE
1	Vistas	393
2	The Generalized Machine	394
3	Direct- and Quadrature-Axis Circuits	395
4	Effect of Rotation	400
5	Torque Equations	401
6	Limitations of the General Equivalent Circuit	402
7	Transformations of Reference Axes	403
8	Symmetrical Components	404
9	Three-Phase to Two-Phase Transformations	405
10	Matrix Transformations	407
11	Impedance Matrix of the Generalized Machine	411
12	Equivalent Circuits	416
	Appendix—Gabriel Kron	421

11

KRON'S GENERALIZED MACHINE

1 Vistas

In the preceding chapters we have visualized, described, and calculated the performance of polyphase induction machines. Especially, there have been developed formulas for calculating all the constants of the equivalent circuit, whereby abstract theories can be translated into specific conclusions.

In this chapter, we shall endeavor to open a window—to overlook the panorama of the entire range of electric machines. For, truly, all rotating electric machines are members of a single family, from the tiny clock motor to the Scherbius steel-mill set, and from the universal motor for hand tools to the 500,000-kw turbine generator.

All these machines behave in accord with immutable laws of physics. In them, electrical, magnetic, and mechanical energies coexist and interact. The reference frames from which these actions are viewed often have accelerated rotation with respect to both their electromagnetic fields and their material structures.

In the early days, theories of performance were built up step by step for each type of machine, on the basis of elementary physical laws and empirical factors. About three decades ago, Gabriel Kron constructed a general theory that encompasses the entire range of machines. To do so, he employed the methods of tensor analysis, the concepts of Einstein's unified field theory, and the geometrical theories of non-Euclidean spaces with torsion, as developed by Weyl and other mathematicians.

However, for every-day engineering, Kron's theories may be considered simply a means of establishing the equivalent circuit for whatever machine or system is of present interest. The general circuits that express his results have been published, and only a little study, with no knowledge of tensors or geometry whatever, is required to select and modify the right circuit to fit any particular case.

Actually to use the circuit, however, it is necessary to determine the numerical values of the circuit constants, that is, to calculate all the resistances and reactances of the machine. This phase of the matter is the subject of Chapters 6 and 7 of this book.

The present chapter, then, enters a little way into the general method of establishing equivalent circuits for other types of machine, whereby the methods of the earlier chapters can be applied to a much wider field of use than polyphase induction machines. In this way, a bridge is provided between the general theories of Kron and the particular theories of this book—a vista of wide scope is opened.

2 The Generalized Machine

To reduce time and effort in deriving the performance equations and circuits of the many different types of rotating machines, it is desirable first to derive the equations of a "primitive" or "generalized" machine that contains the basic elements common to all types. Then, we can establish

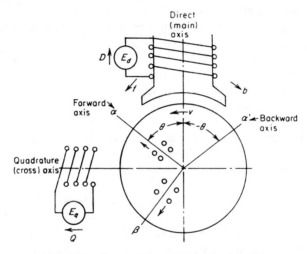

Fig. 11.1. D and Q Axes of Rotating Machine.

methods for reducing and transforming these general equations to fit the various types of windings and connections, so that the equations for any particular machine can be quickly established.

For these purposes, we shall base the analysis on the primitive a-c machine of Kron, Fig. 11.1. The magnetic structure and windings of the stator are assumed to be different along the polar, or "direct" axis and the interpolar, or "quadrature" axis, which are 90° apart in space. The machine is assumed to be symmetrical about each of these axes. For simplicity, a 2-pole machine is shown, but the analysis applies equally well to any number, P, of pole pairs, if the angle θ is measured in electrical degrees, equal to P times the mechanical degrees. It is assumed also that all the magnetic flux densities are proportional to the mmfs, so that both can be superposed without error, i.e., saturation is neglected. In practice, saturation is allowed for by using different values of the flux/mmf ratios, or reactances, to suit the degree of saturation existing under particular conditions.

The stator windings are connected to stationary terminals, and the rotor windings are connected to slip rings or commutator bars in contact with stationary sliding brushes. Although three-phase windings are used almost universally, the analysis is carried out for a two-phase machine, as this permits referring the currents to the D and Q axes directly, thus giving equations much simpler than those for three-phase machines. After the two-phase performance has been determined, it is a simple matter to transform the results to three phase, as will be shown.

In the actual machine, the currents are localized in slots or interpolar spaces, and the corresponding air-gap mmf distributions can be resolved into fundamental, $2P$-pole, sine waves and an infinite series of harmonics with multiples of $2P$ poles, as shown in Sect. 3.7. For the purposes of this analysis, only the fundamental waves are considered. When necessary, the space harmonics of mmf can be taken into account by adding separate equations and circuit branches, as in Fig. 9.6.

At standstill, a magnetic flux in the direct axis does not link the quadrature axis windings, and vice versa, so that the equations for the two axes are independent. However, when a rotor coil is displaced by an angle θ from the direct axis, it is linked by some of the quadrature axis flux, and the direction of this flux through the coil reverses as θ passes through 0. Hence, when rotation occurs, the flux of each axis creates "speed voltages" in the windings of the other axis. These speed voltages are in time phase with the fluxes that produce them, and are proportional to the speed, while the "transformer voltages", or reactance drops, that are produced by the fluxes in the windings of their own axes are 90° out of phase with them, and are proportional to the frequency. At synchronous speed, the proportionality factors become equal, so that for a given flux and winding the speed and transformer voltages are also equal.

To deal with all these relations, first, we shall derive equivalent circuits to represent the complete machine, following a somewhat intuitive procedure due to Kron. Later, we shall derive the complete equations by formal use of matrices and transformation procedures, and thence derive the same circuits as before.

3 Direct- and Quadrature-Axis Circuits

We shall assume four windings in each axis, two each in the stator and the rotor, whose parameters are designated by the subscripts $1d, 2d, 3d, 4d$, and $1q, 2q, 3q, 4q$. At standstill, the equivalent circuit for one axis, Fig. 11.2, is found by adding branches for the two additional windings to the transformer circuit of Fig. 2.4. Each winding has a self-reactance, X_c, chiefly in the coil ends, and a slot reactance, X_s. Part of X_s is due to slot leakage flux that links only the winding that creates it, but another part links both stator (or rotor) windings. Thus, in the figure, X_{1s} represents flux that links winding 1 only, while X_{2s} represents flux that links both windings 1 and 2. With the circuit as shown, windings 1 and 4 are in the bottoms of the slots, and so have larger total reactances than windings 2 and 3, which are located in the upper parts of the slots, next to the air gap. The magnetizing reactance, X_m, due to the air-gap flux, links all the

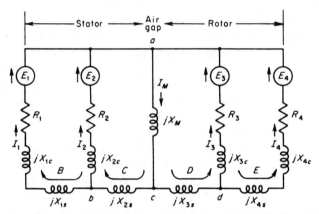

FIG. 11.2. Equivalent Electric Circuit for One Axis of Four-Winding Machine.

windings in one axis. Independent voltages, E_1, E_2, E_3, E_4 are assumed to be impressed on the four windings.

It is helpful, next, to place the D- and Q-axis circuits one above the other, with a common impedanceless branch between, as shown in Fig. 11.3. This is an adequate representation of the entire machine, so long as it does not rotate. To deal with the effects of rotation, it is desirable to resolve the stationary, alternating, mmfs of the two axes into pairs of equal, constant magnitude, and oppositely revolving sine wave mmfs, as we did in Sect. 3.10. And, it is desirable to refer all the electrical quantities to two sets of synchronously rotating axes, one forward and the other backward, which we shall designate as the f and b axes.

Following the procedure of Sect. 3.10; if I_d is the rms value of the D-axis stator current, and taking $\theta = 0$ at the direct axis:

(11.1) $\quad \sqrt{2} I_d \cos P\theta \cos \omega t = (I_d/\sqrt{2})[\cos(P\theta - \omega t) + \cos(P\theta + \omega t)]$

or $\qquad I_d = (I_f + I_b)/\sqrt{2},$

where I_f and I_b are the *constant* (in time) forward and backward field currents. The denominator of Eq. 11.1 is $\sqrt{2}$ rather than 2, because we

DIRECT- AND QUADRATURE-AXIS CIRCUITS

need to preserve the same energy values when we change the reference axes. The rms value of I_d is $1/\sqrt{2}$ times its maximum, because it alternates in time; but the rms values of I_f and I_b are the same as their peaks, since they are unchanging in time. The reference axes for both I_f and I_b are taken

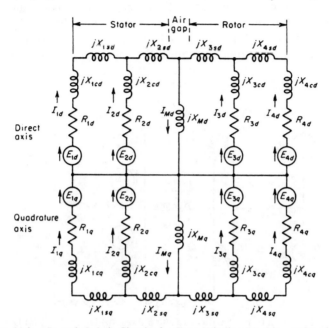

FIG. 11.3. *D-* and *Q*-Axis Circuits for Four-Winding Machine at Standstill.

along the stator *D*-axis at the instant I_d is a maximum. Fig. 11.1 shows the relations between the different axes, and the sign conventions adopted.

Similarly, in the *Q* axis:

$$\sqrt{2}I_q \sin P\theta \omega t = -(I_q/\sqrt{2})[\cos(P\theta - \omega t) - \cos(P\theta + \omega t)]$$

or

(11.2) $$I_q = -j(I_f - I_b)/\sqrt{2}.$$

The $-j$ appears in Eq. 11.2 since we have taken the *D* axis as the reference for I_f and I_b, and the *Q* axis is displaced 90° ahead of this. From Eqs. 11.1 and 11.2, we find the rotating field currents and (by analogy) the voltages, to be:

(11.3) $$I_f = (I_d + jI_q)/\sqrt{2}, \qquad I_b = (I_d - jI_q)/\sqrt{2}$$

(11.4) $$E_f = (E_d + jE_q)/\sqrt{2}, \qquad E_b = (E_d - jE_q)/\sqrt{2}.$$

These last equations show that the f and b quantities are the sums and differences of the d and j times the q quantities, indicating that we can obtain the desired f and b circuits by superposing the D and Q circuits. Fig. 11.4 indicates the procedure. To form the first terms of Eq. 11.4, we draw two D-axis circuits one above the other with a common impedance-less branch at (a), and impress on each of these the voltage $E_d/\sqrt{2}$ in the upward direction. To form the second terms, we draw two Q-axis circuits in the same way at (b), but we impress the voltage $jE_q/\sqrt{2}$ upward on the

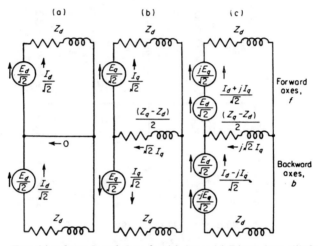

FIG. 11.4. Transition from D and Q to f and b Axes: (a) Direct Axes, (b) Quadrature Axes, (c) Superposed Axes.

upper circuit and downward on the lower circuit, in accordance with the signs of the second terms in Eq. 11.4. Before these two pairs of circuits can be combined, we need to make the impedances (Z_d) the same in the upper and lower branches, which carry D- and Q-axis current; and transfer the differences between Z_q and Z_d to the common return branch, which carries only Q-axis currents. This brings $(Z_q - Z_d)/2$ into the common branch of the Q-axis circuits, since it carries twice as much current as the other branches. Superposing the two pairs of circuits gives the combined circuit (c) in Fig. 11.4, the currents and voltages of which agree with Eqs. 11.3 and 11.4. The derivation is carried through in equation form in this way:

(11.5) $$E_d = (E_f + E_b)/\sqrt{2} = I_d Z_d = Z_d(I_f + I_b)/\sqrt{2}$$

and

(11.6) $$E_q = -j(E_f - E_b)/\sqrt{2} = I_q Z_q = -jZ_q(I_f - I_b)/\sqrt{2},$$

DIRECT AND QUADRATURE-AXIS CIRCUITS

so that

(11.7) $\quad E_f + E_b = Z_d(I_f + I_b)$ and $E_f - E_b = Z_q(I_f - I_b)$

and

(11.8) $\quad\quad E_f = (Z_d + Z_q)I_f/2 + (Z_d - Z_q)I_b/2$

(11.9) $\quad\quad E_b = (Z_d - Z_q)I_f/2 + (Z_d + Z_q)I_b/2.$

The terms in the last two equations can be regrouped to separate the self- and mutual impedances, when they are seen to agree exactly with the circuit equations of Fig. 11.4 (c):

(11.10) $\quad\quad E_f = Z_d I_f + [(Z_q - Z_d)/2](I_f - I_b)$

(11.11) $\quad\quad E_b = Z_d I_b - [(Z_q - Z_d)/2](I_f - I_b).$

When Fig. 11.4 (c) is expanded by replacing Z_d and Z_q by their components and showing all four windings instead of only two, we obtain Fig. 11.5, which is the correct rotating-field equivalent circuit for the generalized four-winding machine at standstill. By symmetry, the

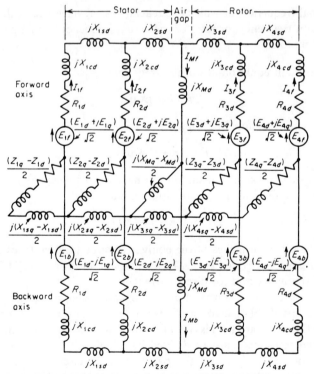

FIG. 11.5. Rotating-Field Circuit for a Four-Winding Machine at Standstill.

differences between the separate winding impedances in each vertical branch are placed in a branch leading off from its midpoint. Also, the differences in the horizontal branch impedances are placed in the common branches into which these are connected. The circuit of Fig. 11.5 is in all respects equivalent to the D- and Q-axis circuit of Fig. 11.3, but it has the advantage that it can be modified readily to give correct performance at all speeds.

If we had taken the Q-axis as the reference instead of the D-axis, Z_q instead of Z_d would have appeared in all the horizontal branches, and the signs of E_b and I_b would have been reversed.

4 Effect of Rotation

The speed of rotation of the forward and backward fields with respect to the stator is $\pm 60f/P$, so that the "speed" voltages induced in the stator (as well as the transformer voltages) are proportional to f. The resistances are independent of frequency and the reactances are proportional to it, neglecting skin effect and hysteresis loss. Let:

F_0 = normal frequency, at which the reactances, X, have been calculated,
F = per unit frequency, or ratio of the actual frequency, f, in a particular winding to F_0.

Then, any circuit equation at normal frequency, F_0,

(11.12) $$E = (R+jX)I$$

becomes, at the per-unit frequency F,

(11.13) $$E = (R+jFX)I.$$

Since we wish to make all calculations at normal frequency, all reactances must appear as jX, and the circuit equations become:

(11.14) $$E/F = [(R/F)+jX]I.$$

Thus, in the stator branches of the equivalent circuit, all voltages and all resistances must be divided by F. This will make the circuit valid for any impressed frequency and for direct current.

When the rotor turns forward at its per-unit speed, v (actual speed is $60fv/P$), its speed relative to the forward field is $(F-v)$ *backward*, and relative to the backward field is $(F+v)$ *forward*. Hence, the rotor impressed voltages and resistance must be divided by $(F-v)$ in all the forward-field circuits, and by $(F+v)$ in all the backward-field circuits. When these changes are made, the circuit of Fig. 11.5 becomes that of Fig. 11.6, which

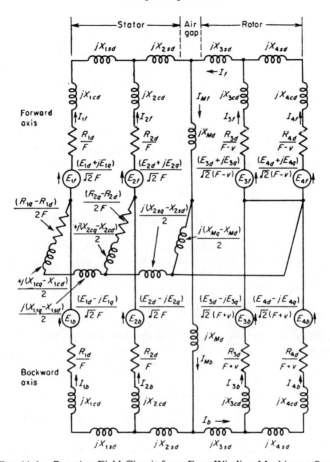

FIG. 11.6. Rotating-Field Circuit for a Four-Winding Machine at Speed.

is the object of our search. This is the equivalent circuit of the generalized a-c electric machine, from which the circuits of all usual types of machines can be derived by appropriate omissions and reconnections.

5 Torque Equations

The total unidirectional torque of the machine is given by the sum of the phasor products for the backward and forward fields of flux-induced voltages by total current:

(11.15) $T = $ real part of $I_{nf}^* E_{nf} + I_{nb}^* E_{nb}$ synchronous watts,†

† Note: I^* is the conjugate of the phasor I. If $I = c + jd$, $I^* = c - jd$.

where I_{nf} and I_{nb} are the currents in the several windings, and E_{nf} and E_{nb} are the rotating-field voltages measured on the equivalent circuit across the secondary "terminals" inside the resistances, as indicated in Fig. 11.6. (The backward-field watts come out with a negative sign.) The term "synchronous watts" means the power corresponding to the torque, if delivered at synchronous speed.

It is sufficient to measure the power in the f and b meshes that are closest to the air gap, by multiplying the sum of all the currents in each field by the air-gap voltage of that field. When there are no impressed secondary voltages ($E_{3f} = E_{3b} = E_{4f} = E_{4b} = 0$), Eq. 11.15 reduces to:

$$(11.16) \quad T = I_{3f}^2 R_{3f}/(F-v) + I_{4f}^2 R_{4f}/(F-v)$$
$$- I_{3b}^2 R_{3b}/(F+v) - I_{4b}^2 R_{4b}/(F+v) \text{ synchronous watts.}$$

There are also alternating torques, due to the interaction of the forward-field flux and the backward-field currents, and vice versa. These are given by.

$$(11.17) \quad T_A = I_f E_b + I_b E_f.$$

The frequency of these torques is $2F$, since they are due to a flux moving at a speed F interacting with a current moving at a speed $-F$. If there are no secondary impressed voltages, Eq. 11.17 reduces to:

$$(11.18) \quad T_A = I_{3f} I_{3b} [R_{3f}/(F-v) - R_{3b}/(F+v)]$$
$$+ I_{4f} I_{4b} [R_{4f}/(F-v) - R_{4b}/(F+v)].$$

If secondary voltages are impressed which have different frequencies from those induced, the entire circuit must be solved twice independently; once with the stator voltages applied and the rotor closed through the impedance of the secondary supply circuit, and again with the rotor voltages impressed and the stator closed through the impedance of the stator supply circuit. The two sets of currents and voltages act independently of each other, except for heating and saturation effects and the presence of alternating torques due to the independent fields slipping by each other (as in a synchronous machine that is "out of step").

6 Limitations of the General Equivalent Circuit

The circuit assumes that a current in either D or Q axis does not create any flux linking the other axis; i.e., the two sets of windings are 90° apart in space. If the angle is not 90° (as in the case of a shaded-pole motor),

Kron's device of a "phase shifter" may be introduced, or the use of equations to supplement the circuit may be resorted to.

The circuit shown in Fig. 11.6 assumes also that the rotor D and Q axes are alike, although the stator D and Q axes are dissimilar (or vice versa). This is because, if both rotor and stator have dissimilar axes, there will be resistances in the rotor midpoint connections which carry both I_f and I_b currents; that is, when v is not zero, they carry currents of two different frequencies. These dual currents will give rise to additional frequencies in both rotor and stator, which require additional circuits for their representation and more complex rules for their calculation. This condition arises in the case of a salient-pole synchronous machine with unbalanced voltages or stator windings. These more complex cases will not be considered here.

Even though Z_d and Z_q in the rotor are assumed to be the same, the rotor values of Z_f and Z_b at speed may be quite different, due to "skin effects" caused by the different frequencies of the f and b fields. The circuit allows these differences to be included.

An induced voltage in the circuit represents a magnetic flux in the machine (volts/frequency gives flux) rather than the actual voltage. However, a current in the circuit corresponds directly to an actual current. Hence, a "power" in the circuit (volts/F times amperes) corresponds to a "torque" in the machine (flux times current). The circuit gives true currents, torques, power factors, etc., but its voltages must be multiplied by the denominators of the resistance elements $[F, (F-v),$ and $(F+v)]$ to give actual values.

The forward- and backward-field values given by the circuit can be converted to D- and Q-axis values by Eqs. 11.1 and 11.2. For two-phase machines, the D and Q values are the actual phase values also. For windings with three or more phases, the transformation to actual phase values is carried out by a similar procedure (known as the *symmetrical components resolution*), as described in Sect. 11.8.

The circuit is valid for steady-state conditions, and also as a basis for the perturbation equations for small oscillations. In the simpler circuits, where the number of loops is the same as the number of original circuit equations, the circuit gives the transient performance equations at constant speed if the per-unit frequency, F, is replaced by $-jp$, where $p = d/\omega t$. This is not true for circuits such as Fig. 11.6, where there are different loops for the components of a single current.

7 Transformations of Reference Axes

The equations of a rotating machine referred to stationary axes usually are non-linear, because the reactance of a coil pulsates between X_d and X_q

as it passes by the D and Q axes. Also, when both forward and backward fields exist (as in single-phase and unbalanced-polyphase machines), the currents on one side of the air gap include components of two different frequencies [$(F-v)$ and $(F+v)$] for which the circuit impedance is not the same. To get around these difficulties, it is necessary to transform the equations to different reference axes. The most important example of this is the $D-Q$ axis transformation, which refers the equations to axes fixed with respect to the salient poles, thus making the reactances constant and the equations linear. The symmetrical components transformation resolves the currents into components that each have a single frequency, thus enabling each component to flow in a separate circuit with the correct impedance. A third important transformation is that from two to three (or from three to two) phases, which simplifies the analysis of salient-pole machines by permitting the use of the D- and Q-axis equations.

In making these transformations, we need to keep the total power the same. For example, in changing from two phases to q phases, the power per phase changes from $1/2$ to $1/q$. Similar transformation equations apply to both current and voltage, so that in general the change from two to q phases changes both currents and voltages in the ratio $\sqrt{(2/q)}$.

All these relations assume that currents are expressed in amperes per phase. If currents are given in per-unit terms, unit current is the rated value, and is 1 regardless of the number of phases.

8 Symmetrical Components

The "symmetrical components" of a system of polyphase currents are found by resolving each phase current into forward- and backward-revolving components (using Eq. 11.1), combining the forward components of all phases by phasor addition into a single "forward-sequence" current, and likewise combining all the backward components into a single "backward-sequence" current. If the sum of all the polyphase currents is not zero, there will be a "zero-sequence" component of current also (Sect. 3.15).

The $2q$ successive phase belts of a q-phase machine are displaced $180/q°$ in space, and their currents (when balanced) are also $180/q°$ apart in time phase. Their contributions to the forward-sequence current are found by rotating their phasor-current values through angles of $180/q$, $360/q$, and $540/q°$, etc., and by phasor addition. Similarly, the backward-sequence current is found by rotating the currents of successive phases through $-180/q$, $-360/q$, and $-540/q°$, etc., and adding. Normally, the phasor sum of all the phase currents is zero; but if there is a neutral connection or a ground fault, their sum may not be zero, and there will be a zero-sequence current, I_0, in addition to the forward and backward currents,

I_f and I_b. These three "symmetrical components" of the currents in a three-phase machine are, therefore:

(11.19) $$I_f = (1/\sqrt{3})(I_A + aI_B + a^2I_C)$$

(11.20) $$I_b = (1/\sqrt{3})(I_A + a^2I_B + aI_C)$$

(11.21) $$I_0 = (1/\sqrt{3})(I_A + I_B + I_C),$$

where
$$a = 1 + j\sqrt{3}/2,$$
$$a^2 = 1 - j\sqrt{3}/2,$$

and
$$a^3 = 1.$$

The inverse transformations are:

(11.22) $$I_A = (1/\sqrt{3})(I_f + I_b + I_0)$$

(11.23) $$I_B = (1/\sqrt{3})(a^2I_f + aI_b + I_0)$$

(11.24) $$I_C = (1/\sqrt{3})(aI_f + a^2I_b + I_0).$$

The factor $1/\sqrt{3}$ enters the equations to keep the total power the same before and after the transformation, or "invariant", as stated above. In general, for a q-phase machine, the factor is $1/\sqrt{q}$.

Since I_0 is the same in all phases, and the sum of three equal phasors 120° apart is zero, the zero-sequence current cannot produce any fundamental field, so it does not contribute to the torque. It can produce fields with 3, 9, etc., times the fundamental number of poles, and thus causes losses. Its reactance drop is important in calculating ground fault currents, but it can be neglected in normal machine operation. The impedances are the same in per-unit terms, whatever the number of phases.

9 Three-Phase to Two-Phase Transformations

The power invariant transformations from three phases (A, B and C) to two phases (α and β) and the inverse are, from Fig. 11.7 (assuming that the A and α axes coincide):

(11.25) $$I_\alpha = \sqrt{(2/3)}(I_A - I_B/2 - I_C/2),$$

(11.26) $$I_\beta = -(1/\sqrt{2})(I_B - I_C),$$

(11.27) $$I_0 = (1/\sqrt{3})(I_A + I_B + I_C),$$

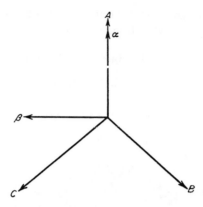

FIG. 11.7. Relations between Two- and Three-Phase Quantities.

and

(11.28) $I_A = \sqrt{(2/3)}\, I_\alpha + \sqrt{(1/3)}\, I_0,$

(11.29) $I_B = [\sqrt{(2/3)}]\,(-I_\alpha/2 - \sqrt{3}\, I_\beta/2) + \sqrt{(1/3)}\, I_0,$

(11.30) $I_C = [\sqrt{(2/3)}]\,(-I_\alpha/2 + \sqrt{3}\, I_\beta/2) + \sqrt{(1/3)}\, I_0.$

The corresponding voltage equations are the same, if we put E in place of I throughout. The total power on two-phase is $2I_\alpha I_\alpha^* R$, which is the same as that on three-phase ($3I_A I_A^* R$), since the current ratio is $I_A/I_\alpha = \sqrt{2/3}$, and the per-unit impedance is the same for two as for three phases.

The α and β axes can be taken as the D and Q rotor axes if desired. It is usually desirable to keep these two sets of axes independent for freedom

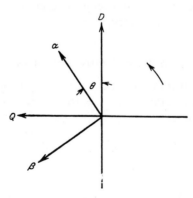

FIG. 11.8. Relations between (Fixed) D and Q Axes and (Moving) α and β Axes.

in analysis. If the α and β axes are displaced $\theta°$ ahead of the D and Q axes (Fig. 11.8), the transformations that leave the power invariant are:

(11.31)
$$I_\alpha = I_d \cos\theta + I_q \sin\theta$$
$$I_\beta = -I_d \sin\theta + I_q \cos\theta$$
$$I_d = I_\alpha \cos\theta - I_\beta \sin\theta$$
$$I_q = I_\alpha \sin\theta + I_\beta \cos\theta.$$

10 Matrix Transformations

Matrix algebra provides a shorthand method of writing equations in several variables, and of carrying through the operations of transforming to different variables and solving simultaneous equations. It is, therefore, extremely useful in performing all the mathematical operations required in the "Dynamic Circuit Theory" that is required for the general analysis of rotating machines and power systems.

In the preceding sections, we developed the general equations and equivalent circuits by somewhat intuitive methods, relying on a clear visualization of the machine for guidance. We shall now follow through with a more orderly procedure, first developed by Kron. This consists of five steps:

1. Establishing the "primitive" machine, in which all its n circuits are closed independently.

2. Writing down the voltage-current equations for all the n circuits, including all mutual inductance terms, and so establishing the "impedance matrix" (z_1) of the machine. This may be done by applying Kirchhoff's laws directly.

3. Making the desired connections and writing down the relations between the currents in the original n circuits and those in the circuits of the particular machine, in the form of a "connection matrix", C.

4. Writing down the corresponding inverse relation between the voltages in the new circuit and those in the original in the form of the conjugate of the transpose (C_t^*)† of the connection matrix.

5. Performing the multiplications indicated by the equation $Z = C_t^* . z . C$, thus obtaining the new impedance matrix for the particular machine.

The inverse voltage change is different from that of the currents, because

† The transpose of a matrix is formed by interchanging the rows and columns. Thus, if

$$C = \begin{array}{|c|c|c|} \hline a & b & c \\ \hline d & e & f \\ \hline g & h & i \\ \hline \end{array} \quad \text{then } C_t = \begin{array}{|c|c|c|} \hline a & d & g \\ \hline b & e & h \\ \hline c & f & i \\ \hline \end{array}$$

the power is the same before and after the reconnection (in a series connection the currents are the same, but the voltages add). For example, if three coils with currents i_1, i_2, i_3 are connected to form only two loops with currents I_1 and I_2, the current equation is $i = C \cdot I$, or:

(11.32)
$$\begin{vmatrix} i_1 \\ i_2 \\ i_3 \end{vmatrix} = \begin{vmatrix} C_{11} & C_{12} \\ C_{21} & C_{22} \\ C_{31} & C_{32} \end{vmatrix} \cdot \begin{vmatrix} I_1 \\ I_2 \end{vmatrix}$$

and the corresponding inverse voltage equation is $E = B \cdot e$, or:

(11.33)
$$\begin{vmatrix} E_1 \\ E_2 \end{vmatrix} = \begin{vmatrix} B_{11} & B_{12} & B_{13} \\ B_{21} & B_{22} & B_{23} \end{vmatrix} \cdot \begin{vmatrix} e_1 \\ e_2 \\ e_3 \end{vmatrix}$$

Since the power is the same, we have:

$$e_1 i_1^* + e_2 i_2^* + e_3 i_3^* = E_1 I_1^* + E_2 I_2^*$$

or

$$e_1(C_{11}I_1 + C_{12}I_2)^* + e_2(C_{21}I_1 + C_{22}I_2)^* + e_3(C_{31}I_1 + C_{32}I_2)^*$$
$$= (B_{11}e_1 + B_{12}e_2 + B_{13}e_3)I_1^* + (B_{21}e_1 + B_{22}e_2 + B_{23}e_3)I_2^*.$$

Since this equation must be true for all values of e and i, the coefficients of like terms must be equal, giving:

(11.34) $\quad C_{11}^* = B_{11}, \quad C_{12}^* = B_{21}, \quad C_{21}^* = B_{12},$

$\quad C_{22}^* = B_{22}, \quad C_{31}^* = B_{13}, \quad C_{32}^* = B_{13},$

which proves that B must be the conjugate of the transpose of C, as stated.

To find the new impedance matrix, we have:

(11.35) $\quad E = C_t^* e = C_t^* z i = C_t^* z C I = Z I,$

so that the new Z is equal to $C_t^* \cdot z \cdot C$, as stated previously.

As a simple example, we shall carry through the development of the equivalent circuit for the three coils of Fig. 11.9 when reconnected as in Fig. 11.10. The current-voltage relation for coil 1 is:

(11.36) $\quad e_1 = (R_1 + pL_{11})i_1 + pL_{12}(i_1 + i_2) + pL_{13}(i_1 + i_3),$

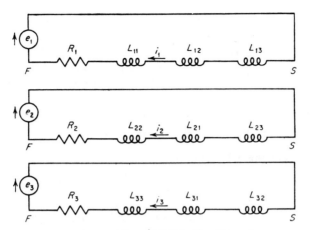

FIG. 11.9. Three-Coil Primitive Network.

where the L values are the self- and mutual inductances of coil 1. The other coils have similar equations, so the "primitive" impedance matrix is given by $e = zi$:

(11.37) $$\begin{vmatrix} e_1 \\ e_2 \\ e_3 \end{vmatrix} = \begin{vmatrix} R_1 + pL_{11} & pL_{12} & pL_{13} \\ pL_{21} & R_2 + pL_{22} & pL_{23} \\ pL_{31} & pL_{32} & R_3 + pL_{23} \end{vmatrix} \cdot \begin{vmatrix} i_1 \\ i_2 \\ i_3 \end{vmatrix}$$

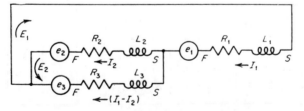

FIG. 11.10. Three-Coil Final Network.

If now we: (1) connect the coils as in Fig. 11.10 with Nos. 2 and 3 in parallel and the two in series with No. 1, and (2) if we choose the new currents to be $I_1 = i_1$, and $I_2 = i_2$ so that $i_3 = I_1 - I_2$, the current equation is:

$$i = C \cdot I$$

or

(11.38)
$$\begin{vmatrix} i_1 \\ i_2 \\ i_3 \end{vmatrix} = \begin{vmatrix} 1 & 0 \\ 0 & 1 \\ 1 & -1 \end{vmatrix} \cdot \begin{vmatrix} I_1 \\ I_2 \end{vmatrix}$$

and the inverse voltage equation is:

$$E = C_t^* \cdot e$$

or

(11.39)
$$\begin{vmatrix} E_1 \\ E_2 \end{vmatrix} = \begin{vmatrix} 1 & 0 & 1 \\ 0 & 1 & -1 \end{vmatrix} \cdot \begin{vmatrix} e_1 \\ e_2 \\ e_3 \end{vmatrix}$$

giving $E_1 = e_1 + e_3$ and $E_2 = e_2 - e_3$. The reversal of sign of e_3 in the E_2 equation means that we have reversed the direction of i_3 with respect to i_2 by the connection of the start of coil 3 to the start of coil 2, but to the finish of coil 1. This requires that we reverse the sign of the mutual inductance terms L_{23} and L_{32} in the impedance matrix. With this change, the multiplication $z \cdot C$ gives:

(11.40)
$$\begin{vmatrix} R_1 + pL_{11} & pL_{12} & pL_{13} \\ pL_{21} & R_2 + pL_{22} & -pL_{23} \\ pL_{31} & -pL_{32} & R_3 + pL_{33} \end{vmatrix} \cdot \begin{vmatrix} 1 & 0 \\ 0 & 1 \\ 1 & -1 \end{vmatrix}$$

$$= \begin{vmatrix} R_1 + pL_{11} + pL_{13} & pL_{12} - pL_{13} \\ pL_{21} - pL_{23} & R_2 + pL_{22} + pL_{23} \\ pL_{31} + R_3 + pL_{33} & -pL_{32} - R_3 - pL_{33} \end{vmatrix}$$

Next, the multiplication $C_t \cdot z \cdot C$ gives:

(11.41)

$$\begin{bmatrix} 1 & 0 & 1 \\ 0 & 1 & -1 \end{bmatrix} \cdot \begin{bmatrix} R_1+pL_{11}+pL_{13} & pL_{12}-pL_{13} \\ pL_{21}-pL_{23} & R_2+pL_{22}+pL_{23} \\ pL_{31}+R_3+pL_{33} & -pL_{32}-R_3-pL_{33} \end{bmatrix}$$

$$= \begin{bmatrix} R_1+pL_{11}+2pL_{13}+R_3+pL_{33} & pL_{12}-pL_{13}-pL_{23}-R_3-pL_{33} \\ pL_{21}-pL_{23}-pL_{31}-R_3-pL_{33} & R_2+pL_{22}+2pL_{23}+R_3+pL_{33} \end{bmatrix}$$

This gives for the equations of the new circuit:

(11.42) $E_1 = [R_1 + p(L_{11}+L_{12}+L_{13}-L_{23})] I_1$
$\qquad\qquad + [R_3 + p(L_{33}+L_{13}+L_{23}-L_{12})](I_1-I_2)$

$E_2 = [R_2 + p(L_{22}+L_{12}+L_{23}-L_{13})] I_2$
$\qquad\qquad + [R_3 + p(L_{33}+L_{13}+L_{23}-L_{12})](I_2-I_1),$

which represents the equivalent circuit of Fig. 11.11. It should be noted that all these impedance matrices are symmetrical about the main diagonal, as the mutual inductances are the same in both directions: $L_{12} = L_{21}$, etc.

11   Impedance Matrix of the Generalized Machine

The four coils of the generalized "primitive" machine, Fig. 11.1, with one coil in each axis of the stator and the rotor, have the impedance matrix:

(11.43)

$$\begin{bmatrix} e_{1d} \\ e_{1q} \\ e_{\alpha} \\ e_{\beta} \end{bmatrix} = \begin{bmatrix} R_{1d}+pL_{dd} & pL_{dq} & pL_{md\alpha} & pL_{md\beta} \\ pL_{dq} & R_{1q}+pL_{qq} & pL_{mq\alpha} & pL_{mq\beta} \\ pL_{md\alpha} & pL_{mq\alpha} & R_{2\alpha}+pL_{aa} & pL_{\alpha\beta} \\ pL_{md\beta} & pL_{mq\beta} & pL_{\alpha\beta} & R_{2\beta}+pL_{\beta\beta} \end{bmatrix} \cdot \begin{bmatrix} i_{1d} \\ i_{1q} \\ i_{\alpha} \\ i_{\beta} \end{bmatrix}$$

To simplify matters, we shall assume: the rotor to be symmetrical, so that $R_{2\alpha} = R_{2\beta}$ and $L_{dq} = 0$; all fluxes to be sinusoidal in time and space; the stator voltages to have a single frequency, f; and the impressed rotor voltages to be zero. The α and β axes of the rotor are assumed to be ahead of the D and Q stator axes by an angle $\theta = vt = (1-s)\omega t$ radians, where v is the rotor speed in electrical radians per second and s is the per-unit slip

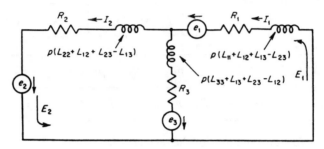

FIG. 11.11. Equivalent Circuit of the Three-Coil Network in Fig. 11.10.

below synchronous speed. $L_{\alpha\alpha}$ and $L_{\beta\beta}$ are the total inductances of the rotor windings in the α and β axes, which vary with θ as the rotor turns, if the stator D and Q axes are not alike. From Fig. 11.1, these inductances are

(11.44) $$L_{\alpha\alpha} = l_2 + L_{md}\cos^2\theta + L_{mq}\sin^2\theta,$$

(11.45) $$L_{\beta\beta} = l_2 + L_{md}\sin^2\theta + L_{mq}\cos^2\theta,$$

(11.46) $$L_{\alpha\beta} = -L_{md}\cos\theta\sin\theta + L_{mq}\sin\theta\cos\theta$$
$$= -\Delta L \sin\theta \cos\theta,$$

where l_2 is the leakage inductance of the rotor winding due to flux that does not cross the air gap, assumed to be the same for the α and β axes, and $\Delta L = L_{md} - L_{mq}$. The sine and cosine factors enter as the square because the flux in the D axis per ampere in the α axis varies as the cosine, and the voltage induced in the α axis by a D-axis flux also varies as the cosine. The mutual inductances between the stator and rotor axes are, from Fig. 11.1:

(11.47) $$L_{md\alpha} = L_{md}\cos\theta, \qquad L_{md\beta} = -L_{md}\sin\theta,$$
$$L_{mq\alpha} = L_{mq}\sin\theta, \qquad L_{mq\beta} = L_{mq}\cos\theta.$$

IMPEDANCE MATRIX OF THE GENERALIZED MACHINE

Making all these substitutions in Eq. 11.43, the impedance matrix becomes:

(11.48)

e_{1d}	$R_{1d}+pL_{dd}$	0	$pL_{md} \cos \theta$	$-pL_{md} \sin \theta$	i_{1d}
e_{1q}	0	$R_{1q}+pL_{qq}$	$pL_{mq} \sin \theta$	$pL_{mq} \cos \theta$	i_{1q}
e_α =	$pL_{md} \cos \theta$	$pL_{mq} \sin \theta$	$R_2+p(l_2+L_{md}\cos^2\theta+L_{mq}\sin^2\theta)$	$-p\Delta L \sin\theta \cos\theta$	i_α
e_β	$-pL_{md} \sin \theta$	$pL_{mq} \cos \theta$	$-p\Delta L \sin\theta \cos\theta$	$R_2+p(l_2+L_{md}\sin^2\theta+L_{mq}\cos^2\theta)$	i_β

To solve these equations and to derive their equivalent circuit, we need to refer them all to the D and Q axes, since this will eliminate the varying reactance terms due to ΔL. (If a machine is asymmetric on both sides of the air gap, these varying terms cannot be eliminated, and they give rise to an infinite series of terms with successively higher frequencies.)

When we transform equations to a new set of reference axes, we keep the same number of currents and voltages, so the transformation matrix must be square. This, together with the rule that the power is invariant, requires that the transformation matrix have a form such that:

(11.49) $\qquad C . C_t^* = $ the unit matrix.

For example, in the case of the symmetrical component transformation (Eqs. 11.9–11.24), the product $C . C_t^*$ is:

$$1/\sqrt{3} \cdot \begin{vmatrix} 1 & a & a^2 \\ 1 & a^2 & a \\ 1 & 1 & 1 \end{vmatrix} \cdot 1/\sqrt{3} \cdot \begin{vmatrix} 1 & 1 & 1 \\ a^2 & a & 1 \\ a & a^2 & 1 \end{vmatrix} = \begin{vmatrix} 1 & 0 & 0 \\ 0 & 1 & 0 \\ 0 & 0 & 1 \end{vmatrix}$$

where $a^* = a^2$, $a^{2*} = a$ and $a^3 = 1$.

Equation 11.31 gives the relations between the α, β, D and Q axes, so that the transformation matrix for the desired change is:

(11.50)

$$C = \begin{array}{c|cccc} & 1d & 1q & 2d & 2q \\ \hline 1d & 1 & 0 & 0 & 0 \\ 1q & 0 & 1 & 0 & 0 \\ \alpha & 0 & 0 & \cos\theta & \sin\theta \\ \beta & 0 & 0 & -\sin\theta & \cos\theta \end{array}$$

and its transpose is:

(11.51)

$$C_t = \begin{array}{c|cccc} & 1d & 1q & 2d & 2q \\ \hline 1d & 1 & 0 & 0 & 0 \\ 1q & 0 & 1 & 0 & 0 \\ \alpha & 0 & 0 & \cos\theta & -\sin\theta \\ \beta & 0 & 0 & \sin\theta & \cos\theta \end{array}$$

As a check, we note that $C \cdot C_t = 1$.

The first step in making the transformation is to multiply $z \cdot C$:

(11.52)

$z \cdot C =$

d_1	q_1	d_2	q_2
$R_{1d}+pL_{dd}$	0	pL_{md}	0
0	$R_{1q}+pL_{qq}$	0	pL_{mq}
$pL_{md}\cos\theta$	$pL_{mq}\sin\theta$	$[R_2+p(l_2+L_{md}\cos^2\theta+L_{mq}\sin^2\theta)]\cos\theta+p\Delta L\sin^2\theta\cos\theta$	$[R_2+p(l_2+L_{md}\cos^2\theta+L_{mq}\sin^2\theta)]\sin\theta-p\Delta L\sin\theta\cos^2\theta$
$-pL_{md}\sin\theta$	$pL_{mq}\cos\theta$	$-[R_2+p(l_2+L_{md}\sin^2\theta+L_{mq}\cos^2\theta)]\sin\theta-p\Delta L\sin\theta\cos^2\theta$	$[R_2+p(l_2+L_{md}\sin^2\theta+L_{mq}\cos^2\theta)]\cos\theta-p\Delta L\sin^2\theta\cos\theta$

IMPEDANCE MATRIX OF THE GENERALIZED MACHINE

At this point we need to recognize that $p = d/dt$, so that each of the terms containing p as a factor is actually two terms, one due to changing flux linkages with fixed current as the rotor turns, and the other due to changing linkages at a fixed rotor position as the current changes. For example:

(11.53) $\quad pL_{md}\cos\theta = (d/dt)(L_{md}\cos\theta) = -vL_{md}\sin\theta + L_{md}\cos\theta\, p.$

Carrying this procedure through for all the terms in Eq. 11.50 gives:

(11.54)

$$z \cdot C =$$

$R_{1d}+L_{dd}p$	0	$L_{md}p$	0
0	$R_{1q}+L_{qq}p$	0	$L_{mq}p$
$L_{md}\cos\theta\, p$ $-vL_{md}\sin\theta$	$L_{mq}\sin\theta\, p$ $+vL_{mq}\cos\theta$	$(l_2+L_{md})(\cos\theta\, p - v\sin\theta) + R_2\cos\theta$	$(l_2+L_{mq})(\sin\theta\, p + v\cos\theta) + R_2\sin\theta$
$-L_{md}\sin\theta\, p$ $-vL_{md}\cos\theta$	$L_{mq}\cos\theta\, p$ $-vL_{mq}\sin\theta$	$-(l_2+L_{md})(\sin\theta\, p + v\cos\theta) - R_2\sin\theta$	$(l_2+L_{mq})(\cos\theta\, p - v\sin\theta) + R_2\cos\theta$

The final step in deriving the new impedance matrix is to multiply this by C_t, giving $Z = C_t \cdot z \cdot C$:

(11.55)

		1d	1q	2d	2q	
e_{1d}		$R_{1d}+L_{dd}p$	0	$L_{md}p$	0	i_{1d}
e_{1q}	=	0	$R_{1q}+L_{qq}p$	0	$L_{mq}p$	i_{1q}
e_{2d}		$L_{md}p$	vL_{mq}	$R_2+(l_2+L_{md})p$	$v(l_2+L_{mq})$	i_{2d}
e_{2q}		$-vL_{md}$	$L_{mq}p$	$-v(l_2+L_{md})$	$R_2+(l_2+L_{mq})p$	i_{2q}

These are the well-known D-Q axis equations for the generalized machine, with salient stator poles and a symmetrical rotor. Reading from left to right in the rotor D-axis equation, the first and third terms give the transformer voltage due to D-axis flux, and the second and fourth terms give the speed voltage due to Q-axis flux. The terms containing v are "joint energy" terms, or "motional inductances".

Under steady-state conditions, the applied voltages are sinusoidal with frequency f in the stator, and the speed is constant, so that we may replace p by $j\omega$ and v by $(1-s)\omega$. Since $j\omega L$ is a reactance, we can replace all the Ls by Xs. The impedance matrix then becomes:

(11.56)

	$1d$	$1q$	$2d$	$2q$	
e_{1d}	R_1d+jX_d	0	jX_{md}	0	i_{1d}
e_{1q}	0	$R_{1q}+jX_q$	0	jX_{mq}	i_{1q}
e_{2d} =	jX_{md}	$(1-s)X_{mq}$	$R_2+j(X_2+X_{md})$	$(1-s)(X_2+X_{mq})$	$\cdot$ i_{2d}
e_{2q}	$-(1-s)X_{md}$	jX_{mq}	$-(1-s)(X_2+X_{md})$	$R_2+j(X_2+X_{mq})$	i_{2q}

12 Equivalent Circuits

These are the equations of the "cross-field theory", which represents the current and voltage relations in the D- and Q-axis circuits, and which considers all currents to flow in the stator at full frequency, ignoring the actual frequencies of the rotor currents. To derive an equivalent circuit for Eq. 11.56, we must rearrange the equations to include only self- and mutual impedances. To do this, we multiply the e_{2q} equation by $j(1-s)$ and add it to the e_{2d} equation:

$$(11.57) \quad e_{2d}+j(1-s)e_{2q} = j[1-(1-s)^2][X_{md}i_{1d}+(X_2+X_{md})i_{2d}] \\ +R_2 i_{2d}+j(1-s)R_2 i_{2q}.$$

Similarly, we multiply the e_{2d} equation by $(1-s)$ and add j times the e_{2q} equation to it:

$$(11.58) \quad (1-s)e_{2d}+je_{2q} = -[1-(1-s)^2][X_{mq}i_{1q}+(X_2+X_{mq})i_{2q}] \\ +jR_2 i_{2q}+(1-s)R_2 i_{2d}.$$

Dividing both equations by $1-(1-s)^2 = s(2-s)$, and regrouping the resistance terms, we find:

$$(11.59) \quad \frac{e_{2d}+j(1-s)e_{2q}}{s(2-s)} = jX_{md}i_{1d}+j(X_2+X_{md})i_{2d}+R_2 i_{2d}/s$$
$$-\frac{(1-s)R_2}{s(2-s)}(i_{2d}-ji_{2q}),$$

$$(11.60) \quad \frac{(1-s)e_{2d}+je_{2q}}{s(2-s)} = X_{mq}i_{1q}-(X_2+X_{mq})i_{2q}+jR_2 i_{2q}/s$$
$$+\frac{(1-s)R_2}{s(2-s)}(i_{2d}-ji_{2q}).$$

These equations represent the equivalent circuit of Fig. 11.12. The fact that the denominator of the resistance in the midpoint connection is $s(2-s)$ shows that the actual rotor currents have two frequencies: s and

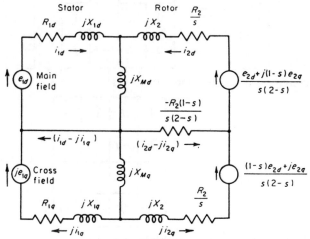

FIG. 11.12. Cross-Field Equivalent Circuit of Generalized Machine.

$2-s$. Therefore, the circuit can not be accurate for a machine with a double-cage or deep-bar rotor, as the skin effect makes the rotor bar resistance very different at the two frequencies, while the circuit provides for a single resistance value only. The impressed rotor voltages, e_{2d} and e_{2q}, must have the frequencies sf and/or $(2-s)f$ if applied to the rotor slip rings, if they are to give useful torque.

In deriving the circuit, we chose counterclockwise rotation, and the Q axis ahead of the D axis (Fig. 11.1). This gave a negative sign to the midpoint resistance term and made the rotor-branch resistance R_2/s. With clockwise rotation, the midpoint resistance becomes positive, and the rotor-branch resistance is $R_2/(2-s)$. The effect of a change in the direction of rotation is merely to substitute $(2-s)$ for s. For forward rotation and balanced voltages in Fig. 11.12, we must have $je_{1q} = e_{1d}$; in which case (with a symmetrical stator) $ji_{1q} = i_{1d}$, the current in the midpoint connection becomes zero, and the four meshes collapse into two, with impressed voltages $2e_{1d}$ and $2e_{2d}/s$.

To obtain an equivalent circuit for the revolving field theory, which correctly allows for rotor skin effect, we need to transform all the D- and Q-axis equations to two sets of axes that rotate forward and backward at synchronous speed with respect to the stator. From Eqs. 11.1 and 11.2, the transformation matrix for making this change is:

(11.61) $\quad C = \begin{vmatrix} i_{1d} \\ i_{1q} \\ i_{2d} \\ i_{2q} \end{vmatrix} = 1/\sqrt{2} \cdot \begin{vmatrix} 1 & 1 & 0 & 0 \\ -j & j & 0 & 0 \\ 0 & 0 & 1 & 1 \\ 0 & 0 & -j & j \end{vmatrix} \begin{vmatrix} i_{1f} \\ i_{1b} \\ i_{2f} \\ i_{2b} \end{vmatrix}$

and the conjugate of the transpose of this is:

(11.62) $\quad C_t^* = 1/\sqrt{2} \cdot \begin{vmatrix} 1 & j & 0 & 0 \\ 1 & -j & 0 & 0 \\ 0 & 0 & 1 & j \\ 0 & 0 & 1 & -j \end{vmatrix}$

From Eqs. 11.56 and 11.61, multiplying $z \cdot C$:

(11.63)

$z \cdot C = 1/\sqrt{2} \cdot \begin{vmatrix} Z_{1d} & Z_{1d} & jX_{md} & jX_{md} \\ -jZ_{1q} & jZ_{1q} & X_{mq} & -X_{mq} \\ j[X_{md} - (1-s)X_{mq}] & j[X_{md} + (1-s)X_{mq}] & Z_{2d} - j(1-s)(X_2 + X_{mq}) & Z_{2d} + j(1-s)(X_2 + X_{mq}) \\ -(1-s)X_{md} + X_{mq} & -(1-s)X_{md} - X_{mq} & -jZ_{2q} - (1-s)(X_2 + X_{md}) & jZ_{2q} - (1-s)(X_2 + X_{md}) \end{vmatrix}$

and $C_t \cdot z \cdot C$ gives:
(11.64)

$$\begin{vmatrix} e_{1f} \\ e_{1b} \\ e_{2f} \\ e_{2b} \end{vmatrix} = \frac{1}{2} \cdot \begin{vmatrix} Z_{1d}+Z_{1q} & Z_{1d}-Z_{1q} & j(X_{md}+X_{mq}) & j(X_{md}-X_{mq}) \\ Z_{1d}-Z_{1q} & Z_{1d}+Z_{1q} & j(X_{md}-X_{mq}) & j(X_{md}+X_{mq}) \\ js(X_{md}+X_{mq}) & js(X_{md}-X_{mq}) & 2(R_2+jsX_2)+js(X_{md}+X_{mq}) & js(X_{md}-X_{mq}) \\ j(2-s)(X_{md}-X_{mq}) & j(2-s)(X_{md}+X_{mq}) & j(2-s)(X_{md}-X_{mq}) & 2[R_2+j(2-s)X_2]+j(2-s)(X_{md}+X_{mq}) \end{vmatrix} \begin{vmatrix} i_{1f} \\ i_{1b} \\ i_{2f} \\ i_{2b} \end{vmatrix}$$

These equations yield the equivalent circuit of Fig. 11.13, which is the same as Fig. 11.6 previously found. The appearance of the factors s and $2-s$ in the rotor f and b reactance terms shows that the actual rotor

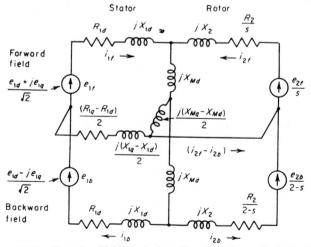

FIG. 11.13. Rotating-Field Equivalent Circuit of Generalized Machine.

frequencies are sf and $(2-s)f$, so that to refer these equations to full frequency, they are divided by s and $2-s$, respectively.

The usual induction machine has a distributed stator winding with uniform air gap, so that $X_{md} = X_{mq}$. Making this change simplifies the circuit considerably.

Summing up, we have derived the basic equations for the generalized machine with salient stator poles and symmetrical rotor by the cross field theory, Eq. 11.56, and by the rotating field theory, Eq. 11.64; and we have developed the corresponding equivalent circuits, Figs. 11.12 and 11.13, respectively. By appropriate modifications, these equations and circuits will give the equations and circuits for all usual specific types of machines

Appendix to Chapter 11

GABRIEL KRON

In view of the importance attached in this book to the work of Gabriel Kron, and the author's belief in its much greater importance in years to come, it is appropriate to give here some perspective of Kron's philosophy.

Gabriel Kron came to the United States in 1921 from Hungary, and enrolled at the University of Michigan, whence he was graduated in 1924 with an electrical engineering degree. After graduation, he was employed first by Robbins & Myers, Inc., then the Lincoln Electric Company, and later by Westinghouse Electric Corporation and Warner Brothers Pictures. His first A.I.E.E. paper, entitled "Generalized Theory of Electrical Machinery", was presented in 1930.

Between employments, he satisfied a long-felt desire for travel by taking a two-year walking trip around the world by way of Samoa, the Fiji Islands, Australia, Indo-China, Siam, India, Arabia, Palestine, and Egypt. The expansive ideas that have marked his later work had their start in meditations on this long hiking trip.

In the early 1930's, these ideas took definite form in the resolve to establish a set of principles that would link together all types of rotating electric machines. He started out to define their relationships as members of a single family on the one hand, and to identify properly this family as a branch of electrodynamics on the other hand. Kron asked himself the questions, "Although all of the existing types of rotating machine seem to have been invented and developed quite independently, is there not a 'universal structure' from which all possible types could be derived by simple laws? Moreover, is there not some basic equation in modern physics that applies equally to all moving electrodynamic systems, be they spinning electrons or rotating machines?" Kron thought that the discovery of such equations, laws, and relationships would facilitate the study and better understanding not only of rotating machines but also of the power systems to which they are connected.

Having a period of freedom from active work in the early 1930's—in common with many others—he returned to Europe and made a systematic study of engineering and mathematical works in pursuit of his objective. In the course of this search, he became interested in non-Euclidean geometry and tensor analysis, and began to see that the methods of treating invariant properties of equations, when transformed into different spaces, could be useful in studying the performance of the various types of rotating machine.

On the basis of these studies, he wrote his classic paper entitled "Non-Riemannian Dynamics of Rotating Electrical Machinery". This he had privately printed in Rumania and sent to a number of friends in the United States. After its informal presentation at an A.I.E.E. conference in New York City, the paper was published in full in the *M.I.T. Journal of Mathematics and Physics* in May 1934. As a direct result of this publication, he joined the Engineering General Staff of the General Electric Company in Schenectady, and since then he has devoted much of his time to the extension of his theories to cover all types of rotating machines and power systems generally. Kron's original hopes are well summed up in a brief talk he gave twenty years ago before a group of central station engineers, from which the following is quoted:

> The opportunities for intellectual adventure in the electric power field are far more promising to the young man with imagination and plenty of courage, than in the fields of electronics and nucleonics. While the latter two fields are already crowded with intellectual adventurers and gold-diggers, the central station field is still a virgin territory for the analyst with the right type of mining tools and a trained eye for valuable minerals. It is intended here to bring out the fact that in spite of the tremendous activity and progress made above ground during the last sixty to seventy years, nevertheless the electrical power engineer has barely ripped the ground open with a wooden plough.
>
> From a strictly analytical point of view, the problems of the central station engineer are in many respects like those of the nuclear physicist or the astronomer. The most obvious similarity between the three types of problems is the large number of units the investigator is forced to deal with. In a uranium atom the behavior of hundreds of spinning electrons and protons has to be analyzed; in a solar system numerous planets, moons and suns influence each other's course. In an integrated electric power system hundreds of rotating machines and stationary transmission lines have to be co-ordinated under operating conditions that defy analysis by conventional tools.
>
> There is, however, another correspondence between the problems of the nuclear physicist, the astronomer and the electric power engineer that is more subtle and, therefore, is appreciated by fewer people. That similarity is the co-existence in a spinning atom, in a whirling planet, as well as within a rotating electric machine, of two types of basic energies; namely, of electromagnetic energies and of mechanical (or gravitational) energies. It is the mutual interrelation of these two, apparently unrelated, types of forces that makes the analysis of these systems so difficult.
>
> During the last two centuries mathematicians and physicists gradually have been developing a method of reasoning, a philosophy of analyzing complex problems with a large number of variables on one hand, and with unusual interrelations between the variables on the other. The codification of two centuries of progress along these lines is known today as "tensor analysis".

APPENDIX 423

It is not a new type of mathematics, it is rather a new type of thinking, a new philosophy of organizing a large number of loosely coupled bits of information into an organic unity. Tensor analysis acts like the mysterious catalyst that combines a large number of lifeless molecules into a living cell. Tensor analysis is a jet-propelled airplane that replaces the horse and buggy in traveling from Schenectady to Los Angeles.

I have been engaged during the last two decades in applying this new type of philosophy to the complex problems confronting the electrical power engineer. In time I came to the conclusion that the most appropriate form in which the engineer can visualize and employ this new philosophy is the already familiar equivalent circuit. It is much easier to interconnect on paper the equivalent circuits of say three interconnected rotating machines undergoing small oscillations than attempt to write the equations of the system. Surprisingly enough it is also possible to interpret even the deepest and most abstruse concepts of modern physics in terms of equivalent circuits that are devised scientifically and not just by brute force, as most conventional equivalent circuits are.

... As has been indicated before, the equations of rotating electric machinery are formally analogous (in the language of tensors) to those used by Einstein and other physicists in studying the atomic nucleus or the stellar galaxies. In fact, the equations of a rotating machine plus a transmission line are far more complicated than those I have as yet seen used by those long-haired physicists or still longer-haired mathematicians.

It must be emphasized that the new physical pictures of damping and synchronizing power flows, superposed on the active and reactive power flows, appear only when the equations are formulated according to the rules of tensor analysis. When the equations are written in a conventional form, the new physical picture is absent. You may laugh in hearing that a really scientific, not just a brute force, old-fashioned, analysis of a synchronous machine implies the introduction of such unearthly concepts as "non-holonomic reference frames" or "multidimensional non-Riemannian spaces", or—please relax—the "Riemann-Christoffel curvature tensor" . . .

... as power systems get bigger and more complex, a need arises with increasing urgency to return for new concepts and new tools to the source itself from which the central station industry originally sprang, namely, to basic electrodynamics. It becomes imperative to return . . . to the basic electrodynamics of the serious researchers in nuclear physics, in stellar dynamics, and in cosmogony. That's where the real progress in electrodynamics has been made during the last half century (page the atomic bomb) and that's where the electrical power engineer must look for new ideas and new inspirations.

The works of Einstein and of his school of collaborators are the beacon lights that show which way electrodynamics is progressing. Whether the central station engineer likes it or not, he will have to start climbing sooner or later along the same road physicists have traveled. What's more, he has no other choice! . . .

The later works of Kron substantiate the hopes expressed in his address. He has succeeded in generalizing the concept of conventional one-dimensional (branch) networks of the electrical engineer by introducing a sequence of multidimensional networks having zero-, two-, three- or higher (p) dimensional elements (so-called p-simplexes). He then interconnects the sequence of p-networks into a variety of *two-phase* (primal and dual) polyhedral networks. These are excited by electromagnetic, magnetohydrodynamic and still more general waves instead of conventional electric currents. The basic building-block of such a dynamic polyhedral network is a p-dimensional magnetohydrodynamic generator, instead of a conventional rotating electric machine. Kron calls the former a "generalized" rotating electric machine, because the same tensorial (and non-Riemannian) equations are valid for both structures.

Since the vertices of an irregular polyhedron can represent the atoms of a polyatomic molecule, the polyhedral waves are analogous to the dynamically self-sustaining dipole-waves and diffracted electromagnetic waves that propagate in a molecule whose atoms are excited by x-rays. Kron can construct his polyhedral models to act also as "self-organizing automata". With the aid of the latter, he has solved highly complex multidimensional physical and information-theoretical (curve-fitting, etc.) problems that could not be solved otherwise. He also hopes to realize physically the polyhedral networks by means of *crystals* and to use the latter eventually as analogue and digital computers. Thus, his tensorial theories have actually led to the representation of a complex polyatomic molecule as a huge multidimensional electrical transmission system—subject to a variety of stability and hunting conditions—just as they were fancied to lead eventually, when still struggling with the intricacies of non-Riemannian electrodynamics. Kron felt that thanks to his careful topological and tensorial organization of single-phase static and two-phase dynamic l-networks, finally he was "in orbit!"

BIBLIOGRAPHY

		Reference to Section
1.	*The Life and Times of Gabriel Kron* or *Walking around the World, and Tensors,* Edited by Philip L. Alger, Mohawk Development Service, Inc., Schenectady, N.Y., 1969.	
2.	*Tensor Analysis of Electric Circuits and Machines,* L. V. Bewley, Ronald Press, 1961.	2
3.	*Circuit Analysis of A-C Power Systems,* Vol. 1, Edith Clarke, John Wiley, 1943.	8
4.	"Induction Motor Damping and Synchronizing Torques," C. Concordia, A.I.E.E. Trans., Vol. 71, Part III, 1952, pp. 358–63.	5
5.	*Matrix Analysis of Electric Networks,* P. Le Corbeiller, Harvard University Press, Cambridge, Mass., 1950.	10
6.	*Tensors in Electrical Machine Theory,* W. J. Gibbs, Chapman and Hall, London, 1952.	1
7.	*Diakoptics and Networks,* Harvey H. Happ, Academic Press, New York, 1970.	
8.	"What Is Tensor Analysis?", B. Hoffman, *Electrical Eng.,* Vol. 57, 1938, pp. 3–9, 61–6, 108–9, 137, 360.	1

BIBLIOGRAPHY

	Reference to Section
9. *Electromechanical System Theory*, H. E. Koenig and W. A. Blackwell, McGraw-Hill, 1961.	2
10. "Generalized Theory of Electrical Machinery," G. Kron, *A.I.E.E. Trans.*, Vol. 49, 1930, pp. 666–83.	1
11. "The Application of Tensors to the Analysis of Rotating Electrical Machinery," G. Kron, *G.E.Rev.*, 1942.	1
12. *Tensors for Circuits*, G. Kron, Dover Press, 1959.	1
13. "Rotating Field Theory and General Analysis of Synchronous and Induction Machines," Y. H. Ku, *I.E.E. Proc.*, Vol. 99, Part IV, 1952, pp. 410–28.	1
14. *Electromagnetic Energy Conversion*, Y. H. Ku, Ronald Press, 1959.	2
15. "Two-Reaction Theory of a General Induction Machine and Its Equivalent Circuit," Y. H. Ku and D. W. C. Shen, *A.I.E.E. Trans.*, Vol. 76, Part III, 1957, pp. 729–34.	3
16. *Electromechanical Power Conversion*, E. Levi and M. Panzer, McGraw-Hill, 1966.	
17. "A Basic Analysis of Synchronous Machines," W. A. Lewis, *A.I.E.E. Trans.*, Vol. 77, Part III, 1958, pp. 436–53.	3
18. "The Tensor Equations of Electrical Machines," J. W. Lynn, *I.E.E. Proc.*, Vol. 102, Part C, 1955, pp. 149–67.	1
19. "A Basic Analysis of the Commutatorless Primitive Machine of Kron," H. Majmudar, *A.I.E.E. Trans.*, Vol. 81, Part III, April 1962, pp. 7–17.	2
20. "Dynamic Circuit Theory and the New Approach to Power," H. K. Messerle, *I.E.E. Elec. and Mech. Eng. Trans.*, Australia, May 1961.	
21. *Dynamic Circuit Theory*, H. K. Messerle, Pergamon Press, London, 1965.	
22. "Two-Reaction Theory of Synchronous Machines," R. H. Park, *A.I.E.E. Trans.*, Vol. 48, 1929, pp. 716–27.	3
23. *Theory of Matrices*, S. Perlis, Addison-Wesley Pub. Co., 1958.	10
24. "Direct and Quadrature Axis Equivalent Circuits of the Synchronous Machine," A. W. Rankin, *A.I.E.E. Trans.*, Vol. 64, 1945, pp. 861–8.	3
25. "Magnetic Revolving Field Coupling," Giovanni Silva, *A.I.E.E. Trans.*, Vol. 71, Part III, 1952, pp. 298–308.	1
26. *Determinants, Matrices, and Tensors*, S. A. Stigant, MacDonald, London, 1959.	10
27. *Electromechanical Energy Conversion*, D. C. White and H. H. Woodson, John Wiley, 1959.	2
28. "The Torque Tensor of the General Machine," Yao-nan Yu, *A.I.E.E. Trans.*, Vol. 81, Part III, Feb. 1963 p. 623–8.	5

CONTENTS—CHAPTER 12

SINGLE-PHASE INDUCTION MOTORS

		PAGE
1	Principal Features	429
2	Stator Windings	430
3	The Capacitor Motor	431
4	Speed Control of Capacitor Motors	440
5	The Pure Single-Phase Motor	447
6	The Dilemma of Single-Phase Motor Theory	449
7	Resistance-Split Phase Motors	458
8	Shaded-Pole Motors	459
9	Repulsion-Induction Motors	461

12

SINGLE-PHASE INDUCTION MOTORS

1 Principal Features

For reasons of simplicity and cost, single-phase power supply is universally preferred for motors rated less than about one horsepower, and is widely used for motors up to about 5 horsepower. A few larger single-phase motors are made for special purposes, such as for auxiliary equipment on a-c locomotives. Single-phase motors are used to drive all kinds of household and office appliances, and for small office equipment.

As shown in Sect. 3.10, the pure single-phase induction motor has no starting torque. It must be started by means of an auxiliary winding, displaced in phase position from the main winding, or some similar device.

Thus, nearly all single-phase induction motors actually are two-phase motors, with the main winding (in the D axis) adapted to carry most or all of the current in operation, and an auxiliary winding (in the Q axis), with a different number of turns, adapted to provide the needed starting torque.

Since the power input pulsates at twice line frequency (Fig. 1.3), all single-phase motors have a double-frequency torque component, which causes slight oscillations in rotor speed and imparts vibration to the motor supports. As noted in Sect. 10.6, many of the larger single-phase motors are provided with elastic mountings to prevent this vibration from causing objectionable noise.

The many forms of single-phase induction motor are grouped into four principal types, depending on the method of starting:

1. *Capacitor Motors.* These have a capacitor in series with the auxiliary winding. There are three varieties: capacitor-start, capacitor-run, and permanent-split capacitor motors. As implied, the first two use a relay or switch to open circuit or reduce the size of the starting capacitor when the motor comes up to speed (Sect. 12.3).

2. *Resistance-Split Phase Motors.* These have a resistance connected in series with the auxiliary winding, which is open circuited by a centrifugal switch or relay when the motor comes up to speed (Sect. 12.7).

3. *Shaded-Pole Motors.* These are used in very small sizes, up to about

50 watts output, chiefly for fan drive. They have salient stator poles, with one-coil-per-pole main windings. The auxiliary winding consists of one (or rarely two) short-circuited coil displaced from the center of each pole (Sect. 12.8).

4. *Repulsion-Induction Motors.* These have distributed rotor windings connected to a commutator, as for a d-c machine, with short-circuited brushes, and with a distributed single-phase stator winding in the D axis only. There are two forms—the repulsion-start motors, which have a centrifugal device that short circuits all the commutator segments when the motor comes up to speed; and the true repulsion-induction motor, which has a buried squirrel cage in the rotor, that limits the no-load speed to a little above synchronism (Sect. 12.9).

2 Stator Windings

For reasons of space economy and ease of machine winding, these small machines are universally provided with concentric stator windings, similar to that shown in Fig. 3.5 for a three-phase machine. In a concentric winding, the outer coils of each pole have a longer span and a higher effectiveness, or pitch factor, while the inner coils have the advantage of a shorter end length. The best performance is obtained by graduating the numbers of turns per pole from a maximum in the outside coil to perhaps half as many in the inner coil, and limiting the width of the phase belt to about 120°. Thus, about one-third of the slots contain coil sides of both windings, while the other two-thirds of the slots, more or less, contain coil sides of only one winding.

Because of its low inherent starting torque, the small number of slots per pole, and the small air gap required for low cost and good running performance, the motor starting performance may be seriously impaired by harmonic fields, as discussed in Sects. 9.3, 9.5, and 9.6. On single phase, the 3rd, 5th, 7th, etc. harmonic fields are all present, with both forward- and backward-revolving components (Eq. 3.29). To prevent these from producing serious crawling tendencies, it is necessary to limit their magnitudes.

The choice of the best numbers of turns in the several coils, therefore, is governed by two distinct considerations, besides that of efficient use of the slot space. The first consideration is to minimize the overall harmonic content of the mmf wave, which means minimizing the ratio of differential leakage (Sect. 7.6) to the magnetizing reactance. The contribution of the kth coil, having m turns, to the leakage reactance is closely proportional to the square of its number of turns, or m^2. The contribution it makes to the magnetizing flux is proportional to its pitch factor, or $\cos kx$, taking $k = 0$ for an outside coil of full pitch, and x for the electrical angle between slots.

Therefore, the ratio of X to X_M is proportional to:

(12.1) $$\frac{X}{X_M} = \frac{K(a^2+b^2+\ldots+m^2+\ldots)}{(a\cos x + b\cos 2x + \ldots + m\cos kx + \ldots)^2},$$

where $a, b, c, \ldots m$, are the numbers of turns in successive coils. For this to be a minimum, the number of turns in each coil must be chosen to make $d\left(\dfrac{X}{X_M}\right) = 0$, or:

(12.2) $$2(ada + bdb + cdc + \ldots)(a\cos x + b\cos 2x + \ldots)$$
$$= 2(a^2 + b^2 + c^2 + \ldots)(da\cos x + db\cos 2x + \ldots).$$

Equating the coefficients of da on the two sides of this equation:

$$2a(a\cos x + b\cos 2x + \ldots) = 2\cos x(a^2 + b^2 + c^2 + \ldots).$$

Equating successive terms on the two sides of this equation:

$$2a^2 \cos x = 2a^2 \cos x, \text{ an identity,}$$
$$2ab \cos 2x = 2b^2 \cos x,$$
$$2ac \cos 3x = 2c^2 \cos x, \text{ etc., giving:}$$

(12.3) $$\frac{a}{\cos x} = \frac{b}{\cos 2x} = \frac{c}{\cos 3x} = \ldots.$$

The same result is found if the coefficients of db, or any other corresponding terms in Eq. 12.2 are equated, instead of the da coefficients.

That is, the lowest leakage factor is obtained when the number of turns in each coil is made proportional to the pitch factor of that coil (Eq. 3.12).

The second consideration is that of minimizing the harmonics which may produce locking or crawling. These are chiefly the 3rd, 5th, and 7th. The nth harmonic may be reduced to zero by choosing the numbers of turns, a, b, c, etc., such that:

$$a\cos nx + b\cos 2nx + c\cos 3nx + \ldots = 0.$$

When there are three or more coils per pole, it is usually possible to keep all the important harmonics down to moderate values in this way, without departing far from the optimum turn numbers given by Eq. 12.3.

3 The Capacitor Motor

The advent about 1930 of good quality, low cost, electrolytic capacitors for low voltage opened a great field for the capacitor motor (first proposed by Steinmetz about 1915). The electrolytic capacitor has a fairly high effective resistance, so that heating severely limits its permissible continuous rating, but it serves excellently for short-time high-current service, as in motor starting. Connecting a capacitor in series with the auxiliary (start) winding of a single-phase motor advances the time phase of the start winding current, so that the angle of split between the main and start winding currents can be made 90°, as in a two-phase motor, or greater if desired. In this way, the starting performance can be made comparable with that of a two-phase motor. Hence, the capacitor motor has almost completely superseded all other types of single-phase induction motor in sizes of one-half horsepower and larger, and it is widely used in the permanent-split capacitor form down to ratings of about 50 watts.

The normal single-phase induction motor has a symmetrical rotor with a short-circuited squirrel-cage winding, Fig. 3.3. It has the same voltage impressed on both the direct- and quadrature-axis windings of the stator. The quadrature-axis (auxiliary) winding, A, has a times as many turns as the winding in the direct axis. The machine has a small air gap, the same in both axes, and has the stator windings distributed in many slots. Therefore, in the equivalent circuit of Fig. 11.13 we have:

(12.4) $\quad E_{1d} = E, \qquad E_{1q} = E/a, \qquad E_{3d} = E_{3q} = 0,$

$$X_{md} = X_{mq} = X_M, \qquad R_{2d} = R_{4d} = \infty,$$

$$Z_{1q} = (1/a^2)(R_{1A} + R_C + jX_{1A} + jX_C), \qquad X_{MA} = a^2 X_M,$$

where $R_C + jX_C$ is the external impedance (of the capacitor) in the circuit of the auxiliary winding (X_C for a capacitor is negative). Since there is only one stator winding in each axis, we shall use the subscript 2 instead of 3 to indicate the secondary winding.

In the equations and circuits of Chapter 11, we have assumed an external voltage to be impressed on each winding, both stator and rotor, and we took the direction of current in each mesh to be the same as that of its impressed voltage. In the case of a squirrel-cage winding, and in most of the cases to be considered, there are no external rotor impressed voltages. Under these conditions, it is appropriate to assume the secondary-current directions to be the same as those produced by the impressed primary voltages. For this reason, in this and in most of the succeeding equations and circuits, the signs of secondary currents are the opposite of those assumed in Chapter 11.

The actual current in the auxiliary winding must be referred to the main-winding turns, and so appears as aI_A in the circuit. We shall consider operation at normal frequency only, so $F = 1$ and v becomes $1 - s$, where s is the per-unit slip. When all these values are placed in the equivalent circuit of Fig. 11.13 the circuit of the capacitor motor is obtained. However, there is one peculiarity of this motor that must be taken into account also. The forward direction of rotation of the capacitor motor is from the Q toward the D axis, contrary to the basic assumption made in deriving the circuit that rotation from D to Q is positive. In the capacitor motor, the current in the A winding (Q axis), being in series with a capacitor, necessarily leads the main-winding current in time phase.

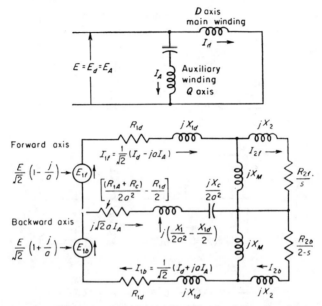

FIG. 12.1. Rotating-Field Equivalent Circuit of Capacitor Motor.

Therefore, the larger of the two rotating fields turns from Q toward D, and this is the direction the motor will turn. In consequence, we should make the further change of reversing the signs of Q-axis voltages and currents. The final result is shown in Fig. 12.1 which is the rotating-field circuit of the capacitor motor. If there is no external impedance and the windings are balanced, this reduces to the circuits shown in Fig. 12.2.

The corresponding equations are:

(12.5) $\quad E_{1f} = I_{1f}(Z_{1d}+Z_f) + (I_{1f}-I_{1b})\left(\dfrac{Z_{1A}+Z_C}{2a^2} - \dfrac{Z_{1d}}{2}\right),$

(12.6) $\quad E_{1b} = I_{1b}(Z_{1d}+Z_b) - (I_{1f}-I_{1b})\left(\dfrac{Z_{1A}+Z_C}{2a^2} - \dfrac{Z_{1d}}{2}\right),$

where Z_f and Z_b are the impedances of jX_M in parallel with

$$\dfrac{R_{2f}}{s}+jX_2 \text{ and } \dfrac{R_{2b}}{2-s}+jX_2, \text{ respectively.}$$

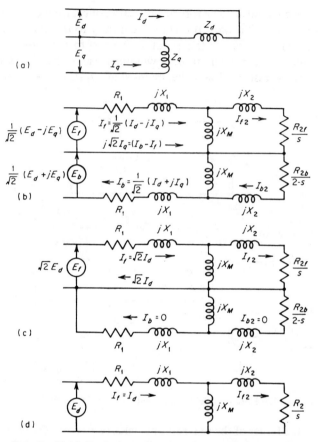

FIG. 12.2. Rotating-Field Equivalent Circuits of Two-Phase Induction Machines: (a) Connection Diagram; (b) Symmetrical Windings, Unbalanced Voltages—Circuit Gives Total Power; (c) Symmetrical Windings, Balanced Voltages—Circuit Gives Total Power; (d) Simplified Circuit for Balanced Two-Phase Operation—Circuit Gives Power of One Phase only.

Adding and subtracting these two equations and dividing by $\sqrt{2}$ give the equations for the currents in the two windings:

(12.7) $\quad E = I_d(Z_{1d}+Z_f) - (I_d + jaI_A)\left(\dfrac{Z_f - Z_b}{2}\right),$

(12.8) $\quad E = I_A(Z_c + Z_{1A} + a^2 Z_b) + (a^2 I_A + jaI_d)\left(\dfrac{Z_f - Z_b}{2}\right).$

At standstill, $s = 1$ and $Z_f = Z_b$, so the last terms of these equations drop out, and the values of I_A and I_d are independent of each other.

THE CAPACITOR MOTOR

To determine the performance of any motor, the values of all the impedances are calculated or are obtained from tests on similar machines, values of the capacitor and turn ratios are selected, and all these are inserted in the circuit. Then, the applied voltage, E, and a value of slip, s, are selected, and the circuit equations are solved just as for any static network. The procedure is repeated for as many values of s as are needed to cover the desired range of operation.

The locked rotor torque of the capacitor motor, from Eq. 11.16 is:

$$(12.9) \qquad T = (I_f^2 - I_b^2)R_2 \text{ synchronous watts.}$$

The winding currents are given by Eqs. 12.7 and 12.8, and the forward and backward currents by Eq. 11.3, with opposite signs from the j terms, because of the changed direction of rotation, as explained above. Assuming that the Q-axis current, aI_A, leads the D-axis current, I_d, by $\theta°$:

$$(12.10) \qquad \sqrt{2}I_f = I_d - jaI_A\underline{/\theta} = I_d + aI_A\sin\theta - jaI_A\cos\theta,$$

$$(12.11) \qquad \sqrt{2}I_b = I_d + jaI_A\underline{/\theta} = I_d - aI_A\sin\theta + jaI_A\cos\theta.$$

The torque then becomes:

$$(12.12) \qquad T = (I_f^2 - I_b^2)R_2 = 2aR_2 I_d I_A \sin\theta \text{ synchronous watts,}$$

and the total locked-rotor current is:

$$(12.13) \qquad I = I_d + I_A\cos\theta + jI_A\sin\theta \text{ amperes.}$$

Best results are obtained by making a considerably greater than one, so reducing the value of I_A required to obtain a given torque, and choosing X_C high enough to make $\sin\theta$ nearly one ($\theta = 60°$ or more). The capacitor cost is proportional to its volt-amperes, $I_A^2 X_C$. As the motor speeds up, the angle θ decreases and the capacitor voltage rises, so that the capacitor must be open-circuited or reduced in size (X_C increased) to avoid over-heating. For continuous duty, Pyranol or oil-filled capacitors are used, as these have lower losses and higher voltage ratings than electrolytic capacitors.

It is sometimes desired to operate a three-phase motor on single phase, with a capacitor in one line, as in Fig. 12.3 (a).

To derive the equations for this arrangement, we express the phase voltages and currents in terms of the sequence voltages and currents:

$(12.14) \quad \sqrt{3}E_A = E_f + E_b + E_0,$ $\qquad (12.17) \quad \sqrt{3}I_A = I_f + I_b,$

$(12.15) \quad \sqrt{3}E_B = a^2 E_f + aE_b + E_0,$ $\qquad (12.18) \quad \sqrt{3}I_B = a^2 I_f + aI_b,$

$(12.16) \quad \sqrt{3}E_C = aE_f + a^2 E_b + E_0,$ $\qquad (12.19) \quad \sqrt{3}I_C = aI_f + a^2 I_b,$

where $a = -\dfrac{1}{2} + j\dfrac{\sqrt{3}}{2}$.

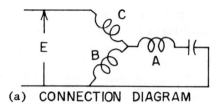

(a) CONNECTION DIAGRAM

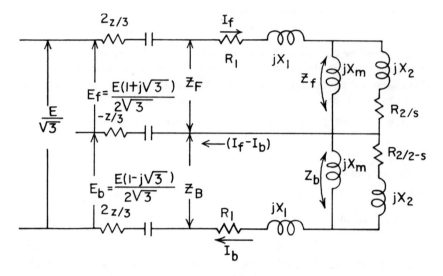

(b) EQUIVALENT CIRCUIT

Fig. 12.3. Connections and Equivalent Circuit for T-Connected Capacitor Motor: (a) Connection Diagram; (b) Equivalent Circuit.

The zero phase sequence current, I_0, is zero, because the Y point is not brought out.

Next, we express the voltages of each phase as the sums of the voltages produced by the currents of the three phases, and the voltages—E'_A, E'_B, and E'_C—induced by the air-gap field. To do this, we recognize that the self (line to neutral) impedance of each phase is $(1/2)(2Z+Z_0)$, where Z is the three-phase impedance of one phase under balanced conditions, and Z_0 is the zero phase sequence impedance of one phase. This gives:

(12.20) $\quad 3E_A = (2Z_1+Z_0+3z)I_A + (Z_0-Z_1)I_B + (Z_0-Z_1)I_C + 3E'_A,$

(12.21) $\quad 3E_B = (Z_0-Z_1)I_A + (2Z_1+Z_0)I_B + (Z_0-Z_1)I_C + 3E'_B,$

(12.22) $\quad 3E_C = (Z_0-Z_1)I_A + (Z_0-Z_1)I_B + (2Z_1+Z_0)I_C + 3E'_C.$

THE CAPACITOR MOTOR

In these equations, we have taken out the air-gap field from the total motor impedance, so that the value of Z is replaced in the equations by Z_1, which is the primary impedance only. z is the external impedance (capacitor) in line A.

Since there is no zero phase sequence field in the rotor, the zero phase sequence voltage in the rotor, E'_0, is zero, and the three air-gap voltages in terms of the sequence voltages are:

(12.23) $\quad \sqrt{3}\,E'_A = E'_f + E'_b; \quad \sqrt{3}\,E'_B = a^2 E'_f + a E'_b; \quad \sqrt{3}\,E'_C = a E'_f + a^2 E'_b.$

The next step is to express the sequence voltages in terms of the phase voltages:

(12.24) $\quad \sqrt{3}\,E_f = \sqrt{3}\,I_f Z_f = E_A + a E_B + a^2 E_C,$

(12.25) $\quad \sqrt{3}\,E_b = \sqrt{3}\,I_b Z_b = E_A + a^2 E_B + a E_C,$

(12.26) $\quad \sqrt{3}\,E_0 = E_A + E_B + E_C.$

Substituting Eqs. 12.20 to 12.23 in Eqs. 12.24 and 12.25, we find expressions for the sequence voltages in terms of the phase currents and impedances:

(12.27) $\quad \sqrt{3}\,E_f = I_A\left[\dfrac{2Z_1+Z_0+3z}{3} + \dfrac{a(Z_0-Z_1)}{3} + \dfrac{a^2(Z_0-Z_1)}{3}\right]$

$$+ I_B\left[\dfrac{Z_0-Z_1}{3} + \dfrac{a(2Z_1+Z_0)}{3} + \dfrac{a^2(Z_0-Z_1)}{3}\right]$$

$$+ I_C\left[\dfrac{Z_0-Z_1}{3} + \dfrac{a(Z_0-Z_1)}{3} + \dfrac{a^2(2Z_1-Z_0)}{3}\right]$$

$$+ E'_A + a E'_B + a^2 E'_C$$

$$= I_A(Z_1+z) + a I_B Z_1 + a^2 I_C Z_1 + \sqrt{3}\,E'_f,$$

(12.28) $\quad \sqrt{3}\,E_b = I_A(Z_1+z) + a^2 I_B Z_1 + a I_C Z_1 + \sqrt{3}\,E'_b,$

(12.29) $\quad \sqrt{3}\,E_0 = 3(Z_0+z)I_A + 3Z_0 I_B + 3Z_0 I_C.$

Finally, we substitute Eqs. 12.17 to 12.19 in Eqs. 12.27 and 12.28, obtaining the sequence equations for the motor:

$$\sqrt{3}\,E_f = \dfrac{I_f}{\sqrt{3}}(Z_1+z+Z_1+Z_1) + \dfrac{I_b}{\sqrt{3}}(Z_1+z+a^2 Z_1+a Z_1) + \sqrt{3}\,E'_f,$$

or

(12.30) $\quad E_f = I_f(Z_1+z/3) + \dfrac{I_b z}{3} + E'_f,$

and

(12.31) $$E_b = I_b(Z_1+z/3)+\frac{I_f z}{3}+E'_b.$$

To derive the equivalent circuit represented by these equations, we rearrange them to a new form:

(12.32) $$E_f = I_f(Z_F+2z/3)+(I_b-I_f)z/3,$$

(12.33) $$E_b = I_b(Z_B+2z/3)+(I_f-I_b)z/3,$$

which gives the equivalent circuit of Fig. 12.3 (b).

In these equations, we have represented the sums of the primary impedance Z_1 and the forward and backward sequence air-gap impedances, Z_f and Z_b, respectively, by Z_F and Z_B.

Referring now to Fig. 12.3 (a) from Eqs. 12.17 and 12.19, the relations between the actual motor currents and voltages and those of the equivalent circuit are:

(12.34) $$I_C = \frac{-1}{2\sqrt{3}}(I_f+I_b)+j\frac{(I_f-I_b)}{2} = \text{main winding current},$$

(12.35) $$I_A = \frac{1}{\sqrt{3}}(I_f+I_b) = \text{auxiliary winding current}.$$

The two mesh equations are:

(12.36) $$E = E_C-E_B = \frac{(a-a^2)}{\sqrt{3}}(E_f-E_b) = j(E_f-E_b) = \text{supply voltage},$$

$$0 = E_A-E_B = \frac{\sqrt{3}}{2}(E_f+E_b)+\frac{E}{2},$$

or

(12.37) $$E = -\sqrt{3}(E_f+E_b).$$

Applying these voltages to the circuit of Fig. 12.3 (b), and solving for the currents, we find, placing

$$M = Z_B(Z_F+2z/3)+Z_F(Z_B+2z/3):$$

(12.38) $$I_f = \frac{-E[Z_B+j\sqrt{3}(Z_B+2z/3)]}{\sqrt{3}M},$$

(12.39) $$I_b = \frac{I_f Z_F+jE}{Z_B} = \frac{-E\left[Z_F-j\sqrt{3}\left(Z_F+\frac{2z}{3}\right)\right]}{\sqrt{3}M}$$

It is worth while to compare the performance of this T connection, Fig. 12.3, to that of the same three-phase winding with the usual two-phase connection, Fig. 12.1. The questions of chief interest are to find which scheme requires the smallest capacitor, and/or the least change in capacitor size as the motor comes up to speed. To determine this, we calculate the values of I_f and I_b for each case, then find what value of z will give perfect phase balance ($I_b = 0$).

The Q-axis winding is one of the three stator phases, so the ratio, a, of its effective turns to those of the two phases in series forming the main winding is $1/\sqrt{3}$. With this value of a, the impedance of the center branch of the circuit of Fig. 12.1 becomes $\dfrac{Z_1}{4} + \dfrac{3z}{2}$, where $z = R_c + jX_c$. We have, then, from Fig. 12.1:

(12.40) $$\sqrt{2}\,E = I_f Z_F + I_b Z_b,$$

(12.41) $$\sqrt{6}\,E = I_b Z_B - I_f Z_F - \left(\frac{Z_1}{2} + 3z\right)(I_f - I_b).$$

These give:

(12.42) $$I_f = \frac{\sqrt{2}\,E\,[2(1 - j\sqrt{3})Z_B + Z_1 + 6z]}{N},$$

(12.43) $$I_b = \frac{\sqrt{2}\,E\,[2(1 + j\sqrt{3})Z_F + Z_1 + 6z]}{N},$$

where

(12.44) $$N = 4Z_F Z_B + (Z_1 + 6z)(Z_F + Z_B).$$

To make $I_b =$ zero, we must have:

(12.45) $z = -(1/6)[Z_1 + 2(1 + j\sqrt{3})Z_F]$, or, approximately,
$|z| = -\tfrac{2}{3} Z_F \underline{/60°}$.

As the motor speed rises, the magnitude and power factor of Z_F increase, so that z should have a considerable resistance at start, and should be a nearly pure capacitance at full speed.

For the T connection, to make $I_b =$ zero, z must be, from Eq. 12.39:

(12.46) $z = -\dfrac{\sqrt{3}}{2}(\sqrt{3} + j)Z_F$, or, approximately, $|z| = -\dfrac{\sqrt{3}}{2}(2Z_F)\underline{/30°}$.

The impedances used in the T circuit are those for one of the three phases, and are, therefore, only half those used in the two-phase circuit, which are for the main axis winding, with two phases in series. Hence, the actual magnitude of z required for perfect phase balance is slightly larger and less resistive for the T connection, giving fewer microfarads. At locked rotor, when Z_F is chiefly reactive, z should be nearly a pure capacitor, while at full speed it should have a considerable negative resistance component. With both connections, z should increase nearly in proportion to Z_F as the speed rises.

Permanent-split capacitor motors, which have a fixed capacitor value and therefore a low starting torque are widely used for fan drive in ratings below one-quarter horsepower, because of their good efficiency, high power factor, quiet operation, and trouble-free operation, without any slip rings or centrifugal devices. By providing taps on the main winding, or adding a series reactance, two or more speeds of operation can be obtained.

As pointed out in Sect. 11.11, the rotor bar resistance, R_{2b}, at $2-s$ frequency may be much higher than at slip frequency, due to induced eddy currents. Since a high value of R_{2b} reduces the maximum torque of a single-phase motor as well as adding to the losses and heating, the deep bars and double squirrel cages that are generally used in polyphase motors (Sect. 8.4) are not used in single-phase motors. Instead, every effort is made to reduce rotor bar skin effect, by using shallow slots and high rotor tooth densities.

However, a rotor with cast-in Alnico bars (Sect. 8.7) has a low effective resistance at low currents, regardless of the frequency, even though it has a high resistance that is proportional to the frequency when the current is above the "pick-up" value for the Alnico. Therefore, if a capacitor-run motor has a capacitor size chosen to hold the induced $2-s$ frequency rotor current below about half the full-load secondary current, and if the Alnico bar pick-up value is at about full-load current, the motor will have a materially lower value of R_{2b} than R_{2f} over the normal operating range. Use of this Saturistor principle should make capacitor-run motors more widely useful, particularly in integral horsepower sizes.

The cross-field circuit, derived from Fig. 11.12, has been widely used for calculating single-phase motor performance, but as pointed out in Sect. 11.11, it does not allow for differences between R_{2f} and R_{2b}. Since these differences can have a major effect on motor performance, the cross-field theory will not be considered here.

4 Speed Control of Capacitor Motors

A fractional horsepower motor has an inherently high secondary resistance (Sect. 2.13), and the full-load slip becomes greater the smaller the motor. Since the torque of a fan varies as the square of the speed, the fan speed can be controlled over a considerable range by merely varying the motor voltage. As the voltage decreases, the speed falls; until the reduced motor torque equals the torque of the fan. The smaller the motor and, therefore, the higher its full-load slip, the greater the range of speed control that can be obtained in this way. At small values of slip, the motor torque rises and the fan torque falls as the speed decreases, giving a stable intersection of the two speed-torque curves. Below the speed of maximum torque, the motor torque also decreases as the speed falls, and at some low speed the two curves become parallel. Below this point the fan speed is unstable.

The permanent-split capacitor motor is most suitable for this service. The capacitor size is chosen to give optimum balance between full-speed performance and cost. The current in the main winding rises rapidly as the speed falls, but the auxiliary winding current is nearly the same at all motor speeds, since it is limited by the high impedance of the capacitor. Hence, if the voltage across the auxiliary winding is maintained at full value, while the voltage on the main winding is varied, satisfactory fan speed control can be obtained with motors up to about 1-hp rating. To obtain $\frac{2}{3}$ speed of the fan, the voltage must be reduced to about half, and the main winding current will increase considerably, when a normal low-resistance squirrel-cage rotor is used. The heating due to the combination of higher current and low speed sets a limit to the permissible speed range.

Figure 12.4 shows a family of speed-torque curves for a 1-hp permanent-split capacitor motor with a normal squirrel-cage rotor, for different voltages on the main winding and a fixed voltage on the auxiliary winding. The torque curve of a fan that requires full motor output at full speed is also shown. The intersections of the fan and motor torque curves determine the voltage required for any desired speed.

The fan and motor torque curves are nearly parallel at speeds below $\frac{2}{3}$ of synchronous, so that the fan speed is quite unstable in this region, unless feedback control is provided. Figure 12.5 shows the corresponding family of speed-current curves, with a dotted curve indicating the locus of the motor current required to operate the fan at different speeds. The motor current at full speed is about 4.5 amperes, and it rises to about 8 amperes at 0.6 speed. For continuous operation at this speed, therefore, the motor will need to be oversize or provided with special cooling means.

Figures 12.6 and 12.7 show similar curves obtained by test when a new rotor with Alnico bars in the tops of the rotor slots was substituted in the same stator. Here, the slot leakage flux across the Alnico increases the reactance, and so reduces the maximum torque at rated voltage to about half the standard value. However, the hysteresis losses in the Alnico contribute torque, so that at low voltage and low speed the torque of the new rotor is actually greater. That is, the Alnico makes the speed-torque curve a great deal flatter.

With this rotor and the same fan, the current at full speed is about 5 amperes, but at $\frac{2}{3}$ speed it is less than 6, as compared with 8 for the

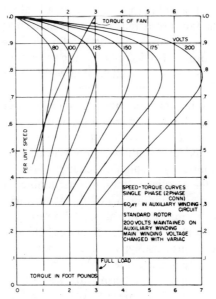

Fig. 12.4. Speed-Torque Curves of Capacitor Motor with Standard Rotor. 200 Volts Maintained on Auxiliary Winding; Main Winding Voltage Changed with Variac.

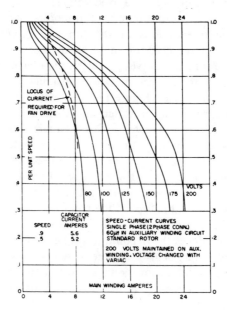

Fig. 12.5. Speed-Current Curves of Capacitor Motor with Standard Rotor. 200 Volts Maintained on Auxiliary Winding; Main Winding Voltage Changed with Variac.

SPEED CONTROL OF CAPACITOR MOTORS

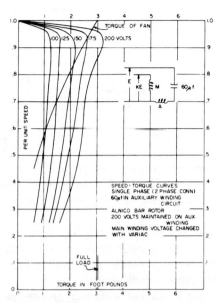

FIG. 12.6. Speed-Torque Curves of Capacitor Motor with Alnico Bar Rotor. 200 Volts Maintained on Auxiliary Winding; Main Winding Voltage Changed with Variac.

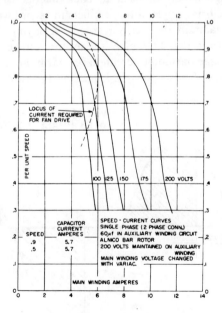

FIG. 12.7. Speed-Current Curves of Capacitor Motor with Alnico Bar Rotor. 200 Volts Maintained on Auxiliary Winding; Main Winding Voltage Changed with Variac.

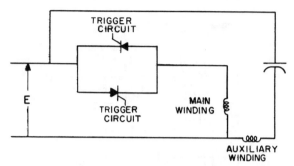

CONNECTION DIAGRAM FOR CAPACITOR MOTOR WITH SCR CONDUCTION ANGLE CONTROL OF MAIN WINDING VOLTAGE

Fig. 12.8. Circuit for Rectifier Supply of Motor Voltage.

standard rotor. And, the motor torque curves intersect the fan torque curve at steeper angles, giving much greater speed stability. Thus, the use of Alnico bars in the rotor markedly improves the motor performance in low-torque adjustable speed applications.

For constant torque loads and whenever precise speeds are required, feedback control is needed. To provide this, it is most convenient to supply the motor through a pair of back to back silicon rectifiers, Fig. 12.8, and to control the firing angle of the rectifiers in accord with a speed signal as required to maintain the desired speed. Delaying the firing of the rectifiers, as shown in Fig. 12.9, gives the motor voltage a large harmonic content. The voltage harmonics create large harmonic currents, since the motor slip for these fields is nearly 1, making the rotor impedance very

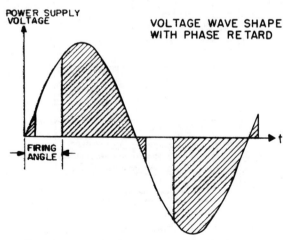

Fig. 12.9. Voltage Wave Shape with Phase Retard.

small. With the standard rotor, the firing angle had to be retarded about 100° to bring the fan speed down to $\frac{2}{3}$; and at this point the motor current was 10 amperes, indicating a harmonic content of about $\sqrt{(10^2-8^2)}=6$ amperes. With the Alnico bar rotor under the same conditions, $\frac{2}{3}$ speed required a firing angle delay of about 90°, and the motor current was only about 6 amperes, nearly the same as with sinusoidal applied voltage.

Figure 12.10 shows the per unit values of the sine and cosine terms of the fundamental voltage for a single-phase supply as functions of the angle of firing delay α and the angle of extinction $180 + \delta$. δ is zero for a unity power factor circuit but approaches 90° for a zero power factor circuit. With a high power factor, the effective voltage falls gradually as α is increased, but with low power factor the voltage does not change at all until a critical firing angle is reached, and then it falls very rapidly. The curves do not apply to a three-phase Y circuit, because at phase angles less than α one of the phases is open and the Y point voltage is halfway between the voltages of the other two terminals. Thus, the Y point voltage jumps each time an SCR fires and all three phases are open when α exceeds 120°.

5 THE PURE SINGLE-PHASE MOTOR

When the auxiliary winding circuit is opened, the capacitor motor becomes a pure single-phase machine, with current in the main winding only, Fig. 12.2 (a). Making $X_C = \infty$ in Fig. 12.1, the equivalent circuit reduces to that of Fig. 12.2 (b) which is the same as that for two identical polyphase motors with opposite phase rotation, connected in series. However, the existence of both forward and backward fields in the same air gap creates a double-frequency torque that would not occur with the independent polyphase motors.

At standstill, the common current creates equal and opposite torques in the two fields, since when $s=1$, $R_2/s = R_2/(2-s)$. If the shaft is turned forward, the forward field secondary resistance, R_2/s, rises and that of the backward field, $R_2/(2-s)$, decreases. Thus, a net forward torque is created, and the motor will accelerate to full speed, if the load torque is not too great. Similarly, if the shaft is turned backward, the motor will come up to full speed backward.

At full speed, the backward field impedance is small, and its torque is correspondingly low, but it nevertheless creates a braking effect that reduces the efficiency well below that of a corresponding polyphase motor. Comparing Fig. 12.2 (b) with Fig. 12.2 (c), it is seen that the single-phase motor performance is similar to that of a polyphase motor with an extra series connected impedance, approximately equal to $\dfrac{R_{2b}}{(2-s)} + j\dfrac{X_2}{2}$. The backward field resistance appearing in the circuit is only $\dfrac{R_{2b}}{2(2-s)}$, but this circuit loss creates a braking torque which must be subtracted from the output, thus doubling its effect.

The best way to understand the pure single-phase motor is to compare it point by point with a similar two-phase motor, Fig. 12.2 (d). In each case, the primary impedance between terminals is $R_1 + jX_1$. In the single-phase case, however, the effective magnetizing and secondary impedances are just half those of the two-phase motor. And, there is an extra series impedance in the single-phase case, due to the backward field. By making these changes in the equations developed for polyphase machines in

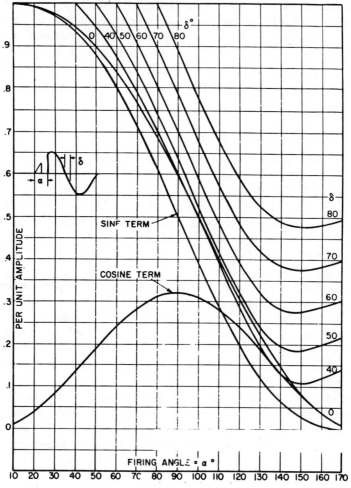

Fig. 12.10. Fundamental Voltage of SCRs with Retarded Firing.

THE PURE SINGLE-PHASE MOTOR

Chapter 5, the same formulas can be used to predict single-phase motor performance. For example, since R_2 is halved, while the total series reactance is nearly the same, and X_M is halved, the maximum torque of a single-phase motor occurs at a slip a little less than half as great as in two-phase operation, by Eq. 5.56.

Figure 12.11 shows the calculated speed-torque-current curves for a two-phase motor with per unit values $R_1 = 0.03$, $R_2 = 0.04$, $X_1 = X_2 = 0.05$, $X_M = 2.0$ in normal two-phase operation and also—with one line open—in single-phase operation.

In the figure, the rated current is taken as unity, and is the same for both single-phase and two-phase operation. Rated torque is taken as the volt-amperes corresponding to the total motor input, and is only half as great for single phase as for two phase. Therefore, the ratio of actual single-phase to two-phase torque is only half as great as that indicated by the numerical ratio of the values shown. The breakdown torque of a two-phase motor is reduced to about 40 per cent when operated on single phase at the same voltage.

6 The Dilemma of Single-Phase Motor Theory

At first sight, it appears remarkable that the magnetic flux in the Q axis of a two-phase induction motor retains nearly its full value when the Q-axis winding is opened, with the motor operating at speed. Since there is no mutual inductance between the D- and Q-axis windings, it seems that the Q-axis flux should disappear when the Q-axis current vanishes, as does in fact occur with the motor at standstill. Just how the Q-axis flux is maintained in single-phase operation constitutes a long standing dilemma of the theory.

When a two-phase motor is running at no load, taking a sine-wave current I_2 in the main phase and $-jI_2$ in the cross phase, Figs. 12.12 (a) and (b), its speed is almost synchronous and the rotor current is very small. The magnetic flux crossing the air gap, due to the stator mmf, Fig. 12.13(b) is a constant-magnitude rotating field with sinusoidal space distribution, except for slight four times frequency ripples as it passes from one phase to another. At any point x on the stator periphery, the mmf (per effective winding turn) due to the D-axis current I is:

(12.40) $$\sqrt{2}\,I_2 \cos x \cos \omega t,$$

and that due to the Q-axis current $-jI_2$ is:

(12.41) $$\sqrt{2}\,I_2 \cos\left(\frac{\pi}{2}-x\right)\cos\left(\omega t-\frac{\pi}{2}\right),$$

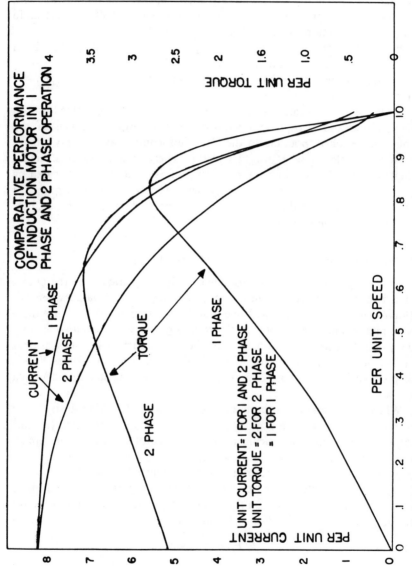

Fig. 12.11. Comparative Performance of Two-Phase Motor with One-Phase and Two-Phase Supply.

THE DILEMMA OF SINGLE-PHASE MOTOR THEORY 449

so that the resultant mmf at the point x at any time t—the sum of Eqs. 12.40 and 12.41—is:

(12.42) $\quad \sqrt{2} I_2 (\cos x \cos \omega t + \sin x \sin \omega t) = \sqrt{2} I_2 \cos(\omega t - x)$,

as indicated in Fig. 12.12 (c).

This 2-pole mmf is rotating at a speed $\omega = 2\pi f$ radians per second, while the rotor is turning in the same direction at a speed $(1-s)\omega$, so that the mmf and its corresponding magnetic flux are slipping forward on the rotor at a speed $s\omega$. The slip s will be less the lower the resistance of the rotor winding, in accordance with the constant-linkage theorem. If this

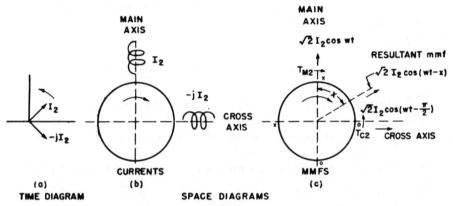

FIG. 12.12. MMF, Time, and Space Relations in Two-Phase Motor.

resistance is assumed to be zero, the flux can not move on the rotor, so that the rotor speed will be exactly synchronous, and $s = 0$.

That is, the rotor current, if any, will then vary in such a way as to provide at each instant the mmf needed to maintain the flux at a constant value, and fixed in position on the rotor. Any variations in the stator mmf from its normal sine-wave shape will produce ripples in the rotor current, but none in the rotor flux.

Fig. 12.13 (a) shows how the primary mmf, due to the main phase of the stator, appears to the rotor as it turns through a half-cycle. At (1), when $t = 0$, the primary mmf at the point Y of maximum flux density on the rotor is a maximum, directed outward. After a quarter-cycle, at (2), it is zero, and after a half-cycle, at (3), it is again a maximum in the outward direction. When Y has turned through the angle ωt away from the stator main axis, the rotor mmf at Y is $\cos \omega t$ times the main-axis mmf at the same instant, or is:

(12.43) $\quad \sqrt{2} I_2 \cos^2 \omega t = \dfrac{I_2}{\sqrt{2}}(1 + \cos 2\omega t)$.

Thus, the mmf of this one stator phase, as seen at Y with the rotor turning at synchronous speed, is unidirectional and pulsating at double frequency between zero and a maximum, as shown by the full line in Fig. 12.13(c). In the two-phase motor, the mmf of the cross phase is exactly similar, but displaced in time phase as shown by the dashed curves in Figs. 12.13(b) and (c). The sum of the two, as seen by the rotor, is a constant mmf, which accordingly produces the constant value of rotor flux needed to balance the impressed stator voltage.

To find the mmf due to the main-phase current at any other point x on the rotor, this mmf at the instant considered must be multiplied by the cosine of the angle between x and the stator main axis at that moment. Hence, as shown in Fig. 12.13 (d), at a point on the rotor θ degrees ahead of Y, the mmf due to the main-phase current at any time t is:

(12.44) $$\sqrt{2}I_2 \cos \omega t \cos(\omega t + \theta) = \frac{I_2}{\sqrt{2}}[\cos \theta + \cos(2\omega t + \theta)]$$

and, the mmf due to the cross-phase current at the same time is:

(12.45) $$\sqrt{2}I_2 \cos(\omega t - 90)\cos(\omega t + \theta - 90) = \frac{I_2}{\sqrt{2}}[\cos \theta - \cos(2\omega t + \theta)].$$

The net mmf at this point on the rotor is the sum of Eqs. 12.44 and 12.45, or:

(12.46) $$\sqrt{2}I_2 \cos \theta,$$

as seen in Fig. 12.13 (d).

Thus, the net mmf is a sine wave that is fixed in rotor position, with its maximum at point Y. Comparing Eq. 12.20 with Eq. 12.42, it is seen that the transfer from stator to rotor is accomplished by merely substituting θ for $\theta - \omega t$, when the rotor speed is synchronous, or $(\theta - s\omega t)$ for $(\theta - \omega t)$ if the slip is s.

If now the cross phase of the stator is opened, it has been found that the motor continues to run at synchronous speed (if the rotor resistance is zero), but the primary current in the main phase changes from I_2 to I_1. Since the voltage is the same, the main-axis flux must be the same, and, to sustain this flux the unidirectional mmf viewed from the rotor must also be the same. Formerly, as shown in Fig.12.13(c), the cross phase supplied half of the unidirectional (net) mmf, so now, to maintain the same total, the main-phase current I_1 must be twice as great as it was on two phase; i.e., $I_1 = 2I_2$, as shown in Figs. 12.14 (a) and (b). In accord with the constant-linkage theorem, a current must then flow in the rotor to cancel the double-frequency alternating component of the primary mmf, and so maintain the flux at its constant value, as also shown in Fig. 12.14 (b).

The mmf due to the induced rotor current must be equal to the difference between the required unidirectional mmf and the varying mmf impressed by the stator main-axis current. This difference at the point Y of maximum rotor flux is given by the dashed curve of Fig. 12.13(b). Evidently, this curve represents a sinusoidal current distribution which varies at twice line frequency *in the rotor*.

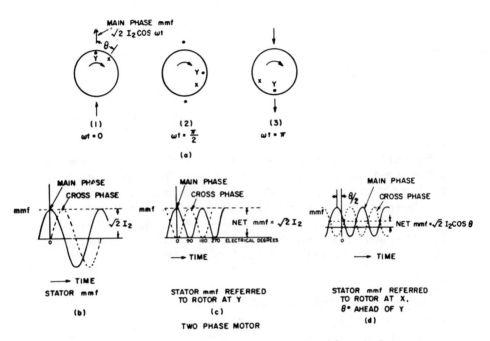

FIG. 12.13. Stator MMF of Two-Phase Motor as Viewed from the Rotor.

At any other point x on the rotor, θ degrees ahead of Y, from Eq. 12.44, the mmf impressed on the rotor by the stator current I_1 is:

$$\frac{I_1}{\sqrt{2}}[\cos\theta + \cos(2\omega t + \theta)]. \tag{12.47}$$

From Eq. 12.46, the necessary mmf at this point to sustain the constant rotor flux is $(I_1/\sqrt{2})\cos\theta = \sqrt{2}\,I_2\cos\theta$, as shown in Fig. 12.14(c). Hence, the rotor mmf at x must be the difference between Eqs. 12.46 and 12.47, or:

$$-\frac{I_1}{\sqrt{2}}\cos(2\omega t + \theta). \tag{12.48}$$

When $\theta = 0$, at point Y, this is $-(I_1/\sqrt{2})\cos 2\omega t$, in agreement with Fig. 12.14(b). When $\theta = 90$ degrees, at the point of zero flux density, it is:

$$\frac{I_1}{\sqrt{2}}\sin 2\omega t = \sqrt{2}I_2 \sin 2\omega t,$$

which is the same as the mmf impressed on the rotor by the cross-phase current in two-phase operation, Fig. 12.13(d).

Therefore, there is a double-frequency current

$$-\frac{I_1}{\sqrt{2}}\cos 2\omega t = -\sqrt{2}I_2 \cos 2\omega t$$

in the rotor axis of maximum flux, and, by Eq. 12.48, there is also a double-frequency current $\frac{I_1}{\sqrt{2}}\sin 2\omega t = \sqrt{2}I_2 \sin 2\omega t$ in the rotor axis of zero flux, 90° ahead of Y.

Figure 12.15 (a) shows the stator mmf in single-phase operation, varying as $\cos \omega t$ in time, and varying in space as $\cos x$, from 1 in the main axis to 0 in the cross axis. Figure 12.15 (b) shows the corresponding mmf of the double-frequency rotor current of Fig. 12.14 (b), referred to the stator. The transfer from rotor to stator was made by substituting x for $(\theta + \omega t)$ in Eqs. 12.47 and 12.48. Therefore, the net mmf referred to the stator—the sum of the mmfs due to the stator and rotor currents—is:

(12.49)

Stator mmf + rotor mmf = net mmf

$$\frac{I_1}{\sqrt{2}}[2\cos x \cos \omega t - \cos(\omega t + x)] = \frac{I_1}{\sqrt{2}}\cos(\omega t - x) = \sqrt{2}I_2 \cos(\omega t - x).$$

This net mmf, as expected, is exactly the same as that of the two-phase motor, Fig. 12.12 (c) and Eq. 12.42. It is proportional to $\cos(\omega t - x)$, which in the cross axis, $x = \pi/2$, takes the values 0, 0.71, 1, 0.71, and 0 at successive time values of 0, 45, 90, 135, and 180°. It, therefore, represents a constant-magnitude mmf, revolving forward in synchronism with the rotor.

On the other hand, the mmf of the double-frequency rotor current, referred to the stator, Fig. 12.15 (b), is proportional to $\cos(\omega t + x)$, which takes the values 0, -0.71, -1, -0.71, and 0 at the times 0, 45, 90, 135, and 180°. This mmf, therefore, revolves backward at synchronous speed with respect to the stator, or backward at twice synchronous speed on the rotor. This is self-evident after all, as the stator frequency must always be the line frequency, and Fig. 12.14 (b) shows that the induced current has double line frequency in the rotor. The mmf speed of twice backward on the rotor,

THE DILEMMA OF SINGLE-PHASE MOTOR THEORY 453

added to the forward rotor speed of one gives a net speed with respect to the stator of one, as required.

When a current of i amperes flows in a circuit, having an inductance L henrys, the voltage across the inductance is $L\,di/dt$, and the power input

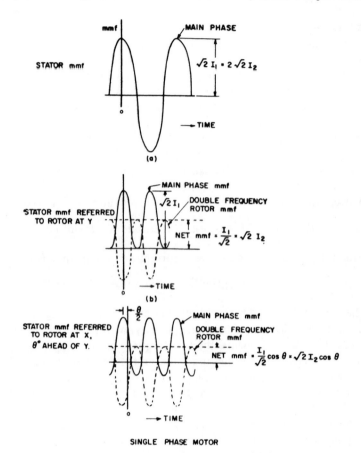

SINGLE PHASE MOTOR

FIG. 12.14. MMF, Time, and Space Relations in Single Phase Motor.

to it is $Li\,di/dt$. The energy stored in the magnetic field is the integral of this times dt, or $Li^2/2$. In terms of the reactance, $X = 2\pi fL$, this energy is $i^2 X/4\pi f$. In terms of the rms volt-amperes EI of the circuit (resistance neglected), the peak energy supplied to the field during each quarter-cycle, while the current is building up, is $Ei/2\pi f$ watt-seconds (Sect. 3.13). In alternate quarter-cycles, while the current is decreasing, this energy is returned to the power source. This constitutes the well-known double-frequency alternating power, or reactive power flow, of any circuit containing reactance.

When the motor operates on two-phase supply, each phase in turn delivers energy directly to the magnetic field in its own axis, the one dying down as the other builds up, so that the combined (rotating) field remains always constant, and the stored field energy also remains the same.

The way this is accomplished may be seen by considering the conditions in a two-phase motor at the instant when the air-gap flux has turned through the angle $x = \omega t$ past the stator main axis, Fig. 12.12 (c). At this moment, the main-axis downward current, shown by ×, flows across the outward cross-axis field, and the cross-axis upward current, shown by o, flows across the main-axis outward field. Thus, there is a torque produced in each axis, proportional to the product of flux and current magnitudes and to the sine of the angle between them. The main-axis current is $\sqrt{2}\,I_2 \cos \omega t$, that of the cross axis is $\sqrt{2}\,I_2 \sin \omega t$, and the air-gap flux is proportional to $\sqrt{2}\,I_2$, at the angle $x = \omega t$.

Hence, the torque due to the main-axis current and the cross-axis flux is:

(12.50) $\qquad T_{M2} = K(\sqrt{2}\,I_2 \cos \omega t)(\sqrt{2}\,I_2 \sin \omega t) = K I_2^2 \sin 2\omega t,$

and the torque due to the cross-axis current and the main-axis flux is:

(12.51) $\quad T_{C2} = -K(\sqrt{2}\,I_2 \sin \omega t)(\sqrt{2}\,I_2 \cos \omega t) = -K I_2^2 \sin 2\omega t.$

The current, flux, and torque directions are indicated in Fig. 12.12 (c). Since all three fields are of the same polarity, with the actual flux located between the two components, the two torques are opposing; and, in fact, their sum is zero. The torques indicated on the figure are those acting on the stator. The main-axis torque tends to turn the stator in the same direction as the field is rotating, so that it represents energy being returned to the line. This is correct, because when sin t is positive the main-axis field is decreasing, and is, therefore, giving up its magnetic energy to the line. An equal amount of energy is being delivered to the cross-field at the same time, so that the total field energy remains constant, and there is no net torque on the rotor (at no load, and neglecting losses).

On single phase, the current in the main phase is twice what it was on two phase, and, therefore, the energy delivered to the main axis in each quarter-cycle is also twice as great on single phase as on two phase. The main-axis flux is the same in both cases, so only half of the energy supplied is needed for the main-axis field, and the other half must go elsewhere. Where this other half goes is explained by the presence of the rotor current in single-phase operation, which offsets half of the stator mmf in the main axis.

THE DILEMMA OF SINGLE-PHASE MOTOR THEORY 455

At the time $x = \omega t$, the torque between the main-axis stator current and the actual air-gap flux is, from Eq. 12.50,

$$(12.52) \quad T_{M1} = K(\sqrt{2}\,I_1 \cos \omega t)\left(\frac{I_1}{\sqrt{2}} \sin \omega t\right)$$

$$= K\frac{I_1^2}{2} \sin 2\omega t = 2KI_2^2 \sin 2\omega t.$$

This is just twice the corresponding torque in two-phase operation, since $I_1 = 2I_2$, while the air-gap flux is unchanged. Equation 12.52 gives the total torque acting on the stator, as there is no cross-axis stator current. Since T_{M1} is twice as large as needed to supply the energy of the main-axis field, half of T_{M1} must be transmitted to the rotor.

At the same time, $\omega t = x$, the double-frequency rotor current in the main axis is flowing across the cross-axis component of the air-gap flux, making a torque applied to the rotor:

$$(12.53) \quad T_{MR1} = -K\left(\frac{I_1}{\sqrt{2}} \cos \omega t\right)\left(\frac{I_1}{\sqrt{2}} \sin \omega t\right)$$

$$= -K\frac{I_1^2}{4} \sin 2\omega t = -KI_2^2 \sin 2\omega t.$$

And, there is a corresponding rotor torque due to the double-frequency rotor current in the cross axis flowing across the main-axis component of the air-gap flux:

$$(12.54) \quad T_{CR1} = -K\left(\frac{I_1}{\sqrt{2}} \sin \omega t\right)\left(\frac{I_2}{\sqrt{2}} \cos \omega t\right)$$

$$= -KI_2^2 \sin 2\omega t.$$

Both T_{MR1} and T_{CR1} are decelerating torques when $\sin 2\omega t$ is positive. T_{CR1} is the same as T_{C2} in two-phase operation, both representing flow of energy from the rotor into the magnetic energy of the cross-field. T_{MR1} represents a flow of energy from the rotor into the main-axis magnetic field, which accounts for the other half of the stator torque, T_{M1}.

Figure 12.16 illustrates what is happening. Referring to Fig. 12.16 (a), at the instant when the rotor flux coincides with the main axis ($\omega t = 0$), the main-axis magnetic energy is a maximum, the cross-field flux and energy are zero, and the rotor mmf is along the main axis. The motor torque is zero, since there is no rotor current in the main-axis field, and no field in the cross axis where the rotor currents are. The power flow from the line is also zero at this instant, since the main axis is a maximum, and, therefore, the voltage induced by it is zero.

One-eighth of a cycle later, Fig. 12.16 (b) ($\omega t = \pi/4$), the rotor flux axis has turned 45°, the main-axis flux has fallen to 71 per cent of its maximum, and the cross-axis field has risen to an equal value. From Fig. 12.15 (b), the rotor mmfs in the main and cross axes are both 71 per cent of their maxima also. Under these conditions, the main-axis rotor current in the cross-axis magnetic field, and the cross-axis rotor current in the main-axis field produce equal decelerating torques on the rotor, slowing it down. Thus, the rotor is giving up kinetic energy, half of which goes into building up the

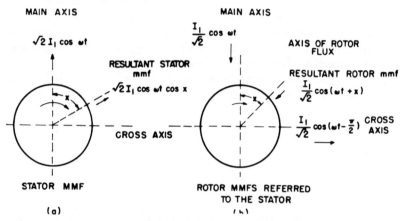

Fig. 12.15. Stator and Rotor MMFs of Single-Phase Motor as Viewed from the Stator.

cross-field energy, and half is delivered to the main field. At the same time, the main-axis field is decreasing, so that its energy, and the equal amount being received from the rotor, flow back to the line, making the total flow into the line the same as that given up by the rotor.

At successive later intervals of one-eighth of a cycle, the torques become zero, reverse, and fall to zero again, accomplishing in this way the transfer of energy from the line to the magnetic fields and back again. Half the energy received from the line in the main axis is delivered to the rotor where it is stored in kinetic energy form as the rotor speeds up, then delivered to the cross-field, by way of the torques produced by the induced rotor current cutting the air-gap flux; and one-quarter of a cycle later the energy is returned to the line by the same method, in reverse.

A convincing proof of the existence of these double-frequency torques in a single-phase motor at no load is given by observing the oscillations of the stator of such a motor when running at no load on a flexible base, or by listening to the twice line frequency noise produced when such a motor is placed directly on a table top, or other resonant surface. A careful

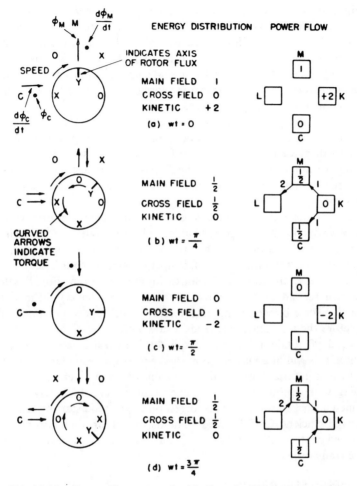

Fig. 12.16. Energy Flow across the Air Gap of a Single-Phase Motor.

comparison of test and calculated values of these oscillations, showing excellent agreement, has been reported in the literature.

7 Resistance-Split Phase Motors

Single-phase induction motors of less than about $\frac{1}{3}$-hp rating, which do not require high starting torque per ampere, are usually of the resistance-split phase type. Typical applications are for washing machines and dishwashers. In these motors, the needed phase angle difference, θ, between the D- and Q-axis currents is obtained by adding resistance in the auxiliary winding circuit instead of capacitance. R_{1d} must be small to obtain good efficiency, so it is not feasible to make $R_{1A}/X_{1A} < R_{1d}/X_{1d}$. Hence, to make θ large enough to obtain adequate torque, it is necessary to make $R_{1A} > R_{1d}$, by using conductors of smaller cross section or higher resistivity in the start winding, and/or to make $X_{1A} < X_{1d}$ by choosing $a < 1$. In practice, it is not feasible to make $\theta > 30°$, approximately, so that the locked-rotor torque per ampere squared, Eq. 12.12, is only about half that obtained in polyphase motors.

Normally, the starting winding is located in the Q axis, 90° away from the main winding, giving winding symmetry. It is possible, however, either to displace the start winding from the Q axis, or to add a third winding, with only a few turns, in the D axis and connect this in series with the start winding. In this way, the effective value of θ may be increased with some additional cost.

Because of its high resistance, the starting winding temperature rises very rapidly with full voltage impressed at locked rotor. With a copper winding, the initial current density may be of the order of 50,000 amps/sq in. or more, giving an apparent initial rate of temperature rise of 40 C/sec or more. Hence, it is essential to open the start winding with a centrifugal switch or relay in a second or two. Most of these motors have built-in thermal protection, or a bimetal temperature relay, which takes the motor off the line before the temperature exceeds a safe limit.

Instead of obtaining the phase split by high resistance in the starting winding, this winding may be connected across the secondary of a current transformer in series with the line. The phase displacement of the starting winding voltage thus obtained enables a phase split of nearly 90° to be obtained at locked rotor, with a correspondingly good starting torque and improved efficiency at full speed. The motor requires no starting switch, and speed control can be obtained by changing taps on the primary winding of the transformer.

8 Shaded-Pole Motors

The lowest cost construction for a single-phase motor is obtained with salient stator poles and permanently short-circuited windings of relatively high resistance linking one tip of each pole, as shown in Fig. 12.17. Normally, the auxiliary winding, called a shading coil, consists of a single uninsulated copper strip encircling about one-third of a pole pitch. Occasionally, two such coils may be used per pole.

With only one coil per pole, the main winding produces an mmf wave with a large harmonic content, even though the pole tips may be chamfered to improve the waveform. The current induced in the shading coil by the D-axis flux creates a counter mmf, reducing the flux density in the area spanned by the coil, and causing this flux to lag in time phase behind the main flux. The air gap under the shaded pole tip may be made smaller than under the rest of the pole, to offset the lower mmf and make the flux densities in the two areas nearly equal. The differences in time- and in space-phase angles of the air-gap fluxes in the shaded and unshaded pole areas create a rotating field, and cause the rotor to turn in the direction from a pole center toward the shaded tip.

As the shading coil has short pitch, it creates large 3rd and 5th harmonic fields as well as a fundamental. Thus, the torques due to the harmonics are

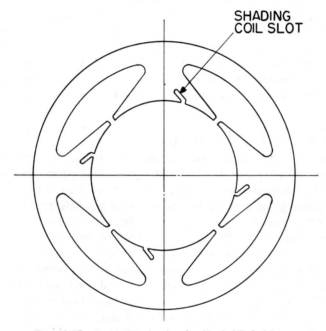

FIG. 12.17. Stator Lamination for Shaded-Pole Motor.

so large that the motor would not accelerate to full speed, if means were not taken to reduce them. For this purpose, it is usual to skew the rotor slots of a shaded-pole motor through 30 or more electrical degrees. And, the rotor resistance is made quite high, to increase the starting torque.

At full speed, the induced currents in the shading coils create a good deal of loss. At best, therefore, the shaded-pole motor performance is much inferior to the other forms of single-phase induction motor, so that it is limited to outputs of about 50 watts or less, and is used chiefly for fan drive.

To obtain a reasonably close check between calculated and test performance, it is necessary to include at least the 3rd and 5th mmf harmonics in the equations. Also, allowance for magnetic saturation is essential, as this is increased by the large skew. Thus, the complete equivalent circuit and equations for the shaded-pole motor are far more complex than for any of the other forms of induction motor, and no consideration will be given to them here.

An especially useful feature of the shaded-pole motor is its high full-load slip, together with a fairly flat torque curve. This enables reduced speed operation of a fan to be obtained, by using a series reactor that can be switched into the line in one or more steps.

Performance similar to the shaded pole motor can be obtained by merely making the air gap larger under one section of the pole, omitting the shading coil. The air gap flux in the pole section with the larger air gap leads in time phase; so the rotor will turn in the direction from the large to the small gap. This "reluctance start" motor has very low starting torque, but should have better efficiency than the shaded pole design.

9 Repulsion-Induction Motors

Before low-cost capacitors became available, repulsion-start-induction-run machines were the most widely used type of single-phase motor in the range from one-third to five horsepower. This motor has a normal direct-current winding and commutator on the rotor with a distributed wound stator having only a single winding, in the D axis. The rotor brushes are permanently short circuited and are displaced from the D axis, so that a rotor current is induced by transformer action. The flux made by this current, which is fixed in position, acts with the main-winding flux to create a torque, which is very high at locked rotor and decreases continually as the speed rises. The speed-torque curve is similar to that of a series d-c motor, so that the motor speed will rise indefinitely until limited by the load torque and losses. To prevent this, these motors are usually provided with a centrifugal mechanism which short circuits the entire commutator when the motor reaches perhaps two-thirds of synchronous speed. Frequently, the mechanism also lifts the brushes off the commutator.

The critical point in the speed-torque curve is that at which the centrifugal mechanism operates, changing the motor from repulsion to induction operation. The torque at this change-over point is usually low but may be adjusted by changing the brush angle.

When operating at speed, the machine behaves like an ordinary squirrel-cage motor, except that with the distributed rotor winding the rotor induced harmonic currents are very small, and the torque curve, therefore, is smooth.

To avoid the centrifugal mechanism, repulsion-induction motors also have been widely used. These have a deeply buried, low-resistance, squirrel cage in the rotor with leakage slots between the bars and the repulsion winding. These slots have such high reactance that the squirrel-cage currents induced at locked rotor are small, and the motor has the normal repulsion characteristic up to about two-thirds speed. At higher speeds the squirrel-cage current increases markedly, and the induction torque becomes larger than the repulsion torque, so that the no-load speed is limited to a little above synchronism. These motors have the disadvantage that the brushes carry current continuously, creating some radio interference.

Both of these types of repulsion-induction motor have fallen into disuse, as they have been replaced by the capacitor motor.

Bibliography

Reference to Section

1. "Test Procedure for Single-Phase Induction Motors," *A.I.E.E. No. 502*, 1958.
2. "The Dilemma of Single-Phase Induction Motor Theory," P. L. Alger, *A.I.E.E. Trans.*, Vol. 75, Part III, 1958, pp. 1045–53. — 5
3. "Single-Phase Induction Motors," P. L. Alger, C. W. Falls, and A. F. Lukens, *Standard Handbook for Electrical Engineers*, Ninth Edition, McGraw-Hill, 1957, pp. 736–43.
4. "Single-Phase Motor Torque Pulsations," P. L. Alger and A. L. Kimball, *A.I.E.E. Trans.*, Vol. 43, 1924, pp. 730–9. — 4
5. "A Capacitor Motor with Alnico Bars in the Rotor," P. L. Alger, R. L. Mester, and J. G. Yoon, *I.E.E.E. Trans.*, Vol. 83, No. 10, October 1964, pp. 989-997.
6. "Self-Excitation of Capacitor Motors," G. Angst, *A.I.E.E. Trans.*, Vol. 71, Part III, 1952, pp. 557–62. — 3
7. "A New Type of Single-Phase Motor," S. R. Bergman, *A.I.E.E. Trans.*, Vol. 43, 1924, pp. 1039–43. — 8
8. "Auxiliary Winding Design for Split-Phase Motors," L. W. Buchanan and N. Maupin, *A.I.E.E. Trans.*, Vol. 79, Part III, 1960, pp. 1183–7. — 6
9. "Induction Motor Theory—Some Elementary Concepts Extending to Supersynchronous Speeds," C. T. Button, *A.I.E.E. Trans.*, Vol. 73, Part IIIA, 1954, pp. 289–93. — 4
10. "Equivalent Circuits and Their Application in Designing Shaded-Pole Motors," S. S. L. Chang, *A.I.E.E. Trans.*, Vol. 70, Part I, 1951, pp. 690–8. — 7
11. "Single-Phase Induction Motors with Multiple Non-Quadrature Winding," S. S. L. Chang, *A.I.E.E. Trans.*, Vol. 75, Part III, 1956, pp. 913–16. — 7
12. "A New Method of Starting Single-Phase Induction Motors," A. K. Chatterjee and V. K. Jain, *A.I.E.E.* Conference Paper No. 62–1433. — 6
13. "The Repulsion-Start Induction Motor," J. L. Hamilton, *A.I.E.E. Trans.*, Vol. 34, 1915, pp. 2443–74. — 8
14. "Equivalent Circuit of the Capacitor Motor," G. Kron, *G.E. Rev.*, Vol. 44, Sept. 1941, pp. 511–13. — 3
15. "Equivalent Circuits of the Shaded Pole Motor with Space Harmonics," G. Kron, *A.I.E.E. Trans.*, Vol. 69, Part II, 1950, pp. 735–41. — 7

16. "Physical Conception of the Operation of the Single-Phase Induction Motor," B. G. Lamme, *A.I.E.E. Trans.*, Vol. 37, 1918, pp. 627–58. 5
17. "The Revolving Field Theory of the Capacitor Motor," W. J. Morrill, *A.I.E.E. Trans.*, Vol. 48, 1929, pp. 614–29. 3
18. "Digital Computer Analysis of the Instantaneous Reversal Transient of a Single Phase Motor," D. W. Novotny and J. J. Grainger, *I.E.E.E. Trans.*, Vol. 83 PAS. 1964, pp. 380-386.
19. "High-Frequency Iron Losses in Fractional Horsepower Motors," V. C. Shaneman, *A.I.E.E. Trans.*, Part III, 1961, p. 579.
20. "Calculation of Shaded-Pole Motor Performance," G. H. Sherer and G. E. Herzog, *A.I.E.E. Trans.*, Vol. 78, Part III, 1959, pp. 1607–10. 7
21. "Equivalent Circuits for Single-Phase Motors," G. R. Slemon, *A.I.E.E. Trans.*, Vol. 74, Part III, 1955, pp. 1335–42. 3
22. "A Theory for Shaded-Pole Induction Motors," F. W. Suhr, *A.I.E.E. Trans.*, Vol. 77, Part III, 1958, pp. 509–15. 7
23. "An Analysis of the Shaded-Pole Motor," P. H. Trickey, *A.I.E.E. Trans.*, Vol. 55, 1936, pp. 1007–14. 7
24. "Performance Calculations on Capacitor Motors by the Cross-Field Theory," P. H. Trickey, *A.I.E.E. Trans.*, Vol. 75, Part III, 1956, pp. 1547–52. 3
25. "MONECA, A New Network Calculator for Motor Performance Calculations," C. G. Veinott, *A.I.E.E. Trans.*, Vol. 71, Part III, 1952, pp. 231–8. 3
26. *Fractional Horsepower Electric Motors*, C. G. Veinott, McGraw-Hill Book Company, Inc., New York, 1939.
27. "Theory and Calculation of the Squirrel-Cage Repulsion Motor," H. R. West, *A.I.E.E. Trans.*, Vol. 43, 1924, pp. 1048–55. 8

CONTENTS—CHAPTER 13

Special Types and Connections of Induction Machines

		PAGE
1	Unbalanced Two-Phase Motors	467
2	Unbalanced Three-Phase Motors	469
3	Dual-Winding Motors	476
4	Transient Currents	480
5	Reluctance Motors	486
6	The Permasyn Motor	488
7	Hysteresis Motors	489
8	Synchronous Induction Motors	490
9	A-C Excited Induction Motors	490
10	Linear Induction Motors	491

13

SPECIAL TYPES AND CONNECTIONS OF INDUCTION MACHINES

1 Unbalanced Two-Phase Motors

If different voltages, E_d and E_q, are applied to the two phases of a capacitor motor, Fig. 13.1 (a), the sequence voltages are:

(13.1) $$E_f = \left(E_d - \frac{jE_q}{a}\right)\bigg/\sqrt{2}; \quad E_b = \left(E_d + \frac{jE_q}{a}\right)\bigg/\sqrt{2}.$$

Putting Z_{1d} for the total impedance in the primary circuit of the D winding, and Z_{1q} for that in the Q winding, Fig. 12.1 becomes Fig. 13.1 (b), which is the equivalent circuit for an unbalanced two-phase motor with unbalanced applied voltages. Making $a = 1$ gives the circuit for equal turn numbers in the two phases, and also making $Z_{1d} = Z_{1q}$ gives the circuit for a symmetrical two-phase motor with unbalanced applied voltages. This is represented by Fig. 13.1 (b), with zero impedance in the midpoint connection.

If the voltages are balanced also, $E_d = jE_q$, and E_b becomes zero, so the circuit reduces to Fig. 13.1 (c). All of these circuits give the total motor power, so in the last case the circuit voltages and currents are $\sqrt{2}$ times the actual values. By dividing E and I by $\sqrt{2}$, the circuit reduces to Fig. 13.1 (d), which is the conventional equivalent circuit of a polyphase motor, Fig. 5.1, giving the power of one phase only.

If in a two-phase induction motor with balanced windings ($a = 1$) but unbalanced external impedances and unbalanced applied voltages, the rotor is single phase (with current in only one axis), the equivalent circuit becomes that of Fig. 13.2 (a), derived from Figs. 9.6 and 13.1. Here, the s frequency rotor current creates, in addition to the normal s frequency forward field, a field turning backward on the rotor at speed s, whose speed on the stator is $1-s-s = 1-2s$; and the $2-s$ frequency rotor current creates an additional field turning forward on the rotor at speed $2-s$, so that its speed on the stator is $1-s+2-s = 3-2s$. These frequencies are indicated by the denominators of the resistance terms in the primary circuit branches which shunt the magnetizing reactances of the two additional rotor fields. At standstill, all the fields represent motor action, and all have the line frequency, so that all the denominators must be $+1$ when

468   SPECIAL TYPES AND CONNECTIONS OF INDUCTION MACHINES

$s = 1$. That is, the relative speed of each field is always measured with respect to its initial direction of rotation, so that the denominator for the backward motor field is $2s - 1$, not $1 - 2s$.

If the stator is also single phase, $Z_q = \infty$, the midpoint connection is open, and the circuit becomes that of Fig. 13.2 (b). Here the $1 - 2s$

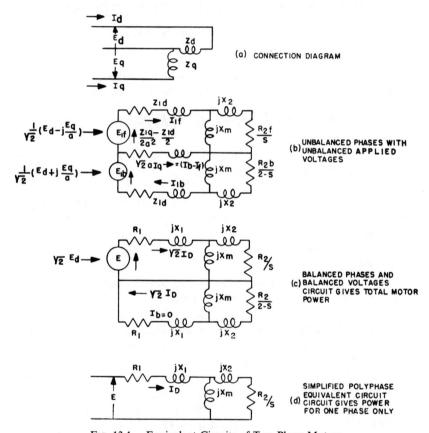

FIG. 13.1.   Equivalent Circuits of Two-Phase Motors.

frequency current in the stator creates an additional backward $1 - 2s$ frequency field, which therefore has a $2 - 3s$ frequency in the rotor. And, the $3 - 2s$ frequency stator current creates an additional $3 - 2s$ frequency forward field, which has a $4 - 3s$ frequency in the rotor. Each of these new fields creates currents in the single-phase windings of both stator and rotor, which in turn produce both forward and backward fields. Thus, there is an infinite series of currents with successively higher frequencies on both sides of the air gap, and the total currents are,

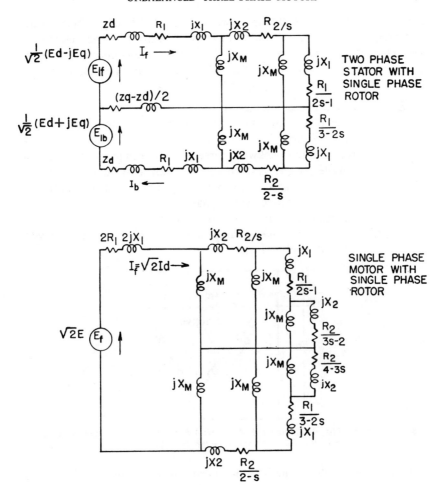

FIG. 13.2. Equivalent Circuits of Two Phase Motor with Single Phase Rotor.

therefore, far from sinusoidal. As mentioned in Sect. 9.3, this single-phase connection creates very high rotor voltages. For this and other reasons mentioned previously, the polyphase stator with single-phase rotor, or Dreyfus motor, is not useful.

2 Unbalanced Three-Phase Motors

To make use of the equivalent circuits presented here, it is necessary to transform the circuits of three-phase machines to rotating field axes. As an example of this process, we shall take a split-winding, three-phase induction motor, with K times as many turns (but the same volume of copper)

470 SPECIAL TYPES AND CONNECTIONS OF INDUCTION MACHINES

in phase A as in phases B and C, which is treated briefly in Sect. 8.10. However, we shall make the additional assumptions that the machine is connected in delta, and that external impedances are connected in series, $K^2 y$ in phase A, and z in phases B and C, as shown in Fig. 13.3. Let:

I_A, I_B, I_C = actual phase currents,
E_A, E_B, E_C = actual phase voltages,
$\quad Z_1$ = three-phase impedance of stator per phase, with normal winding turns,
Z_f, Z_b = forward and backward three-phase impedances of rotor and parallel magnetizing circuit, with normal winding turns,
$\quad K$ = ratio of A phase turns to normal (B and C phase) turns,
$K^2 y$ = external impedance in A phase,
$\quad z$ = external impedance in B and C phases,
Z_0 = zero-phase sequence impedance of stator per phase with normal winding turns.

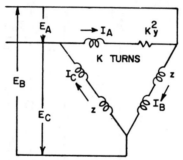

FIG. 13.3. Three-Phase Δ-Connected Motor with K Times as Many Turns in Phase A as in B or C; External Impedances in Each Phase.

Then, from Eq. 7.10, the line-to-neutral impedance of one phase with normal winding turns is:

$$\tfrac{1}{3}[2(Z_1+Z_f)+Z_0],$$

and the mutual impedance between any two phases is:

$$\tfrac{1}{3}(Z_0-Z_1-Z_f).$$

The impressed phase voltages, assumed balanced, are (Fig. 3.16):

(13.2) $\quad E_A = E,$

(13.3) $\quad E_B = E\,\underline{/240} = E\left(-\dfrac{1}{2}-j\dfrac{\sqrt{3}}{2}\right) = a^2 E,$

(13.4) $\quad E_C = E\,\underline{/120} = E\left(-\dfrac{1}{2}+j\dfrac{\sqrt{3}}{2}\right) = aE.$

UNBALANCED THREE-PHASE MOTORS

For the assumed winding, Fig. 13.3, the current-voltage relations are:

(13.5) $\quad 3E_A/K = K(3y+2Z_1+Z_0)I_A+(Z_0-Z_1)I_B+(Z_0-Z_1)I_C+3E'_A/K,$

(13.6) $\quad 3E_B = K(Z_0-Z_1)I_A+(3z+2Z_1+Z_0)I_B+(Z_0-Z_1)I_C+3E'_B,$

and

(13.7) $\quad 3E_C = K(Z_0-Z_1)I_A+(Z_0-Z_1)I_B+(3z+2Z_1+Z_0)I_C+3E'_C,$

where E'_A, E'_B, and E'_C are the voltages due to the air-gap fields. That is, the first terms of the above equations are the voltages consumed in the stationary and external motor circuits, the final terms are the remaining voltages that are impressed on the rotor.

All that the rotor "sees" are the air-gap fields due to the stator ampere turns. Since the zero-phase sequence stator currents produce no fundamental-frequency air-gap fields, only forward and backward fundamental fields exist in the air gap and rotor. Hence, from Fig. 3.13 and Sect. 11.8, we can substitute:

(13.8) $\quad \sqrt{3}\, E'_A/K = E'_f+E'_b = I_f Z_f+I_b Z_b,$

(13.9) $\quad \sqrt{3}\, E'_B = a^2 E'_f+a E'_b = a^2 I_f Z_f+a I_b Z_b,$

(13.10) $\quad \sqrt{3}\, E'_C = a E'_f+a^2 E'_b = a I_f Z_f+a^2 I_b Z_b.$

The applied voltages and currents referred to rotating field axes, remembering that they must all be expressed in terms of normal winding turns, are:

(13.11) $\quad \sqrt{3}\, E_A/K = E_f+E_b+E_0,$

(13.12) $\quad \sqrt{3}\, E_B = a^2 E_f+a E_b+E_0,$

(13.13) $\quad \sqrt{3}\, E_C = a E_f+a^2 E_b+E_0,$

(13.14) $\quad \sqrt{3}\, E_f = E_A/K+a E_B+a^2 E_C,$

(13.15) $\quad \sqrt{3}\, E_b = E_A/K+a^2 E_B+a E_C,$

(13.16) $\quad \sqrt{3}\, E_0 = E_A/K+E_B+E_C,$

(13.17) $\quad \sqrt{3}\, K I_A = I_f+I_b+I_0,$

(13.18) $\quad \sqrt{3}\, I_B = a^2 I_f+a I_b+I_0,$

(13.19) $\quad \sqrt{3}\, I_C = a I_f+a^2 I_b+I_0.$

472   SPECIAL TYPES AND CONNECTIONS OF INDUCTION MACHINES

To transform the three-phase equations to rotating field axes, we substitute Eqs. 13.5 to 13.13 in Eqs. 13.14 to 13.16, giving:

(13.20) $\sqrt{3}\,E_f = K(y+Z_1)I_A + a(z+Z_1)I_B + a^2(z+Z_1)I_C + \sqrt{3}\,I_f Z_f,$

(13.21) $\sqrt{3}\,E_b = K(y+Z_1)I_A + a^2(z+Z_1)I_B + a(z+Z_1)I_C + \sqrt{3}\,I_b Z_b,$

(13.22) $\sqrt{3}\,E_0 = K(y+Z_0)I_A + (z+Z_0)I_B + (z+Z_0)I_C.$

Substituting Eqs. 13.17 to 13.19 in these equations and dividing by $\sqrt{3}$ throughout:

(13.23) $E_f = (y+Z_1+Z_f)I_f + \left(\dfrac{z-y}{3}\right)(I_f - I_0) + \left(\dfrac{z-y}{3}\right)(I_f - I_b),$

(13.24) $E_b = (y+Z_1+Z_b)I_b + \left(\dfrac{z-y}{3}\right)(I_b - I_0) + \left(\dfrac{z-y}{3}\right)(I_b - I_f),$

(13.25) $E_0 = (Z_0+y)I_0 + \left(\dfrac{z-y}{3}\right)(I_0 - I_f) + \left(\dfrac{z-y}{3}\right)(I_0 - I_b).$

These last three equations are represented by the equivalent circuit of Fig. 13.4. If the external impedances were different in all three phases, no equivalent circuit could be made to represent the equations, due to the lack of D- and Q-axis symmetry. If $y = z$, the circuit breaks down into two independent circuits. One is the usual circuit for a balanced two-phase machine with unbalanced applied voltages, Fig. 13.1 (b). The other is the

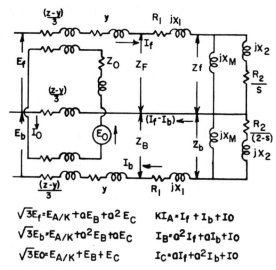

FIG. 13.4.   Equivalent Circuit of Motor of Fig. 13.3, Rotating Field Axes.

UNBALANCED THREE-PHASE MOTORS 473

single-phase zero-sequence circuit. Thus, when there is no external impedance, the motor of Fig. 13.3 behaves exactly as if it were a balanced three-phase motor, with voltages E_A/K, E_B, and E_C impressed across the three phases.

Assuming balanced impressed voltages, as in Eqs. 8.107 to 8.109:

(13.26) $$E_A = E,$$

(13.27) $$E_B = a^2 E = \left(-\frac{1}{2} - j\frac{\sqrt{3}}{2}\right) E,$$

(13.28) $$E_C = aE = \left(-\frac{1}{2} + j\frac{\sqrt{3}}{2}\right)\sqrt{2}.$$

From Eqs. 13.13 to 13.16:

(13.29) $$E_f = \frac{(2K+1)E}{\sqrt{3}K},$$

(13.30) $$E_b = \frac{-(K-1)E}{\sqrt{3}K},$$

(13.31) $$E_0 = \frac{-(K-1)E}{\sqrt{3}K}.$$

These agree with Eqs. 8.120 to 8.122.

Let, for convenience:

(13.32) $Z_F = Z_1 + Z_f =$ total impedance of forward field on three-phase with normal turns,

(13.33) $Z_B = Z_1 + Z_b =$ total impedance of backward field on three-phase with normal turns.

Substituting in Eqs. 13.23 to 13.25 and solving for the currents:

(13.34) $$I_b = \frac{\sqrt{3}E(z+Z_0)(z-Ky+Z_F-KZ_F)}{\left\{K[3Z_F Z_B Z_0 + (y+2z)(Z_F Z_B + Z_F Z_0 + Z_B Z_0) \atop + z(z+2y)(Z_F+Z_B+Z_0) + 3yz^2]\right\}},$$

(13.35) $$I_f = \frac{\sqrt{3}E + I_b(z+Z_B)}{z+Z_F},$$

(13.36) $$I_0 = \frac{I_b(z+Z_B)}{z+Z_0}.$$

474 SPECIAL TYPES AND CONNECTIONS OF INDUCTION MACHINES

At standstill, $Z_F = Z_B = Z$, and Eqs. 13.34 to 13.36 become:

(13.37) $\quad I_b = \dfrac{\sqrt{3}\,E(z+Z_0)(Z-KZ+z-Ky)}{K[3Z^2 Z_0 + (y+2z)(Z^2+2ZZ_0) + z(z+2y)(2Z+Z_0) + 3yz^2]}$,

(13.38) $\quad I_f = \dfrac{\sqrt{3}\,E}{z+Z} + I_b$,

(13.39) $\quad I_0 = \dfrac{I_b(z+Z)}{z+Z_0}$.

Placing $K=1$ and either $y=0$, $z=0$, or $y=z$ gives the equations of a balanced three-phase motor with an external impedance in one, two, or three phases.

Taking, for simplicity, $y=0$, these give, from Eqs. 13.17 to 13.19:

(13.40) $\quad KI_A = \dfrac{E[(K+z)Z_0 - (K-1)Z + 3z]}{K[3ZZ_0 + z(2Z+Z_0)]}$,

(13.41) $\quad I_B = \dfrac{-E[(K+2)Z_0 + 2(K-1)Z]}{2K(2Zz + 3ZZ_0 + Z_0 z)} - \dfrac{j\sqrt{3}\,E}{2(Z+z)}$,

(13.42) $\quad I_C = \dfrac{-E[(K+2)Z_0 + 2(K-1)Z]}{2K(2Zz + 3ZZ_0 + Z_0 z)} + \dfrac{j\sqrt{3}\,E}{2(Z+z)}$.

When $z=0$, these reduce to Eqs. 8.114 to 8.116.

The actual zero-sequence current in the stator windings at locked rotor with $y=0$ is:

(13.43) $\quad I'_0 = \tfrac{1}{3}(I_A + I_B + I_C) = \dfrac{-E[(K-1)(K+2)Z_0 + (2K+1)Z - 3z]}{K^2[3ZZ_0 + z(2Z+Z_0)]}$.

This is the current circulating in the delta. It must be clearly distinguished from I_0, Eqs. 8.124 and 13.39, which is the apparent zero-sequence current derived by referring the impressed air-gap mmf to normal winding turns. This latter value creates a third harmonic air-gap field, which produces an asynchronous crawling tendency at one-third speed.

If the motor is Y-connected, Eqs. 13.5 to 13.25 are still true, so that the equivalent circuit of Fig. 13.4 is valid for Y- as well as Δ-connected windings. However, the line currents must add up to zero:

(13.44) $\quad\quad\quad\quad\quad\quad I_A + I_B + I_C = 0.$

Hence, by analogy with Eq. 13.16:

(13.45) $\quad\quad\quad\quad \sqrt{3}\,I_0 = KI_A + I_B + I_C = (K-1)I_A$

and, from Eq. 13.17:

(13.46) $\quad\quad\quad \sqrt{3}(I_f + I_b) = 3KI_A - \sqrt{3}\,I_0 = (2K+1)I_A.$

UNBALANCED THREE-PHASE MOTORS

To find the phase voltages, E_A, E_B, E_C, with Y connection and balanced impressed voltages, $\sqrt{3}\,E$, we shall assume:

(13.47) $\qquad E_A - E_C = \sqrt{3}\,E\,\underline{/330}$,

(13.48) $\qquad E_C - E_B = \sqrt{3}\,E\,\underline{/90}$,

(13.49) $\qquad E_B - E_A = \sqrt{3}\,E\,\underline{/210}$.

From Eqs. 13.14 to 13.16:

(13.50) $\quad \sqrt{3}(E_C - E_B) = (a - a^2)E_f + (a^2 - a)E_b = j3(E_f - E_b)$,

so that

(13.51) $\qquad E_f - E_b = E$.

Also:

(13.52) $\quad 2E_A - E_B - E_C = \sqrt{3}\,E(\underline{/330} - \underline{/210}) = 3E$.

From Eqs. 13.11 to 13.13:

(13.53) $\quad \sqrt{3}(2E_A - E_B - E_C) = (2K+1)(E_f + E_b) + 2(K-1)E_0 = 3\sqrt{3}\,E$.

Substituting Eqs. 13.23 to 13.25 in 13.51 and 13.53:

(13.54) $\quad E_f - E_b = E = (z + Z_F)I_f - (z + Z_B)I_b$,

(13.55) $\quad 3\sqrt{3}\,E = (2K+1)[I_f(Z_F + y) + I_b(Z_B + y)] + 2(K-1)I_0(Z_0 + y)$
$\qquad\qquad\qquad\qquad\qquad\qquad + (z - y)(I_f - I_0) + (z - y)(I_b - I_0)$.

From Eqs. 13.45 and 13.46:

(13.56) $\qquad I_f + I_b = \dfrac{(2K+1)I_A}{\sqrt{3}} = \dfrac{(2K+1)I_0}{K-1}$.

Solving these three equations for I_f, I_b, and I_0:

(13.57) $\quad I_f = \dfrac{\sqrt{3}\,E}{M}[(K+2)(2K+1)Z_B + 3(K+1)z + 3K^2 y + (K-1)^2 Z_0]$,

(13.58) $\quad I_b = \dfrac{-\sqrt{3}\,E}{M}[(K-1)(2K+1)Z_F - 3Kz + 3K^2 y + (K-1)^2 Z_0]$,

(13.59) $\quad I_0 = \dfrac{(K-1)\sqrt{3}\,E}{M}[(K+2)Z_B - (K-1)Z_F + 3z]$,

where:

(13.60) $\quad M = \{(2K+1)^2 Z_F Z_B + 2(K^2 + K + 1)z(Z_F + Z_B) + 3z^2$
$\qquad\qquad\qquad\qquad + [3K^2 y + (K-1)^2 Z_0](Z_F + Z_B + 2z)\}$.

These give for the phase currents:

(13.61) $$I_A = \frac{3E}{M}[(K+2)Z_B - (K-1)Z_F + 3z],$$

(13.62) $$I_B = \frac{-\sqrt{3}E}{2M}[\sqrt{3}[(K+2)Z_B - (K-1)Z_F + 3z] + j\{(2K+1) \\ \times [(K-1)Z_F + (K+2)Z_B] + 3z + 3K^2 y + 2(K-1)^2 Z_0\}],$$

(13.63) $$I_C = \frac{-\sqrt{3}E}{2M}[\sqrt{3}[(K+2)Z_B - (K-1)Z_F + 3z] - j\{(2K+1) \\ \times [(K-1)Z_F + (K+2)Z_B] + 3z + 3K^2 y + 2(K-1)^2 Z_0\}].$$

At standstill, $Z_F = Z_B = Z$, and with $y = z = 0$, these last three equations reduce to Eqs. 8.132 to 8.134.

It is interesting to see whether Fig. 13.4 is consistent with the circuits derived earlier, such as Fig. 12.2 (b). In deriving Fig. 13.4 we assumed the axis of phase A as the reference, in order to obtain a symmetrical form for the equations. For a single-phase line-to-line connection, and with $K = 1$, the voltage E is impressed across phases A and B in series reversed, giving $E = E_A - E_B$. Substituting this in Eqs. 13.14 to 13.16 does not enable E_f or E_b to be determined. However, by taking a reference axis 30° ahead of phase A, that is, halfway between A and $-B$, Eqs. 13.14 and 13.15 become:

$$\sqrt{3}E_f = E_A\underline{/30°} + E_B\underline{/150°} + E_C\underline{/270°},$$

$$\sqrt{3}E_b = E_A\underline{/330°} + E_B\underline{/210°} + E_C\underline{/90°},$$

giving: $$\sqrt{3}(E_f + E_b) = \sqrt{3}(E_A - E_B) = \sqrt{3}E,$$

or $$E_f + E_b = E.$$

For this connection $I_C = 0$. If y and $z = 0$ also, from Fig. 13.4:

$$E = I_f(Z_F + Z_B) = I_f[2(R_1 + jX_1) + Z_f + Z_b].$$

Since $I_f = I_b$, we have:

$$I_A = -I_B = I_f\underline{/330} + I_b\underline{/30} = \sqrt{3}I_f = \frac{E}{Z_F + Z_B}.$$

This checks with the circuit of Fig. 12.2 (b), allowing for a 2 to 1 difference in the definition of unit impedance. In that case, the line-to-line single-phase impedance was defined as $(Z_F + Z_B)/2$, without reference to per phase values; whereas in Fig. 13.4 the impedance per phase is taken as the unit, making the line-to-line impedance twice as great, or $Z_F + Z_B$.

3 Dual-Winding Motors

Many machines are wound with two independent circuits, which may be connected in series for high voltage or in parallel for low voltage. Sometimes, as for part-winding starting, or for double-winding generators, the two half windings have independent external circuits, with different voltages and currents. To derive equations and circuits for the analysis of these dual-winding machines, we may start with the equivalent circuit of a four-winding machine, Fig. 11.6.

Limiting our consideration to three-phase machines with symmetrical windings and uniform air gaps, so that $Z_d = Z_q$, with normal frequency,

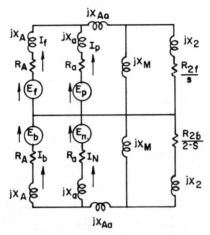

Fig. 13.5. Elementary Equivalent Circuit of Dual-Winding Machine.

$F = 1$, and with a single short-circuited rotor winding, so that $Z_4 = \infty$ and $E_{3f} = E_{3d} = 0$, this circuit reduces to that of Fig. 13.5. The two stator windings are designated as A, B, C, and a, b, c, respectively. The forward-sequence voltages are designated by E_f for the A winding and E_p for the a winding, and the backward-sequence voltages are E_b and E_n.

However, there are two things missing in this circuit. In the first place, since there are three phases, the impressed voltages may include zero-sequence components, E_0 and E_z. Since zero-sequence currents produce no fundamental air-gap fluxes, they do not induce any rotor voltages nor create any torque. Therefore the zero-sequence circuits are independent single-phase circuits which have no influence on the forward- and negative-sequence circuits, so long as the three-phase windings are balanced.

In the second place, the circuit only allows for a mutual slot reactance between the stator windings A and a. That is, the reactance X_{Aa} in the figure has a voltage drop proportional to the sum of I_A and I_a. In actual machines, however, the A and a windings normally are located in different phase belts, so that the air-gap flux made by either winding alone includes large harmonic fields which are cancelled out when $I_A = I_a$. That is, if

Table 13.1

Test Procedure and Equations for Determining Harmonic Impedance Values for Dual-Winding Motors

Connection		Locked Rotor Impedance Equations
One leg	I	*Line to Neutral* $(1/3)[2(Z_1+Z_H+Z_F)+(Z_0+Z_{0H})]$
Two legs in series aiding	II	$(2/3)(2Z_1+Z+4Z_{0F})$
Two legs in series bucking	III	$(2/3)(2Z_1+Z_0+2Z_{0H}+4Z_H)$
Three legs in series Δ	IV	*Open Delta* $3(Z_0+Z_{0H})$
Six legs in series Δ	V	$6Z_0$
Two half phases in series Y	VI	*Line to Line* $2(Z_1+Z_H+Z_F)$
Four half phases in series Y aiding	VII	$4(Z_1+2Z_F)$
Four half phases in series Y bucking	VIII	$4(Z_1+2Z_H)$
Full winding, half phases in parallel aiding	IX	$2[Z_1+2Z_H/2-Z_H+Z_F]=Z_1+2Z_F$
Half phases in Δ	X	$(2/3)(Z_1+2Z_H-Z_H+Z_F)$ $=(2/3)(Z_1+Z_H+Z_F)$
Half winding in Δ	XI	*Three Phase (on per phase basis)* *Winding Connected to 3-Phase Supply* $Z_1+2Z_H-Z_H+Z_F=Z_1+Z_H+Z_F$
Full winding with half phases in parallel Δ	XII	$Z_1/2+Z_F$
Full winding with half phases in series aiding Δ	XIII	$2Z_1+4Z_H-4Z_H+4Z_F=2Z_1+4Z_F$

DUAL-WINDING MOTORS 479

winding A has fractional pitch and is located in alternate poles, the air-gap mmf wave will be made up of components like those in Fig. 3.10, and will include a large 2nd harmonic. The a winding will create an exactly similar but opposite harmonic field, if $I_A = I_a$. If $I_A = -I_a$, however, the harmonic field will be doubled. Hence, the voltage across the mutual reactance due to these air-gap field harmonics is zero if $I_A = I_a$, is $I_A X_H$ if $I_a = 0$, and is $4I_A X_H$ if $I_A = -I_a$.

We must arrange the circuit to give these values. This is done by placing a self-impedance $2Z_H$ in each winding, and a mutual impedance $-Z_H$ in both windings.

With these additions, the circuit becomes that of Fig. 8.19, that was used in the part-winding analysis of Sect. 8.9. The values of Z_H, Z_{0H}, and the other impedances of the circuit can be found by calculation, using the methods of Chapters 7 and 9, or by test, as indicated in Table 13.1. Impedance measurements of this kind can be made on any dual-voltage motor, with 6 or 9 leads brought out. In this way, very useful information about harmonic fields can be obtained, as affected by numbers of poles, winding pitch, rotor construction, etc.

Figure 13.6 shows speed-torque curves of a 4-pole motor with normal

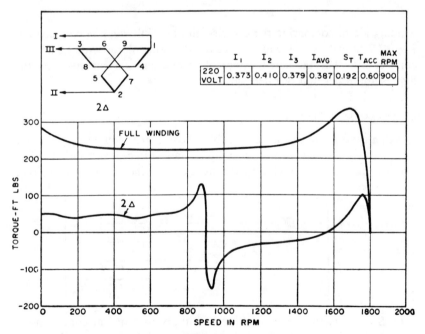

FIG. 13.6. Speed-Torque Curves of 4-Pole Motor with Double-Delta and Full Winding Connections.

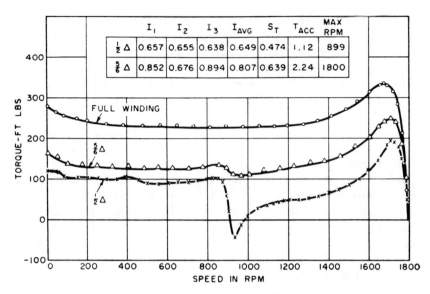

FIG. 13.7. Speed-Torque Curves of 4-Pole Motor with $\tfrac{1}{2}\Delta$, $\tfrac{5}{6}\Delta$, and Full Winding Connections.

winding pitch, comparing the double-delta and full-winding connections. Fig. 13.7 shows speed-torque curves for the same motor with $\tfrac{1}{2}\Delta$ and $\tfrac{5}{6}\Delta$ connections. The tabulated values of torque and current given on the figures are in per unit of rated values.

4 Transient Currents

As an example of a transient problem, consider a double squirrel-cage polyphase motor, Fig. 13.8, when a three-phase voltage is suddenly applied. From inspection of the circuit, the steady-state line current and voltage relation is:

$$(13.64) \quad \frac{E}{F} = I_1 \left[\frac{R_1}{F} + jX_1 + \cfrac{1}{\cfrac{1}{jX_M} + \cfrac{1}{jX_2 + \cfrac{1}{\cfrac{F-v}{R_B} + \cfrac{F-v}{R_A + j(F-v)X_A}}}} \right].$$

By Sect. 11.6, the same equation will apply to transient conditions at constant motor speed, if F is replaced by $-jp$, where $p = \dfrac{d}{\omega\, dt}$. The voltage across a reactance carrying a current I is, under steady-state conditions:

(13.65) $$jfXI = j2\pi f_0 FLI = j\omega FLI,$$

where f_0 = normal frequency, F = ratio of actual frequency to f_0, and X = reactance at normal frequency. Under transient conditions, the reactive voltage is:

(13.66) $$L\frac{di}{dt} = \omega pLI = pXi,$$

and these two equations become identical when $F = -jp = -j\dfrac{d}{\omega\,dt}$. The ω in the denominator is introduced in order to retain X rather than L values in the equations to be derived. (This is the same as if t were measured in

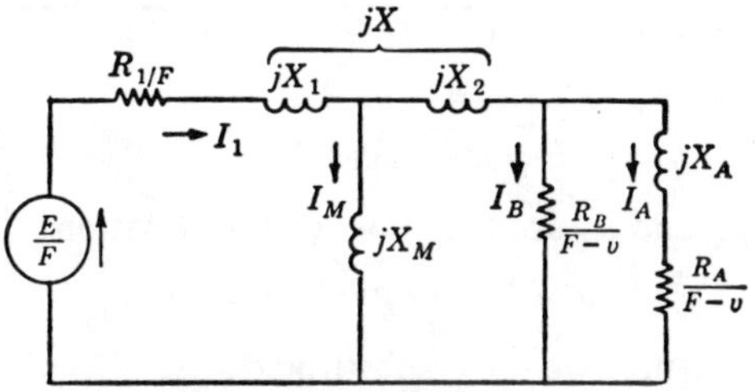

Fig. 13.8. General Equivalent Circuit of Double Squirrel-Cage Polyphase Motor.

radians of angular motion at normal frequency, instead of in seconds, and ω were omitted.)

Considering first a suddenly applied voltage, $E = \sin \omega t$, with the motor at rest, and neglecting both the magnetizing current and the primary resistance (for simplicity), the transient equation is obtained by making $R_1 = 0$, and $X_M = \infty$, and putting $v = 0$ and $F = -jp$ in Eq. 13.64, giving:

(13.67) $$E(t) = i_1(t)\left[pX + \frac{1}{\dfrac{1}{R_B} + \dfrac{1}{R_A + pX_A}}\right].$$

The expressions $i(t)$ and $E(t)$ denote that the current and voltage are now varying functions of time.

Putting $E(t) = E \sin \omega t$, this is:

(13.68) $$i_1(t) = \frac{(R_A + R_B + pX_A)E \sin \omega t}{p^2 XX_A + p[X(R_A + R_B) + X_A R_B] + R_A R_B}.$$

To solve Eq. 13.68, we note, first, that a fraction whose denominator is an algebraic expression of nth order, and whose numerator is an algebraic expression in the same variable, but of a lower order than , can be resolved into the sum of n fractions, each having a first-order denominator. For it may be easily verified that:

$$\text{(13.69)} \qquad \frac{f(p)}{F(p)} = \frac{f(p_1)}{(p-p_1)F'(p_1)} + \frac{f(p_2)}{(p-p_2)F'(p_2)} + \ldots,$$

where $p_1, p_2, \ldots$ are the n roots of the nth-order equation $F(p) = 0$. As a simple example, consider the fraction:

$$\frac{x+1}{x^2-5x+6}.$$

Since the two roots of $x^2-5x+6 = 0$ are 3 and 2, and since the first derivative of x^2-5x+6 is $2x-5$, Eq. 13.69 applied to the above fraction gives the identity:

$$\frac{x+1}{x^2-5x+6} = \frac{3+1}{(x-3)(6-5)} + \frac{2+1}{(x-2)(4-5)}$$

$$= \frac{4}{x-3} - \frac{3}{x-2}.$$

Therefore, if p_1 and p_2 are the roots of the equation:

$$\text{(13.70)} \qquad p^2 X X_A + mp + R_A R_B = 0,$$

where, for convenience, we have substituted:

$$\text{(13.71)} \qquad m = X(R_A+R_B)+X_A R_B,$$

Eq. 13.68 may be written:

$$\text{(13.72)} \quad i_1(t) = \frac{(R_A+R_B+p_1 X_A)E \sin \omega t}{(p-p_1)(m+2p_1 XX_A)} + \frac{(R_A+R_B+p_2 X_A)E \sin \omega t}{(p-p_2)(m+2p_2 XX_A)}.$$

The line current, therefore, consists of two components, due to the different decrement rates of the currents in the two branches of the squirrel cage. Each of these components can be considered separately. Taking the p_1 term, multiplying through by $(p-p_1)$, and substituting $\dfrac{d}{\omega\, dt}$ for p, we obtain:

$$\text{(13.73)} \qquad \frac{di_1}{dt} = \frac{\omega(R_A+R_B+p_1 X_A)E \sin \omega t}{m+2p_1 XX_A} + \omega p_1 i_1.$$

From Sect. 2.9, we know that the solution of Eq. 2.37 (taking $\theta = 0$):

$$\text{(2.37)} \qquad \frac{di}{dt} = \frac{E \sin \omega t}{L} - \frac{Ri}{L}$$

is:

$$\text{(2.42)} \qquad i = \frac{E}{Z}[\sin(\omega t - \alpha) + e^{-Rt/L} \sin \alpha],$$

where

$$Z = \sqrt{(R^2 + \omega^2 L^2)} \text{ and } \sin\alpha = \frac{\omega L}{Z}.$$

By analogy, therefore, we can write down the solution of Eq. 13.73, by merely substituting the corresponding values in Eq. 2.42.

For R/L in Eq. 2.42, we put $-\omega p_1$, and for L we put

(13.74) $$L = \frac{m + 2p_1 X X_A}{\omega(R_A + R_B + p_1 X_A)}.$$

The same procedure applied to the second term of Eq. 13.72 will give the second component of the current, and adding the two will give the sought-for value of the total current.

Before carrying this through, we shall substitute the values of p_1 and p_2 found by solving Eq. 13.70:

$$p_1 = \frac{1}{2XX_A}[-m + \sqrt{(m^2 - 4R_A R_B X X_A)}],$$

$$p_2 = \frac{1}{2XX_A}[-m - \sqrt{(m^2 - 4R_A R_B X X_A)}].$$

Since $4R_A R_B X X_A$ is small in comparison with m^2, these reduce to:

(13.75) $$p_1 = \frac{-R_A R_B}{m}, \text{ approximately,}$$

(13.76) $$p_2 = \frac{-m}{XX_A} + \frac{R_A R_B}{m}, \text{ approximately.}$$

Substituting these values in the indicated solution of Eq. 13.72, the total current is found to be:

(13.77)

$$i_1 = \frac{\left(R_A + R_B - \frac{R_A R_B X_A}{m}\right)}{\left(m - \frac{2R_A R_B X X_A}{m}\right)\sqrt{(1 + p_1^2)}}[E \sin(\omega t - \alpha_1) + E e^{\omega p_1 t} \sin\alpha_1]$$

$$- \frac{\left(R_A + R_B - \frac{m}{X} + \frac{R_A R_B X_A}{m}\right)}{\left(m - \frac{2R_A R_B X X_A}{m}\right)\sqrt{(1 + p_2^2)}}[E \sin(\omega t - \alpha_2) + E e^{\omega p_2 t} \sin\alpha_a],$$

where $\sin\alpha_1 = 1/\sqrt{(1 + p_1^2)}$ and $\sin\alpha_2 = 1/\sqrt{(1 + p_2^2)}$.

By setting up the equations for i_A and i_B in the same way as for i_1, the separate currents in the upper and low squirrel-cage bars, whose sum is i_1, are found to be:

(13.78)
$$i_A = \frac{R_B}{\left(m - \frac{2R_A R_B X X_A}{m}\right)\sqrt{(1+p_1^2)}} [E\sin(\omega t - \alpha_1) + Ee^{\omega p_1 t}\sin\alpha_1]$$
$$- \frac{R_B}{\left(m - \frac{2R_A R_B X X_A}{m}\right)\sqrt{(1+p_2^2)}} [E\sin(\omega t - \alpha_2) + Ee^{\omega p_2 t}\sin\alpha_2],$$

and

(13.79)
$$i_B = \frac{R_A X(R_A + R_B)}{(m^2 - 2R_A R_B X X_A)\sqrt{(1+p_1^2)}} [E\sin(\omega t - \alpha_1) + Ee^{\omega p_1 t}\sin\alpha_1]$$
$$- \frac{R_B[R_A X^2 + R_B(X+X_A)^2]}{(m^2 - 2R_A R_B X X_A)\sqrt{(1+p_2^2)}} [E\sin(\omega t - \alpha_2) + Ee^{\omega p_2 t}\sin\alpha_a].$$

Similar procedures will give the transient currents for any other assumed conditions.

It is evident that the transient currents given by the second and fourth terms of Eq. 13.79 will coexist with the steady-state air-gap fluxes and currents, and, therefore, will produce additional torques, lasting for a few cycles after the switching operation. These torques may be large compared to normal torque, providing an additional reason for designing motors with liberal shaft sizes. However, the squirrel-cage resistance of a polyphase induction motor is seldom less than 0.02 per unit, while the per unit value of leakage reactance is seldom higher than 0.20, so that the slower decrement, ωp, is nearly always less than $0.1(2\pi f)$. On this basis, the time constant of the current decay curve, or time before the transient has fallen to $1/e$, or 36 per cent, of its initial value, is seldom longer than two cycles.

As a second case, we shall consider the motor to be running at a speed $v = 1 - s$, when the voltage $E\sin\omega t$ is suddenly applied. As the presence of the v terms increases the order of the equation by one, we shall consider a single squirrel cage only, and shall again neglect R_1. Putting $R_2 + jX_2$ for $R_A + j(X_2 + X_A)$ in Fig. 13.8, making $R_1 = 0$, $R_B = \infty$, and $F = -jp$, Eq. 13.64 becomes:

(13.80) $$E\sin\omega t = -jp_1(t)\left(jX_1 + \frac{1}{\frac{1}{jX_M} + \frac{-jp-v}{R_2 + (p-jv)X_2}}\right),$$

or

(13.81) $$i_1(t) = \frac{[R_2 + (p-jv)(X_M + X_2)]E\sin\omega t}{p[(p-jv)(X_M X_1 + X_M X_2 + X_1 X_2) + R_2(X_M + X_1)]}.$$

By Eq. 13.69, this is:

$$i_1(t) = \frac{[R_2 - jv(X_M + X_2)] E \sin \omega t}{pR_2(X_M + X_1) - jv(X_M X_1 + X_M X_2 + X_1 X_2)} \tag{13.82}$$
$$+ \frac{[R_2 + (p_1 - jv)(X_M + X_2)] E \sin \omega t}{(p - p_1)[(2p_1 - jv)(X_M X_1 + X_M X_2 + X_1 X_2) + R_2(X_M + X_1)]},$$

since the two roots of the denominator of Eq. 13.81 are 0 and p_1, where:

$$p_1 = \frac{-R_2(X_M + X_1)}{X_M X_1 + X_M X_2 + X_1 X_2} + jv. \tag{13.83}$$

By Eq. 2.42, the solution of Eq. 13.81 is, if $i_1 = 0$ when $t = 0$:

$$i_1 = \frac{[R_2 - jv(X_M + X_2)] E(1 - \cos \omega t)}{R_2(X_M + X_1) - jv(X_M X_1 + X_M X_2 + X_1 X_2)} \tag{13.84}$$
$$- \frac{R_2 X_M^2 E [\sin(\omega t - \alpha) + e^{\omega p_1 t} \sin \alpha]}{\left\{(X_M X_1 + X_M X_2 + X_1 X_2)[-R_2(X_M + X_1) \atop + jv(X_M X_1 + X_M X_2 + X_1 X_2)]\sqrt{(1 + p_1^2)}\right\}},$$

where $\sin \alpha = 1/\sqrt{(1 + p_1^2)}$.

The first term of Eq. 13.84 is a fully offset alternating current. The time decrement of the offset term does not appear, because we neglected R_1 in deriving it. The second term is an alternating current that is initially fully offset, but the offset has an oscillating decrement, except when $v = 0$.

If $R_A = R_2$, $X_A + X_1 = X$, $R_B = \infty$ in Eq. 13.77, and $v = 0$, $X_M = \infty$ in Eq. 13.84, the two equations become identical and equal to:

$$i_1 = \frac{E}{X\sqrt{(1 + p_1^2)}} [\sin(\omega t - \alpha) + e^{\omega p_1 t} \sin \alpha], \tag{13.85}$$

where $p_1 = -R_2/X$ and $\sin \alpha = X/\sqrt{(R^2 + X^2)}$. This is the familiar transient performance equation for a simple R, X circuit.

An interesting special case arises when a capacitance is inserted in series with the stator, making X_1 negative. If the capacitance is sufficiently large, the value of p_1 from Eq. 13.83 becomes positive, and the oscillating second term of Eq. 13.82 increases indefinitely, until limited by magnetic saturation or losses. By putting $-X_C + X_1$ for X_1 in Eq. 13.83, and equating the real term to zero, the two limiting values of X_C between which the decrement is positive are found to be:

$$X_C > \frac{X_M X_1 + X_M X_2 + X_1 X_2}{X_M + X_2} \quad \text{and} \quad X_C < X_M + X_1. \tag{13.86}$$

486 SPECIAL TYPES AND CONNECTIONS OF INDUCTION MACHINES

Between these two values of capacitance the motor will be self-excited, with an alternating current of its own natural frequency superposed on the normal power current. This phenomenon can be cured by shunting the capacitor with a resistance of the proper value.

5 Reluctance Motors

In a growing number of motor applications it is desired to maintain exact synchronous speed. To do this, without the complications of rotor windings, brushes, and slip rings, reluctance motors are widely used. Such a motor has a normal induction motor stator and a squirrel-cage

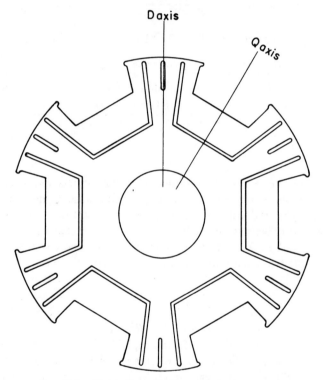

Fig. 13.9. Rotor Cross-Section of Reluctance Motor.

rotor from which sectors of the periphery have been cut away to form salient poles, as in Fig. 13.9.

With a sinusoidal air gap mmf distribution, the flux density and, therefore, the stored magnetic energy, increase markedly when the rotor is turned from the position of low to high permeance in the axis of peak mmf. In accord with the procedure of Sect. 3.3, the total air-gap magnetic energy can be calculated by integrating the square of the flux density times the differential of the air-gap volume around the entire periphery. And, the reluctance torque can be found by differentiating this energy with respect to the angle α between the axis of peak mmf and that of maximum air-gap permeance. There will be several harmonics of considerable amplitude in the flux density wave, so that a formal evaluation of the torque by the procedure outlined yields expressions that are too complex for consideration here. Since the torque depends on the rate of change of the square of the density, the torque-angle curve may be quite far from sinusoidal. To obtain a high peak torque it is desirable to make the ratio of X_d to X_q as high as possible, and also to make the steepness of transition from X_d to X_q as great as possible.

Early forms of reluctance motor had salient poles only, but recent designs have added narrow slits in the rotor laminations, parallel to the desired rotor flux paths, as shown in Fig. 13.9. These slits markedly increase the rate of change of permeance with angle of displacement, and give superior performance. Motors of this kind are now built in industrial sizes up to 150 hp or so. With such extreme slitting as that shown, reliance is placed on the cast-in aluminum squirrel-cage bars in the interpolar spaces to withstand centrifugal forces, and the permissible rotor speed is quite limited.

The equivalent circuit of a reluctance motor can be derived from that of the generalized machine with salient stator poles, Fig. 11.13. To change this circuit from stator to rotor saliency, we short-circuit the voltages applied to the salient side of the air gap, and apply instead forward and backward voltages to the nonsalient side. Also, we interchange the subscripts 1 and 2, to put the secondary on the rotor where the salient poles are. Reversing the circuit also, to put the stator on the left as in previous figures, we obtain the upper half of Fig. 13.10. The voltage e_f is applied where e_{2f}/s was in Fig. 11.13. Since the stator frequency is now the rated value, both e and R_1 are divided by 1 instead of by s. This forward field induces a rotor voltage of frequency s, so the secondary resistances are divided by s instead of by 1. The rotor dissymmetry introduces the differential impedance, $(R_\beta - R_\alpha)/2s + j(X_{2\beta} - X_{2\alpha})/2$, in the rotor midpoint connection, and the differential magnetizing reactance, $j(X_{m\beta} - X_{m\alpha})/2$, in the stator midpoint connection, as shown. Here $R_\alpha + jX_{2\alpha}$ and $R_\beta + jX_{2\beta}$ are the secondary impedances in the direct- and quadrature-axes of the rotor, respectively. $X_{m\alpha}$ and $X_{m\beta}$ are the magnetizing reactances in the two rotor axes.

If the rotor was uniform around the periphery, the midpoint impedances would disappear, and the upper two meshes in the figure would be short circuited, so they too would disappear from the circuit. With the dissymmetry present, the induced rotor s frequency current has both a forward and a backward component. The backward s frequency component creates a stator frequency of $1-s-s$, or $1-2s$. Since this is backward, the stator resistance in the upper branch of the circuit is divided by $2s-1$, just as in the case of the induction motor with single-phase rotor, Fig. 13.2.

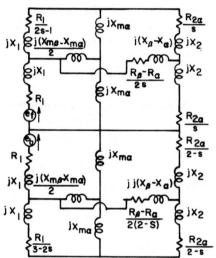

FIG. 13.10. Rotating-Field Equivalent Circuit of Reluctance Motor with Unbalanced Voltages.

Thus, the upper half of Fig. 13.10 provides for only the forward-sequence component of the stator applied voltage. To provide for the backward-sequence component, if any, a duplicate circuit is required. Since the backward field induces a rotor frequency $2-s$, and a reflected stator frequency $3-2s$, which are different from the frequencies induced by the forward field, the two halves of the complete circuit are completely independent, as indicated by the zero impedance in their common branches.

When operating in synchronism, with balanced applied voltages, the bottom half of the circuit drops out. Also, the rotor current paths are open circuited, as $s = 0$, so the circuit reduces to the two upper right-hand meshes only.

6 The Permasyn Motor

Instead of relying on the salient pole effect, it is possible to use permanent magnets in the rotor, located radially below the rotor slots at a depth sufficient to keep them out of the rotor flux paths during the starting period.

Since only a fraction of the total flux required to balance the impressed voltage penetrates below the rotor bars of a squirrel-cage motor at locked rotor, a thin ring of steel below the slots will suffice to carry the rotor flux in starting. If the $2P$-pole permanent magnet center is placed inside this outer thin ring of laminations, the rotor flux will be entirely outside the magnets at locked rotor, but will gradually penetrate them as the motor speeds up and the flux increases, until at full speed the magnet flux will cross the air gap and convert the machine into a synchronous motor. These "Permasyn" motors have remarkably high power factor, since the magnets supply most of the required magnetizing current. However, the low flux capacity of permanent magnets and the mechanical limitations imposed by the low tensile strength and brittleness of the material limit these motors to fractional or small integral horsepower ratings.

7 Hysteresis Motors

Still another widely used form of small synchronous motor employs a ring of permanent magnet material in the rotor, in the path of the air-gap flux. Ordinarily, the hysteresis motor has no rotor slots or conductors, and the outer periphery of the rotor is the solid (non-laminated) magnet material. At locked rotor, the air-gap mmf creates a corresponding rotor flux wave that is displaced peripherally by a considerable angle, just as was described in Sect. 1.10, in considering the Saturistor. This displaced flux wave creates a torque, which remains substantially constant all the way up to synchronous speed. At no load, the angle of flux lag decreases until it is just sufficient to overcome the friction and windage.

One of the difficulties with these motors is that eddy currents are induced in the solid rotor surface by the flux pulsations due to stator slot openings. Since the inherent torque is low, these losses may be a large fraction of the output. To avoid this difficulty, the stator may be made with closed slots, and the winding may be inserted from the outside, the stator yoke or outer ring of laminations being slipped on after the winding is in place.

Another problem is that the magnetizing current is very high, as the permanent magnet material is directional and its permeability is very low in the cross direction of flux. If the ring of permanent magnet material is deep enough to provide a considerable torque, flux will flow both radially and peripherally, requiring high ampere turns. If it is thin, the flux may be all radial, but the energy available will be low, especially as it is difficult to fabricate a ring with the preferred direction of magnetization radial. For such reasons as these hysteresis motors are limited to fractional horsepower sizes, generally only a few watts. Much better use of permanent magnet materials in induction motors can be obtained by the Saturistor principle, with slot-embedded Alnico bars, that carry the rotor leakage flux only, as described in Sect. 8.7. However, such motors have the normal slip of an induction motor, and are not synchronous.

8 SYNCHRONOUS INDUCTION MOTORS

For large machines, synchronous speed operation can be obtained by supplying direct current to the rotor winding after the motor has come up to speed. The rotor current will then be constant, independent of the load, up to the breakdown torque limit. If the motor has a normal air gap, so that the no-load current in induction motor operation is only about half rated current; and if the d-c rotor current in synchronous operation is equivalent to rated current, the motor at no load will supply a leading current to the power system about equal to the normal magnetizing current. However, to obtain a breakdown torque of 150 per cent of rated or more, the rotor current must be equivalent to at least 1.5 times rated current, and the corresponding no-load current will be high. Thus, a field current regulator is required if such a synchronous induction motor is to have reasonable power factor and overload ability.

To remedy this, the air gap may be made larger than normal, but then the required rotor current will be even larger, and an oversized motor will be required. A salient-pole rotor can provide a great many more ampere turns, with less cost, than a rotor with a distributed winding. And, a squirrel-cage winding located in the pole faces of a salient-pole rotor can provide good starting performance with lower cost than a wound-rotor motor. For these reasons, salient-pole synchronous motors are used in preference to wound-rotor induction motors whenever the starting duty is light and continuous operation at full speed is desired, and the synchronous induction motor is now very rarely used.

9 A-C EXCITED INDUCTION MOTORS

By connecting the slip rings of a wound-rotor motor to a polyphase voltage source of the same frequency as the stator voltage, with opposite phase rotation, two similar but independent rotating fields will be produced, and these will come into synchronism at twice normal synchronous speed. This doubly-fed motor, having full frequency on both sides of the air gap, has very little or negative damping torque, and therefore is subject to severe speed oscillations. Also, during acceleration, when the speed reaches normal synchronism, the frequency of the currents induced in the rotor by the stator field will become zero, and this is also true for the currents induced in the stator by the rotor field, so that the motor will operate as an induction motor with double torque at this point, and will not accelerate above it. To make the doubly-fed motor useful, therefore, it is necessary to provide auxiliary means to bring it above half the operating speed of twice normal synchronism, and also to provide some kind of mechanical damper to prevent speed oscillations.

The Schrage motor has been widely used to provide stepless speed adjustment, with a constant speed characteristic and good efficiency. This motor has two independent rotor windings, one connected through slip rings to a polyphase a-c supply, and the other to a commutator. The stator has a distributed winding also. There are two independently movable sets of brushes on the commutator, and the stator winding is connected across these. When the two brush sets are together, on the same commutator bars, the stator voltage is zero, and the motor operates as an ordinary induction motor, with inverted construction. As the two brush sets are moved apart, a slip frequency voltage of increasing magnitude is impressed on the stator, which may either add to or oppose the induced voltage, causing the motor speed to either rise or fall, depending on the direction of brush shift. By adjusting the position at which the two brush sets come together, the power factor may be varied also.

Thus, this brush-shifting type of motor has highly useful characteristics. However, in commutating alternating currents, there is a voltage proportional to frequency induced in the coils that are short-circuited by the brushes, and this limits the permissible horsepower per pole to quite low values, and also limits the permissible range of speed variation. Also, the requirement for two sets of brushes, with provisions for their motion, make the motor cost and maintenance higher than induction or d-c machines of similar capacity.

The present availability of low-cost and high-efficiency silicon rectifiers enables direct-current power to be obtained far more economically than formerly, and this is giving a strong impetus to the use of d-c motors wherever adjustable speed with good efficiency is desired. Hence, the Schrage motor is now very little used in the United States.

Much work is now being done to develop silicon rectifier frequency changers, which may be used either to supply adjustable frequency to a squirrel-cage induction motor, or to convert the slip frequency power of a wound rotor motor back to the line frequency. In this way, wide-range speed control of induction motors for vehicle and other drives may become feasible, with high efficiency and without commutators or slip rings, but it is too early at the present writing (1964) to judge how successful this will be.

10 Linear Induction Motors

By making a motor straight instead of circular, as if a normal motor was cut in two by a plane through the shaft, and the two halves were flattened out, linear motors can be made, that will produce straight line motion. Such a motor may have 5, or 7½, or any other number of "poles," as the magnetic flux can complete its path through the air spaces at the ends of the laminations. The associated extra reactance and losses can be tolerated, and they will be small if the core extends over 8 or more pole pitches.

One of the earliest proposals of this nature was made by P. Trombetta to provide an electric hammer. By opening and closing contactors, the squirrel-cage element of the motor was driven up and down, providing a great deal more impact than that due to gravity alone. Troubles with the needed contactors made this scheme impractical at the time. Other designs have been made for electric catapults that shoot a projectile out at the end of a long linear stator, and the idea has been considered also for launching planes from a short runway. And, a number of linear motors have been made to pump liquid metals, such as mercury, the metal serving as the squirrel cage.

The most widely considered use for linear motors is to provide transportation. For railway use it has been proposed that a vertical strip of aluminum be placed along the center of the track, and that the moving car or locomotive have two lengths of laminated stator placed opposite each other, with the stationary aluminum strip between them. Then, the strip serves as a squirrel cage, and the linear motor will drive the car at a speed corresponding to 2 pole pitches times the cycles per second of the power frequency, less the slip. Such a motor has a large air gap, made up of the thickness of the aluminum strip plus clearances, and the clearance must be quite large if the track is curved.

A more promising scheme is to use the linear motor for moving baggage and small vehicles along tracks in airport terminals or other crowded areas. Here, the "stators" may be placed half a car length apart in the center of the track, and the vehicles will have laminations with embedded squirrel-cage windings located just above the stators, each "slider" being the full length of the vehicle, so it will always overlap at least one stator. To reduce the magnetic pull, and also the high stator current in the absence of a slider, the stators may be made two sided, and the sliders would then be aluminum strips attached below the vehicle, to move between the stators. The vehicles do not need any power supply, so problems of switching from one track to another and current collection are removed.

If the linear motor is used to accelerate a car or projectile, the pole pitches of the fixed stators can be lengthened, or the frequencies supplied to them can be increased, in successive track sections, to match the increasing speed of the car. There are very many possible arrangements, so it seems certain that some important applications for linear motors will be found.

A very different kind of a-c excited linear motor is coming into use, to provide short-stroke reciprocating motion. The simplest form is a solenoid, with a single coil encircling a magnetic core and a moving plunger, as shown at A in Fig. 13.11. Usually, the plunger is tapered, to make the air gap shorter than the stroke, and the magnetic circuit is completed through parasitic air gaps at the sides, as shown. The magnetic force is positive for both plus and minus currents, so the force has twice the line frequency, and there are 120 complete strokes per second with a 60-cycle supply. To pull the plunger back on the first stroke, and to conserve the (large) kinetic energy of the moving parts, a spring, tuned to resonance at the double frequency, is attached to the plunger. In operation, the magnetic force is nearly in phase with the velocity, with the peak occurring at midstroke.

A serious difficulty with Design A is that, since the air gaps on each side are magnetically in parallel, the slightest inequality causes the flux to shift to the short gap side, creating a strong side pull, and making the plunger "seize" at the side unless restrained by a close-fitting guide bearing. This trouble can be minimized by providing balancing coils on the two sides, that are connected in series bucking, as indicated in the figure. Any inequality of the fluxes induces a circulating current, that restores equality, to the extent permitted by the circuit resistance.

Operation at the line frequency can be obtained by connecting a diode in series with the coil. The current then consists of fully-displaced sinusoid, so that the force consists chiefly of a line frequency sine wave, superposed on a steady one-way force, which is offset by biasing the spring. By using two duplicate motors, connected to the same plunger in opposite directions, and using reversed diodes, so that the two units operate on opposite half cycles of voltage, a pure sine wave line current is obtained, and also a sine wave single frequency force. The magnetic energy due to the d-c component of coil current is not lost, but is transferred to and fro between the two motors. On this account, the power factor of the two units together is much higher than that of one alone.

Another form of linear motor, shown at B in Figure 13.11, has two "working" air gaps magnetically in series, and no parasitic gaps. This requires two coils, which should be connected in parallel, to serve as balancing coils as well as providing the main mmf. An alternative construction, shown at C, has only one coil, but two plungers. This has four working gaps, with adjacent gaps in series, but the center gaps in parallel, so that balancing coils may be needed on the outer legs.

A quite different form of motor, shown at D, can be used for single frequency operation only, as with only a-c excitation the plunger is pulled to midstroke position, and locks there. By providing either a permanent magnet or a d-c coil, as indicated in the figure, the flux is polarized, as with a diode. The air-gap flux density on one side of the slot is proportional to the d-c, plus a-c mmf, and on the other side to the d-c minus the a-c, creating a single frequency force that moves the plunger up and down.

Design D may be made circular or rectangular. The circular shape is preferable for mechanical reasons, but it is difficult to laminate the magnetic paths to allow the flux to move both along and across the line of motion, as required to avoid eddy-current losses. Thus, for small motors, the circular shape seems preferable, but for large sizes the rectangular form seems better.

The force equations for all these motors can readily be derived by the methods cited in sections 1.6 and 1.7. The stored magnetic energy in air with a density of 1 weber per square meter is 0.398 newton meter, so that a motion dS of the plunger, that changes the flux density in dV from B_1 to B_2 is caused by a force of (13.87) $F = 0.398 \, (B_2^2 - B_1^2) \, dV/dS$ newtons, where dV is the change in volume of the magnetic field due to the motion dS.

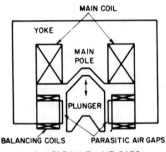

A. PARALLEL AIR GAPS

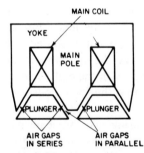

C. TWIN PLUNGERS - SINGLE COIL

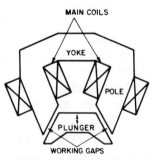

B. SERIES AIR GAPS - TWIN COILS

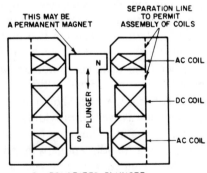

D. POLARIZED PLUNGER

LINEAR MOTORS FOR RECIPROCATING MOTION
FIG. 13.11

Bibliography

		Reference to Section
1.	"Transient Overspeeding of Induction Motors," R. W. Ager, *A.I.E.E. Trans.*, Vol. 60, 1941, pp. 1030–2.	4
2.	"Switching Transients in Wound Rotor Induction Motors," P. L. Alger and Y. H. Ku, *A.I.E.E. Trans.*, Vol. 73, Part IIIB, 1954, pp. 19–27.	4
3.	"The Synchronous Double-Fed Induction Machine," R. E. Bedford, *A.I.E.E. Trans.*, Vol. 75, Part III, 1956, pp. 1486–91.	9
4.	"Induction-Type Synchronous Motors," L. H. A. Carr, *The Jour. of the I.E.E.*, Vol. 60, 1922, pp. 165–74.	8
5.	"The Synchronous Induction Motor and Its Application," L. H. A. Carr, *The Metropolitan-Vickers Gazette*, Vol. 16, June 1936, pp. 136–40.	8
6.	"An Analysis of the Hysteresis Motor," M. A. Copeland and G. R. Slemon, *I.E.E.E. Trans.*, April 1963, pp. 34–41.	7
7.	"Characteristics of Induction Motors with Permanent Magnet Excitation," J. F. H. Douglas, *A.I.E.E. Trans.*, Vol. 78, Part III, 1959, pp. 221–5.	6
8.	*The Theory of the Polyphase Induction Motor with a Single-Phase Rotor*, L. A. Dreyfus, Ernst, Berlin, 1909.	1
9.	"The Görges Phenomenon," H. L. Garbarino and E. T. B. Gross, *A.I.E.E. Trans.*, Vol. 69, 1950, pp. 1569–74.	1
10.	"Polyphase Induction Motors with Unbalanced Rotor Connections," B. N. Garudachar and N. L. Schmitz, *A.I.E.E. Trans.*, Vol. 78, Part III, 1959, pp. 199–204.	1
11.	"Steady-State Performance of Reluctance Machines," V. B. Honsinger, *I.E.E.E. PWR Trans.*, Paper 70 TP12, 1970.	
12.	"The Inductances L_d and L_q of Reluctance Machines," V. B. Honsinger, *I.E.E.E. PWR Trans.*, Paper 70 TP193, 1970.	
13.	"Transient Torques in Squirrel-Cage Induction Motors, with Special Reference to Plugging," E. S. Gilfillan, Jr., and E. L. Kaplan, *A.I.E.E. Trans.*, Vol. 60, 1941, pp. 1200–9.	4
14.	"Transient Analysis of Rotating Machines and Stationary Networks, by Means of Rotating Reference Frames," Y. H. Ku, *A.I.E.E. Trans.*, Vol. 70, 1951, pp. 943–54.	4
15.	"Linear Induction Motors," E. R. Laithwaite, *Proc. I.E.E.*, Vol. 104A, 1957, p. 461.	
16.	"Application of Linear Induction Motor to High Speed Transport," E. R. Laithwaite and F. T. Barwell, *Proc. Inst. Mech. Eng.*, Vol. 181, Part 3G, 1967.	
17.	"An Analysis of the Induction Motor," S. J. Levine, *A.I.E.E. Trans.*, Vol. 54, 1935, pp. 526–9.	2
18.	"Equivalent Circuits of Reluctance Machines," Chi-Yung Lin, *A.I.E.E. Trans.*, Vol. 71, 1952, Part III, pp. 1–9.	5
19.	"Transient Conditions in Electric Machinery," W. V. Lyon, *A.I.E.E. Trans.*, Vol. 42, 1923, pp. 157–76.	4
20.	"Analysis of Unsymmetrical Machines," W. V. Lyon and C. Kingsley, Jr., *A.I.E.E. Trans.*, Vol. 55, 1936, pp. 471–6.	2
21.	"Reclosing Transients in Induction Motors with Terminal Capacitors," F. P. deMello and G. W. Walsh, *A.I.E.E. Trans.*, Vol. 79, Part III, 1960, pp. 1206–13.	6

	Reference to Section
22. "Permanent Magnet Excited Synchronous Motors," F. W. Merrill, *A.I.E.E. Trans.*, Vol. 73, Part IIIB, 1954, pp. 1754–9.	6
23. "Parametric Pump-Down of Synchronous Machine Oscillations," D. W. Novotny and N. L. Schmitz, *I.E.E.E. Trans.*, Oct. 1963, pp. 652–7.	9
24. "Equivalent Circuits for Induction and Hysteresis Motors," D. O'Kelly, *I.E.E.E. Power Group Trans.*, Paper 70 TP9, 1970.	
25. "Induction Motors on Unbalanced Voltages," H. R. Reed and R. J. W. Koopman, *A.I.E.E. Trans.*, Vol. 55, 1936, pp. 1206–13.	2
26. "The Hysteresis Motor," H. C. Roters, *A.I.E.E. Trans.*, Vol. 66, 1947, pp. 1419–30.	7
27. *Electromagnetic Devices*, H. C. Roters, John Wiley & Sons, New York, 1941.	
28. "The Stabilized Doubly Fed Synchronous Induction Machine," N. L. Schmitz and V. D. Albertson, *I.E.E.E. Trans.*, Vol. 83, No. 8, August 1964, pp. 859-864.	
29. "The Equivalent Circuit of the Schrage Motor," C. L. Sheng, *A.I.E.E. Trans.*, Vol. 73, Part IIIA, 1954, pp. 114–18.	9
30. "A Linear Induction Motor with Discrete Pole-Pitch Grading," by R. A. C. Slater, *Advances in Machine Tool Design and Research*, Pergamon Press, 1967, pp. 91-113.	
31. "Equivalent Circuits for Cylindrical Rotor, Reluctance, and Salient Rotor Synchronous Machines," G. R. Slemon, *A.I.E.E. Trans.*, Vol. 81, Part III, Aug. 1962, pp. 219–24.	5
32. "Theory of Hysteresis Motor Torque," B. R. Teare, *A.I.E.E. Trans.*, Vol. 59, 1940, pp. 907–12.	7
33. "The Electric Hammer," P. Trombetta, *A.I.E.E. Trans.*, Vol. 41, 1922, pp. 1269-80.	
34. "Self-Excitation of Induction Motors," C. F. Wagner, *A.I.E.E. Trans.*, Vol. 58, 1939, pp. 47–51.	4
35. "Self-Excitation of Induction Motors with Series Capacitors," C. F. Wagner, *A.I.E.E. Trans. Supplement*, Vol. 60, 1941, pp. 1241–7.	4
36. "Transient Starting Torques in Induction Motors," A. M. Wahl and L. A. Kilgore, *A.I.E.E. Trans.*, Vol. 59, 1940, pp. 603–7.	4
37. "Electromechanical Transient Performance of Induction Motors," C. N. Weygandt and S. Charp, *A.I.E.E. Trans.*, Vol. 65, pp. 1000–9.	4
38. "Operation of Three-Phase Induction Motors on Unbalanced Voltages," J. E. Williams, *A.I.E.E. Trans.*, Vol. 73, Part IIIA, 1954, pp. 125–32.	2

CONTENTS—CHAPTER 14

Rating and Application of Induction Motors

		PAGE
1	The Purpose of a Rating System	499
2	General-Purpose Motors	500
3	Starting Current Limitations	503
4	Temperature Limitations	503
5	Special-Purpose Motors	508

14

RATING AND APPLICATION OF INDUCTION MOTORS

1 The Purpose of a Rating System

The nameplate rating of a motor is a measure of its working abilities for whose integrity the entire electrical industry, as well as the makers of the particular motor, is responsible. The rating together with the standards to which the motor was designed implies a host of different qualities built into the motor, including starting, continuous rating, and overload torque; temperature endurance; high potential strength; low starting current; and other factors. The American motor standards, developed by the cooperative efforts of makers, users, and disinterested experts, define these various qualities, and also set normal values for many of them. The Induction Motor Test Procedure, prepared by the Institute of Electrical and Electronics Engineers, provides approved means for measuring all these motor characteristics.

Such standards are invaluable in the maintenance of a proper variety of motor types and sizes, for the fair comparison of competitive designs, and the economical handling of power supply, control and installation problems. The selected horsepower ratings and mounting dimensions, in preferred number steps (Table 6.1), the starting current limits, normal overloads, preferred voltages, and time ratings allow motor users and power suppliers to design standard equipment and to carry in stock bases, pulleys, fuses, overload relays, etc., with complete assurance that they will fit any standard make of motor. The resultant cost and time savings through increased production of fewer varieties, advance planning, warehouse stocking, and simplified paper work are immense.

The purpose of the motor rating, therefore, is to define accurately its useful range of performance, in a way most intelligible to the user, and in a way also that can be accurately verified by simple test procedures. The performance of gas engines, pumps, turbines, and other mechanical apparatus is normally limited by mechanical considerations in the self-contained unit. The continuous output rating is, therefore, very little below the maximum momentary capacity, and users do not expect to load such machines appreciably beyond their ratings, even briefly. Electric motors, on the other hand, may carry large short-time overloads. For motors, there are two non-mechanical factors of importance, the electric power supply system, and the temperature limitations of the electrical insulation, which must be taken into account.

The rating of an induction motor includes six major variables: the supply voltage, number of phases, and frequency of the power supply, and the motor horsepower, speed, and temperature rise. The unit of output generally employed is the horsepower (746 watts), which together with the rated speed determine the torque the motor can deliver. Besides these factors, the rating specifies the type of service, whether continuous, intermittent, or varying duty, for which the motor is intended.

The full-load (line) current of any three-phase motor is given by the formula:

$$(14.1) \quad \text{Full-load current} = \frac{746 \times \text{hp rating}}{1.73 \times \text{eff} \times \text{pf} \times \text{voltage}} \text{ amperes.}$$

For single-phase, the factor 1.73 becomes unity, and for two-phase it is 2.

2 General-Purpose Motors

The prime purpose of a motor is to deliver torque (Eq. 3.11). Recognizing that service reliability, and flexibility in application, are of major importance, American Standards require that a polyphase motor for general-purpose use shall have a maximum, or breakdown, torque not less than 200 per cent of its nameplate rating, and for single-phase motors the breakdown torque shall not be less than 175 per cent. Somewhat greater values are specified for motors in the smaller sizes, as given in Table 14.1.

A "general-purpose motor" is any open motor having a continuous 40 C rating and designed, listed, and offered in standard ratings with standard operating characteristics and mechanical construction for use under usual service conditions, without restriction to a particular application.

These motors are designed to operate successfully when the voltage variation does not exceed 10 per cent above or below normal, and the frequency does not vary more than 5 per cent either way from normal; the sum of the two variations at any one time not exceeding 10 per cent.

Usual service conditions are: (*a*) when and where the temperature of the cooling medium does not exceed 40 C; (*b*) where the altitude does not exceed 3,300 ft (1,000 meters); (*c*) where the location or atmospheric conditions do not interfere with ventilation; and (*d*) where the motor is solidly mounted with a drive arrangement in accord with good standard practice.

The torque values given in Table 14.1 are the upper limits of the range of application; i.e., they are the highest the user can rely upon, as well as the lowest the manufacturer must provide.

Table 14.1

Locked-Rotor and Breakdown Torques of Three-Phase General-Purpose Motors

(Per Cent of Full-Load Torque)

Synchronous Speed	3,600		1,800		1,200		900	
Torque	Break-down	Locked Rotor	Break-down	Locked Rotor	Break-down	Locked Rotor	Break-down	Locked Rotor
Horsepower Rating								
½	350*	—	325*	—	300*	—	225	140
¾	325*	—	300*	—	275	175	220	135
1	325*	—	300	275	265	170	215	135
1½	250	175	280	250	250	165	210	130
2	240	170	270	235	240	160	210	130
3	230	160	250	215	230	155	205	130
5	215	150	225	185	215	150	205	130
7½	200	140	215	175	205	150	200	125
10	200	135	200	165	200	150	200	125
15	200	130	200	160	200	140	200	125
20, 25, 30	200	130	200	150	200	135	200	125
40	200	125	200	140	200	135	200	125
50, 60	200	120	200	140	200	135	200	125
75	200	105	200	140	200	135	200	125
100	200	100	200	125	200	125	200	125
125	200	100	200	110	200	125	200	120
150	200	100	200	110	200	120	200	120
200	200	100	200	100	200	120	200	120

* Approximate.

The basis for choosing the value of 200 per cent for the breakdown torque of the standard polyphase motor is roughly this:

First, it is desired to be certain that the motor will continue to deliver its normal output during temporary periods of low voltage. Thermal overload protective relays for general-purpose motors are normally set to take the motor off the line if the current rises above a value corresponding roughly to full load at 85 per cent of normal voltage. It is highly desirable to have the motor continue to carry its load down to this voltage level. Since the breakdown torque varies as the square of the voltage, this requires this torque at rated voltage to be at least 1.38 times the rated torque.

Also, it is a requirement of the American Standards that every general-purpose motor shall be able to operate continuously without injury at an output equal to 115 per cent of its rating. This 1.15 ratio, called the "service factor", provides a margin to allow for unforeseen service conditions. It permits the motor to carry loads of 115 per cent of the rating in cases where the conditions are accurately known in advance. A somewhat larger service factor is required for smaller motors (Table 14.2).

Finally, a factor of about 1.25 is allowed to provide for occasional overloads and unforeseen contingencies in motor operation, and to ensure a liberal starting torque with little sacrifice of efficiency.

The product of these three factors, $1.38 \times 1.25 \times 1.15 =$ roughly 2.0.

Progressively higher starting and breakdown torques are standard for the smaller motors for two chief reasons:

(a) Loads driven by small motors generally need higher starting torques than those of large motors, because of relatively greater friction (as in shaft seals), cost limitations on the use of unloaders, more frequent starting, etc.

(b) The inherently larger ratio of no-load to full-load current of the smaller motors gives them a smaller per cent increase in current on overload. For example, the line current of a ½-hp single-phase motor will increase to only about 1.5 times rated value at twice its rated torque, whereas the corresponding factor is about 2 for a three-phase motor of 20 hp or larger. This means that a greater percentage overload can be

TABLE 14.2

SERVICE FACTORS OF SINGLE-PHASE AND THREE-PHASE GENERAL-PURPOSE MOTORS AND LOCKED-ROTOR CURRENTS OF THREE-PHASE GENERAL-PURPOSE MOTORS

Rated Horsepower	Service Factor	Locked-Rotor Current Three Phase, at 220 Volts 60 Hertz* Amperes
⅛ and less	1.40	—
⅙	1.35	—
¼	1.35	—
⅓	1.35	—
½	1.25	20
¾	1.25	25
1	1.25	30
1½	1.15	40
2	1.15	50
3	1.15	64
5	1.15	92
7½	1.15	127
10	1.15	162
15	1.15	232
20 and up	1.15	Approx. 14.5 amp/hp

Locked-rotor currents of motors designed for voltages other than 220 volts are inversely proportional to the voltage.

* Hertz = cycles per second.

carried by a small single-phase motor than by a large polyphase motor, for a given increase in temperature rise.

Along with the greater breakdown-torque values, and for the same reasons, the service factors for the small motors are increased above the 1.15 value standard for the larger motors, as shown in Table 14.2.

The torque values given in the American Standards have evolved over many years of experience and discussion, so that the above explanation shows the general trend of thought, rather than any exact process.

The starting torque values, also given in Table 14.1, are based on logical relations between starting and breakdown torques from a design viewpoint, and on experience in the successful application of motors.

3 Starting Current Limitations

It is assumed that general-purpose motors in nearly every case will be started at full voltage. Under this condition, the torques provided will bring any usual load up to full speed within 5 to 10 seconds, and yet will not be high enough to overstress the shaft or couplings. With reduced voltage starting, the torque being reduced as the square of the voltage, only loads requiring relatively low torque can be started. In general, any increase in the allowable starting current allows a lower secondary resistance to be used, giving a higher operating efficiency for a given starting ability.

The starting current of a motor causes a dip in the supply voltage, and it also determines the requirements for overcurrent protective devices. To avoid ambiguity, and facilitate measurement, the term "locked-rotor current" is used for all guaranteed values, in place of the loosely defined term, "starting current". It is desirable from the user's viewpoint to hold this current as low as possible, but this handicaps the designer. The presently accepted values, given in Table 14.2, are a compromise between the needs of the power supplier, the user, and the designer.

As shown by Eq. 5.62, the ratio of per cent locked-rotor current to per cent breakdown torque is nearly a constant for any type of induction machine, and is a little greater than 2 for usual polyphase motors. Therefore, a per unit breakdown torque of 2 plus requires a per unit locked-rotor current of more than 5. The current values given in Table 14.2 are, therefore, in close accord with the torque values in Table 14.1.

4 Temperature Limitations

The life of organic insulations, such as paper, varnishes, enamels, and other "Class A" materials, is reduced progressively as the temperature increases. Long ago it was agreed that a hot-spot temperature of 105 C is the highest value compatible with the satisfactory life of these insulations in electric motors (Sect. 6.4). Taking 40 C as the normal ambient temperature, and allowing 15 C differential between the hottest spot and the temperature as measured by thermometer, give an allowable temperature rise of $105-40-15 = 50$ C by thermometer measurement. When the temperature is determined by the resistance method, the hot-spot allowance is considered to be only 5 C, so that a 60 C rise by resistance is generally accepted as the equivalent of 50 C by thermometer.

The advent of many new synthetic insulating materials, with greater temperature endurance and good mechanical properties, has removed the chief objection to the use of higher temperatures, and therefore Class B insulation became the accepted basis for standard motors in 1966. At the same time the uncertain thermometer method of measuring temperature was replaced by the resistance method. Thus, the present motor standards allow 80 C rise, plus 40 C ambient, plus 5 C hot-spot allowance, giving 125 C for continuous operation, and 90 C rise for (short-time) operation at the service factor load of 115% of rating. Still higher temperatures are permissible for Class F and H insulation systems, that are used for some special machines.

These higher temperatures make it necessary to provide better overload protection, and more careful application to ensure that reasonable loads are not exceeded. To understand the problems that come with higher temperatures, it is helpful to review the reasoning that was the basis for temperature limits when Class A insulation was standard.

A large volume of test data, supported by the known laws of chemical reaction rates, has shown that the mechanical properties of organic insulation deteriorate at a rate which doubles for each 8 C to 12 C increase in temperature (an 8 C rule is used for organic insulation oil-immersed, while 10 C and 12 C are appropriate for Class A and Class B insulation, respectively, in air). Hence, for each 10 C that the motor temperature increases above the limiting value of 105 C, the expected insulation life is roughly halved. Figure 14.1 shows a calculated curve of hot-spot temperature rise versus load, in continuous operation, for a typical general-purpose induction motor, taking the rise to be 50 C at rated load, and assuming constant resistance. The dotted line, rising more steeply, indicates the effects of temperature creep, or the cumulative rise in temperature caused by the progressive increase in the resistivity of copper as the temperature rises. Temperature rise is assumed proportional to losses. This indicates that a load of 150 per cent of nameplate, or 130 per cent of the service factor rating, will give a hot-spot temperature rise of roughly 110 C in continuous service, or 60 C above normal. As 60 divided by 10 is 6, the expected service life at 150 per cent load by the rules given above is only $1/2^6$, that is, about 1/64 as long as for operation at the nameplate rating. On the other hand, if the motor operated at this sustained load for only one-tenth of its total time in service, the expected life would be of the order of $1/[0.9+0.1(64)]$ times normal, or about 15 per cent of normal. High

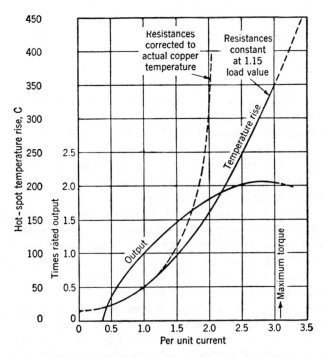

Fig. 14.1. Per Unit Output and Steady-State Temperature Rise for Typical General-Purpose Induction Motor, Showing Temperature Creep due to Increase of Copper Resistivity with Temperature.

sustained overloads are rapidly destructive, but occasional short-time overloads of moderate value can be tolerated without greatly decreasing the motor life.

From this analysis, it appears that the overall cost of an open-type motor will not be greatly reduced by designing for operation at a higher temperature rise. If a new insulation gave the same life at 20° higher temperature, and the rise at rated load was, therefore, increased from 50 to 70°, the rise at 1.50 load would be increased from 110 C to $\frac{70}{50} \times 110 = 154$ C, plus some more to allow for temperature creep, or roughly 50 C more than for the original motor. That is, the overload rise is increased some 30 C more than the rise at rated load, giving an apparent reduction of the life on overload to only $(\frac{10}{30})^2$, or, say, 10 per cent of the original motor's life on overload. Unless the new insulation has a temperature increase for half life materially higher than the 10° figure, at the same time that its life at normal temperature is improved, the hotter motor will have a materially lower overload capacity. Also, there are knotty problems of bearing design, lubricants, and thermal expansion stresses to be solved, which become progressively more difficult as the temperature rises.

In spite of these problems, the reduced size and weight of the Class B motor were found to outweigh the difficulties of higher temperature. However, the torque requirements were reduced slightly for the smaller ratings, and better methods of temperature protection have been adopted for the new designs. The new designs are better able to withstand high ambient temperatures and enclosed conditions.

In the interests of economy, it is desirable to use the same standard motor for as many different applications as possible. Consequently, it is customary to apply general-purpose motors (continuous rated) whenever:

(*a*) The peak momentary overloads do not exceed about 75 per cent of the breakdown torque.

(*b*) The root-mean-square value of the motor losses over any extended period does not appreciably exceed the losses at the service factor rating.

(*c*) The duration of any overload is not long enough to bring the momentary peak temperature above a reasonable value from the safety viewpoint, about 100 C rise for Class B insulation.

Assurance that all these limits will be adhered to is secured in practice by selecting overcurrent protection devices with the proper time delay adapted to each motor, which will permit the motor to make full use of its torque and thermal abilities, but will cut it off the line when either limit is exceeded. Normally, the protective devices cut the motor off when the current exceeds normal for a considerable time, as indicated by the lower (shaded) curve in Fig. 14.2. When the motor is stalled, the current usually is about 600 per cent, and the relay operates in about 15 seconds.

The conventional method has been to use bimetal heater strips in series with two of the motor lines, located on the control panel. At best, however, the temperature of such a heater varies considerably from the actual motor winding temperature, so that this method either does not give full protection, or does not allow full use of the motor's capacity. There is, therefore, a growing trend toward the use of built-in "Thermo-Tectors" that are inserted in or cemented on the end windings in direct contact with the conductors of motors with random windings. Figure 14.3 shows the construction of one of these devices. A snap action contact is opened when

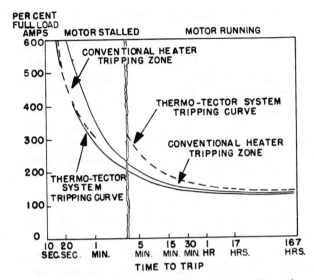

Fig. 14.2. Time-Current Tripping Curves for Motor Protection.

the thermal expansion of an outer tube in contact with the windings exceeds a certain amount. The contact is mounted on an inner rod made of a low expansion metal, and separated by an air space from the outer tube. When the temperature rises rapidly, the inner rod temperature lags behind that of the tube, so the contact opens quickly. When the temperature rises slowly, the rod and tube temperatures are nearly the same, and the contact will not open until the winding temperature is a good deal higher than in the case of a rapid temperature rise. This temperature anticipation feature enables the Thermo-Tector to open the holding coil of the line contactor

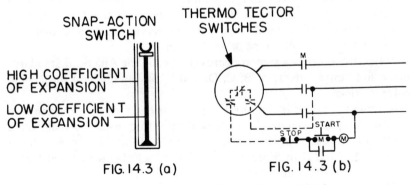

Fig. 14.3. Thermo-Tector Construction and Circuit.

at just the right time to avoid damage to the motor, whatever the duty cycle may be.

In general, the dimensions of a drip-proof general-purpose motor are fixed by considerations of torque, efficiency, noise, and mechanical requirements. The ventilating system is then designed to provide sufficient cooling to hold the temperature rise at rated load down to the allowable value. The situation would be quite different if the American Standards required only a very limited breakdown torque, such as 160 instead of 200 per cent. Such a low-overload motor would be very much closer to the breakdown torque, when operating at its rating, and might naturally have a temperature rise some $(\frac{200}{160})^2$, or 1.5 times as great as for the present standard designs.

It cannot be too strongly emphasized that the liberal system of rating now employed, permitting wide latitude in motor application, and giving assurance of reliable operation under a wide variety of service conditions, is dependent on a conservative temperature rise at the rating. A good way to judge motors of widely different designs is to compare their temperature rises at the same per cent of their breakdown-torque values.

5 Special-Purpose Motors

The philosophy of the American Standards is to recommend a limited number of standard designs, and to require that motors so rated shall strictly adhere to the standard guarantees, but at the same time to permit wide freedom for use of other motor types at the option of the manufacturer or the user, so long as such motors are distinguished by clearly defined ratings.

A special-purpose motor is a motor with special operating characteristics or special mechanical construction, or both, designed for a particular application and not falling within the definition of a general-purpose or definite-purpose motor. And a definite-purpose motor is any motor designed, listed, or offered in standard ratings with standard operating characteristics or mechanical construction for service conditions other than usual, or for a particular type of application.

Polyphase squirrel-cage motors are classified in the American Standards under five design letters: A, B, C, D, and F, having different torque and current values.

A Design A motor is a squirrel-cage motor designed to withstand full-voltage starting, which develops a locked-rotor torque not less than, and a breakdown torque greater than, those given in Table 14.1, with a slip at rated load of less than 5 per cent, and has a locked-rotor current higher than those given in Table 14.2.

These high-starting-current designs are limited to services where it is imperative to have unusually low values of copper loss for efficiency or heating reasons. In the totally enclosed, fan-cooled motors, for example, cost and application reasons make it desirable to keep the same frame size as for the open motors of the same ratings. To accomplish this, it is sometimes necessary to use a reduced number of stator winding turns, so obtaining a lower resistance, at the expense of a lower reactance and, therefore, a high starting current.

Design B motors are similar to Design A, except that they have breakdown torques not less than those given in Table 14.1, and starting current

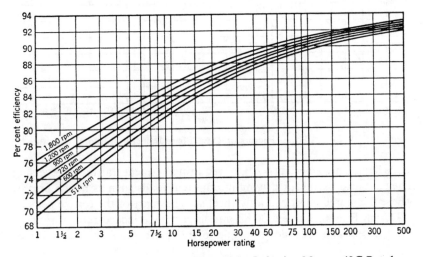

FIG. 14.4. Normal Efficiencies of Three-Phase Induction Motors, 40C Rated.

values not exceeding those given in Table 14.2. The general-purpose motors come in this category. Figures 14.4 and 14.5 show typical efficiency and power factor values for these machines (40 C rated).

Design C motors have higher locked-rotor and lower breakdown torques than A or B, as given in Table 14.3, but have locked-rotor current values the same as for Design B motors (Tables 14.1 and 14.2), and a slip at rated load of less than 5 per cent.

These motors are designed with double squirrel cages or equivalent rotor constructions (Sect. 8.6), and are used for applications requiring high breakaway torque or rapid acceleration.

Design D motors have locked-rotor torques of 275 per cent of rated torque, with locked-rotor currents in accord with Table 14.2, and have a slip at rated load of 5 per cent or more.

These motors have high-resistance rotor windings, and are used for

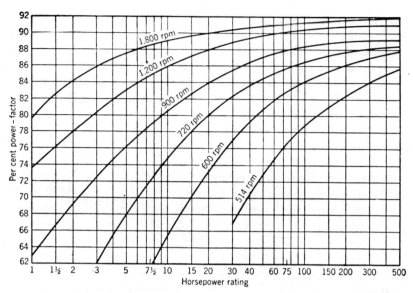

Fig. 14.5. Normal Power Factors of Three-Phase Induction Motors, 40C Rated.

loads having high inertia (Wk^2), or requiring frequent starting or reversing. They are also useful for driving intermittent loads such as punch presses, where the stored energy in a flywheel supplies part of the torque peaks.

Design F motors (rated 30 hp and larger) have locked-rotor torques of 125 per cent and breakdown torques of 135 per cent of their rated values,

TABLE 14.3

BREAKDOWN AND LOCKED-ROTOR TORQUES OF DESIGN C SQUIRREL-CAGE MOTORS

(Per Cent of Full-Load Torque)

Synchronous Speed	1,800		1,200		900	
Torque	Break-down	Locked Rotor	Break-down	Locked Rotor	Break-down	Locked Rotor
Horsepower Rating						
3	—	—	225	250	200	225
5	200	250	200	250	200	225
7½, 10	190	250	190	225	190	200
15	190	225	190	200	190	200
20 and up	190	200	190	200	190	200

The values given are the upper limits of the range of application; i.e., they are the highest the user can rely upon, as well as the lowest the manufacturer must provide.

SPECIAL-PURPOSE MOTORS

with a locked-rotor current of only 9 amp per hp. These have very limited applications.

Motors are also classified according to their mechanical protection and methods of cooling (Sect. 6.2). The American Standards define open, drip-proof, splash-proof, semi-protected, protected, fully protected, totally enclosed non-ventilated, totally enclosed fan-cooled, explosion-proof, dust-explosion-proof, water-proof, and pipe-ventilated machines. Any one of these can be obtained at a price, and each has its own field of usefulness. Every effort is made, however, to apply either the open (normally built to be substantially drip-proof) or the totally enclosed fan-cooled design in preference to the others, as these two types have the highest production and, therefore, the least cost and shortest procurement times.

A third classification of motors is in accordance with the required duty cycle. General-purpose motors have a single "continuous" rating, and are designed for steady operation at rated load for an indefinite period—of days or even years without shutdown. For loads requiring repeated short-time operation, motors with definite time ratings, of 5, 15, 30, or 60 minutes, are available. A 30-min rated motor, for example, will normally have the same torques, locked-rotor currents, and temperature rise ratings as the comparable continuous rated motor, but its temperature rise will exceed the standard value if the rated load operation continues beyond the specified 30-min period, or if this load cycle is repeated at too brief intervals. The definite time rating gives no information on the permissible frequency of repetition of the load cycle, as this depends on the balance between the losses and heat-storage capacity of the motor, and upon the rate of cooling.

For this reason, there are also provided motors with intermittent or varying duty ratings. These ratings define a repetitive load cycle of time on and time off which the motor can repeat indefinitely without exceeding the normal temperature rise.

For loads requiring continuous operation at reduced speeds, and for the most severe starting duty, wound-rotor motors with external rheostats are employed. Or, multispeed motors with pole changing can be used. Very large motors, in ratings of more than 500 hp, are generally of wound-rotor construction, to provide speed adjustment and easily controlled starting with low starting current.

Large wound-rotor motors may be provided with an auxiliary commutator machine on the motor shaft (Scherbius or Kramer system), which supplies a slip-frequency voltage to the rotor (Sect. 8.14). In this way, the motor speed can be adjusted to any desired value, without loss of efficiency, the slip losses being returned to the line or the shaft through the commutator machine. Only a moderate departure from normal motor speed is feasible, as cost and performance of the commutator machine rapidly become worse as the slip frequency increases.

All these motor types can be provided with open, protected, or enclosed construction, with or without gears, or brakes, or clutches, in vertical or horizontal mounting, etc. Also, innumerable modifications and combinations of such motors can be provided, with special control to take care of almost any conceivable time cycle of speed and torque variations.

The polyphase induction motor is indeed the industrial electric motor par excellence, with lower cost, greater reliability, and more adaptability than any other, so long as reasonably high speed, and predominantly constant speed performance, in moderate power ratings are desired. It gives way to the commutator machine, however, when very low speeds, frequent reversing, or high accelerating torques are wanted. And it is excelled by the synchronous machine for power-generation purposes, for the largest powers, and for drives at low and constant speed.

For ratings below 1 hp, a single-phase power supply is generally more economical. In this very extensive fractional-horsepower motor field, the single-phase induction motor reigns supreme.

BIBLIOGRAPHY

	Reference to Section
1. *American Standard, Rotating Electric Machinery*, A.S.A. C50.	1
2. "General Principles upon which Temperature Limits are Based in the Rating of Electric Machines and Apparatus," I.E.E.E. No. 1.	1
3. "Standards for Motors and Generators," *N.E.M.A. Pub. No. MG1*.	1
4. "Rating of General-Purpose Induction Motors," P. L. Alger and T. C. Johnson, *A.I.E.E. Trans.*, Vol. 58, 1939, pp. 445–56.	2
5. "Rating of High-Temperature Induction Motors," P. L. Alger and H. A. Jones, *A.I.E.E. Trans.*, Vol. 64, 1945, pp. 300–2.	4
6. "Base for Class A Random Wound Motor Insulation Life," R. L. Balke and D. R. Blake, *A.I.E.E. Trans.*, Vol. 78, Part III, 1959, pp. 660–5.	4
7. "Application of Capacitors for Power-Factor Improvement of Induction Motors," W. C. Bloomquist and W. K. Boice, *A.I.E.E. Trans.*, Vol. 64, 1945, pp. 274–8.	2
8. "Insulation Systems for Random Wound Motors Evaluated by Motorette Tests;" T. J. Gair, *A.I.E.E. Trans.*, Vol. 73, Part IIIB, 1954, pp. 1702–6.	4
9. "The Rating of Electrical Machinery and Apparatus," R. E. Hellmund, *A.I.E.E. Trans.*, Vol. 58, 1939, pp. 499–503.	1
10. "Duty Cycles and Motor Rating," L. E. Hildebrand, *A.I.E.E. Trans.*, Vol. 58, 1939, pp. 478–83.	2

BIBLIOGRAPHY

	Reference to Section
11. "Squirrel-Cage Induction Motors," D. B. Hoseason, *The Jour. of the I.E.E.*, Vol. 66, 1928, pp. 410–31.	2
12. "Thermal Co-ordination of Motors, Control, and Their Branch Circuits on Power Supplies of 600 Volts and Less," B. W. Jones, *A.I.E.E. Trans.*, Vol. 61, 1942, pp. 483–7.	4
13. "Thermal Relationships in Induction Motors," W. J. Martiny, R. M. McCoy, and H. B. Margolis, *A.I.E.E. Trans.*, Vol. 80, Part III, 1961, pp. 66–77.	4
14. "Inspection and Tests of Explosion-Proof Motors," A. H. Nuckolls, *A.I.E.E. Trans.*, Vol. 55, 1936, pp. 151–8.	5
15. "New NEMA Fractional-Horsepower Motor Standards," C. P. Potter, *A.I.E.E. Trans.*, Vol. 66, 1947, pp. 508–10.	1
16. "Transient Stalled Temperature Rise of Cast Aluminum Squirrel Cage Rotors," G. M. Rosenberry, Jr., *A.I.E.E. Trans.*, Vol. 74, Part III, 1955, pp. 819–24.	4
17. "The Rating and Application of Motors for Refrigeration and Air Conditioning," P. H. Rutherford, *A.I.E.E. Trans.*, Vol. 58, 1939, pp. 519–23.	2
18. "Problems Confronting the Designer in Starting Large Motors," C. E. Schlecker and C. M. Lathrop, *A.I.E.E. Trans.*, Vol. 70, Part III, 1951, pp. 1351–3.	3
19. "The Development of Standards for Insulation Systems and Insulating Materials," M. L. Schmidt, *A.I.E.E. Trans.*, Vol. 77, Part III, 1958, pp. 747–9.	4

PAPERS BY PHILIP LANGDON ALGER

1. "The Development of Low-Starting Current Induction Motors," *G. E. Rev.*, Vol. 28, 1925, pp. 499–508.
2. "A Comparison of the Efficiency of Synchronous Machines as Determined by Various Methods," *G. E. Rev.*, Vol. 29, 1926, pp. 765–74.
3. "Synchronous Condensers," *A.I.E.E. Trans.*, Vol. 47, 1928, pp. 124–34.
4. "The Calculation of the Armature Reactance of Synchronous Machines," *A.I.E.E. Trans.*, Vol. 47, 1928, pp. 493–513.
5. "Induction Motor Performance Calculations," *A.I.E.E. Trans.*, Vol. 49, 1930, pp. 1055–67.
6. "Progress in Engineering Knowledge During 1933," *G. E. Rev.*, Vol. 37, 1934, pp. 4–10.
7. "Progress in Engineering Knowledge During 1934," *G. E. Rev.*, Vol. 37, 1934, pp. 551–62.
8. "Progress in Engineering Knowledge During 1935," *G. E. Rev.*, Vol. 38, 1935, pp. 546–57.
9. "Some Remarks on Engineering Education," *G. E. Rev.*, Vol. 39, 1936, p. 316.
10. "Progress in Engineering Knowledge During 1936," *G. E. Rev.*, Vol. 39, 1936, pp. 572–87.
11. "Another Tensor Anniversary," *G. E. Rev.*, Vol. 40, 1937, p. 170.
12. "Progress in Engineering Knowledge During 1937," *G. E. Rev.*, Vol. 41, 1938, pp. 72–88.
13. "Progress in Engineering Knowledge During 1938," *G. E. Rev.*, Vol. 42, 1939, pp. 58–77.
14. "Progress in Engineering Knowledge During 1939," *G. E. Rev.*, Vol. 43, 1940, pp. 60–88.
15. "Progress in Engineering Knowledge During 1940," *G. E. Rev.*, Vol. 44, 1941, p. 85.
16. "Progress in Engineering Knowledge During 1941," *G. E. Rev.*, Vol. 45, 1942, p. 82.
17. "A.I.E.E. Proposes Dual Rating of Electric Apparatus," *Ind. Standardization*, Vol. 13, 1942, p. 40.
18. "The Relation of Ethics to Human Progress," *The Scientific Monthly*, 1942, p. 166.
19. "Synergy," *G. E. Rev.*, Vol. 46, 1943, p. 89.
20. "Using Hydrogen to Save Coal," *The Scientific Monthly*, April 1944, p. 273.
21. "Raise Faculties," *G. E. Rev.*, Vol. 47, 1944, p. 7.
22. "Civic Responsibilities of Engineers," *A.I.E.E. El. Eng.*, Vol. 64, 1945, p. 251.
23. "Quality Control Through Product Testing," *A.I.E.E. El. Eng.*, Vol. 65, 1946, p. 11.
24. "Progress in Engineering Knowledge in 1945," *G. E. Rev.*, Vol. 49, 1946, p. 9.
25. "The Significance of Electrical Progress," *The Engineers' Digest*, July 1946, p. 326.
26. "Progress in Engineering Knowledge in 1946," *G. E. Rev.*, Vol. 50, 1947, p. 12.
27. "Engineering and Quality Control," *A.I.E.E. El. Eng.*, Vol. 66, 1947, p. 16.
28. "The Growing Importance of Statistical Methods in Industry," *G. E. Rev.*, Vol. 51, 1948, p. 11.
29. "Standard Temperatures of Reference for Efficiency Calculations," *A.I.E.E. Trans.*, Vol. 70, Part II, 1951, pp. 2006–07.
30. "Late Trends in Polyphase Induction Motors," *Indus. & Power*, Vol. 64, 1953, pp. 75–79.
31. "The Magnetic Noise of Polyphase Induction Motors," *A.I.E.E. Trans.*, Vol. P.A.S–73, 1954, pp. 118–25.
32. "Unfinished Business of Engineering Ethics," *A.I.E.E. El. Eng.*, Vol. 75, 1956, pp. 552–54.
33. "Performance Calculations for Part-Winding Starting of Three-Phase Motors," *A.I.E.E. Trans.*, Vol. P.A.S–75, 1956, pp. 1535–43.
34. "The Growing Importance of Ethics to Engineers," *A.I.E.E. C.P.*, pp. 57–1103.
35. "Induced High-Frequency Currents in Squirrel-Cage Windings," *A.I.E.E. Trans.*, Vol. P.A.S.–76, 1957, pp. 724–29.
36. "The Dilemma of Single-Phase Induction Motor Theory," *A.I.E.E. Trans.*, Vol. P.A.S.–75, 1958, pp. 1045–55.

37. *Mathematics for Science and Engineering*, Second Edition, McGraw-Hill Book Co., 1969.
38. *Induction Machines*, Second Edition, Gordon and Breach Science Publishers, 1970.

PAPERS WITH CO-AUTHORS

1. "Magnetic Flux Distribution in Annular Steel Laminae," A. E. Kennelly, *A.I.E.E. Trans.*, Vol. 36, 1917, pp. 1113–56.
2. "Induction Motor Core Losses," R. Eskergian, *A.I.E.E. Jour.*, Vol. 39, 1920, pp. 906–20.
3. "A New Power-Factor Slide Rule," H. W. Samson, *G. E. Rev.*, Vol. 25, 1922, pp. 455–57.
4. "Shaft Currents in Electric Machines," H. W. Samson, *G. E. Rev.*, Vol. 27, 1924, pp. 188–98.
5. "Shaft Currents in Electric Machines," H. W. Samson, *A.I.E.E. Trans.*, Vol. 43, 1924, pp. 235–45.
6. "Single-Phase Motor Torque Pulsations," A. L. Kimball, *A.I.E.E. Trans.*, Vol. 43, 1924, pp. 730–39.
7. "Double Windings for Turbine Alternators," E. H. Freiburghouse and D. D. Chase, *A.I.E.E. Trans.*, Vol. 49, 1930, pp. 226–44.
8. "The Development of Electrical Machinery in the United States," F. D. Newbury, *Proc. I.E.C.*, Paris, 1932, pp. 485–532.
9. "Rating of General-Purpose Induction Motors," T. C. Johnson, *A.I.E.E. Trans.*, Vol. 58, 1939, pp. 445–59.
10. "Induction Machines. Alternating-Current Commutator Motors," M. N. Halberg, *Standard Handbook for Electrical Engineers*, Seventh Edition., McGraw-Hill, N.Y., 1941, pp. 694–748.
11. "Progress in Engineering Knowledge During 1942," J. Stokely, *G. E. Rev.*, Vol. 46, 1943, p. 90.
12. "Progress in Engineering Knowledge During 1943," J. Stokely, *G. E. Rev.*, Vol. 47, 1944, p. 9.
13. "Progress in Engineering Knowledge During 1944," J. Stokely, *G. E. Rev.*, Vol. 48, 1945, p. 9.
14. "Rating of High-Temperature Induction Motors," H. A. Jones, *A.I.E.E. Trans.*, Vol. 64, 1945, pp. 300–02; 458.
15. "The Air Gap Reactance of Polyphase Machines," H. R. West, *A.I.E.E. Trans.*, Vol. 66, 1947, pp. 1331–43.
16. "A Derivation of Heaviside's Operational Calculus Based on the Generalized Functions of Schwartz," J. J. Smith, *A.I.E.E. Trans.*, Vol. 68, Part II, 1949, pp. 939–46.
17. "Split-Winding Starting of Three-Phase Motors," H. C. Ward, Jr. and F. H. Wright, *A.I.E.E. Trans.*, Vol. 70, Part I, 1951, pp. 867–73.
18. "Finite Representation of Impulse Functions in Solving Differential Equations," J. J. Smith, *A.I.E.E. Elec. Eng.*, Vol. 70, 1951, p. 153.
19. "Use of the Null-Unit Function in Generalized Integration," J. J. Smith, *J. Franklin Inst.*, Vol. 253, 1952, pp. 235–50.
20. "Energy Flow in Induction Machines," W. R. Oney, *G. E. Rev.*, Vol. 56, 1953, pp. 56–60.
21. "Short-Circuit Capabilities of Synchronous Machines for Unbalanced Faults," R. F. Franklin, C. E. Kilbourne, and J. B. McClure, *A.I.E.E. Trans.*, Vol., P.A.S.–72, 1953, pp. 394–404.
22. "Double and Triple Squirrel Cages for Polyphase Induction Motors," J. H. Wray, *A.I.E.E.*

Trans., Vol. P.A.S.–72, 1953, pp. 637–45.
23. "Switching Transients in Wound Rotor Induction Motors," Y. H. Ku, *A.I.E.E. Trans.*, Vol. P.A.S.–73, 1954, pp. 19–27.
24. "Speed-Torque Calculations for Induction Motors with Part Windings," Y. H. Ku and C. H. T. Pan, *A.I.E.E. Trans.*, Vol P.A.S.–73, 1954, pp. 151–60.
25. "Torque-Energy Relations in Induction Machines," W. R. Oney, *A.I.E.E. Trans.*, Vol. P.A.S.–73, 1954, pp. 259–64.
26. "Turbine Generator Stator Winding Temperatures at Various Hydrogen Temperatures," C. E. Kilbourne and D. S. Snell, *A.I.E.E. Trans.*, Vol. P.A.S.–74, 1955, pp. 232–51.
27. "New Method for Part-Winding Starting of Polyphase Induction Motors," L. M. Agacinsky, *A.I.E.E. Trans.*, Vol. P.A.S.–74, 1955, pp. 1455–62.
28. "Calculation of the Magnetic Noise of Polyphase Induction Motors," E. Erdelyi, *Acoustical Society of Am. Jour.*, Vol. 28, 1956, pp. 1063–67.
29. "Speed Control of Induction Motors Using Saturable Reactors," Y. H. Ku, *A.I.E.E. Trans.*, Vol. P.A.S.–75, 1956, pp. 1335–41.
30. "Better Part-Winding Starting," J. Sheets, *Power Industry*, October 1957, p. 14.
31. "Induction Machines," C. W. Falls and A. F. Lukens, *Standard Handbook for Electrical Engineers*, Ninth Edition, McGraw-Hill, 1957, pp. 704–48.
32. "Single-Phase Induction Motors," C. W. Falls and A. F. Lukens, *Standard Handbook for Electrical Engineers*, Ninth Edition, McGraw-Hill, 1957, pp. 735–43.
33. "Stray-Load Losses in Polyphase Induction Machines," G. Angst and E. J. Davies, *A.I.E.E. Trans.*, Vol. P.A.S.–78, 1959, pp. 349–57.
34. "Saturation Factors for Leakage Reactance of Induction Motors," P. D. Agarwal, *A.I.E.E. Trans.*, Vol. P.A.S.–79, 1960, pp. 1037–42.
35. "Electric Machines and System Engineering Courses," O. I. Franksen, *A.I.E.E. Special Pub.*, S–128, April 1961, pp. 62–71.
36. "Electromechanical Energy Conversion," E. Erdelyi, *Electro-Technology*, September 1961.
37. "Energy Conversion in Magnetic Devices," E. Erdelyi, *Electro-Technology*, June 1962.
38. "Stepless Starting of Wound Rotor Induction Motors," Jalaluddin, *A.I.E.E. Trans.*, Vol. App. & Ind.–81, 1962, 262–71.
39. "A New Wide-Range Speed Control System for Induction Motors," E. A. De Meo and H. C. Ward, Jr., *I.E.E.E. C.P.* pp. 63–364.
40. "Saturistors and Low Starting Current Induction Motors," G. Angst and W. M. Schweder, *A.I.E.E. Trans.*, P.A.S.–82, 1963, pp. 291–98.
41. "A Projects Laboratory for Electrical Engineering Students," L. P. Winsor, *I.E.E.E. Trans.*, Vol. Educ.–6, 1963, 9–12.
42. "A Capacitor Motor with Alnico Bars in the Rotor," R. L. Mester and J. G. Yoon, *I.E.E.E. Trans.*, Vol. 83, 1964, pp. 989–97.
43. "A Self-Excited Synchronous Generator for Isolated Power Supply," R. S. Barton, *I.E.E.E. Trans.*, Vol. 83, 1964, pp. 1002–06.
44. "Speed Control of Wound Rotor Motors with SCRs and Saturistors," W. A. Coelho and M. R. Patel, *I.E.E.E. Trans.*, Vol. I.G.A.–4, 1968, pp. 477–85.
45. "Subsynchronous Motors," V. Hamata, *I.E.E.E. P.A.S. C.P.* 72–001–01, 1972.
46. "Equivalent Circuits for Double Squirrel-Cage Induction Motors," M. Ivanes and M. Poloujadoff, *I.E.E.E. P.A.S. C.P.* 74–222–26, 1974.
47. "A Saturistor Motor for Pump Drive with SCR Speed Control," I. V. Ingvarsson and W. R. Oney, *I.E.E.E. Trans.*, P.A.S.–95, May / June 1976, pp. 766–72.

DISCUSSIONS BY PHILIP LANGDON ALGER

1. "A Method for Separating No-Load Losses in Electrical Machinery," C. J. Fechheimer, *A.I.E.E. Trans.*, Vol. 39, 1920, pp. 291–308.
2. "Magnetic Flux Distribution in Transformers," K. B. McEachron, *A.I.E.E. Trans.*, Vol. 41, 1922, pp. 247–61.
3. "Continuous Current Generator for High-Voltage," S. R. Bergman, *A.I.E.E. Trans.*, Vol. 42, 1923, pp. 910–14.
4. "Gaseous Ionization in Built Up Insulation—Part II," J. B. Whitehead, *A.I.E.E. Trans.*, Vol. 43, 1924, pp. 116–25.
5. "Eddy Current Losses in Armature Conductors," R. E. Gilman, *A.I.E.E. Trans.*, Vol. 43, 1924, pp. 246–51.
6. "Surface Iron Losses with Reference to Laminated Materials," T. Spooner and I. F. Kinnard, *A.I.E.E. Trans.*, Vol. 43, 1924, pp. 262–81.
7. "Short Circuits of A-C Generators," C. M. Laffoon, *A.I.E.E. Trans.*, Vol. 43, 1924, pp. 356–73.
8. "Theory of Three-Circuit Transformers," A. Boyajian, *A.I.E.E. Trans.*, Vol. 43, 1924, pp. 508–29.
9. "Theory and Calculation of the Squirrel Cage Repulsion Motor," H. R. West, *A.I.E.E. Trans.*, Vol. 43, 1924, pp. 1048–57.
10. "Single-Phase Induction Motor," L. M. Perkins, *A.I.E.E. Trans.*, Vol. 44, 1925, pp. 40–52.
11. "Squirrel-Cage Induction Motor Core Losses," T. Spooner, *A.I.E.E. Trans.*, Vol. 44, 1925, pp. 155–63.
12. "No-Load Copper Eddy Current Losses," T. Spooner, *A.I.E.E. Trans.*, Vol. 45, 1926, pp. 231–39.
13. "The Ratings of Electrical Machines as Affected by Altitude," C. J. Fechheimer, *A.I.E.E. Trans.*, Vol. 45, 1926, pp. 354–64.
14. "Motor Band Losses," T. Spooner, *A.I.E.E. Trans.*, Vol. 45, 1926, pp. 364–68.
15. "Starting Characteristics and Control of Polyphase Squirrel-Cage Induction Motors," H. M. Norman, *A.I.E.E. Trans.*, Vol. 45, 1926, pp. 369–82.
16. "The Cross-Field Theory of Alternating Current Machines," H. R. West, *A.I.E.E. Trans.*, Vol. 45, 1926, pp. 466–74.
17. "Synchronous Machines: Part I— An Extension of Blondel's Two-Reaction Theory," R. E. Doherty and C. A. Nickle, *A.I.E.E. Trans.*, Vol. 45, 1926, pp. 912–26; 946.
18. "Synchronous Machines: Part II— Steady State Power-Angle Characteristics," R. E. Doherty and C. A. Nickle, *A.I.E.E. Trans.*, Vol. 45, 1926, pp. 927–47.
19. "Mechanism of Breakdown of Dielectrics," P. L. Hoover, *A.I.E.E. Trans.*, Vol. 45, 1926, pp. 983–97.
20. "Magnetomotive Force Wave of Polyphase Windings With Special Reference to Subsynchronous Harmonics," Q. Graham, *A.I.E.E. Trans.*, Vol. 46, 1927, pp. 19–29.
21. "Starting Performance of Synchronous Motors," H. V. Putnam, *A.I.E.E. Trans.*, Vol. 46, 1927, pp. 39–59.
22. "Additional Losses of Synchronous Machines," C. M. Laffoon, *A.I.E.E. Trans.*, Vol. 46, 1927, pp. 84–100.

DISCUSSIONS

23. "Reduction of Armature Copper Losses," I. H. Summers, *A.I.E.E. Trans.*, Vol. 46, 1927, pp. 101–11.
24. "Static Stability Limits and the Intermediate Condenser Station," C. F. Wagner and R. D. Evans, *A.I.E.E. Trans.*, Vol. 47, 1928, pp. 94–123.
25. "Recent Improvements in Turbine Generators," S. L. Henderson and C. R. Soderberg, *A.I.E.E. Trans.*, Vol. 47, 1928, pp. 549–78.
26. "Improvements in Insulation for High-Voltage A-C Generators," C. F. Hill, *A.I.E.E. Trans.*, Vol. 47, 1928, pp. 845–52.
27. "220-Kv Transmission Line for the Conowingo Development," P. H. Chase, *A.I.E.E. Trans.*, Vol. 47, 1928, pp. 900–08.
28. "Squirrel-Cage Rotors with Split Resistance Rings," H. Weichsel, *A.I.E.E. Trans.*, Vol. 47, 1928, pp. 929–43.
29. "Condenser Motor," B. F. Bailey, *A.I.E.E. Trans.*, Vol. 48, 1929, pp. 596–606; 630–32.
30. "Fundamental Theory of the Capacitor Motor," H. C. Specht, *A.I.E.E. Trans.*, Vol. 48, 1929, pp. 607–13; 630–32.
31. "Revolving Field Theory of the Capacitor Motor," W. J. Morrill, *A.I.E.E. Trans.*, Vol. 48, 1929, pp. 614–29; 630–32.
32. "Line-Start Induction Motors," C. J. Koch, *A.I.E.E. Trans.*, Vol. 48, 1929, pp. 633–44.
33. "No-Load Induction Motor Core Losses," T. Spooner and C. W. Kincaid, *A.I.E.E. Trans.*, Vol. 48, 1929, pp. 645–55.
34. "Insulation Tests of Electrical Machinery Before and After Being Placed in Service," C. M. Gilt and B. L. Barns, *A.I.E.E. Trans.*, Vol. 48, 1929, pp. 656–65.
35. "Increased Voltages for Synchronous Machines," C. M. Laffoon, *A.I.E.E. Trans.*, Vol. 49, 1930, pp. 213–25.
36. "Telephone Interference from A-C Generators Feeding Directly on Line with Neutral Grounded," J. J. Smith, *A.I.E.E. Trans.*, Vol. 49, 1930, pp. 798–809.
37. "Damper Windings for Water-Wheel Generators," C. F. Wagner, *A.I.E.E. Trans.*, Vol. 50, 1931, pp. 140–52.
38. "Calculation of Synchronous Machine Constants— Reactances and Time-Constants Affecting Transient Characteristics," L. A. Kilgore, *A.I.E.E. Trans.*, Vol. 50, 1931, pp. 1201–14.
39. "Determination of Synchronous Machine Constants by Test-Reactances, Resistances, and Time Constants," S. H. Wright, *A.I.E.E. Trans.*, Vol. 50, 1931, pp. 1331–51.
40. "Measurement of Stray Load Loss in Polyphase Induction Motors," C. J. Koch, *A.I.E.E. Trans.*, Vol. 51, 1932, pp. 756–63.
41. "Adequate Wiring of Buildings, an Essential for Good Illumination," G. H. Stickney and W. Sturrock, *A.I.E.E. Trans.*, Vol. 51, 1932, pp. 993–1000.
42. "Higher Steam Temperatures and Pressures— A Challenge to Engineers," M. D. Engle and I. E. Moultrop, *A.I.E.E. Trans.*, Vol. 52, 1933, pp. 630–46.
43. "Reactive Power Concepts in Need of Clarification," A. E. Knowlton, *A.I.E.E. Trans.*, Vol. 52, 1933, pp. 744–47; 779–801.
44. "Reactive and Fictitious Power," V. G. Smith, *A.I.E.E. Trans.*, Vol. 52, 1933, pp. 748–51; 779–801.
45. "Operating Aspects of Reactive Power," J. A. Johnson, *A.I.E.E. Trans.*, Vol. 52, 1933, pp. 752–57; 779–801.
46. "Power, Reactive Volt-Amperes, Power Factor," C. L. Fortescue, *A.I.E.E. Trans.*, Vol. 52, 1933, pp. 758–62; 779–801.
47. "Reactive Power and Power Factor," W. V. Lyon, *A.I.E.E. Trans.*, Vol. 52, 1933, pp. 763–70; 779–801.
48. "Notes on the Measurement of Reactive Volt-Amperes," W. H. Pratt, *A.I.E.E. Trans.*, Vol. 52, 1933, pp. 771–801.

49. "Irregular Windings in Wound Rotor Induction Motors," R. E. Hellmund and C. G. Veinott, *A.I.E.E. El. Eng.*, Vol. 53, 1934, pp. 342–46; 1316–17.
50. "Stray Load Loss Test on Induction Machines," T. H. Morgan and P. M. Narbutovskih, *A.I.E.E. El. Eng.*, Vol. 53, 1934, pp. 286–90; 1317–18.
51. "Insulation Resistance of Armature Windings," R. W. Wieseman, *A.I.E.E. El. Eng.*, Vol. 53, 1934, pp. 1010–21; 1407–08.
52. "Split Winding Transformers," D. D. Chase and A. N. Garin, *A.I.E.E. El. Eng.*, Vol. 53, 1934, pp. 914–22; 1411–12.
53. "Industry Demands and Engineering Education," L. W. W. Morrow, *A.I.E.E. El. Eng.*, Vol. 53, 1934, pp. 518–22; 1413.
54. "Engineers of the Next Generation," C. F. Hirshfeld, *A.I.E.E. El. Eng.*, Vol. 53, 1934, pp. 857–59; 1413.
55. "Encouraging Initiative in the Engineering Student," C. L. Dawes, *A.I.E.E. El. Eng.*, Vol. 53, 1934, pp. 910–14; 1413.
56. "Engineering Education Is Meeting the Challenge," H. W. Bibber, *A.I.E.E. El. Eng.*, Vol. 53, 1934, pp. 1356–59; Vol. 54, 1935, p. 757.
57. "Characteristics of a Group of Engineers," T. Spooner, *A.I.E.E. El. Eng.*, Vol. 53, 1934, pp. 1571–76; Vol. 54, 1935, p. 757.
58. "On the Schooling of Engineers," A. Dow, *A.I.E.E. El. Eng.*, Vol. 53, 1934, pp. 1589–91; Vol. 54, 1935, p. 757.
59. "Measurement of Noise from Power Transformers," A. P. Fugill, *A.I.E.E. El. Eng.*, Vol. 53, 1934, pp. 1603–08; Vol. 54, 1935, p. 437.
60. "Measurement of Noise from Small Motors," C. G. Veinott, *A.I.E.E. El. Eng.*, Vol. 53, 1934, pp. 1624–28; Vol. 54, 1935, p. 437.
61. "Standardization of Noise Meter," R. G. McCurdy, *A.I.E.E. El. Eng.*, Vol. 54, 1935, pp. 14–15; 437.
62. "Quieting Substation Equipment," E. J. Abbott, *A.I.E.E. El. Eng.*, Vol. 54, 1935, pp. 20–26; 437.
63. "Noise Measurements for Engineering Purposes," B. G. Churcher, *A.I.E.E. El. Eng.*, Vol. 54, 1935, pp. 55–65; 437.
64. "Armature Leakage Reactance of Synchronous Machines," L. A. March and S. B. Crary, *A.I.E.E. El. Eng.*, Vol. 54, 1935, pp. 378–81; 1116.
65. "Insulation for High Voltage Alternators," C. M. Laffoon and J. F. Calvert, *A.I.E.E. El. Eng.*, Vol. 54, 1935, pp. 624–31; Vol. 55, 1936, p. 179.
66. "Reactance of End Connections," J. F. H. Douglas, *A.I.E.E. El. Eng.*, Vol. 56, 1937, pp. 257–59; 1312.
67. "Characteristic Constants of Single-Phase Induction Motors— Part I, Air-Gap Reactances," W. J. Morrill, *A.I.E.E. El. Eng.*, Vol. 56, 1937, pp. 333–38; 1312.
68. "End-Winding Inductance of a Synchronous Machine," B. H. Caldwell, Jr., *A.I.E.E. El. Eng.*, Vol. 56, 1937, pp. 455–61; 474–75; 1312.
69. "Vibration Measurements," C. D. Greentree, *A.I.E.E. El. Eng.*, Vol. 56, 1937, pp. 706–10; 1305–06.
70. "Dynamic Balancing of Small Gyroscope Motors," O. E. Esval and C. B. Frische, *A.I.E.E. El. Eng.*, Vol. 56, 1937, pp. 729–34; 1305–06.
71. "Approximating Potier Reactance," S. Beckwith, *A.I.E.E. El. Eng.*, Vol. 56, 1937, pp. 813–18; 1312.
72. "Temperature Aging Characteristics of Class A Insulation," J. J. Smith and J. A. Scott, *A.I.E.E. Trans.*, Vol. 58, 1939, pp. 445–59.
73. "Determination of Temperature Rise of Induction Motors," E. R. Summers, *A.I.E.E. Trans.*, Vol. 58, 1939, pp. 459–67; 472–78.
74. "The Hydrogen-Cooled Turbine Generator," D. S. Snell, *A.I.E.E. Trans.*, Vol. 59, 1940,

pp. 35-50.
75. "Fractional-Slot and Dead-Coil Windings," M. G. Malti and F. Herzog, *A.I.E.E. Trans.*, Vol. 59, 1940, pp. 782-94; 1079-80.
76. "Temperature Survey of the United States," J. J. Smith and H. W. Tenney, *A.I.E.E. Trans.*, Vol. 59, 1940, pp. 769-75; 1090-92.
77. "The Application of Class B Insulation to Auxiliary-Type D-C Motors in Severe Duty Service," F. A. Compton, Jr., *A.I.E.E. Trans.*, Vol. 59, 1940, pp. 828-34; 1098.
78. "Dynamoelectric Amplifier for Power Control," E. F. W. Alexanderson, M. A. Edwards, and K. K. Bowman, *A.I.E.E. Trans.*, Vol. 59, 1940, pp. 937-39; 1136.
79. "Some Problems in the Standardization of Temperature Ratings of Fractional-Horsepower Motors," C. G. Veinott, *A.I.E.E. Trans.*, Vol. 59, 1940, pp. 1055-61; 1218-21.
80. "A Study of Short-Time Ratings and Their Application to Intermittent Duty Cycles," R. E. Hellmund and P. H. McAuley, *A.I.E.E. Trans.*, Vol. 59, 1940, pp. 1050-55; 1221-24.
81. "Performance Calculations on Capacitor Motors— the Revolving Field Theory," P. H. Trickey, *A.I.E.E. Trans.*, Vol. 60, 1941, pp. 73-76; 662-63.
82. "The Service Factor Rating of Arc-Welding Generators and Transformers," R. C. Freeman and A. U. Welch, *A.I.E.E. Trans.*, Vol. 60, 1941, pp. 137-41; 726.
83. "The Use of Auxiliary Impedances in the Single-Phase Operation of Polyphase Induction Motors," R. W. Ager, *A.I.E.E. Trans.*, Vol. 60, 1941, pp. 494-99; 661.
84. "Complete Analysis of Motor Temperature Rise," F. Felix and H. G. Jungk, *A.I.E.E. Trans.*, Vol. 60, 1941, pp. 578-86; 684.
85. "Rotor-Bar Currents in Squirrel Cage Induction Motors," J. S. Gault, *A.I.E.E. Trans.*, Vol. 60, 1941, pp. 784-91; 1327-28.
86. "Classification and Coordination of Short-Time and Intermittent Ratings and Applications," R. E. Hellmund, *A.I.E.E. Trans.*, Vol. 60, 1941, pp. 792-98; 1323-25.
87. "Current Rating and Life of Cold-Cathode Tubes," G. H. Rockwood, *A.I.E.E. Trans.*, Vol. 60, 1941, pp. 901-03; 1390-91.
88. "A New Transformer for Base-Load Stations," P. Sporn and H. V. Putnam, *A.I.E.E. Trans.*, Vol. 60, 1941, pp. 916-18; 1358-61.
89. "Evening Courses at Graduate Levels— a challenge to Colleges of Engineering," R. Beach, *A.I.E.E. Trans.*, Vol. 61, 1942, pp. 88-94; 430.
90. "Temperature Rise Values for D-C Machines," A.I.E.E. Comm. Report, *A.I.E.E. Trans.*, Vol. 68, Part I, 1949, pp. 206-18.
91. "Differential Leakage of the Different Patterns for Fractional Slot Winding," M. M. Liwschitz, *A.I.E.E. Trans.*, Vol. 68, Part II, 1949, pp. 1129-32.
92. "Circuit Analysis Method for Determination of A-C Impedances of Machine Conductors," D. S. Babb and J. E. Williams, *A.I.E.E. Trans.*, Vol. 70, Part I, 1951, pp. 661-66.
93. "An Experimental Study of Induction Machine End-Turn Leakage Reactance," E. C. Barnes, *A.I.E.E. Trans.*, Vol. 70, Part I, 1951, pp. 671-79.
94. "Magnetic Revolving Field Coupling," G. Silva, *A.I.E.E. Trans.*, Vol. P.A.S.-71, 1952, pp. 298-308.
95. "The Economics of High-Temperature Dry-Type Transformers," L. C. Whitman, *A.I.E.E. Trans.*, Vol. P.A.S.-71, 1952, pp. 377-85.
96. "Characteristics of Overlapping Joints in Magnetic Circuits," T. D. Gordy and H. L. Gabarino, *A.I.E.E. Trans.*, Vol. P.A.S.-71, 1952, pp. 386-92.
97. "New Large Short-Circuit Testing Generators," C. E. Kilbourne, *A.I.E.E. Trans.*, Vol. P.A.S.-71, 1952, pp. 829-36.
98. "Thermal Endurance of Silicone Magnet Wire Evaluated by Motor Test," W. J. Bush and J. F. Dexter, *A.I.E.E. Trans.*, Vol. P.A.S.-73, 1954, pp. 1005-10.
99. "The Resistance of Twisted Segmental Amortisseur Bars," J. C. White, *A.I.E.E. Trans.*, Vol. P.A.S.-73, 1954, pp. 1282-88.

DISCUSSIONS

100. "New Performance Standards for Electrical Insulation of Rotating Machines," A.I.E.E. Comm. Report, *A.I.E.E. Trans.*, Vol. P.A.S.-73, 1954, pp. 1542–46.
101. "Permanent-Magnet Excited Synchronous Motors," F. W. Merrill, *A.I.E.E. Trans.*, Vol. P.A.S.-73, 1954, pp. 1754–60.
102. "Effect of Skew on Induction Motor Magnetic Fields," C. E. Linkous, *A.I.E.E. Trans.*, Vol. P.A.S.-74, 1955, pp. 760–65.
103. "Computation of Skin Effect in Bars of Squirrel-Cage Rotors," M. M. Liwschitz-Garik, *A.I.E.E. Trans.*, Vol. P.A.S.-74, 1955, pp. 768–71.
104. "Application of Statistics of Motor Testing," J. L. Oldenkamp, *A.I.E.E. Trans.*, Vol. P.A.S.-74, 1955, pp. 815–18.
105. "Equivalent Circuit for the Concatenation of Induction Motors," Y. H. Ku, *A.I.E.E. Trans.*, Vol. P.A.S.-74, 1955, pp. 1214–19.
106. "Ten Part-Winding Arrangements in Sample 4-Pole Induction Motor," J. J. Courtin, *A.I.E.E. Trans.*, Vol. P.A.S.-74, 1955, pp. 1248–54.
107. "Equivalent Circuits for Single-Phase Motors," G. R. Slemon, *A.I.E.E. Trans.*, Vol. P.A.S.-74, 1955, pp. 1335–43.
108. "Single-Phase Induction Motor Noise Due to Dissymmetry Harmonics," D. F. Muster and G. L. Wolfert, *A.I.E.E. Trans.*, Vol. P.A.S.-74, 1955, pp. 1365–72.
109. "Double-Energy Conversion in an Air Gap— A Novel Asynchronous Frequency Changer," W. LaPierre and J. Y. Louis, *A.I.E.E. Trans.*, Vol. P.A.S.-74, 1955, pp. 1373–76.
110. "The Penetration of Electromagnetic Radiation into Ferromagnetic Material," C. A. Adams, *A.I.E.E. Trans.*, Vol. P.A.S.-75, 1956, pp. 468–81.
111. "Synchronous-Motor Starting Performance Calculation," J. C. White, *A.I.E.E. Trans.*, Vol. P.A.S.-75, 1956, pp. 772–78.
112. "Calculation of Windage-Noise Power Level in Large Induction Motors," M. E. Talaat, *A.I.E.E. Trans.*, Vol. P.A.S.-76, 1957, pp. 46–55.
113. "A Basic Analysis of Synchronous Machines," W. A. Lewis, *A.I.E.E. Trans.*, Vol. P.A.S.-77, 1958, pp. 436–56.
114. "Calculation of Induction Motor Torque and Power," W. H. Middendorff, *A.I.E.E. Trans.*, Vol. P.A.S.-77, 1958, pp. 1055–59.
115. "Torque and Speed Control of Induction Motors Using Saturable Reactors," J. F. Szablya, *A.I.E.E. Trans.*, Vol. P.A.S.-77, 1958, pp. 1676–82.
116. "Eddy-Current Losses in Solid and Laminated Iron," P. D. Agarwal, *A.I.E.E. Trans.*, Vol. Comm. & Electr.-78, 1959, pp. 169–81.
117. "Functional Life Tests of Apparatus as Compared with Insulation Material Tests," M. L. Manning, *A.I.E.E. Trans.*, Vol. Comm. & Electr.-78, 1959, pp. 1107–11.
118. "Polyphase Induction Motors With Unbalanced Rotor Connections," B. N. Garudachar and N. L. Schmitz, *A.I.E.E. Trans.*, Vol. P.A.S.-78, 1959, pp. 199–205.
119. "Theory of End-Winding Leakage Reactance," V. B. Honsinger, *A.I.E.E. Trans.*, Vol. P.A.S.-78, 1959, pp. 417–26.
120. "Differential Leakage of 3-Phase Winding with Consequent Pole Connection," C. H. Lee, *A.I.E.E. Trans.*, Vol. P.A.S.-78, 1959, pp. 759–64.
121. "Voltage Harmonics of Salient Pole Generators Under Balanced 3-Phase Loads— I," D. Ginsberg and A. L. Jokl, *A.I.E.E. Trans.*, Vol. P.A.S.-78, 1959, pp. 1573–80.
122. "Equations for Induction Motor Slip," V. B. Honsinger, *A.I.E.E. Trans.*, Vol. P.A.S.-78, 1959, pp. 1621–25.
123. "A New Brushless D-C Excited Rotating Field Synchronous Motor," G. M. Rosenberry, Jr., *A.I.E.E. Trans.*, Vol. App. & Ind.-79, 1960, pp. 136–39.
124. "Dynamic Circuit Theory," H. K. Messerle, *A.I.E.E. Trans.*, Vol. P.A.S.-79, 1960, pp. 1–12.

125. "Calculation of Stray Load Losses in D-C Machinery," E. Erdelyi, *A.I.E.E. Trans.*, Vol. P.A.S.–79, 1960, pp. 129–38.
126. "End Component of Zero-Sequence Reactance of A-C Machines," R. T. Smith, *A.I.E.E. Trans.*, Vol. P.A.S.–79, 1960, pp. 259–64.
127. "Equivalent Circuits and Performance of Canned Motors," P. D. Agarwal, *A.I.E.E. Trans.*, Vol. P.A.S.–79, 1960, pp. 635–42; 1301–02.
128. "Saturation Harmonics of Polyphase Induction Machines," C. H. Lee, *A.I.E.E. Trans.*, Vol. P.A.S.–80, 1961, pp. 597–603.
129. "The Torque Tensor of the General Machine," Yao-nan Yu, *A.I.E.E. Trans.*, Vol. P.A.S.–81, 1962, pp. 623–29.
130. "Derating of Polyphase Induction Motors Operated with Unbalanced Line Voltages," M. M. Berndt and N. L. Schmitz, *A.I.E.E. Trans.*, Vol. P.A.S.–81, 1962, pp. 680–86.
131. "An Analysis of the Hysteresis Motor— Analysis of the Idealized Machine," M. A. Copeland and G. R. Slemon, *I.E.E.E. Trans.*, Vol. P.A.S.–82, 1963, pp. 34–42.
132. "The Development of Torque in Slotted Armatures," A. S. Langsdorf, *I.E.E.E. Trans.*, Vol. P.A.S.–82, 1963, pp. 82–88.
133. "Improved Starting Performance of Wound-Rotor Motors Using Saturistors," C. E. Gunn, *I.E.E.E. Trans.*, Vol. P.A.S.–82, 1963, pp. 298–302.
134. "Saturation Factors for Leakage Reactance of Induction Motors with Skewed Rotors," G. Angst, *I.E.E.E. Trans.*, Vol. P.A.S.–82, 1963, pp. 716–25.
135. "The Effect of Voltage Waveshape on the Performance of a 3-Phase Induction Motor," G. C. Jain, *I.E.E.E. Trans.*, Vol. P.A.S.–83, 1964, pp. 561–66; 744–45.
136. "Engineering Education at a Crossroad," J. W. Rittenhouse, *I.E.E.E. Trans.*, Vol. P.A.S.–83, 1964, pp. 817–26.
137. "Spatial Harmonic Magnetomotive Forces in Irregular Windings and Special Connections of Polyphase Windings," C. G. Veinott, *I.E.E.E. Trans.*, Vol. P.A.S.–83, 1964, pp. 1246–53.
138. "Pulling-Into-Step of Reluctant Synchronous Machines," K. Burian, *I.E.E.E. Trans.*, Vol. P.A.S.–84, 1965, pp. 349–56; 528.
139. "A Single-Phase Induction Motor with One Distributed Winding," D. D. Hershberger and J. L. Oldenkamp, *I.E.E.E. Trans.*, Vol. P.A.S.–87, 1968, pp. 1862–66.
140. "The GM High Performance Induction Motor Drive System," P. D. Agarwal, *I.E.E.E. Trans.*, Vol. P.A.S.–88, 1969, pp. 86–93.
141. "Refinements in I.E.E.E. No. 117 Test Procedure for Evaluation of Life Expectancy of Random Wound Motor Insulation Systems," I.E.E.E.-W.G. Report, *I.E.E.E. Trans.*, Vol. P.A.S.–88, 1969, pp. 258–66.
142. "Direct Simulation of A-C Machinery Including Third Harmonic Effects," R. J. W. Koopman and F. C. Trutt, *I.E.E.E. Trans.*, Vol. P.A.S.–88, 1969, pp. 465–74.
143. "Tangential Forces in Squirrel-Cage Induction Motors," B. Heller and A. L. Jokl, *I.E.E.E. Trans.*, Vol. P.A.S.–88, 1969, pp. 484–92.
144. "The Quadrature-Axis Equivalent Circuit of the Synchronous Machine with a Grill," S. B. Jovanovski, *I.E.E.E. Trans.*, Vol. P.A.S.–88, 1969, pp. 1620–24.
145. "Magnetic Fields, Eddy Currents, and Losses, Taking the Variable Permeability into Account," K. Oberretl, *I.E.E.E. Trans.*, Vol. P.A.S.–88, 1969, pp. 1646–57.
146. "Effects of Axial Slits on the Performance of Induction Machines with Solid Iron Rotors," P. K. Rajagopalan and V. B. Murty, *I.E.E.E. Trans.*, Vol. P.A.S.–88, 1969, pp. 1695–1709.
147. "Two-Speed Single-Winding Shaded-Pole Single-Phase Induction Motors," G. S. Rao, V. V. Sastry, and P. V. Rao, *I.E.E.E. Trans.*, Vol. P.A.S.–89, 1970, pp. 1308–21.
148. "A New Equivalent Circuit for Double-Cage Induction Motors," A. K. Bandyopadhyay, *I.E.E.E. Trans.*, Vol. P.A.S.–89, 1970, pp. 1540–45.

INDEX

ADAMS, C. A., x, 222
Air-gap energy flow, 255
Air-gap field, 246
Air-gap flux distribution, 83
Air-gap fringing coefficients, 184
Air-gap reactance, 246
Alnico, bars, x, 283, 441
  B-H curve, 29
ANGST, G., x
Application of motors, 499
Armature reaction, 93
Armature windings, 76
ARNOLD, E., 222
ASA Standards, 501

Bearings, 172
BEHREND, B. A., 126
BENNETT, A. A., 184
BERGMAN, S. R., x
BEWLEY, L. V., x, 15, 66
Blondel diagram, 38
Braking, dynamic, 156
  with Saturistor, 158
Brush-shifting motor, 491

Capacitor motor, 429, 431
  permanent-split, 441
  T-connected, 436
CARR, L. H. A., 222
CARTER, F. W., 182
CHAPMAN, F. T., 209
Circle diagram, 44, 114
Circuit calculations, 5, 133
Circuit, generalized, 140, 417, 419
  response, 52
  simple, 5
  steady-state analysis, 43
  transient analysis, 46, 49
Circulating current losses, 340
CLYMER, C. C., 320
Cogging, 346
Concatenation, 296, 302
Constant linkage theorem, 49
Core losses, 118
Coupling coefficient, 40
Crawling, asynchronous, 331

Crawling, synchronous, 355
Cross-field circuit, 417
Current loading, 72
Current, locked rotor, 502

Deep-bar factors, 270, 276
Deep-bar rotor, 265
Design, art of, 3
  process of, 165
  progress in, 5, 170
Dielectric constant, 18, 176
Dilemma of single phase motor, 449
Dimensional analysis, 60
  quantities, 63
Dirac function, 57
Dissymmetry effects, 382
Distribution factor, 80
Double squirrel cage, 277
Doubly-fed motor, 490
Dreyfus motor, 308
Dual-winding motors, 476
Dynamic braking, 156

Electromechanical analogies, 63
Electrons, 12, 14, 64
Electrostatic capacity, 18
  motors, 67
Energy flow, 64, 252, 458
  air-gap, 254
Energy storage, electrostatic, 18
  magnetic, 17
Equations, of motion, 63
  transient, 46, 49
Equivalent circuit calculations, 133
Equivalent circuit, capacitor motor, 433 436
  cross-field, 417
  D and Q axis, 395
  double squirrel-cage motor, 278, 48
  dual-winding, 291, 477
  general, 41
  generalized, 140, 417, 419
  importance, 42
  ladder, 283
  polyphase induction motor, 97, 114, 132, 338

526 INDEX

Equivalent circuit, reluctance motor, 488
  single-phase motor, 434
  transformer, 41
  two-phase motor, 434, 468
Equivalent circuit formulas, 159

Field, harmonic, 85
  of single coil, 83
  revolving, 82
Flux distribution, 246
  pulsation, 183
Foote scheme, 318
Force waves, 377
Fourier analysis, 84
Fractional-horsepower motors, 165
Frame, cast iron, 167
  dimensions, 168
  enclosures, 172, 511
  standard N.E.M.A., 169
  steel plate, 167
  structure, 166
  vibration, 178
Fringing coefficients, 183, 184

Generalized machine, 394, 411
General purpose motors, 498
GREEN, C. F., 182

Harmonic fields, 85, 325
Harmonics, mmf, 85, 325, 339
  permeance, 341
Heating in acceleration, 262
Heating limitations, 503
HEAVISIDE, O., 54
High frequency losses, 119, 340
Homopolar generator, 16
Hysteresis motor, series, 283
  shunt, 489

Idle bars, 272, 283
Inductance, 23, 25
  motional, 416
Induction generator, 148
Induction, law of, 3
Induction motor, polyphase, 3, 105
  single phase, 429
  utility of, 107
Insulation, classes, 171, 504
  temperature, life of, 504
Integral horsepower motors, 165
Interspersing, 333

Kelvin's law, 20
KOCH, C. J., x
Krämer drive, 320
KRON, G., ix, x, 393, 403, 421

LAMME, B.G., 16
Laplace equation, 182
  transform, 57
Linear Motors, 491
Locking, 346
Losses, copper, 10, 122, 189
  core, 118, 177, 192
  friction and windage, 119, 194
  high frequency, 119, 340
  slip, 122, 189
  stray-load, 123, 193, 339
LYON, W. V., x

MACMILLAN, C. 149, 333
Magnetic calculations, 11
  core, 177
  energy, 17
  forces, 19
  noise, 365
  structure, 71, 108
  unbalanced pull, 356
Magnetization curve, 12, 187
Magnetizing current, 93, 181
Magnetomotive force, 72, 86
  geometrical pattern of, 92
Magnetomotive force waves, 72
Matrix, connection, 407
  impedance, 411
  transformations, 407
MAXWELL, H., v, x, 149
Motors, application of, 499
  brush-shifting, 491
  capacitor, 429, 431
  dilemma of single phase, 449
  doubly-fed, 490
  Dreyfus, 308
  dual-winding, 476
  efficiencies of, 509
  electrostatic, 67
  fractional-horsepower, 165
  general purpose, 500
  hysteresis, 283, 489
  integral horsepower, 165
  linear, 491
  permanent-split capacitor, 441
  Permasyn, 488
  polyphase induction, 3, 105
  power factors of, 508
  rating of, 169, 499
  reluctance, 486
  reluctance start, 461

INDEX

Motors, repulsion induction, 461
  Saturistor, 283
  Schrage, 491
  shaded pole, 459
  single-phase, 447
  single-phase induction, 429
  special purpose, 506
  split-phase, 458
  subsynchronous, 356
  synchronous induction, 490
  T-connected capacitor, 436
  three-phase, unbalanced, 469
  two-phase, 449
  two-phase, unbalanced 467
  wound rotor, rheostatic control, 306
  wound rotor, with Saturistor, 309
  wound rotor, with single-phase rotor, 308

N.E.M.A. frames, 169
Noise calculations, 372
Noise, magnetic, 365
Noise spectrum, 366, 379
Null-unit function, 54

Offset current, 51

Part-winding starting, 288
Permeance waves, 328, 342
Permasyn motor, 488
Per unit notation, 43
Per unit quantities, 43
Phase shifter, 403
Phasor, 6
  conjugate of, 9
Phasor diagram, induction motor, 113
  transformer, 38
Pitch factor, 80
Pole changing, 302
Polyphase induction machines
  circle diagram, 114
  cross section, 77
  equivalent circuit of, 97, 114, 132
  formulas for, 159
  load tests, 122
  losses, 188
  no-load tests, 117
  performance curves of, 137
  phasor diagram, 113
  rating, 170, 499
  starting performance, 116
  torque, 109
  utility, 107
Power factor, 8

Power-factor chart, 150
Power output, 75
Power, reactive, 8, 9
  real, 8
Poynting vector, 254

Rating of motors, 169, 499
Reactance, air-gap, 213, 244
  belt leakage, 228
  calculations, 23
  coil end, 233
  differential, 209, 216
  incremental, 122, 279
  leakage, 23, 46
  magnetizing, 201
  peripheral leakage, 232, 251
  slot, 201, 208
  zero-phase sequence, 206
  zigzag, 209, 216, 222, 335
Reactors, 23, 26
Rectifiers, silicon, 491
Reluctance motor, 486
Reluctance start motor, 461
Repulsion induction motor, 461
Resistance calculations, 10, 190
Resistivity, 10
Rotor, deep-bar, 265
  double squirrel-cage, 277
  high-impedance, 263
  idle-bar, 272, 283
  single-phase, 308
  wound, 306

Saturable
  reactor, 27
  resistor, 30
Saturation effects, 120, 185
Saturistors, x, 28, 158, 309, 314
Scherbius drive, 318
SCHMIDT, M. L., x
Schrage motor, 491
SCHWARTZ, LAURENT, 54, 57
SCHWEDER, W. M., x
Selsyns, power, 306
Service factor, 502
Shaded pole motor, 459
Shafts, 172, 359
Silicon rectifiers, 319, 444, 491
Single phase motor, 447
Skew, 227, 335, 345, 349
Skin effect, 111, 269, 417
Slip, 109
SMITH, J. J., 54
Sound intensity, 365

INDEX

Sound levels, 367
  measurement, 366
  radiation, 370
Special purpose motors, 508
Speed control, of capacitor motors, 440
  with auxiliary machines, 314
Split-phase motor, 458
Squirrel-cage, deep bar, 265
  double, 277, 481
  idle-bar, 272, 283
Standards, ASA, 501
Starting currents, 503
  double delta, 479
  multicircuit, 302
  part-winding, 288
  reduced voltage, 287
  split-winding, 296
  Y-delta, 288
Starting performance, 116
STEINMETZ, C. P., 131
STIEL, WILHELM, 327
Stray losses, 123, 193, 339
Subsynchronous motor, 356
Superposition theorem, 52
Symmetrical components, 100, 404
Synchronous induction motor, 490

Temperature, creep, 505
  limitations, 503
  rise, 171
TESLA, NIKOLA, 107, 126
Test,
  code, 499
  load, 122
  locked rotor, 119
  no-load, 117
  running light, 117
  stray-load loss, 123
Thermotector, 507
Three-phase motor, unbalanced, 469
Thyristors, 314, 319
Torque, 73, 109
  breakdown, 153, 501
  dynamic braking, 156
  equations, 73
  locked rotor, 159, 501
  locking, 346
  pulsation, 382, 456
Transformations, matrix, 407
  symmetrical components, 405

Transformations, three-phase to two-phase, 405
Transformer, 37
  circle diagram of, 45
  equivalent circuit of, 41
  maximum output of, 46
  phasor diagram of, 38
  regulation of, 44
  transient performance, 57
Transient analysis, 49
  equations, 54, 480
Transients in current build up, 65
Triple squirrel cage, 283
Two-phase motor, 449
  unbalanced, 467

Units, systems of, 62
Unbalanced pull, 356
Unbalanced windings, 467, 469

Ventilation, 171
Vibration of core, 178, 327, 366, 369
Virtual displacement principle, 20
Voltage, calculations, 14
  generated, 80
  induced, 14
  ripples, 327, 383
  speed, 15, 132, 415
  transformer, 15, 415

WEST, H. R., x, 222
WIESEMAN, R. W., 183
Windings, analysis, 96
  armature, 76, 179, 430
  double-layer, 77
  irregular, 231
  multispeed, 302
  single layer, 78
Wound rotor motor, rheostatic control, 306
  with Saturistor, 309
  with single-phase rotor, 308

Zero-phase sequence circuit, 473
Zero-phase sequence reactance, 206
Zigzag leakage, 209, 216, 222, 335